# 23rd Symposium on Physics of Switching Arc 2019



Nove Mesto na Morave, Czech Republic
9 - 13 September 2019

ISBN: 978-1-7138-0316-4

**Printed from e-media with permission by:**

Curran Associates, Inc.
57 Morehouse Lane
Red Hook, NY 12571

**Some format issues inherent in the e-media version may also appear in this print version.**



Printed with permission by Curran Associates, Inc. (2020)



Brno University of Technology
c/o Milada Bartlová - Department of Physics
Technical 2848/8, Zabovresky, 61600
Brno, Czech Republic

Phone: +420 541 143 431
Fax: +420 541 143 133

bartlova@feec.vutbr.cz

**Additional copies of this publication are available from:**

Curran Associates, Inc.
57 Morehouse Lane
Red Hook, NY 12571 USA
Phone: 845-758-0400
Fax: 845-758-2633
Email: curran@proceedings.com
Web: www.proceedings.com

# TABLE OF CONTENTS

## ISSUE 1

## ISSUE 2

## ISSUE 3

doi:10.14311/ppt.2019.1.1
*Plasma Physics and Technology* **6**(1):1–6, 2019 

# METALLIC NANOPARTICLES GENERATION BY REPETITIVE PULSED LASER FOR APPLICATIONS IN BIO-MEDICINE

N. Restuccia[a,*], L. Silipigni[a], M. Cordaro[b], and L. Torrisi[a]

[a] *Dipartimento di Scienze Matematiche e Informatiche, Scienze Fisiche e Scienze della Terra, MIFT, Università di Messina, V.le F. Stagno d'Alcontres 31, 98166 S. Agata, Messina, Italy*

[b] *Dip.to di Scienze Chimiche, Biologiche, Farmaceutiche e Ambientali, Università di Messina, Italy*

[*] nrestuccia@unime.it

**Abstract.** An Nd:YAG pulsed laser operating at the 1064 nm wavelength, the 3 ns pulse duration, the $10^{10}$ W/cm$^2$ intensity and the 10 Hz repetition rate is employed to irradiate biocompatible metallic targets based on Au, Bi and Ag placed in water. The laser-matter interaction produces nanometric spherical particles. The concentration of the solution with nanoparticles is controllable by the laser parameters, the ablative emission process, the irradiation time and the water's volume. Generally, nanoparticles of about 10 nm in size and concentrations of the order $(0.1 \div 10)$ mg/ml are prepared to be injected in cell cultures or in living systems (mice). The nanoparticles introduction in the extra and intracellular cellular liquids improves the bio-imaging of the tissue and organs by using fluorescence techniques. Moreover, if these nanoparticles are concentrated in tumour cells, they make possible high efficiency radio-therapy and thermal-therapy treatments, as it will be presented and discussed.

**Keywords:** Laser ablation in water; nanoparticles; imaging; therapy.

## 1. Introduction

Recently nanoparticles (NPs) are more and more used for many applications in different scientific fields, such as physics, chemistry, materials science and bio-medicine. Advancements have been obtained in the applications of heavy metallic elements nanoparticles in the field of biological and medical applications with the aim to enhance both the diagnostic imaging contrast, by using these nanoparticles such as contrast medium, and the targeting of diseased tissues owing to the increase in the effective atomic number before they are submitted to radiotherapy [1]. By using nanoparticles as markers to control the metabolic phases of internalization in the cells, nanotechnology applied to life sciences has led to a better understanding of the cellular and molecular processes and factors that cause a given disease. By means of different techniques, such as the electronic microscopy, the optical absorption spectroscopy and the X-ray spectroscopy, it is possible to follow the uptake and the decay of nanoparticles often functionalized with specific bio-molecules.

Among all the diagnostic imaging modes, the X-ray imaging techniques are the most used for their simplicity, high spatial resolution and low cost. The improvement of these techniques strongly depends on the development of new X-ray contrast agents. Up to now, the clinically used contrast medium consists of small iodine molecules in liquid solutions that allow highlighting the functionality of particular organs. However, the iodine medium can cause serious side effects due to excessive repeated doses necessary to achieve a good image contrast. Recently, metallic nanoparticles (NPs) have been used as contrast agents for the radiological imaging [2]. To this reason there is a growing interest in the development of alternative contrast agents with a high atomic number and a high X-ray absorption, such as for example silver, gold and bismuth nanoparticle solutions.

A special interest is devoted to the use of functionalized nanoparticles able to cross the cell membranes and to diffuse in the intra-cellular liquid up to the nucleus [3]. The aim is that to introduce high atomic number (Z) nanoparticles in the cancer cells to enhance the electronic density of the medium and the possibility to release higher quantities of energy coming from radiations used for radiotherapy (X-rays, electron and ion beams) [4].The main purpose of radiotherapy is to maximize the radiation dose delivered to the tumors and at the same time minimize the level of damage to surrounding healthy tissues by reducing the dose they receive.

A complementary approach to potentially achieve this goal involves the localization of high atomic number (Z) elements and compounds in the target tumor tissue before irradiation. The presence of high Z atoms in the target leads to an increase in the cross-section of the photoelectric effect (it strongly depends on the Z value) and to the generation of low-energy free radicals (photoelectrons and Auger electrons). These low-energy free radicals have high rates of linear energy transfer (LET) and, consequently, can cause DNA damage through the generation of reactive oxygen species. NPs have several advantages over iodine because they have a higher Z-number and therefore a higher probability of photoelectric interactions.

Silver, gold and bismuth are biocompatible metals that can be obtained as nanometric particles in biological solutions. They have high electron density, low toxicity and they easily tend to chemically bind to specific bio-molecules in order to be vehicled trough the blood flux. Their production by laser ablation in water is consolidated by years and it is due to the ablation of atoms from the solid surface, to their energetic interaction in water, inside the plasma generated at the solid-liquid interface, and to their nucleation in nanoparticles with a dimension depending on the laser parameters, irradiation conditions and nature of the used liquid. The mass absorption coefficients of X-ray and the stopping power for electrons and ions versus the energy show a significant increase on the values for such metallic elements with respect to the light biological tissues composition, constituting high contrast medium, when introduced in biological environment, and target elements in cancerous biological tissues for appropriate radiotherapy expositions.

## 2. Materials and Methods

Metallic nanoparticles of Ag, Au and Bi were prepared by using the technique of the pulsed laser ablation in liquids. A Nd:YAG laser was employed at the 1064 nm wavelength, the 3 ns pulse duration, the 100 mJ maximum pulse energy, the 1 mm$^2$ focused spot and the 10 Hz repetition rate. The metal, in the form of sheets of about 2 cm$^2$ surface and 1 mm thickness, was placed on the bottom of a glass beaker. A volume of (5 ÷ 10) ml of distilled water was employed to generate the solution. Generally the metal surface was placed 4 mm under the free level of the water in the container. The laser light was perpendicular to the free water surface and it was focused on the metallic target surface. In order to avoid generation of deep craters during the repetitive laser shots, the beaker glass was placed on a rotating and sliding platform to be able to hit the fresh surface.

The ablation yield, in terms of ablated mass per laser pulse, increased about linearly with the laser fluence (J/cm$^2$) and has been controlled by the laser fluence and the solution concentration, because the increment of the laser fluence enhances the ablation yield while an higher concentrated solution absorbs laser energy and decreases the ablation yield [2]. Generally, for the repetition rate, we used exposition times of the order of (10 ÷ 30) min, i.e. a number of laser shots within $6 \times 10^3$ and $1.8 \times 10^4$. A scheme of the used experimental set-up is reported in Figure 1a. Figure 1b shows a photo of the experimental apparatus.

The metallic nanoparticles can be produced by different laser wavelengths, however, the IR radiation was chosen to have higher ablation yield with respect to visible and UV radiations, in agreement with the literature [5].

By using a 70 mJ pulse energy and an irradiation time of 10 min and covering the metallic sheet surface with 4 mm liquid, 10 ml in volume, the ablated mass was of 1.2 mg, 1.0 mg and 3.8 mg, for Ag, Au and Bi, respectively. The IR laser ablation approximately is inversely proportional to the boiling point of the elements, which is 961°C, 1063°C and 271°C for Ag, Au and Bi, respectively. In such conditions the produced solution will have a concentration of 1.2 mg/10 ml, 1 mg/10 ml and 3.8 mg/10 ml for Ag, Au and Bi, respectively.

The production of nanoparticles in a liquid at high concentration promotes their interaction coalescence producing agglomerated and micrometric particles that may precipitate in the solution.

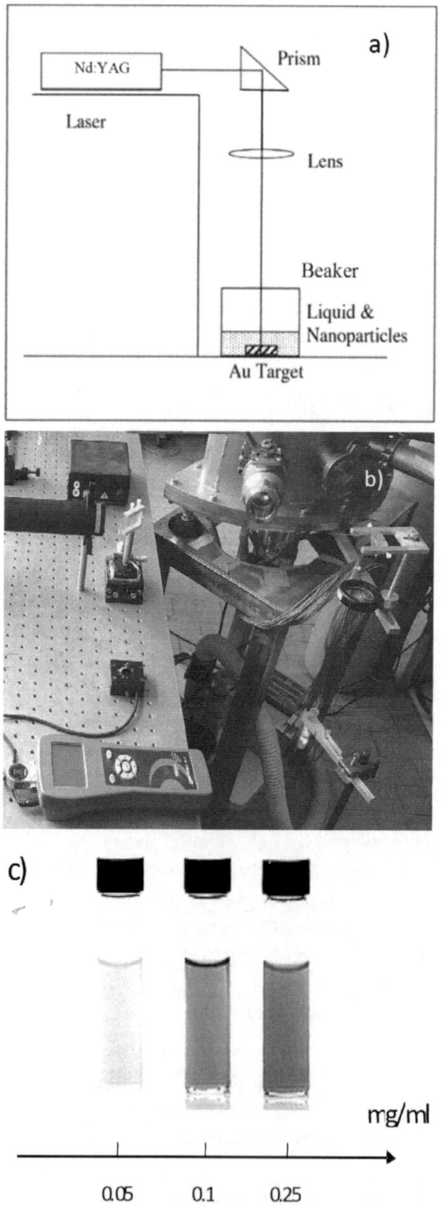


*Figure 1. Laser Ablation in liquid: (a) scheme of the experimental set up, (b) its photo and (c) colour of the solution containing Au NPs in different concentrations.*

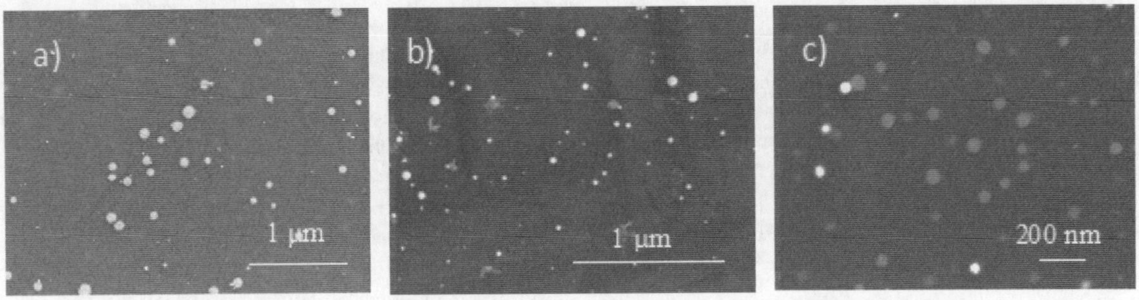


*Figure 2. SEM images of nanoparticles of Ag (a), Au (b) and Bi (c) prepared by laser ablation in water.*

Such effect increases with the particles concentration, time and temperature. In order to reduce the nanoparticles coalescence effect a biocompatible surfactant (cap agent) was added, in the form of sodium citrate ($Na_3C_6H_5O_7$), by using a 1 $\mu$g/10 ml of surfactant concentration in water. In this way we have stabilized the nanoparticles and avoided a fast their coalescence. The color of the solution changes both with the particles nature and their concentration: it is grey for Ag and Bi solutions and red for gold and these colors become more and more intense as the concentration of the solution increases. In particular, Figure 1c shows the colour changes with the Au NPs concentration in distilled water.

Different physical techniques were employed to characterize the produced nanoparticles such as scanning electron (SEM) microscopy, energy dispersive X-ray (EDX) characteristic fluorescence by using electron ionization, X-ray diffraction (XRD) to obtain crystallographic information, and absorbance spectroscopy in the UV- Vis-NIR regions.

The measurements of optical absorbance due to surface plasmon resonance (SPR) analysis were obtained using transmitted light through the solution versus the wavelength and it is affected by scattering effects due to the thin layer crossed in the 4 mm inner cuvette size.

Measurements of uptake and decay of the used solutions containing nanoparticles were performed in living systems (mice) by using the X-ray fluorescence of the element induced by a photon ionization of a 45 kV X-ray tube (Bruker-in vivo MS-FX PRO system).

## 3. Results

The SEM images of the metallic nanoparticles have indicated their spherical shape and a dimension ranging between about 5 nm and 50 nm in the three cases. The dimensions distribution, the shape and the study of little nanoparticles aggregates was performed by using a 20 kV SEM to observe nanoparticles deposited on a silicon surface. Typical SEM images are reported in Figure 2 for silver (a), gold (b) and bismuth (c). The prolonged laser irradiation up to about 30 min determines two size distributions of nanoparticles in the liquid, one at about 20-30 nm diameter and another at about 5-10 nm diameter. This is due to the laser absorption in the high concentrated solution that produces further ablation and fragmentation of nanoparticles in smaller dimensions.

The presence of the metallic nanoparticles in the solution induces surface plasmon resonance (SPR) absorption effects when it is illuminated in the Vis-NIR wavelength regions. In fact, the solution shows absorption bands at wavelengths depending strongly on the particles shape, dimensions and concentration and on the nature of liquid medium. The resonant absorption is due to the dipole induction by light and to electron oscillations on the surface of metallic nanoparticles absorbing energy from the incident light.

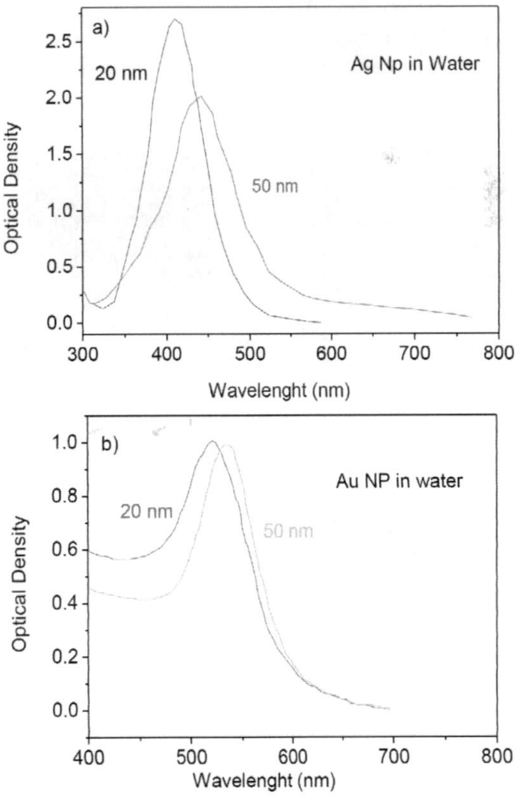


*Figure 3. Typical optical absorbance spectra for Ag (a) and Au (b) NPs, as a function of wavelength and the spherical nanoparticles diameter.*

Figure 3 reports two plots relative to the typical absorption bands for Ag (a) and Au (b) NPs versus the wavelength. It is possible to observe that for spherical particles with 20 nm diameter the absorption SPR resonance occurs at about 400 nm for silver and 520 nm for gold, and that increasing the NPs dimension the absorption peak shifts towards higher wavelengths. For bismuth the peak is less intense than for Ag and Au and occurs at about 270 nm, according to the literature [6].

The peak shape is typical of Ag, Au and Bi NPs in distilled water showing the SPR band peak and laterally the absorption values vs. wavelength due to other mechanisms. It depends strongly on the nanoparticle nature, size and concentration and on the type of liquid, as reported in the literature [7].

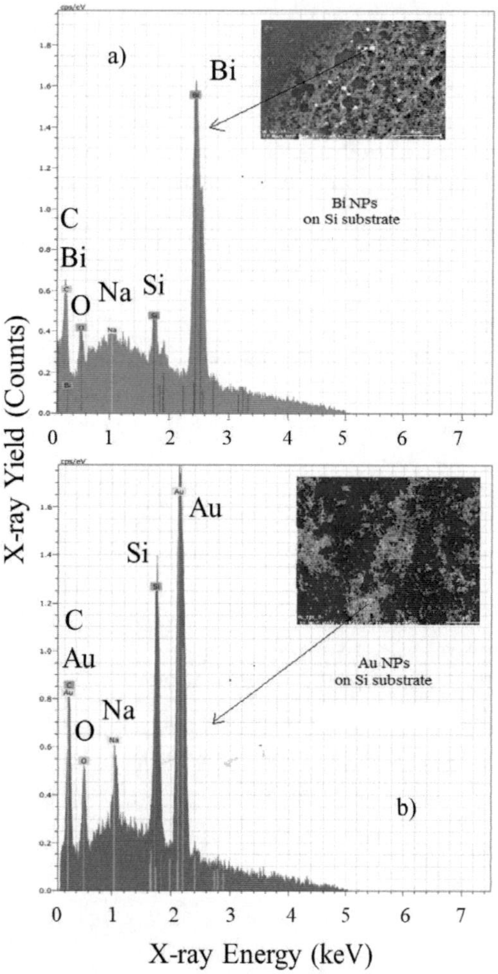


*Figure 4. EDX spectra for Bi (a) and Au (b) NPs deposited on a Si substrate: the insets show the analyzed portion.*

Figure 4 illustrates EDX spectra, carried out during SEM analysis with a 5 keV electron beam and relative to a group of Bi-NPs (a) and Au-NPs (b) reported in the SEM image of the inset. The Bi NPs EDX spectrum (Figure 4a) displays the characteristic M-lines of Bi, at about 2.4 keV (M$\alpha$), the K-line of Si due to the silicon substrate, the O and C K-lines derived from contamination of the substrate. The signal from Na comes from the surfactant compounds used at a very low concentration to stabilize the Bi NPs. The O low presence indicates that the Bi is metallic and it is very little oxidized, as confirmed by our Raman spectroscopy, XRD, and XPS spectroscopy analyses and in agreement with the literature [8, 9].

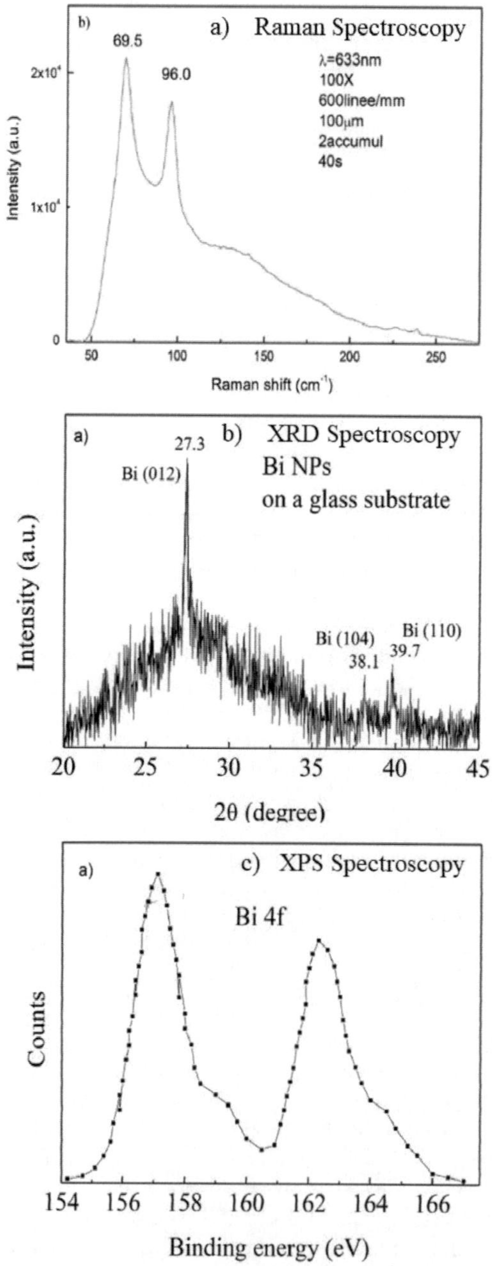


*Figure 5. Raman spectroscopy (a), X-ray diffraction analysis (b) and X-ray photoelectron spectroscopy (c) of the prepared Bi-NPs films.*

In particular, the characteristic first order Raman modes of rhombohedral bismuth are visible in the Bi NPs film Raman spectrum together with a broad shoulder attributable to a very slight oxidation of the Bi NPs (Figure 5a). The absence of considerable oxidized phase for the Bi NPs is also evident in the XRD (Figure 5b) and XPS (Figure 5c) spectra. In fact, in the former the typical peaks of the rhombohedral structure of metallic bismuth are only present, while in the latter the Bi 4f and O1s core-level regions are characterized by two asymmetrical peaks due to the Bi $4f_{7/2}$ and $4f_{5/2}$ spin-orbit components for Bi metal and by a low intensity signal for the O1s line.

The insertion of the nanoparticles in the cell cultures and in tissues and organs of living systems (mice) was performed by using a microsyringe to inject small quantitative of the solution in the extracellular liquids of the culture or in the mice's organ. Their transport, accumulation (uptake) and decay was monitored by using the X-ray fluorescence yield induced by 5 keV electrons from a SEM in the case of cells and by an X-ray tube operating at 45 kV in the case of living mice [10].

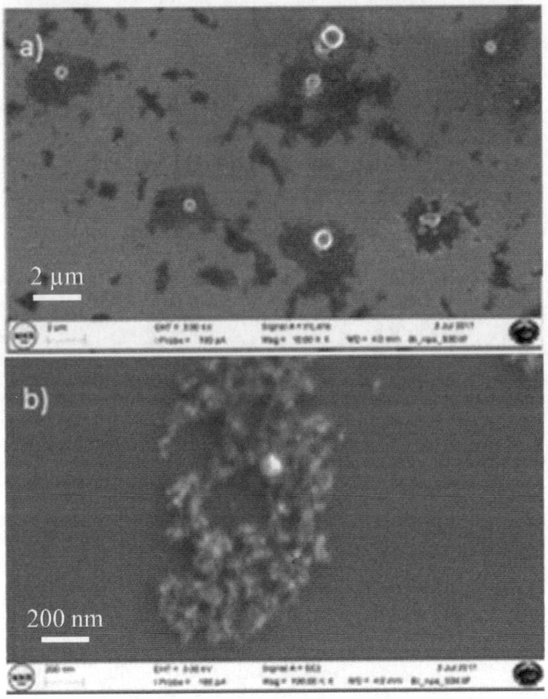


*Figure 6. SEM photos of the Bi-NPs deposited (white spots) together the physiological extracellular liquid on a silicon surface at low (a) and high magnification (b).*

Figure 6a shows a typical SEM investigation of Bi NPs in a physiological solution injected in the cell culture. It was observed that the Bi NPs have the possibility to go inside the cell, crossing its membrane, only if the Bi NPs dimension is lower than about 25 nm; in fact Bi NPs with higher diameters ($\sim$ 50 nm), as the one shown in Figure 6b, may be stopped on the cell membrane. Moreover the results indicate that the Bi nanoparticles are very well contrasted with respect to the surrounding low Z biological environment.

The use of Bi NPs as contrast medium in living tissues has been tested in mice by injecting the prepared biocompatible solution in some organs and by observing the Bi fluorescence image induced by the Bruker X-ray tube.

Figure 7a shows the image when the solution was injected into the heart. The image displays very well the organ (observed in dark grey scale in the X-ray transmission) and the nearest blood irrigated tissues appear as very bright due to the X-ray induced fluorescence. Figure 7b reports the image when the solution was directly injected in the Sx kidney. The picture illustrates very well the organ (observed in greyscale in the X-ray transmission) and the nearest blood irrigated tissues are very bright due to the X-ray induced fluorescence. The obtained medical images are well contrasted and the interested organ shows an high spatial resolution of the zones where the Bi-NPs are localized. The aim of the study was to distinguish the interested organ in order to observe it in detail before that a successive radiotherapy treatment can be applied to the diseased tissues [11]. For comparison, Figure 7c reports the X-ray image without the use of Bi-NPs.

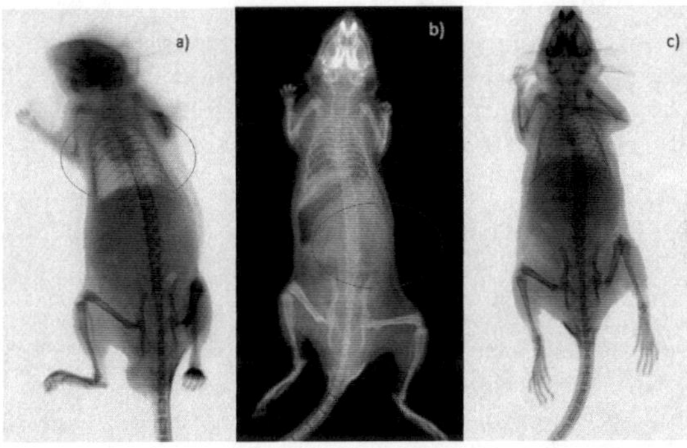


*Figure 7. X-ray fluorescence image showing the Au-NPs for an injection of solution in mice heart (a) and Sx kidney (b). For comparison the image without use of Bi NPs (c).*

## 4. Discussion and Conclusions

Our results have showed that Ag, Au and Bi NPs have good biocompatibility and safety and do not affect the normal functioning of the organs. Cells (in culture) are alive after injection of the solution into their extracellular liquid and mice are also alive after the NPs injection in the blood through the tail vein or in some organs.

Measurements have been performed on the uptake and the time decay of the NPs vs. time, demonstrating that a fast uptake ($\sim 1 \div 2$ hr) and a slowly decay ($\sim 12 \div 24$ hr) occurs, depending on the nature of the used nanoparticles and on the particular organ of the mice. The contrasts of the fluorescence images are good and improve with the NP concentration injected in the living system.

Thanks to this type of approach it will be possible, also guided by the literature [12] to design, to functionalize and to direct the nanoparticles in the affected tumour organs and, after the diagnostics concerning their spatial localization, to proceed to their irradiation with ionizing radiations for treatments based on the traditional radiotherapy (X-rays and electrons) and on the innovative one (protons and carbon ions). In this direction we are investigating more, overall by preparing, by laser ablation, different kinds of biocompatible nanoparticles, with diverse size and shape, and by taking into consideration the application of X-ray radiotherapy whose ionization cross section increases with the atomic number such as $Z^{4\div5}$. Of course, because of the interdisciplinary research aims, experts of physics, biology, chemistry and pharmacology are involved in this research.

## Acknowledgements

This research was supported by University of Messina Research & Mobility 2016 Project (project code RES_AND_MOB_2016_TORRISI) no. 74893496. Many thanks to Dr. G. Costa of MIFT Department for the support given to the preparation of this paper.

## References

[1] L. Torrisi. Radiotherapy improvements by using au nanoparticles. *Recent patents on Nanotechnology*, 9(2):114–125, 2015. doi:10.2174/1872210509021508191533355.

[2] L. Torrisi, N. Restuccia, S. Cuzzocrea, I. Paterniti, D. Ielo, S. Pergolizzi, M. Cutroneo, and L. Kovacik. Laser-produced au nanoparticles as x-ray contrast agents for diagnostic imaging. *Gold Bulletin*, 50(1):51–60, 2017. doi:10.1007/s13404-017-0195-y.

[3] L. Torrisi. Evaluation of the radiotherapy and proton therapy improvements using gold nanoparticles. *Gold Bulletin*, 50(4):299–311, 2017. doi:10.1007/s13404-017-0216-x.

[4] L. Torrisi. Gold nanoparticles enhancing protontherapy efficiency. *Recent patents on Nanotechnology*, 9(1):51–60, 2015. doi:10.2174/1872210509999141222224121.

[5] L. Torrisi and A. Torrisi. Laser ablation parameters influencing gold nanoparticle synthesis in water. *Radiation Effects and Defects in Solids*, 173(9-10):729–739, 2018. doi:10.1080/10420150.2018.1528598.

[6] J. Fang, K. L. Stokes, J. A. Wiemann, W. L. Zhou, J. Dai, F. Chen, and C. J. O'Connor. Microemulsion-processed bismuth nanoparticles. *Materials Science and Engineering: B*, 83(1-3):254–257, 2001. doi:10.1016/S0921-5107(01)00528-1.

[7] M. Maciulevičius, A. Vinčiūnas, M. Brikas, A. Butsen, N. Tarasenka, N. Tarasenko, and G. Račiukaitis. On-line characterization of gold nanoparticles generated by laser ablation in liquids. *Physics Procedia*, 41:531–538, 2013. doi:10.1016/j.phpro.2013.03.112.

[8] L. Torrisi, L. Silipigni, N. Restuccia, S. Cuzzocrea, M. Cutroneo, F. Barreca, B. Fazio, G. Di Marco, and S. Guglielmino. Laser-generated bismuth nanoparticles for applications in imaging and radiotherapy. *Journal of Physics and Chemistry of Solids*, 119:62–70, 2018. doi:10.1016/j.jpcs.2018.03.034.

[9] F. Xia, X. Xu, X. Li, L. Zhang, L. Zhang, H. Qiu, W. Wang, Y. Liu, and J. Gao. Preparation of bismuth nanoparticles in aqueous solution and its catalytic performance for the reduction of 4-nitrophenol. *Industrial & Engineering Chemistry Research*, 53(26):10576–10582, 2014. doi:10.1021/ie501142a.

[10] L. Torrisi, N. Restuccia, and I. Paterniti. Gold nanoparticles by laser ablation for x-ray imaging and protontherapy improvements. *Recent patents on Nanotechnology*, 12(1):59–69, 2018. doi:10.2174/1872210511666170609093433.

[11] P. K. Jain, I. H. El-Sayed, and M. A. El-Sayed. Au nanoparticles target cancer. *nano today*, 2(1):18–29, 2007. doi:10.1016/S1748-0132(07)70016-6.

[12] K. Huang, H. Ma, J. Liu, S. Huo, A. Kumar, T. Wei, X. Zhang, S. Jin, Y. Gan, P. C. Wang, S. He, X. Zhang, and X. Liang. Size-dependent localization and penetration of ultrasmall gold nanoparticles in cancer cells, multicellular spheroids, and tumors in vivo. *ACS nano*, 6(5):4483–4493, 2012. doi:10.1021/nn301282m.

doi:10.14311/ppt.2019.1.7
*Plasma Physics and Technology* **6**(1):7–9, 2019 

# OUTPUT CURRENT CONTROL SYSTEM OF A HIGH VOLTAGE ELECTRIC PULSE GENERATOR FOR PLASMA EXCITATION

D.V. Godun*, S.V. Bordusau, G.P. Budzko

*Belarusian State University of Informatics and Radioelectronics, P. Brovki 6, 220013, Minsk, Belarus*

* godundv@mail.ru

**Abstract.** A control and pulse discharge current limiting system integrated into an AC/DC converter and pulse modulator of a high voltage pulse generator have been developed. The peculiarity of such system's operation is the stabilization of the power supplied to the discharge and the correction of the width of output electric pulses towards decrease upon reaching the specified pulsed current amplitude value. The system enables the pulse generator to work in the modes close to the "short circuited load" mode. In this case the driving module of a composite IGBT key performs the correction of the working pulse width and blocks the pulse generator operation if needed. The suggested circuit design solutions allow using the generator in a wide range of electric plasma-forming parameters' modes and working with various types of vacuum gas discharge systems.

**Keywords:** output current, control, electric pulses, plasma load.

## 1. Introduction

It is known that for working with plasma load in order to protect a generator of electric pulses from the avalanche current flowing through the open crossover of the power IGBT key, it is necessary to provide high speed electronic protection system [1, 2]. It is due to the fact that unstable plasma load may lead to the change of its electric characteristics (as well as because of spontaneous transition into another kind of gas discharge). The speed and the value of the growing current, flowing across the power key, are directly determined by the conductivity and volume of the formed discharge zone. The stability of output electric pulse generator characteristics can be achieved with the specific circuit solutions and a modern component base.

The most important factor influencing the effective operation of a power pulse generator is the right choice of the key IGBT transistor. The key block must have the necessary frequency properties enabling it to perform commutation of the load at the specified frequencies, provided the rate of power pulses rise corresponds to the required. Non-compliance of the frequency properties of the selected key can cause a breakdown of a transistor, distortion of the power pulse fronts, the emergence of inflowing currents and the increase of the formed plasma discharge reactivity [3].

The most perspective concept of designing a power key for a universal pulse plasma discharge systems is the use of high speed discrete keys with sequential switching [4]. Such circuit solution allows to get the required steepness of the formed pulses fronts at frequencies up to 60 kHz while switching pulses currents up to 80 A.

Because of the peculiarities of a pulse generator application for synthesizing an unstable plasma, it is desirable to embed a system of pulse current control into an output generator circuit. Such system should provide the required time for safety shutdown and provide the function of a controlled turn off threshold. To reduce the number of false protection block's triggering, the preliminary amplifier should be based on a separate low noise operational amplifier and provide the selection of the desired signal with the help of passive filtering elements.

## 2. Main part

In reference source [5] some circuit solutions of designing a powerful high voltage electric pulse generator with the functions of stabilizing the current discharge as well as the voltage at a discharge gap are presented.

In the designed high voltage pulse generator, to provide a breakdown of gas discharge medium at different work pressure, control and a mode of stabilizing the voltage amplitude applied to the gas discharge gap are provided. The availability of such option in the pulse generator allows to control the amplitude of pulses in the interval from -600 to -1200 V with the function of limiting the current in the load circuit. It allows to abandon the additional high voltage igniting devices. It makes the cost of a generator significantly lower, simplifies the construction and improves its operation reliability. The availability of an output pulse voltage control makes it possible to use the pulse generator for different discharge systems.

The discharge current stabilization mode is required to move smoothly through the voltage-current characteristic of the pulse discharge. It is also necessary to take care of limiting the pulse power consumed by the non-stationary plasma discharge. After the gas breakdown the pulse generator allows to work in different stationary modes, even close to the mode

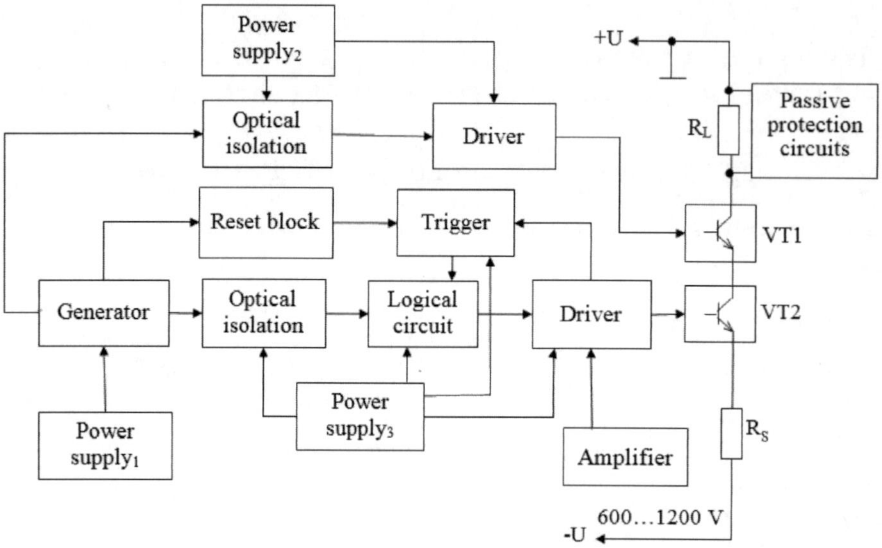


*Figure 1. Structural diagram of a driver device with the function of pulse discharge current limiting*

of forming an arc discharge. The implemented circuit design solutions provide a decreased probability of damaging the processed samples due to the formation of a great number of micro-arcs in a limited volume [6]. The mentioned effect can be eliminated by increasing the off-duty factor of power pulses both manually and automatically. Note, the system, in this case, increases the amplitude value of current pulses and this, in practice, ensures maintaining constant level of the pulse power applied to the discharge. The implementation of the stabilization mode of the discharge pulse current allows a significant reduction in the pulse current growth rate through the power key when operating in non-stationary modes. This is one of the most important advantages of the system under study.

An additional barrier of protecting power semiconductor devices from admissible pulse current excess is an integrated protection system built in a driver module.

A structural diagram of the designed driver module with the function of pulse current limiting is shown in Fig.1.

As it has been mentioned above, a composite pulse key on the discrete IGBT transistors connected in series is the most preferable variant of designing a pulse modulator. Because of a standard construction of the pulse modulator to control the upper key at VTI transistor, let's analyze the functioning of the lower key based on the VT2 transistor with its functional peculiarities.

The generator forms one channel of program-controlled pulses that get simultaneously through the optical converters block to the driver module of the upper key and through the logical circuit to the driver module of the lower key. The driver forms the signals controlling the VT1 transistor. In a normal mode of system operation, the two keys must open in-phase. The logical circuit is implemented for converting signals into the required type according to the developed operation algorithm. After processing in the block of logical conversion, the pulses get to the input of the lower key driver module. The lower IGBT transistor is controlled in the same way as the upper key. In the presented structural diagram, the formed plasma non-stationary discharge is indicated as $R_L$. The voltage controlled AC/DC converter with the function of stabilizing the output power through $R_L$, the composite key and the current sensor is connected to the ground. The signal, proportional to the value of the flowing current, is taken from the sensor $R_S$ and is passed to the low noise amplifier with a controlled amplification coefficient. The use of passive correction circuits in the scheme of a low-noise amplifier of provides the filtration and conversion of the signal which subsequently is passed to the input of the current control of lower key driver module. The driver's micro circuit at this input differentiates two modes – overload and short circuit mode. All pulse current control systems are included into driver's microcircuit. According to the technical characteristics of the driver's microcircuit the time of the safety shutdown does not exceed 3 microseconds. This has been proved by practical experiments. When in an overload mode, the driver makes the lower key control pulse narrower, thus, limiting the average value of the pulse current flowing through the composite transistor. In the short circuit mode, the driver module blocks the operation of the lower IGBT transistor, forms an error signal which is passed to the trigger circuit and, further, a blocking signal of module operation is formed. The withdrawal of the system from this state is possible on an user's external signal, using the reset block included in the system.

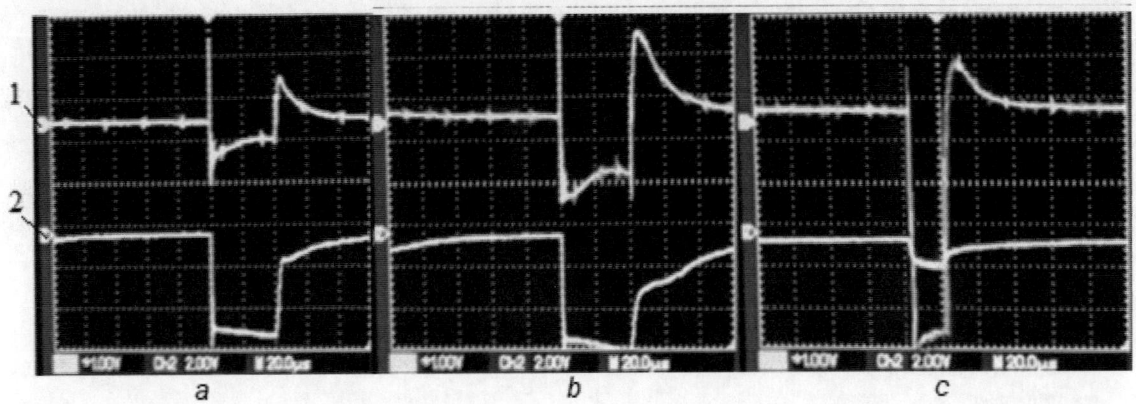


*Figure 2. Oscillograms of current pulses and voltage at plasma load at f=5 kHz; 1- signal from current pulse; 2- pulse voltage signal*

The presented device has been implemented and tested as part of the high voltage pulse generator. The examples of the developed generator's output parameters control system while working with a non-stationary plasma load for the output pulses' frequency f=5 kHz are given in Fig.2.

Fig.2a shows the pulse current and voltage oscillograms at plasma load. Fig.2b shows the change of current and voltage pulses when the consumed power of the discharge increases. Fig.2c shows the oscillogaram of current and voltage pulses when the pulse power supply source works in the mode of discharge current stabilization close to the short circuit mode. One can see that when the generator works for plasma load using the function of stabilization of the power applied to the discharge, it is possible to maintain different current and voltage modes in the load, i.e. to move along the voltage-current characteristic of the discharge.

## 3. Conclusion

A system for the output current control of a high-voltage electric pulse generator has been developed and tested with the formation of a non-stationary plasma discharge. It has been experimentally established that the reaction time of the system to the excess of pulse value of the current flowing through the non-stationary discharge does not exceed 3 microseconds. Taking into account that the system already includes the circuit of current negative feedback which provides the stabilization of AC/DC converter's output power, the rate of current growth in the load circuit in emergency situations significantly decreases and the reaction time of the system is sufficient to protect the key IGBT transistors from overloading.

### Acknowledgements

This research has been supported by Belarusian State Program of Scientific Research "Science of physical materials, new materials and technologies", assignment 3.5.02

### References

[1] C. Yong-Nong and K. Chih-Ming. Design of plasma generator driven by high-frequency high-voltage power supply. *Journal of Applied Research and Technology*, 11(2):225–234, 2013. doi:10.4028/www.scientific.net/amm.284-287.1136. doi:10.4028/www.scientific.net/AMM.284-287.1136.

[2] J. E. Harry. *Introduction to Plasma Technology: Science, Engineering, and Applications*. Wiley-VCH, 2010.

[3] A. Polishchuk. Selection of key transistors for hard-switching converters. *Power Electronics*, (2):22–25, 2004.

[4] A. Wintrich, U. Nicolai, W. Tursky, and R. T. Problems of parallel and serial igbt connection. part 2. sequential work of the igbt. *Power Electronics*, (5):34–40, 2013.

[5] D. Godun, S. Bordusov, and A. Dostanko. The software-controlled generator of high-voltage impulses for plasma technological application. *Acta Technica*, (56):332–337, 2011.

[6] D. Godun, S. Bordusov, and A. Dostanko. Systems of electronic overcurrent protection in pulse power generator operating on plasma load. *Acta Polytechnica*, 53(2):152–154, 2013.

doi:10.14311/ppt.2019.1.10
*Plasma Physics and Technology* **6**(1):10–14, 2019 

# EXPERIMENTAL STUDY ON DYNAMIC WELDING OF AgW60, CuW60 AND Cu CONTACT MATERIALS IN SYMMETRICAL AND ASYMMETRIC PAIRINGS AT DC 270 V/200 A

Liao Yuan[a], Zhang Dandan[a], He Mengbing[a], Makoto Hasegawa[b], Li Zhenbiao[a,*]

[a] *State Key Laboratory of Advanced Electromagnetic Engineering and Technology, School of Electrical and Electronic Engineering, Huazhong University of Science and Technology, Wuhan 430074, China*

[b] *Chitose Institute of Science and Technology, Chitose, Hokkaido, 066-8655, Japan*

[*] zbli@hust.edu.cn

**Abstract.** Experiments were conducted for AgW60, CuW60 and Cu contacts and their different pairings at DC 270 V/200 A in a surrounding atmosphere of carbon dioxide or nitrogen. Welding resistances, average welding forces and relationships between welding forces and the number of operations required for welding were measured and analyzed. The anti-welding ability of the three materials in their symmetric pairings in carbon dioxide and nitrogen can be ranked as AgW60>Cu>CuW60. The "intermediate effect" occurs in the anti-welding properties of the asymmetric pairings. The relationships between the number of operations required for welding and the average welding force is approximately in an inverse proportional function.

**Keywords:** surrounding atmosphere, contact material, DC, welding, welding force, AgW60, CuW60, Cu.

## 1. Introduction

Switching capabilities in large DC load conditions while realizing miniaturization and long life are strongly required for switching devices such as relays and contactors [1]. For example, solar power generation facilities and energy storage batteries becomes popular increase year by year, leading to larger demands for high voltage/current switches. Relay contacts are subjected to heavier loads more and more, and their failures are sometimes inevitable. Dynamic welding is one of the most serious failures. With high voltage/current levels, different contact materials and ambient atmospheres have important influences on anti-welding abilities of contacts for improving reliabilities and electrical life of miniaturized high voltage/current relays. However, at present, there are few experimental studies on effects of different contact materials and their matching forums as well as ambient atmospheres on welding characteristics of DC contacts.

In this paper, dynamic welding characteristics were studied and analyzed with AgW60, CuW60 and Cu contacts under DC resistive 270 V/200 A in carbon dioxide or nitrogen with 9 kinds of symmetrical and asymmetrical material pairings.

## 2. Experimental method

Fig.1 shows a schematic diagram of the experimental circuit in this study. The main circuit included a DC regulated power source E providing an output voltage of 270 V, a load resistor R of 1.35 Ω, and a pair of electrical contacts K which was set on the simulation test device. The voltage $U_N$ between the pair of contacts K was measured with a voltage divider 1, while the voltage $U_M$ was measured with another voltage divider 2.

A schematic diagram of the electrical contact simulation device K is shown in Figure 2. The contacts are normally closed, and the electromagnet provides a breaking force, while the spring provides a closing force. A force sensor is connected to one side of the static contact. At each contact operation, the welding will be cooled and solidified after the material is melted, resulting in a different degree of welding. The breaking force at the moment of separation of the two contact surfaces in the breaking operation is taken as the welding force $F_W$ of that time.

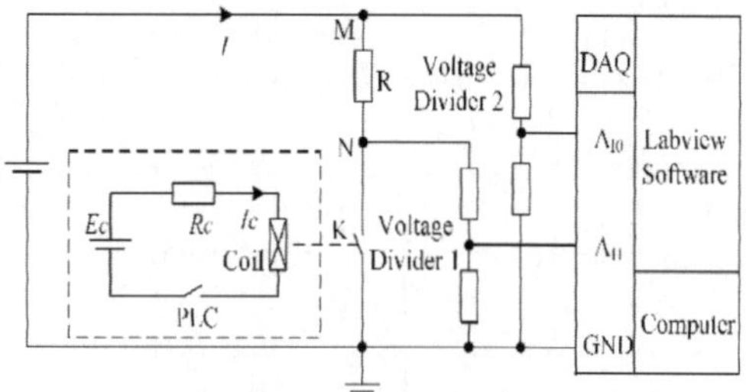


*Figure 1. A schematic diagram of the experimental circuit.*

AgW60, CuW60 and Cu contact materials were studied. One make-and-break operation was set as 0.7 s ON and 2.3 s OFF. Under the experimental conditions, because the division strategy is relatively high, the setting is the contact turn-on time is much shorter than the break-off time, which can ensure more operation times and facilitate the study of various electrical properties of the contacts. Magnetic blow-out was created at the magnetic flux density of 110 mT with a pair of magnets. Before the experiment, the contact simulation device was placed in a vacuum simulation chamber as a whole, and carbon dioxide or nitrogen gas with a pressure of $1 \times 10^5$ Pa was charged. The two samples 1# and 2# were tested for each pair of contacts in both carbon dioxide and nitrogen. In the experiment, when the first welding occurred, the electrodes were manually disconnected and the experiment was resumed up to the third welding incidence. For the convenience of description, the number of operations required for each of the three welding incidences were recorded as$N_1$,$N_2$,and $N_3$,respectively (each counting was started from 1).

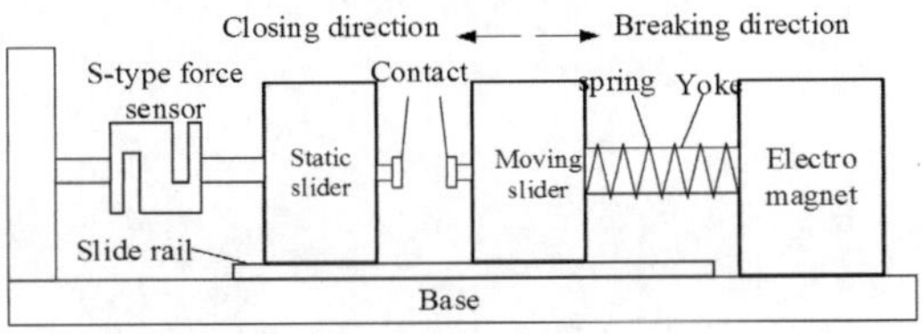


*Figure 2. Electrical contacting simulator.*

## 3. Experimental results and analysis

### 3.1. Contact welding resistance

Welding resistance of contacts is defined by the number of operations when welding occurs for the first time. The larger the number of operations for the first welding, the higher the welding resistance of contacts. In this study, the average number of operations $\overline{N}_1$ for the results of the two samples 1# and 2# is used to determine the welding resistance capability. The calculation formula of $\overline{N}_1$ is:

$$\overline{N}_1 = (N_1(1) + N_1(2))/2 \tag{1}$$

where $N_1(j)(j = 1,2)$ is the number of operations when the $j$-th sample of contact pairs is first welded.

Tables 1 and 2 show the results of the average first welding values $\overline{N}_1$. The higher the value of $\overline{N}_1$, the higher the welding resistance capability. The welding resistance capabilities in carbon dioxide and nitrogen are not significantly different from each other, while the contact materials have strong influences. For example, AgW60 and CuW60 have smaller values of $\overline{N}_1$ in nitrogen than in carbon dioxide, which means their welding resistance in nitrogen is lower than that in carbon dioxide. However, Cu shows the opposite tendency. The order of the welding resistances of these materials in symmetric pairings in carbon dioxide and nitrogen is: AgW60>Cu>CuW60.

The welding resistance capabilities in the asymmetric pairings in carbon dioxide and nitrogen are mostly "intermediate effect". Specifically, when the materials A and B are asymmetrically paired, their welding resistance (corresponding to the number of operations during the first fusion welding, recorded as $N_{A-B}$ ) is between the anti-welding abilities of the symmetrical pairings (recorded as $N_{A-A}$, $N_{B-B}$) where $A - A$ or $B - B$ indicates symmetrical pair (i.e., $N_{A-A} > N_{A-B} > N_{B-B}$ ). In the total of twelve asymmetric pairs in carbon dioxide and nitrogen, the "worst effect" ($N_{A-B} < N_{A-A}, N_{A-B} < N_{B-B}$) and "optimal effect" ($N_{A-B} > N_{A-A}, N_{A-B} > N_{B-B}$ occur only once. The specific results are shown in Table 3.

| Cathode(-) / Anode(+) | AgW60 | CuW60 | Cu |
|---|---|---|---|
| AgW60 | 1262 | 57 | 100 |
| CuW60 | 142 | 6 | 113 |
| Cu | 29 | 9 | 27 |

*Table 1. The average first welding $\overline{N}_1$ value of contact material in carbon dioxide.*

| Cathode(-) / Anode(+) | AgW60 | CuW60 | Cu |
|---|---|---|---|
| AgW60 | 428 | 15 | 249 |
| CuW60 | 4 | 2 | 59 |
| Cu | 22 | 5 | 153 |

*Table 2. The average first welding $\overline{N}_1$ value of contact material in nitrogen.*

According to the following formula, the number of operations $\overline{N}_k$ required for each welding in the carbon dioxide and nitrogen can be calculated:

$$\overline{N}_k = (N_3(1) + N_3(2))/6 \tag{2}$$

$\overline{N}_3(j)(j = 1,2)$ is the number of operations for the third welding incidence of the $j$-th sample. Table 4 shows the obtained results for $\overline{N}_k$.

Comparing Table 4 with Tables 1 and 2, it can be seen that as the number of operations increases, the number of operations between the consecutive welding incidences is reduced, indicating weaker welding resistance. In Table 4, the number of operations required for each welding in carbon dioxide and nitrogen is obtained by diving the number of operations in the third welding by 3. Among the 18 sets of values, 15 sets shows a smaller value than the first welding.

### 3.2. Average welding force of contacts

The average value of the welding force of all the breaking operations before the first welding was recorded as $\overline{F}_{w1}$ Table 5 and Table 6 show their results in carbon dioxide and nitrogen, respectively. The similar average welding force value of all the effective operations up to the third welding can be called as the total average welding force, and recorded as $\overline{F}_w$. Tables 7 and 8 show their results in carbon dioxide and nitrogen, respectively.

| Gases | Pairing material | Worst effect | Intermediate effect | Optimal effect |
|---|---|---|---|---|
| | AgW60, CuW60 | | AgW60(+)& CuW60(+) | |
| Carbon dioxide | AgW60, Cu | | AgW60(+)&Cu(+) | |
| | CuW60, Cu | | Cu(+) | CuW60(+) |
| | AgW60, CuW60 | | AgW60(+)&CuW60(+) | |
| Nitrogen | AgW60, Cu | Cu(+) | AgW60(+) | |
| | CuW60, Cu | | CuW60(+)&Cu(+) | |

Note: The pairing form of each polarity effect is only labeled with the anode material, and the cathode is another material.

*Table 3. Three polar effects of anti-welding ability corresponding to asymmetrical mating.*

| | | $CO_2$ | $N_2$ |
|---|---|---|---|
| Symmetric pairing | AgW60 | 489 | 190 |
| | CuW60 | 6 | 4 |
| | Cu | 20 | 109 |
| Asymmetric pairing | AgW60(+)CuW60(-) | 188 | 12 |
| | AgW60(-)CuW60(+) | 81 | 2 |
| | AgW60(+)Cu (-) | 383 | 115 |
| | AgW60(-)Cu (+) | 10 | 13 |
| | CuW60(+)Cu (-) | 68 | 51 |
| | CuW60(-)Cu (+) | 9 | 1 |

*Table 4. The value of average number of operations $\overline{N}_k$ required for each weld.*

Table 9 shows the three polar effects of total average welding force when the contacts are asymmetrically paired in carbon dioxide and nitrogen.

Table 5 - Table 8 shows that as the number of operations increases, the average welding force of the contacts decreases. The three materials in Tables 7 and 8 have a total of 18 pairings, of which the total average welding force of the 16 pairs is lower than the average welding force before the first welding in Tables 5 and 6.

Table 9 shows that in carbon dioxide and nitrogen, the "worst effect", "intermediate effect" and "optimal effect" of the asymmetric average pairing of AgW60, CuW60 and Cu materials exist.

## 3.3. Relationship between welding force and the number of operations during welding

By correlating the data in Tables 1 and 2 with the data in Tables 5 and 6, relationships between the average number of operations $\overline{N}_1$ in the first welding and the average welding force $\overline{F}_{w1}$ before the first welding can be shown as in Fig. 3 in log scale. Similarly, by correlating the data in Tables 4 with the data in Tables 6 and 7, relationships between the average number of operations required for each welding incidence and the total average welding force can be shown as in Fig.4 in log scale.

Fig. 3 shows that the larger the average welding force $\overline{F}_{w1}$ before the first welding, the smaller the average number of operations $\overline{N}_1$ in the first welding, and the relationship between them is approximately an inverse proportional function (power function). The fitting relationship in Fig.4 is $y = 28.263x^{-0.989}$, the goodness of fit $R^2 = 0.8812$.

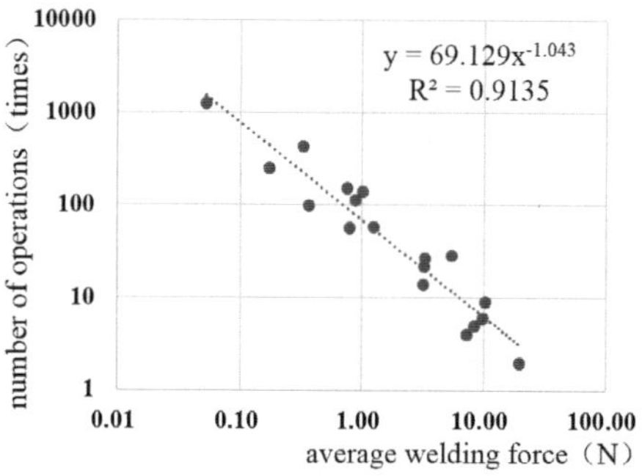


*Figure 3. The relationship between $\overline{N}_1$ and $\overline{F}_{w1}$.*

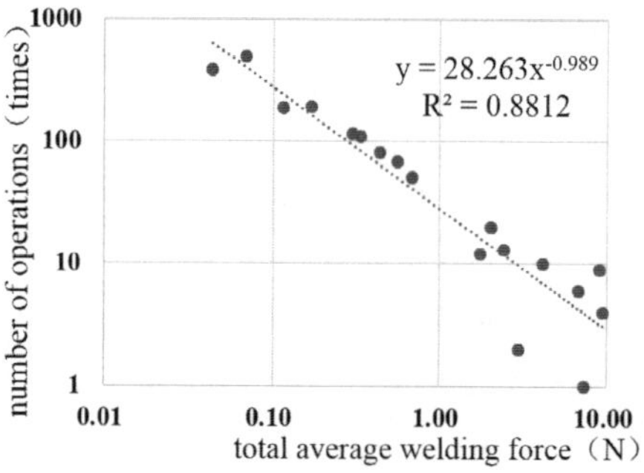


*Figure 4. The relationship between $\overline{N}_k$ and $\overline{F}_w$.*

## 4. Discussions

The occurrence of welding during the closing or breaking process of the contact material mainly comprises two processes: Process 1 is a heating process, reflecting the time during which the material per unit area of the contact surface melts after the arc energy is injected, thereby estimating the welding tendency of the contact material. Process 2 is a cooling process that reflects the area of fusion and the amount of welding force. The welding tendency and the welding force of the integrated contact are related to the welding resistance of a certain contact material [2]. In this paper, as the number of operations increases, the average welding force of the contacts decreases,

| Gases | Pairing material | Worst effect | Intermediate effect | Optimal effect |
|---|---|---|---|---|
| Carbon dioxide | AgW60, CuW60 | | AgW60(+)&CuW60(+) | |
| | AgW60, Cu | Cu(+) | | AgW60(+) |
| | CuW60, Cu | Cu(+) | | CuW60(+) |
| Nitrogen | AgW60, CuW60 | | AgW60(+)&CuW60(+) | |
| | AgW60, Cu | Cu(+) | AgW60(+) | |
| | CuW60, Cu | | CuW60(+)&Cu(+) | |

*Table 9. Three polar effects of anti-welding ability corresponding to asymmetrical mating.*

| Anode(+) \ Cathode(-) | AgW60 | CuW60 | Cu |
|---|---|---|---|
| AgW60 | 0.05 | 0.78 | 0.36 |
| CuW60 | 1.00 | 9.84 | 0.88 |
| Cu | 5.45 | 10.25 | 3.29 |

*Table 5. Average welding force $\overline{F}_{w1}$ before the first welding in carbon dioxide [N].*

| Anode(+) \ Cathode(-) | AgW60 | CuW60 | Cu |
|---|---|---|---|
| AgW60 | 0.33 | 3.20 | 0.17 |
| CuW60 | 7.25 | 19.40 | 1.23 |
| Cu | 3.25 | 8.36 | 0.75 |

*Table 6. Average welding force $\overline{F}_{w1}$ before the first welding in nitrogen [N].*

| Anode(+) \ Cathode(-) | AgW60 | CuW60 | Cu |
|---|---|---|---|
| AgW60 | 0.07 | 0.12 | 0.04 |
| CuW60 | 0.04 | 6.70 | 0.56 |
| Cu | 4.11 | 8.95 | 2.01 |

*Table 7. Total average welding force $\overline{F}_{w}$ in carbon dioxide [N].*

| Anode(+) \ Cathode(-) | AgW60 | CuW60 | Cu |
|---|---|---|---|
| AgW60 | 0.17 | 1.75 | 0.30 |
| CuW60 | 2.94 | 9.45 | 0.68 |
| Cu | 2.42 | 7.24 | 0.34 |

*Table 8. Total average welding force $\overline{F}_{w}$ in carbon dioxide [N].*

but the welding resistance of the contacts is weakened. The reason may be that the number of operations is increased, the heat generation is increasing, leading to higher temperature. This will lead to a shorter time required for the contact material to melt, and thus, welding becomes likely to occur, resulting in the weaker welding resistance of the contacts.

The surface conditions directly affects the resistance of the contact [3]. Comparing Fig. 3 and Fig. 4, the goodness of fit of the power function for the total average welding force and the average number of operations required for each weld is lower than the goodness of fit of the power function for the average welding force before the first welding and the average number of operations during the first welding. The reason may be: as the number of operations increases:

1. Under the action of arc heat, the lower melting point metals such as Ag and Cu in the two-phase pseudoalloys such as AgW60 and CuW60 flow under the action of electromagnetic force, and vaporize even splashing, so that the composition of the contact surface is no longer uniform, locally become a single-phase Ag, Cu or W skeleton;
2. The pseudo-alloy containing tungsten forms a silver tungstate $Ag_2WO_4$ or tungsten oxide $WO_3$ when it is working in carbon dioxide [4];
3. The surface of the contact is ablated by the arc to produce different degrees of defects.

## 5. Conclusion

In this paper, the following conclusions can be drawn:

1. The welding resistance of the three materials in the symmetric pairing of carbon dioxide and nitrogen is ranked as AgW60>Cu>CuW60. The "intermediate effect" occurs in the welding resistance of asymmetric pairing.
2. As the number of operations increases, the number of operations between each welding incidence is reduced, indicating weaker welding resistance.
3. As the number of operations increases, the average welding force of the contacts decreases.
4. The relationship between the average number of operations of the first welding of the contact and the average welding force before the first welding is approximately in an inverse proportional function. The relationship between the number of operations required for each occurrence of welding and the total average welding force is approximately in an inverse proportional function.

### References

[1] Y. Shiba, Y. Morishita, S. Kaneko, S. Okabe, H. Mizoguchi, and S. Yanabu. Study of DC circuit breaker of $H_2$-$N_2$ mixture gas for high voltage. *Ieee Transactions on Power & Energy*, 174(2):9–17, 2008. doi:10.1541/ieejpes.128.1407.

[2] X. Liu. A new mathematical model of dynamic welding for electric contacts. *Transactions of China Electrotechnical Society*, pages 81–84, 2002.

[3] F. Meng, J. Gao, Y. Wang, L. Qiao, and L. I. Rui. The study on process and properties of Ag-W contact material. *Materials Review*, pages 321–324, 2006.

[4] F. Q. Dou, J. Zheng, Z. H. Ren, and L. Jian. Study on the arc-erosion properties of AgW and $AgSnO_2$ used in the simulated conditions of automobile relays. *Electrical Engineering Materials*, pages 6–9, 2009.

doi:10.14311/ppt.2019.1.15
*Plasma Physics and Technology* **6**(1):15–18, 2019 

# INVESTIGATIONS ON THE EFFECT OF THE NOZZLE MATERIAL ON THE INTERRUPTION CAPABILITY OF A MEDIUM VOLTAGE LOAD BREAK SWITCH

M. Bendig[a,*], T. Krampert[a], N. Götte[a], A. Kalter[b], M. Schaak[b], K. Ermeler[c]

[a] *Institute for High Voltage Technology, RWTH Aachen University, Schnikelstr. 2, 52056 Aachen, Germany*
[b] *Siemens AG, Energy Management Division, Carl-Benz-Str. 22, 60386 Frankfurt am Main, Germany*
[c] *Siemens AG, Energy Management Division, Nonnendammallee 104, 13629 Berlin, Germany*
[*] bendig@ifht.rwth-aachen.de

**Abstract.** In the process of substituting sulphur hexafluoride in medium voltage load break switches by atmospheric gases, the inferior arc quenching capabilities of possible substitutes have to be compensated. By introducing a polymer nozzle into the switching gap of a load break switch, the interruption capability can be enhanced as the ablated nozzle material changes the composition of the arc plasma. In this contribution the interruption capability of a model load break switch is investigated using different nozzle materials. The results show a good interruption capability when using polypropylene and polyamide 6.6 for high blowing pressures. Polytetrafluorethylene shows good results across a wide blowing pressure range. Polylactide has the lowest interruption capability among the polymers in this work.

**Keywords:** load break switch, nozzle material, ablation, alternative gases.

## 1. Introduction

Medium voltage load break switches (LBS) are widely used switchgear in the secondary distribution grid. They serve as combined load and disconnector switches and are capable of interrupting load currents in the range of a few hundred amps. Due to the space limitations in urban areas LBS are mainly built into metal-enclosed switchgear filled with the insulation gas sulphur hexafluoride ($SF_6$). $SF_6$, with its high dielectric strength and superior arc quenching capabilities, allows for a compact and reliable switchgear design. Nevertheless, $SF_6$ has a global warming potential of 23500 carbon dioxide ($CO_2$) mass equivalents, which makes it the most potent greenhouse gas known. Possible atmospheric substitutes like air or $CO_2$ have been investigated in previous works. As their arc quenching capability is inferior to $SF_6$, additional measures like an axial arc blowing have to be applied [1, 2]. One way of enhancing the interruption capability of a gas switch is by introducing a gassing polymer nozzle into the switching gap. If a switching arc in drawn between the contacts, the polymer ablates and the released gas is changing the composition of the plasma. It has been shown, that polymers that release either fluorine or hydrogen are good candidates to be used as a nozzle in LBS [3, 4].

This work analyses the influence of the applied nozzle material on the interruption capability of a model load break switch. The investigated polymers are polytetrafluorethylene (PTFE), as a polymer releasing fluorine as well as polypropylene (PP), polyamide 6.6 (PA6.6) and polylactide (PLA) as polymers releasing hydrogen. All tests were performed using a nitrogen - carbon dioxide mixture as quenching and insulation gas at an elevated filling pressure of $p = 1800\,\mathrm{hPa}$ absolute to enhance the electric strength.

## 2. Experimental method

To evaluate the switching performance the thermal interruption capability as well as the rate of the dielectric recovery are determined individually. By using this method, the effects of the used nozzle material can be analyzed for the different phases.

### 2.1. Thermal interruption limit

A successful thermal interruption is dependent on the current steepness shortly before current zero (CZ) as well as on the voltage steepness shortly after CZ. In this work the critical current steepness $\mathrm{d}i/\mathrm{d}t_{\mathrm{crit}}$ is used to quantify the thermal interruption capability. For the determination the synthetic test circuit shown in Figure 1 is used. The high current circuit provides one half oscillation of a sinusoidal current with a value of $I_{\mathrm{rms}} = 630\,\mathrm{A}$ and a frequency of $f_1 \approx 50\,\mathrm{Hz}$. Shortly before the current reaches its natural CZ the thyristor in the high voltage circuit is triggered and a current with an amplitude in the range of a few ten amps and a frequency of $f_2 \approx 1000\,\mathrm{Hz}$ is superimposed over the test current. Using this superposition method, the current steepness at CZ can be varied independent of the test current amplitude. When the current is interrupted successfully a recovery voltage rises across the

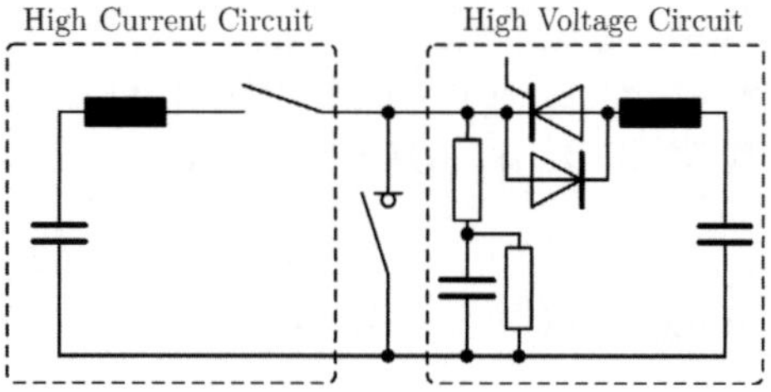


*Figure 1. Test circuit to determine the thermal interruption capability*

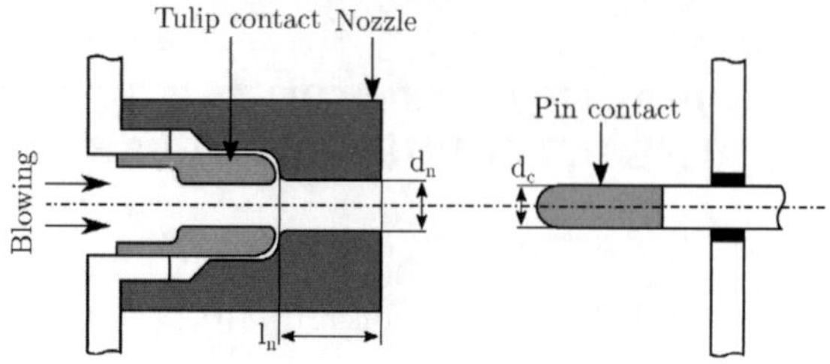


*Figure 2. Section of the model load break switch*

| Property | PTFE | PP | PA6.6 | PLA |
|---|---|---|---|---|
| $\rho$ [g/cm$^3$] | 2.16 | 0.91 | 1.15 | 1.15 |
| $T_L$ [°C] | $\approx 330$ | $\approx 160$ | 260 | 215 |
| Color | Gray | Gray | Natural | White |
| Vapor | F | H | H | - |

*Table 1. Properties of the investigated polymers*

open contacts. To set the rate of rise of the recovery voltage (RRRV) a parallel resistor with $R_p = 220\,\Omega$ is used. This resistor is chosen in a way that the RRRV corresponds to the RRRV of a standardized test for $U_N = 24\,\mathrm{kV}$ when a current steepness according to the steepness of a 630 A test at a frequency of $f = 50\,\mathrm{Hz}$ is set. As the parallel resistor is fixed, the RRRV changes when the current steepness is changed.
At the beginning of a test series a low current steepness ($\mathrm{d}i/\mathrm{d}t$) is chosen and five successful interruption tests are performed. Afterwards the $\mathrm{d}i/\mathrm{d}t$ is increased and another five interruptions are performed. This procedure is repeated until all five interruptions for one $\mathrm{d}i/\mathrm{d}t$ fail. Afterwards the measured $\mathrm{d}i/\mathrm{d}t$s and RRRVs are plotted into a diagram and a linear regression is performed to make sure that all tests have the right ratio between $\mathrm{d}i/\mathrm{d}t$ and RRRV. In a next step the critical RRRV is determined as the mean value between the lowest failed test and the next lowest successfull test. Afterwards, the critical current steepness is determined as the product of the regression slope and the critical RRRV. By using this method, inaccurancies from the low signal to noise ratio of the current signal around CZ are avoided.

### 2.2. Dielectric recovery

The dielectric recovery of the model switch is determined with the test method and circuit described in [5]. The test object is stressed with a half oscillation of a current with a value of $I_{\mathrm{rms}} = 630\,\mathrm{A}$. Afterwards, the breakdown voltage at a certain delay after CZ is determined by applying a steep impulse voltage to the test object. This procedure is repeated for several delay times to get a time resolved course of the breakdown voltage after current zero. For each delay time a minimum of four breakdown tests is performed.

## 3. Test setup

A sectional view of the model switch is shown in Figure 2. The contact system consists of a combination of a tulip and a pin contact made from a tungsten-copper-alloy with a weight ratio of 80/20. The maximum contact distance is $s = 58{,}5\,\mathrm{mm}$ and the contact diameter of the pin is $d_c = 10\,\mathrm{mm}$. The switching arc is cooled by an axial blowing through the tulip contact. The blowing is initiated by a magnetic valve, which is connected to an external gas volume. The blowing pressure is measured upstream of the nozzle using a Pitot probe and is given as the total pressure drop across the tulip and nozzle. All investigated pressures in this work lead to a subsonic flow. The tulip contact and part of the contact distance is surrounded by a polymer nozzle with a nozzle throat diameter of $d_n = 11\,\mathrm{mm}$ and a nozzle throat length of $l_n = 24\,\mathrm{mm}$. This configuration was identified as a good parameter set in previous work [1]. The investigated nozzle materials in this work are shown in Table 1. The density $\rho$ as well as the melting point $T_L$ vary across the different materials. While the used PTFE, PP and PLA are colored, the PA6.6 is used in its natural ivory color. If ablated by the arc, the resulting vapor mainly consists of fluorine for PTFE and hydrogen for PP and PA6.6 [4, 6, 7]. For PLA no data was found on the ablation behavior. All nozzles are milled except of the PLA nozzle, which is 3D printed. Due to the manufacturing the internal surface as well as the structure of the material differs from the other nozzles.
The model switch is operated by a pneumatic piston drive with a stroke of 90 mm and an average opening speed of $v_{\mathrm{m}} = 4{,}5\,\mathrm{m\,s^{-1}}$. The stroke is recorded during the opening process using a laser triangulation measurement system.

## 4. Results

The thermal interruption capability is tested for nozzles made from PTFE, PP, PA6.6 and PLA. The dielectric recovery is only analyzed for PTFE, PP and PA 6.6. Additionally the mass losses due to ablation are determined for all investigated materials.

### 4.1. Thermal interruption capability

The critical current steepnesses for the different materials are shown in Figure 3 for different blowing pressures. The lines are drawn for better visualization. For the maximum blowing pressure and when using PP or

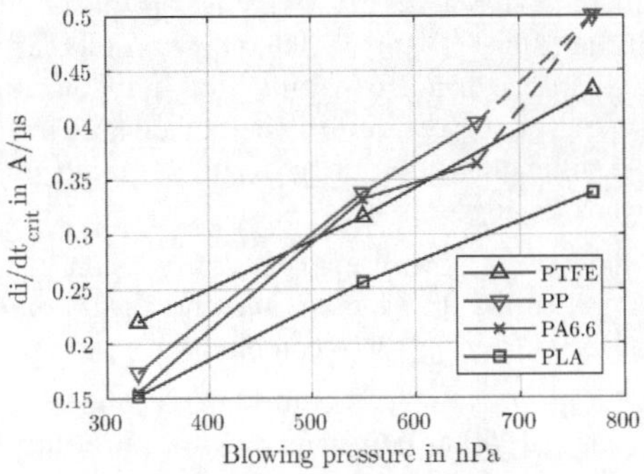


*Figure 3. Thermal interruption capability for different nozzle materials and blowing pressures*

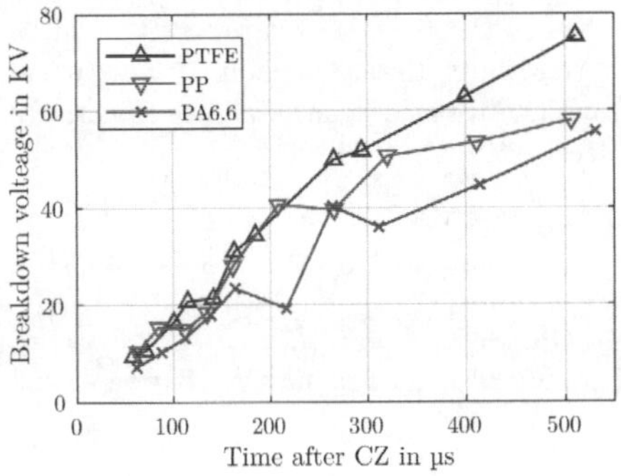


*Figure 4. Dielectric recovery after current zero for different nozzle materials*

PA6.6 no thermal failure occurs with the maximum steepness of the test circuit $di/dt_{max} = 0{,}5\,\mathrm{A\,\mu s^{-1}}$. Dashed lines indicate that the critical steepness is above this level.

For all investigated nozzle materials the critical current steepness rises with increasing blowing pressure. However, there are differences in the pressure dependence. When using PTFE or PLA as a nozzle material, the increase in interruption capability rises nearly parallel for a rising blowing pressure, with PTFE showing significantly higher critical current steepnesses. When using either PP or PA6.6 the critical current steepnesses are low for low blowing pressures but the increase with pressure is much higher than for PTFE or PLA.

## 4.2. Dielectric recovery

The time resolved breakdown voltages after CZ are shown in Figure 4 for PTFE, PP and PA6.6. PLA was not investigated due to its low thermal interruption capability. The plotted values represent the mean values of at least four breakdowns at one specific delay time. The blowing pressure is set to $p = 330\,\mathrm{hPA}$ for all tests.

For all investigated nozzles a strong increase in breakdown voltage occurs after current zero. In the first 170 µs the difference between the nozzle materials is low. Afterwards PA6.6 shows a slight decrease in

| Material | PTFE | PP | PA6.6 | PLA |
|---|---|---|---|---|
| Ablation [mg] | 3 | 5.8 | 5.2 | 11.5 |

*Table 2. Ablated mass per 630 A half oscillation for the investigated nozzle materials*

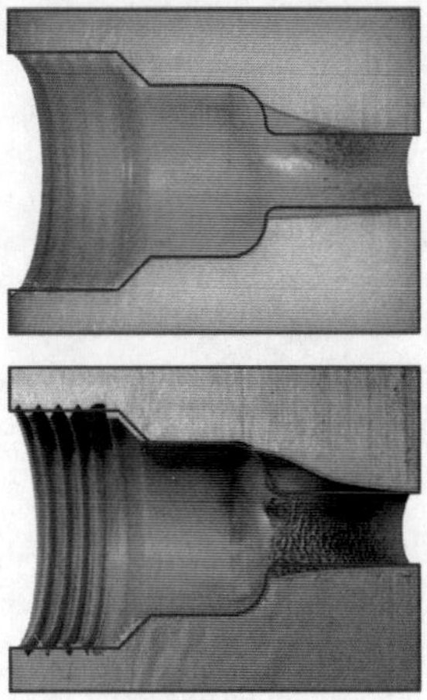

*Figure 5. Photos of the PA 6.6 (top) and the PP (bottom) nozzle after 250 switching operations*

dielectric strength followed by a slight decrease of the breakdown voltage of PP. Local minima are expected to result from statistical scatter. From roughly 300 µs on, the highest breakdown voltages are achieved when using a PTFE nozzle followed by PP and PA6.6.

## 4.3. Nozzle ablation

To compare the ablation behavior of the different nozzle materials the nozzles are weighed after production and after 100 consecutive interruption tests. Then the average weight loss per 630 A half oscillation is calculated. In Table 2 the resulting ablation masses are given. The ablation rates show large deviations among the different materials. Due to its high sublimation temperature PTFE shows the lowest weight loss. The weight losses of PP and PA6.6 are in the same range, despite of their different melting temperatures. However, as the density of PP is lower than of PA6.6 the volumetric losses are higher. PLA shows by far the highest mass losses. This might be due to the 3D printed structure being not as dense as the other nozzles. Furthermore, the significantly higher roughness inside the nozzle throat might lead to increased ablation during switching.

On all nozzles except from the PTFE nozzle a soot formation can be observed after the tests. Photos of the PP and the PA 6.6 nozzle after 250 switching operations are shown in Figure 5. The original geometry is higlighted with a blue curve, while the actual geometry is highlighted by a red curve. The stronger volumetric loss of the PP nozzle is clearly visible. As the arc has a preferred footpoint on the top of the tulip, the maximum ablation occurs there.

# 5. Discussion

As the different nozzle materials show different effects in the thermal and the dielectric phase the two phases are discussed individually.

## 5.1. Thermal phase

The ablated vapor of PP and PA6.6 mainly consists of hydrogen. Due to its high thermal conductivity an efficient arc cooling can be achieved. However, the ablated masses are almost twice as high as for PTFE, which leads to a higher pressure rise inside the nozzle due to ablation. It is assumed, that this higher pressure rise leads to a nozzle clogging at lower blowing pressure, resulting in an insufficient flow field and therefore a reduced interruption capability. At higher blowing pressures the pressure rise due to ablation is lower than the external pressure leading to a sufficient flow. The high hydrogen content results in an increased interruption capability in comparison to PTFE.

Due to the lower ablated mass when using PTFE, the effective flow field is more efficient which results in a good interruption capability. As the ablated gas is very similar to dissociated $SF_6$, an admixture of ablated PTFE to the plasma increases the interruption capability as well [7]. As PLA has a significantly higher ablation rate, the flow is insufficient even at high external blowing pressures.

## 5.2. Dielectric phase

As the experiments for determining the dielectric recovery are performed with a blowing pressure of $p_b = 330\,\text{hPa}$, similar effects can be assumed to those in the thermal phase. This leads to PP and PA6.6 having an insufficient cooling and therefore also a slower recovery than PTFE. Additionally, the ablated gas almost completely dissociates in vicinity of the arc. Due to the absence of the highly reactive fluorine conductive carbon black might be formed in the recombination process leading to a lower breakdown voltage and therefore a slower dielectric recovery.

To isolate the effect of carbon black formation, experiments with a higher blowing pressure should be conducted. However, as the recovery voltage at medium voltage load interruption is not as demanding as i.e. during fault interruption, the dielectric recovery in the latter part is sufficient for all investigated cases. Only the first 120 µs might be critical.

# 6. Conclusions

The thermal interruption capability as well as the rate of the dielectric recovery were experimentally determined for a model load break switch with an axial arc blowing using different nozzle materials. The following results were obtained:

- □ PTFE shows a good thermal interruption capability for all investigated blowing pressures as well as a good dielectric recovery, due to its low ablation rate.
- □ Polymers that release hydrogen in high parts during arcing, like PP and PA6.6, show an increased thermal interruption capability for high blowing pressures. However, due to their higher ablation rate, insufficient cooling may result from clogging at lower blowing pressures.
- □ The dielectric recovery may be slower when using PP or PA6.6 due to clogging and the formation of carbon black during the recombination process.
- □ 3D printed PLA is not a suitable nozzle material, as the high ablation rate leads to a low interruption capability.

Summarizing, PTFE is a good candidate for a nozzle material in load break switches using alternative insulation gases. If sufficient blowing pressures are available using PP or PA6.6 can result in an even higher thermal interruption capability. The disadvantage is the higher nozzle wear as well as the formation of soot after switching.

## Acknowledgements

This work was supported by the Federal Ministry for Economic Affairs and Energy on the basis of a decision by the German Bundestag.

The authors would like to thank Tim Ballweber for his assistance in this work.

## References

[1] M. Bendig, N. Götte, T. Krampert, A. Schnettler, A. Kalter, and M. Schaak. Investigations on the switching capability of medium voltage load break switches in an alternative quenching gas. In *Proc. of the 22nd Int. Conf. on Gas Discharges and their Applications*, 2018.

[2] N. Sasaki Aanensen, E. Jonsson, and M. Runde. Air-flow investigation for a medium-voltage load break switch. *IEEE Trans. Power Delivery*, 30(1):299–306, 2015. doi:10.1109/TPWRD.2014.2334360.

[3] D. Gonzales, H. Pursch, and F. Berger. Experimental investigation of the interaction of interrupting arcs and gassing polymer walls. In *57th Holm Conference on Electrical Contacts*, 2011. doi:10.1109/HOLM.2011.6034774.

[4] H. Taxt, K. Niayesh, and M. Runde. Medium voltage load current interruption in presence of ablating polymer material. *IEEE Trans. Power Delivery*, 33(5):2535–2540, 2018. doi:10.1109/TPWRD.2018.2803165.

[5] M. Bendig, N. Götte, T. Krampert, A. Schnettler, A. Kalter, and M. Schaak. A method to determine the rate of he dielectric recovery in a medium voltage load break switch with a free burning switching arc. In *Proc. of the 22nd Int. Conf. on Gas Discharges and their Applications*, 2018.

[6] P. Andre. Composition and thermodynamic properties of ablated vapours of PMMA, PA6-6, PETP, POM and PE. *J. Phys. D: Appl. Phys.*, 29(7):1963–1972, 1996. doi:10.1088/0022-3727/29/7/033.

[7] M. Kriegel. *Einfluss des Düsenmaterials auf das Ausschaltverhalten von SF6-Selbstblasschaltern.* Dissertation, RWTH Aachen University, 1999.

doi:10.14311/ppt.2019.1.19
*Plasma Physics and Technology* **6**(1):19–22, 2019 

# CONTROL OF DIFFUSE VACUUM ARC USING AXIAL MAGNETIC FIELDS IN COMMERCIAL HIGH VOLTAGE SWITCHGEAR

S. Giere[b], T.Heinz[b], A. Lawall[a,*], C. Stiehler[a], E. D. Taylor[a], S. Wethekam[b]

[a] *Siemens AG, Smart Infrastructure, Rohrdamm 88, 13623 Berlin, Germany*
[b] *Siemens AG, Gas and Power, Nonnendammallee 104, 13623 Berlin, Germany*
[*] andreas.lawall@siemens.com

**Abstract.** During the development of a commercial vacuum interrupter for application in HV switchgear at a rated voltage of 145 kV, we investigated the behavior of vacuum arcs controlled by axial magnetic fields (AMF). AMF arc control is already extensively used in medium voltage (1-52 kV) applications, the key difference is the 2-3 times larger contact gap and the corresponding reduction of the AMF strength for HV applications. We conducted several stress tests with short circuit currents up to 40 kA, thus not only testing the interrupting capability, but also the electrical endurance of such a contact system. We also investigated the dielectric behavior of the vacuum interrupter by testing the capacitive switching duty. Despite the large number of operations, the contacts showed a minimal amount of contact erosion and damage and demonstrated behavior very similar to the extensive experience with MV vacuum interrupters. In line with simulation results, we conclude that even at high contact gaps and currents, a diffuse vacuum arc was maintained which distributed the arc energy evenly over the contacts.

**Keywords:** High Voltage, Vacuum Interrupter, Vacuum Arcs.

## 1. Introduction

Vacuum interrupters (VI) have been used in high voltage (HV) applications above 52 kV from the start of practical VI manufacturing in the late 1950's and early 1960's. Early circuit breaker designs successfully achieved ratings up to 145 kV/40 kA [1] and 132 kV/3500 MVA [2] in type tests, and achieved some limited field use. However, the circuit breakers were extremely complex, using multiple VI in series [1], often with complex voltage grading systems [2]. The complexity and cost of these circuit breakers limited their use to very isolated examples. A concerted focus on HV VI development in Japan resulted in the field installation of approximately 8300 circuit breakers using VI from the late 1970's to 2010 [3]. 99% of these circuit breakers were for 72 kV and lower system voltages. Many of these designs, especially in later years, used a single VI or at most two VI in series instead of the previous multiple VI systems. Nevertheless, even in Japan only a very small percentage of the installed base of HV circuit breakers used vacuum.

In the 2000's, a broader interest in the use of VI for HV circuit breakers arose, primarily driven by legal burdens and restrictions on the use of $SF_6$. This renewed focus centered on applications for system voltages of 110-170 kV. Interest appeared in Europe [4, 5], China [6], and Korea [7]. These were primarily R&D projects; however, a series of type tests were performed on a single-break VI design for 126 kV/40 kA from a Chinese university, which however did use $SF_6$ as an external insulation [8].

Very early in the development of VI, it was thought that holding off the voltage would be straightforward, but current interruption would be difficult. This resulted in some early designs for >72 kV applications as load break switches and capacitor switches [9]. Instead, while high current interruption was manageable, enabling generator breaker applications up to 100 kA [10] and several specialized designs for high currents up to 200 kA [11–13] application to voltages above the medium voltage (MV) range faced several challenges. One fundamental problem was the general dependence of the breakdown voltage on the contact gap, which scales as roughly the square root of the contact separation [14]. In the MV range, this leads to very moderate contact gaps that are very compatible with various arc control techniques using the magnetic field generated by the short-circuit current [14, 15]. HV (high voltage) applications, particularly for system voltages >=110 kV, require longer contact gaps that creates challenges for arc control [16] and prompted consideration of sophisticated contact opening parameters [17]. In addition, a high voltage across an open metal contact gap in vacuum can produce x-ray emission, with increasing risk of x-ray emission as the voltage increases [18].

A concerted effort lead to the successful complete type testing of a commercial $SF_6$-free vacuum circuit breaker for 145 kV/40 kA. The tests included verification of the complete range of requirements for practical circuit breaker applications. This achievement built on considerable efforts in the field experience of 72 kV circuit breakers with VI [19], control of x-ray emission

| | Arc voltage | Normal current | Short-circuit breaking current |
|---|---|---|---|
| Vacuum, AMF-contact | 30 – 40 Volts | 10,000 operations | 50 operations |
| $SF_6$, self-compression | Several Kilovolts | 6,000 operations | 12 operations |

*Table 1. Comparison of typical breaking performance for vacuum and* $SF_6$.

from HV VI [20], and modeling of the arc behavior [16].

## 2. Experiments

The 3AV1 passed all relevant tests to ensure the reliability of a fully type tested product.

In comparison to a gas circuit breaker, the stress on vacuum interrupter contacts during switching operations with normal and short-circuit currents is significantly lower. The short arcing times and the low arc voltage of vacuum interrupters help to reduce the contact wear inside of the vacuum interrupter.

This results in a significantly higher number of breaking operations for vacuum interrupters compared to gas circuit breakers, enabling the design of vacuum interrupters capable to switch currents of several kA up to a million times, e.g., in tap changer applications [14].

In order to determine the high switching capability of our commercial vacuum interrupter for 145 kV/40 kA, we have performed arc voltage and arcing times investigations in extended high-power test series. In the latter, we demonstrate that the stresses of several high-power type tests can be handled with a single vacuum interrupter, and the electrical endurance performance is superior to established commercial high-voltage $SF_6$ switchgear.

Table 1 compares the arc voltage during a short circuit current interruption using our 145 kV high-voltage vacuum and a commercial 145 kV $SF_6$ gas circuit breaker. Whereas the intensively cooled plasma column of a $SF_6$ circuit breaker produces several kilovolts of arc voltage, the vacuum arc voltage is of the order of some 10 V only, resulting in a two orders of magnitude lower arc energy dissipated in the interrupter unit of the circuit breaker.

In Fig. 1, we compare minimum arcing times of our tests with a commercial 145 kV $SF_6$ gas circuit breaker with our 145 kV vacuum circuit breaker for a number of high-power test duties according to international standards (e.g. IEC 62271-100, etc.). Whereas for the $SF_6$ circuit breaker, minimum arcing times of 12–16 ms are observed, the minimum arcing times for the vacuum circuit breaker are in the range of 4–6 ms only. This results in significantly shorter arc durations for our vacuum circuit breaker, i.e. a second effect reducing the arc energy impact on the contact system of the interrupter unit compared to $SF_6$ gas circuit breakers.

In our experiments, we found that these effects result in pratically neglibible erosion (see section 3) of the vacuum interupter unit during type test for high-voltage circuit breakers according to international standards. As an example, during the development of our 145 kV/40 kA vacuum circuit breaker, we have performed the series of type tests with a single vacuum interruper given in Table 2. We performed 65 high-power tests including 17 tests T100a and T100s, which represent the highest thermal stress for the contacts of the standard type tests according to interantional standards.

For commercial 145 kV self-compression gas circuit breakers, the contact system and nozzle show a pronounced erosion during switching operations with rated short-circuit currents, so that the interrupter unit needs to be replaced after about 12 operations. As indicated in Table 1, the number of high-current operations for our vacuum interrupter in the high-power test series was significantly higher.

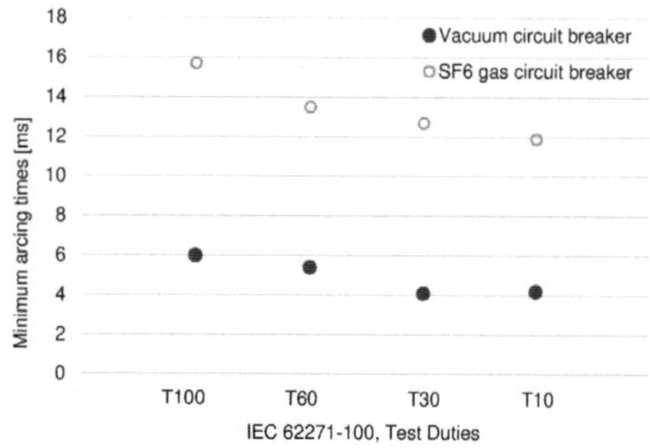


*Figure 1. Comparison of minimum arcing times between* $SF_6$ *and vacuum in type test duties.*

| test duty | Current [kA] | Operations |
|---|---|---|
| T10 | 4 | 9 |
| T30 | 12 | 7 |
| T60 | 24 | 9 |
| L75 | 30 | 8 |
| OP2 | 6 - 10 | 9 |
| T100a | 40 | 4 |
| T100s(b) | 40 | 13 |
| Verification | 40 | 3 |
| Integrity Check | 6 - 17 | 3 |
| **Sum** | | **65** |

*Table 2. Summary of test duties conducted with a single vacuum interrupter.*

*Figure 2. Picture of one contact surface of the opened VI after all current interruptions.*

We have also successfully performed line– and cable–charging capacitive switching tests with a capacitive voltage factor $k_c = 1.4$ where we achieved the highest capacitive switching class (very low probability of restrike) C2. This demeonstrated that high voltage vacuum circuit breakers are not limited to capacitive switching class C1 [8].

All tests were concluded with verfication tests, integrity tests and condition checks according to the specific international standard being tested (eg. IEC 62271-100, etc.).

## 3. Experimental Results

After all tests, which were part of the realized type test procedure, the vacuum interrupter was opened and inspected. Figure 2 shows a picture of the contact plate surface. We found that the contacts suffered almost no wear which indicates that their lifetime is significantly higher than tested in our extended test series.

A suitably dimensioned AMF contact system is essential for low contact erosion and ensures a high number of successful breaking operations. In summary, we can state that for appropriately dimensioned AMF contacts contact wear seem to play an almost negligible role. After the thermal load for the confirmed breaking operations, the loss of contact material is negligible. Therefore, we can conclude that contact wear of 145 kV vacuum interrupters does not need to be monitored during the electrical lifetime of the circuit breaker.

## 4. Simulation Results

The simulations are based on a two-temperature magneto-hydrodynamic model of the arc plasma. The AMF produced by the current path in the contacts

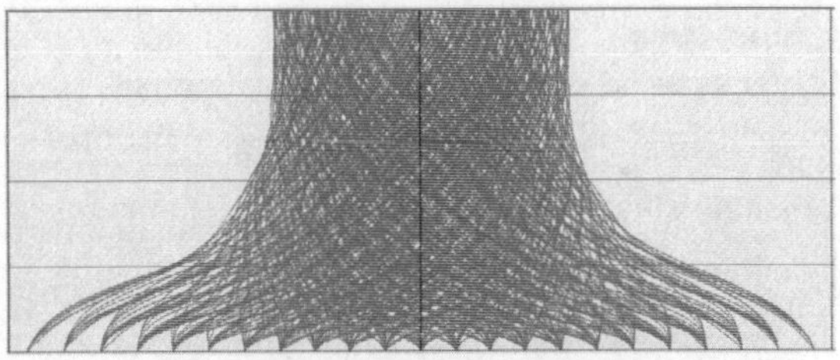

*Figure 3. Plot of the streamlines of the ion velocity of ions emerging from the cathode (side-view). The cathode is located at the bottom, the anode is located at the top. Shown is the full contact diameter and contact gap.*

is first calculated using standard FEM-tools. Then a smoothed approximation of this AMF is used as input for the plasma model. Details of this model can be found in Refs [21] and [22].

The arc plasma is simulated within a cylindrical volume between a pair of disc-shaped electrodes. Plasma length and disc diameter correspond to the contact gap length and the contact diameter of our commercial circuit breaker. The applied current was 40 kA (RMS).

Figure 3 shows a side-view of the streamlines of the ions emerging from the cathode. Although a slight contraction of the arc can be seen in this result it still remains in a diffuse state. For the design of the contacts it is crucial to establish enough AMF to keep at least such a partly diffuse state of the arc. So the simulation model proved to be a useful tool to explore the contact designs for the high gaps needed in this application.

## 5. Conclusion

Vacuum interrupters are the dominant technology for circuit breakers in medium voltage applications based on the long electrical life at load and short-circuit currents, as well as being maintenance-free and offering minimal contact erosion during operation. Issues with the relatively long contact gap required for HV applications initially prevented the penetration of vacuum into HV circuit breakers and switches, aside from a few specialized applications. Extensive efforts in arc modeling guided the successful development of AMF arc control techniques for these long contact gaps. Type testing verified the performance of VI for practical HV circuit breaker applications. Further high-power testing demonstrated that VI in HV keep the advantages seen in MV applications, such as very long electrical life and minimal contact erosion.

### Acknowledgements

We like to thank Dr. Norbert Wenzel and Dr. Andreas Geisler for their support concerning the numerical arc model.

## References

[1] P. G. Slade et al. The development of a vacuum interrupter retrofit for the upgrading and life extension of 121 kV–145 kV oil circuit breakers. *IEEE Trans. Power Del.*, 6(3):1124/1131, 1991. doi:10.1109/61.85857.

[2] L. Falkingham and M. Waldron. Vacuum for HV applications - perhaps not so new? - thirty years service experience of 132 kV vacuum circuit breaker. *Proc. 22nd ISDEIV*, 1:200–203, 2006. doi:10.1109/DEIV.2006.357267.

[3] R. Smeets et al. The impact of the application of vacuum switchgear at transmission voltages. *Cigre Working Group A3.27; Cigre Technical Report*, (589), 2014.

[4] R. Renz. High voltage vacuum interrupters - technical and physical feasibility versus economical efficiency. *Proc. 22nd ISDEIV*, 1:257–262, 2006. doi:10.1109/DEIV.2006.357281.

[5] X. Godechot, S. Chakraborty, A. Girodet, and P. Vinson. Design and tests of vacuum interrupters for high voltage circuit breakers. *Proc. 26th ISDEIV*, 1:417–420, 2014. doi:10.1109/DEIV.2014.6961708.

[6] Z. Liu, J. Wang, S. Xiu, Z. Wang, S. Yuan, L. Jin, H. Zhou, and R. Yang. Development of high-voltage vacuum circuit breakers in China. *IEEE Trans. Plasma Sci*, 35(4):856–865, 2007. doi:10.1109/TPS.2007.896929.

[7] J. Ryu, Y.-G. Kim, J. Choi, and S. Park. The experimental research of 170 kV VCB using single-break vacuum interrupter. *Proc. 25th ISDEIV*, 2:493–496, 2012. doi:10.1109/DEIV.2012.6412563.

[8] X. Yao, J. Wang, Y. Geng, J. Yan, Z. Liu, J. Yao, and P. Liu. Development and type test of a single-break 126-kV/40-kA-2500-A vacuum circuit breaker. *IEEE Trans. Power Del.*, 31(1):182, 2016. doi:10.1109/TPWRD.2015.2456033.

[9] H. C. Ross. Vacuum power switches: 5 years of field application and testing. *Trans. AIEE Part III: Power App. Syst.*, 71:758, 1961. doi:10.1109/AIEEPAS.1961.4501132.

[10] H. Urbanek, K. R. Venna, and N. Anger. Vacuum circuit breakers - promising switching technology for pumped storage power plants up to 450 MVA. *Proc. ICEPE 4th Int. Conf. Electr. Power Equip. - Switching Technol.*, page 107, 2017.

[11] S. Yanabu, S. Souma, T. Tamagawa, S. Yamashita, and T. Tsutsumi. Vacuum arc under an axial magnetic field and its interrupting ability. *Proc. IEE*, 126(4):313, 1979. doi:10.1049/piee.1979.0079.

[12] T. Bonicelli, A. DeLorenzi, D. Hrabal, R. Piovan, E. Saches, E. Sapietro, and S. R. Shaw. The European development of a full scale switching unit for the ITER switching and discharging networks. *Fusion Eng. Design*, 75:193, 2005. doi:10.1016/j.fusengdes.2005.06.225.

[13] A. Zamengo. Operational experience of the 50 kA–35 kV RFX-mod DC-current interruption system. *Proc. 28th ISDEIV*, 2:543–546, 2018.

[14] P. G. Slade. *The Vacuum Interrupter: Theory Design and Application*. CRC Press, Boca Raton, 2008.

[15] R. Renz. Vacuum Interrupters. In *Vacuum Electronics: Components and Devices*, chapter 8. Springer, 2008.

[16] N. Wenzel, A. Lawall, U. Schümann, and S. Wethekam. Combined experimental and theoretical study of constriction threshold of large-gap AMF vacuum arcs. *Proc. 26th ISDEIV*, 1:193, 2014.

[17] J. Kusserow and R. Renz. Method for opening the contact gap of a vacuum interrupter. *U.S. Patent 7,334,319*, 2008.

[18] J. Yan, Z. Liu, Y. Geng, S. Zhang, and Y. Zhang. Investigation on X-radiation for 126 kV vacuum interrupters. *Plasma Sci. Technol.*, 18(5):577, 2016. doi:10.1088/1009-0630/18/5/22.

[19] J. Brucher, S. Giere, C. Watier, A. Hessenmüller, and P. E. Nielsen. 3AV1FG - 72.5 kV prototype vacuum circuit breaker (case study with pilot customers). *44th Int. Conf. Large High Voltage Electric Systems*, A3:101, 2012.

[20] S. Giere, T. Heinz, A. Lawall, C. Stiehler, E. D. Taylor, N. Wenzel, and S. Wethekam. X-radiation emission of high-voltage vacuum interrupters: Dose control under testing and operating conditions. *Proc. 28th ISDEIV*, pages 523–526, 2018.

[21] W. Hartmann, A. Hauser, A. Lawall, R. Renz, and N. Wenzel. The 3D numerical simulation of a transient vacuum arc under realistic spatial AMF profiles. *Proc. 24th ISDEIV, Braunschweig*, 2:285–288, 2010. doi:10.1109/DEIV.2010.5625791.

[22] N. Wenzel, S. Kosse, A. Lawall, R. Renz, and W. Hartmann. Numerical simulation of multi-component arcs in high-current vacuum interrupters. *Proc. 25th ISDEIV*, 2:321–324, 2012. doi:10.1109/DEIV.2012.6412518.

doi:10.14311/ppt.2019.1.23
*Plasma Physics and Technology* **6**(1):23–26, 2019 

# NOZZLE WEAR AND PRESSURE RISE IN HEATING VOLUME OF SELF-BLAST TYPE ULTRA-HIGH PRESSURE NITROGEN ARC

F. Abid[a,*], K.Niayesh[a], N. Støa-Aanensen[b]

[a] *Department of Electric Power Engineering, Norwegian University of Science and Technology, 7491, Trondheim, Norway*

[b] *SINTEF Energy Research, 7465, Trondheim, Norway*

[*] fahim.abid@ntnu.no

**Abstract.** This paper reports on experiments with ultra-high pressure nitrogen arcs in a self-blast type switch design. The effect of filling pressure on nozzle mass loss and pressure-rise in the heating volume is investigated. An arc current peak of 130 A at 190 Hz, and a fixed inter-electrode gap of 50 mm are used. The arc burns inside a polytetrafluoroethylene nozzle with a gas outflow vent in the middle. Three different nitrogen filling pressures were tested: 1, 20 and 40 bar, which also covers the supercritical region. To study the effect of vent size on blow pressure near current zero, three different vent dimensions were investigated. The energy deposited in the arc increases with the filling pressure. It is observed that the pressure-rise in the heating volume is linked to the filling pressure, while the vent size plays a crucial role in the blow pressure near current zero. The nozzle mass loss per unit energy deposited in the arc is found to be independent of the filling pressure.

**Keywords:** ultrahigh pressure arc, nozzle wear, supercritical fluid, arc discharge.

## 1. Introduction

The interruption chamber of a medium voltage (MV) gas circuit breaker is generally filled at atmospheric pressure or at a slightly elevated pressure of a few bar. Due to safety reasons, these MV gas circuit breakers are almost never filled with extremely high pressure. For gas circuit breakers, the interruption performance is believed to be related to filling pressure due to the increase of gas flow rate near current zero (CZ) [1]. Switchgears placed at the seabed for subsea power transfer applications have the potential to be filled with extremely high filling pressure. By filling the interrupting chambers of such switchgears at the surrounding pressure, the complexity and cost of the encapsulation can be substantially reduced.

When the pressure and temperature of a gas exceed a critical point, the gas enters into the supercritical (SC) region. High density, high heat conductivity, high diffusivity, absence of vapour bubbles and self-healing properties are some of the unique features of an SC fluid. For gas circuit breakers, the properties of SC fluids are believed to be in favour of improved current interruption performance [2]. For nitrogen ($N_2$), the critical point is 126 K and 33.5 bar. Nitrogen is chosen in this work due to its environmentally benign nature, good insulation strength and low critical pressure.

There are few publications on gas discharges in ultra-high pressure medium in general, among which almost none of them cover high energy arc discharges typical for power switching applications [2, 3]. Recently, some efforts have been made to investigate arc discharges in SC $N_2$ at tens of bars filling pressure. The arc voltage is reported to increase with filling pressure, without any abrupt change during the transition from gas to SC state [4]. A higher arc voltage results in an increased energy deposition in the arc channel. Moreover, the arc radius decreases as the filling pressure increases [5].

To investigate the effect of gas flow for very high pressure arc discharges, a self-blast arrangement is adopted in this paper. A self-blast switch design is studied in a fixed electrode arrangement where the arc burns inside a polytetrafluoroethylene (PTFE) nozzle with a gas outflow vent in the middle. During the high current phase, the PTFE nozzle will ablate. Some of the ablated mass escapes through the vent and the rest causes a pressure rise in the heating volume. The compressed gas in the heating volume creates a backflow of relatively cold gas near CZ to cool the arc for current interruption. To study the effect of vent size on blow pressure near CZ, three different vent dimensions are investigated. Experiments are conducted at three different $N_2$ filling pressures: 1 bar (atmospheric pressure), 20 bar and 40 bar (supercritical region). In this paper, the effect of filling pressure on mass loss of the PTFE nozzle and consecutive pressure rise in the heating volume is reported.

## 2. Experimental setup

The electrical setup is shown in Figure 1(a). The test circuit consists of the charging and discharging sections of a 7.2 µF high voltage (HV) capacitor bank, $C$. The capacitor bank is charged through a diode-resistance unit ($R_C$–$D_C$) from a HV source. Once the capacitor bank is charged to a predefined voltage, it is disconnected from the grid by switch $S_C$. Afterwards,

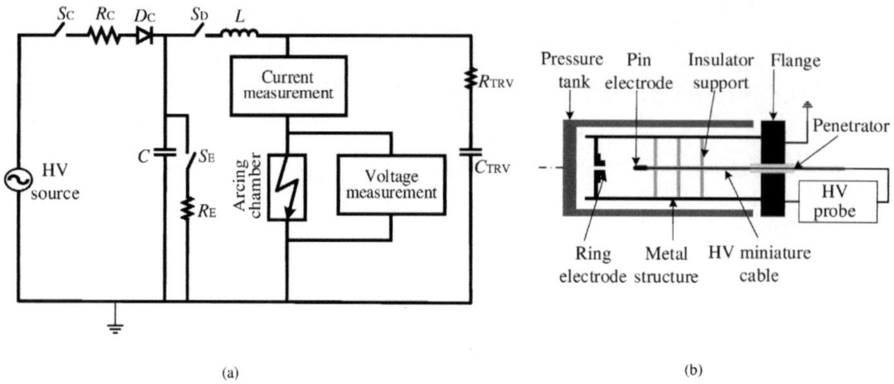


*Figure 1. Experiment setup, (a) electrical circuit, (b) interior connection of the pressure tank.*

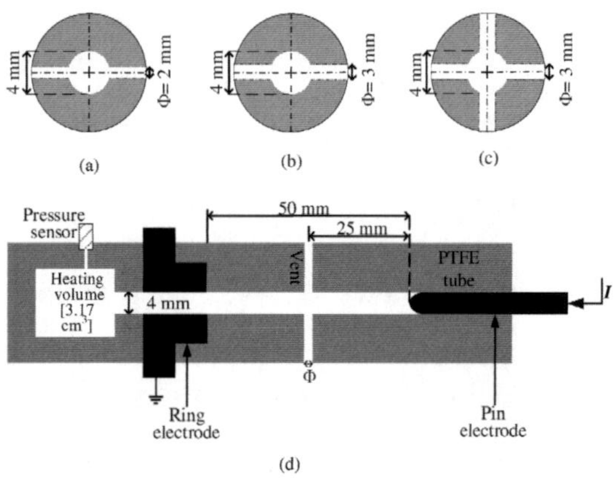


*Figure 2. Cross sectional view of the vent of three designs. (a) design* ***a****, (b) design* ***b****, (c) design* ***c****. Electrode arrangement, nozzle, heating volume, (d) Longitudinal view.*

the switch $S_D$ is closed which results in a current flow through the inductor, $L$ and further through a copper ignition wire (40 µm diameter) inside the arcing chamber. The inductance of $L$, and the charging voltage of $C$ are kept constant to generate a current amplitude of 130 A in the first half cycle at a frequency of 190 Hz for all the experiments conducted. As the arc resistance varies with the filling pressure, the damping of the $LC$ oscillation also changes. Hence, only the first half cycle is considered for all cases. A parallel resistive-capacitive branch ($R_{TRV}$–$C_{TRV}$) is placed across the arc to generate a controlled initial rate of rise of recovery voltage (IRRRV) of 43 V/ µs after CZ, see Figure 1(a).

A pressure tank of 15.7 litres rated for 500 bar, is used as arcing chamber. The interior connection of the pressure tank is shown schematically in Figure 1(b). A HV miniature cable is inserted into the pressure tank using a cable penetrator. The arc burns between the pin and the ring electrode. The arc is initiated by melting of a copper ignition wire. To investigate the effect of gas flow on the arc based on self-blast principle, the arc is ignited inside a 4 mm diameter PTFE nozzle with vents in the middle. The electrode arrangement, nozzle and the heating volume are shown schematically in Figure 2(d). A heating volume of 3.17 cm$^3$ is attached on the ring electrode where the ablated gas is stored during the high current phase and generates a back-flow of the stored gas near CZ. In this paper, three different vent sizes are investigated: 2 opposite holes of $\Phi = 2$ mm (design **a**), 2 opposite holes of $\Phi = 3$ mm (design **b**), and 4 opposite holes $\Phi = 3$ mm (design **c**), see Figure 2(a-c). Three different filling pressures are investigated in this paper: 1 bar, 20 bar and 40 bar, which also covers the supercritical region. For each filling pressure all three designs are tested with a new nozzle.

A HV probe measures the arc voltage across the electrodes while a current shunt is used to measure the arc current. A piezoelectric pressure sensor is mounted on the heating volume to measure the pressure rise, as can be seen in Figure 2(d). The pressure sensor is recess-mounted, and a vinyl electric tape is used to protect it from the thermal blast. All the data are sent to the control room via optical fibre links and stored in a digital oscilloscope with a sampling rate of ten samples per microsecond. For mass loss measurement, weight measurement with an accuracy of 0.1 mg is carried out before and after five tests of all the new nozzles.

## 3. Results and discussions

A typical measurement of the arc voltage, arc current, and pressure rise in the heating volume for design **a** at 40 bar filling pressure is shown in Figure 3. The ignition wire melts at approximately 0.2 ms marked by a voltage peak. The pressure in the heating volume starts building from approximately 0.5 ms. The pressure rises to a maximum of 13.5 bar at approximately 0.8 ms. Afterwards, the back-flow of stored gas from heating volume starts and flows out completely at approximately 2.9 ms. The CZ occurs approximately

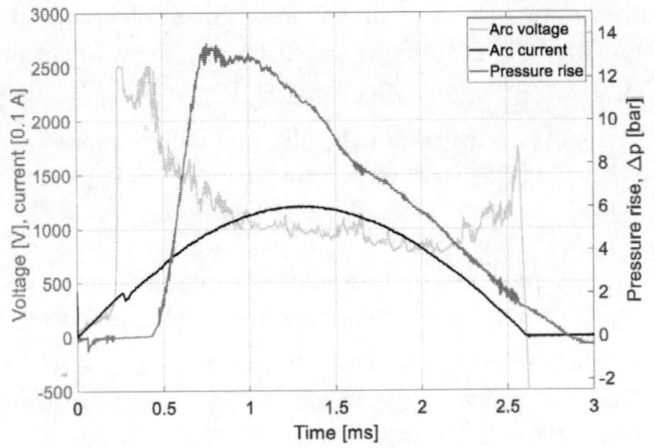


*Figure 3. Arc voltage, current and pressure rise in heating volume for design **a** at a filling pressure of 40 bar.*

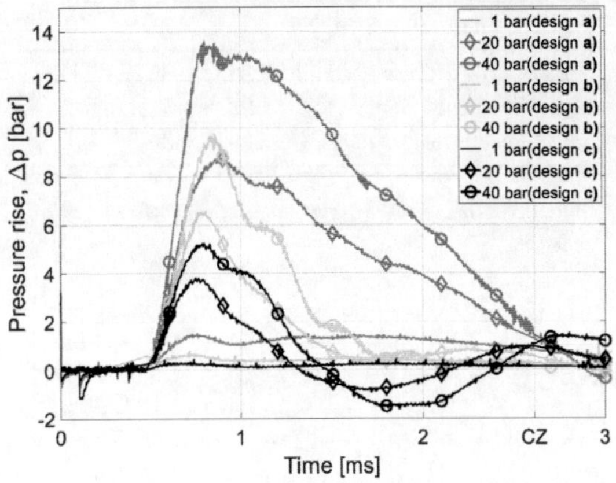


*Figure 4. Pressure rise in the heating volume as a function of filling pressure for different designs.*

| **Filling pressure** [bar] | 1 | 20 | 40 |
|---|---|---|---|
| Design **a** | 0.81 | 1.31 | 1.64 |
| Design **b** | 0.27 | 0.53 | 0.28 |
| Design **c** | 0.15 | 0.96 | 1.07 |

*Table 1. Over-pressure in the heating volume at CZ [bar].*

at 2.61 ms where the current is interrupted by the self-blast test switch. The over-pressure in the heating volume at CZ in this case is approximately 1.64 bar. The back-flow of stored gas in heating volume near CZ cools the arc, which can be seen by the extinction peak of the arc voltage.

### 3.1. Pressure rise in heating volume

The measured over-pressures in the heating volume for three designs at three different filling pressures are plotted in Figure 4. The measured pressure-rise in the heating volume is different for different vent dimensions. As the hole size in the vent is increased, more ablated PTFE vapour leaves the arcing zone through the vent instead of contributing to the pressure build up in heating volume. As a result, design **a** shows higher pressure rise in comparison with design **b** and **c** for any specific filling pressure.

The vent dimension also plays a crucial role in the outflow of the stored gas. The heated $N_2$ and PTFE vapour flows out more slowly in design **a** compared to other designs. In contrast to design **a**, the stored gas leaves completely in design **b** at approximately 1.75 ms. As a result the over-pressure at CZ is low for design **b**. For design **c** however, a negative pressure rise in the heating volume is observed at 20 bar and 40 bar filling pressure, indicating a gas intake through the hole.

The filling pressure also has a significant effect on the pressure-rise in the heating volume. For all designs, the highest pressure rise is observed for 40 bar filling pressure, followed by 20 bar and 1 bar. The rate of change of over-pressure in the heating volume around CZ determines the gas flow rate, hence, the effective cooling of the arc. The measured over-pressures at CZ for all the cases tested are presented in Table 1.

### 3.2. Mass loss of PTFE nozzle

Five tests were conducted for each filling pressure and for each design. Among them, the only design that resulted in successful current interruption in the first two CZ was design **a**. The current interruption in design **b** and **c** happened after first, second or several half cycles. Consequently, only design **a** is considered for mass loss analysis.

As the filling pressure increases, the energy deposition in the arc increases too as a result of the increased arcing voltage. From the measured arc voltage and arc current, the energy deposited in each test is calculated until the current is interrupted. At 1 bar, a total nozzle mass loss of 9.2 mg was measured after five tests, while the total energy deposition of five tests was approximately 313 J. For 20 bar and 40 bar, the mass losses measured after 5 tests were 31.1 mg (851 J) and 42.3 mg (1186 J), respectively. Afterwards, the mass loss per unit energy deposited in the arc at different filling pressure is calculated, which is shown in Table 2. It can be seen that although the overall mass loss in PTFE nozzle increases with the filling pressure, the mass loss per unit energy deposition is not strongly dependent on the gas pressure.

### 3.3. Current interruption performance

The number of successful interruptions after first half cycle among the five tests conducted for each case are plotted in Figure 5. For design **a** at 1 bar pressure, the current was interrupted 4 times among the 5 tests conducted. The number of successful interruption increased to 5 out of 5 for both 20 bar and 40 bar filling pressure using design **a**. For design **b**, however, the interruption performance deteriorated for all filling pressures. Interruption performance is observed to

| Filling pressure [bar] | 1 | 20 | 40 |
|---|---|---|---|
| Total energy deposited [J] | 313 | 851 | 1186 |
| Mass loss [mg] | 9.2 | 31.1 | 42.3 |
| Mass loss/energy [mg/kJ] | 29.4 | 36.5 | 35.7 |

*Table 2. Nozzle mass loss as a function of filling pressure.*

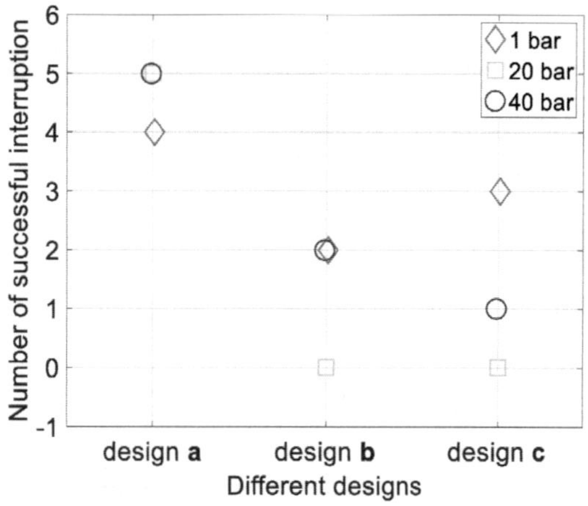


*Figure 5. Number of successful interruption among the five tests conducted at different filling pressures for three designs.*

deteriorate as the filling pressure increases for design **c**.

For a successful current interruption, effective cooling of the arc is necessary. The arc resistance build-up just before CZ is a key parameter for a successful current interruption [6]. As the gas filling pressure is increased, the arc radius decreases, while the temperature of the arc column increases. Without the sufficient flow at a higher filling pressure (i.e., 20 bar, 40 bar), the arc resistance fails to build up before CZ resulting in a poor interruption performance. The rate of change of overpressure near CZ is high in design **a** for all the filling pressures tested in comparison to other designs, see Figure 4. The gas flow due to high rate of change of over-pressure near CZ results in an improved cooling of the arc and hence, better interruption performance for design **a**. The filling pressure also affects the flow field near CZ [7]. Without sufficient cooling of the arc, the high filling pressure can deteriorate the interruption performance, in this case design **b**, and **c**. This paper underpins the importance of gas flow near CZ for ultra-high pressure $N_2$ arc for current interruption application.

## 4. Conclusions

The mass loss of PTFE nozzle, pressure rise in the heating volume and interruption performance for arc burning at different filling pressure of nitrogen are reported in this paper. To study the effect of vent size on blow pressure near current zero, three different vent dimensions are studied. Based on the experimental results following conclusions have been drawn.

1. The pressure rise in the heating volume increases with the filling pressure. The outlet vent dimension plays a decisive role in both the pressure rise in the heating volume and the back-flow near CZ. With proper designing the interruption performance can be improved for ultra-high pressure nitrogen arcs.

2. Total nozzle mass loss is found to increase with the filling pressure. A high filling pressure also constitutes for a high energy deposition in the arc as a result of high arcing voltage. Nevertheless, the mass loss per unit energy deposition in the arc is found to be less dependent on filling pressures.

## Acknowledgements

This work is supported by the Research Council of Norway, project number 280539.

## References

[1] W. Hermann and K. Ragaller. Theoretical description of the current interruption in HV gas blast breakers. *IEEE Transactions on Power Apparatus and systems*, 96(5):1546–1555, 1977. doi:10.1109/T-PAS.1977.32483.

[2] J. Zhang, E. van Heesch, F. Beckers, A. Pemen, R. Smeets, T. Namihira, and A. Markosyan. Breakdown strength and dielectric recovery in a high pressure supercritical nitrogen switch. *IEEE Transactions on Dielectrics and Electrical Insulation*, 22(4):1823–1832, 2015. doi:10.1109/TDEI.2015.005013.

[3] T. Kiyan, A. Uemura, B. C. Roy, T. Namihira, M. Hara, M. Sasaki, M. Goto, and H. Akiyama. Negative DC prebreakdown phenomena and breakdown-voltage characteristics of pressurized carbon dioxide up to supercritical conditions. *IEEE transactions on plasma science*, 35(3):656–662, 2007. doi:10.1109/TPS.2007.896774.

[4] F. Abid, K. Niayesh, E. Jonsson, N. S. Støa-Aanensen, and M. Runde. Arc voltage characteristics in ultrahigh-pressure nitrogen including supercritical region. *IEEE Transactions on Plasma Science*, 46(1):187–193, 2018. doi:10.1109/TPS.2017.2778800.

[5] F. Abid, K. Niayesh, and N. S. StÃÿa-Aanensen. Ultrahigh-pressure nitrogen arcs burning inside cylindrical tubes. *IEEE Transactions on Plasma Science*, 47(1):754–761, Jan 2019. doi:10.1109/TPS.2018.2880841.

[6] M. Seeger, B. Galletti, R. Bini, V. Dousset, A. Iordanidis, D. Over, N. Mahdizadeh, M. Schwinne, P. Stoller, and T. Votteler. Some aspects of current interruption physics in high voltage circuit breakers. *Contributions to Plasma Physics*, 54(2):225–234, 2014. doi:10.1002/ctpp.201310067.

[7] Z. Guo, X. Li, Y. Zhang, X. Guo, and J. Xiong. Investigation on the influence of gas pressure on $CO_2$ arc characteristics in high-voltage gas circuit breakers. *Plasma Physics and Technology*, 4(1):95–98, 2017.

doi:10.14311/ppt.2019.1.27
*Plasma Physics and Technology* **6**(1):27–30, 2019 

# THERMAL PLASMA OF ELECTRIC ARC DISCHARGE BETWEEN Cu–Cr COMPOSITE ELECTRODES

A. Veklich[a,*], V. Boretskij[a], M. Kleshych[a], S. Fesenko[a], A. Murmantsev[a], A. Ivanisik[a], O. Khomenko[b], O. Tolochyn[b], M. Bartlova[c]

[a] *Taras Shevchenko National University of Kyiv, 64/13, Volodymyrska str., Kyiv, Ukraine*
[b] *Frantsevich Institute for Problems of Materials Science NAS of Ukraine, 3, Krzhizhanivsky str., Kyiv, Ukraine*
[c] *Brno University of Technology, Czech Republic, bartlova@feec.vutbr.cz*
[*] an.n.veklich@gmail.com

**Abstract.** This work deals with investigations of model plasma source realised as electric arc discharge in gas atmosphere between Cu–Cr composite electrodes. Radial distributions of temperature and electron density in arc plasma column were studied by optical emission spectroscopy. The content of electrode metals' vapours in discharge was calculated on the base of experimentally obtained plasma parameters as initial data. So, in this way the erosion properties of such composition can be determined.

**Keywords:** optical emission spectroscopy, plasma of arc discharge, erosion of surface electrode, composite materials, electrical contact.

## 1. Introduction

Composite materials on copper base with addition of refractory metals, particularly chromium, have advanced applications in electric power industry. Namely, such composition are mostly used as contact material in vacuum arc-quenching chambers of high and middle loads systems [1, 2]. The erosion properties of abovementioned material play key role, which determined the efficiency of circuits' commutation performance.

It is well-known that heterogeneous compound materials of two or more components with different melting and boiling points, such as Cu–Cr, are mostly manufactured by sintering and/or infiltration techniques. The compositions between 75/25% and 50/50% by weight of Cu–Cr contact material has proved optimum behavior in many respects for circuit-breakers [1–3]. Namely, high breaking capacity, gettering effect, smooth surface after arcing with subsequent high electrical strength, moderate chopping current of such composition are just those properties which are required in this kind apparats [1].

There are some investigations of plasma produced by high current vacuum arcs between Cu–Cr contacts in literature. Namely, in paper [3] was shown, that microparticles can be produced during interruption of a high current vacuum arc between such contacts. It is the reason, which causes the reduce of the insulation performance of vacuum interrupters. The authors concluded, that the contact material has a great effect on the particle generation patterns. It was found that it is much more difficult to form a melting pool on a Cu–Cr contact surface with a higher proportion of chromium. The refractory material can suppress the flow of metal liquid and, consequently, fewer particles can be produced from it.

In paper [4] the spatially resolved optical emission spectroscopy was used to study plasma parameters of high-current vacuum arcs between Cu–Cr contacts. It was found that the plasma layer in front of the anode is not in local thermal equilibrium (LTE). The two-temperature plasma model is suggested by authors to use in study of such non-equilibrium plasma areas.

The aim of this paper is investigation of plasma parameters of model source realised as electric arc discharge in air atmosphere between Cu–Cr composite electrodes. In assumptions of LTE the optical emission spectroscopy techniques are developed to study the content of electrode metals' vapours in discharge, which can be calculated on the base of experimentally obtained plasma parameters as initial data.

The results of calculation of metal component concentration in arc discharge plasma can be used and applied to development and optimization of novel composite materials.

## 2. Experimental setup

### 2.1. Measurements of temperature

The arc was ignited between the end surfaces of vertically-oriented non-cooled copper-chromium electrodes. Electrodes were produced by powder metallurgy in the form of a rectangular parallelepiped with a cross section $3.5 \times 3.5$ mm. The Cu–Cr composite was formed at a temperature of $750°C$ in a volume ratio of 65% Cr and 35% Cu. Discharge gap was of 8 mm and arc current 3.5 A.

Radial profiles of emission intensities in the average cross section of the free burning in air electric arc plasma were investigated. Technique of one-pass tomographic recording of the spatial distribution of spectral line intensities was used [5]. Monochromator MDR-12 with 3000-pixels CCD linear image sensor Sony ILX526A (B/W) accomplished fast scanning of spatial distribution of radial intensity. Due to the instability of the discharge, statistical averaging of the recorded spatial distributions of the radiation characteristics was carried out.

Because side-on (lateral) observation of plasma object was realized in proposed experimental setup, it is necessary to use Abel inversion [6] for obtaining of local values of intensity.

The radial distributions of plasma temperature were determined in the middle section of discharge gap. The techniques of Boltzmann plot in the assumption of local thermodynamic equilibrium (LTE) were used. In this study spectral lines Cu I 510.5 nm, 515.3 nm, 521.8 nm, 570 nm, 578.2 nm, 793.3 nm, 809.3 nm and Cr I 435.1 nm, 458 nm, 464.6 nm, 487 nm, 532.8 nm, 540.9 nm were used.

It must be noted, that in our previous investigation [7] of electric arc discharge between Cu–Cr electrodes with addition of 1% of W the realisation of LTE in plasma column was found at the same arc current. It is a reason of the application of such model in the frame of this study.

## 2.2. Electron density measurements

The electron density was obtained from electric conductivity, which can be calculated by solution of energy balance equation (Elenbaas-Heller) in assumption of LTE in plasma. Preliminary, the measurement of electric field has been carried out in positive plasma column of arc discharge. So, radial distribution of electrical conductivity can be calculated in the following manner:

$$\sigma(r) = -\frac{1}{rE^2}\left[\frac{\mathrm{d}}{\mathrm{d}r}\left(r\lambda(T)\frac{\mathrm{d}T}{\mathrm{d}r}\right)\right]. \qquad (1)$$

It must be noted, that radiation loss term was neglected in Eq.1 according to results of work [8]. The author found, that the arc radiation losses are insignificant for current up to 30 A.

It must be noted as well, that according to results, obtained in [9], the metal vapour does not noticeably change the thermal conductivity of air as plasma-forming gas in the investigated temperature range. Naturally, we took into account the temperature dependency of air thermal conductivity.

At the next step of investigation the radial profiles of electron density in arc discharge were obtained from calculated profiles of electrical conductivity. Since the mobility of electron is much higher than the ion mobility, the ion current in plasma can be neglected. Then plasma electrical conductivity can be written as:

$$\sigma(r) = eN_e(r)\mu_e(r), \qquad (2)$$

where $e$ is the elementary electrical charge, $N_e(r)$ - electron density distribution, $\mu_e$ - electron mobility.

Thus, electron densities can be obtained from Eq.1-2. Peculiarities of performed calculation of electron density in plasma and the cross sections determination are discussed in detail in papers [10, 11].

## 2.3. Calculation of plasma composition

The calculation of the component composition of the plasma in the air atmosphere with the impurities of copper and chromium vapours is performed on the base of the preliminary determined radial distributions of the temperature $T(r)$ and the electron density $N_e(r)$ (Fig.1). In addition, the radial profiles of two spectral lines intensities of different elements were used. Namely, copper and chromium atoms spectral lines intensities were measured to obtain the necessary in calculation the ratio of these element concentrations.

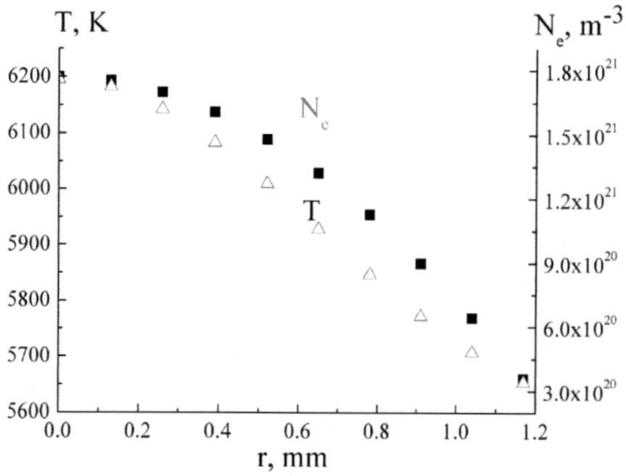


*Figure 1. Radial distributions of plasma temperature and electron density of arc discharge between composite Cu–Cr electrodes at current 3.5 A.*

To calculate the component composition on the base of the aforementioned parameters, the following set of equations is solved:

1. four Saha equations for atoms of copper, chromium, nitrogen and oxygen:

$$S_i(T, N_e) = \frac{N_e N_i^+}{N_i}, \quad i \to \text{Cu, Cr, N, O}; \qquad (3)$$

2. two equations for dissociation of nitrogen and oxygen molecules:

$$D(T) = \frac{N_i^2}{N_{i2}}, \quad i \to \text{N, O}; \qquad (4)$$

3. electroneutrality equation:

$$N_e = \sum N_i^+, \quad i \to \text{Cu, Cr, N, O}; \qquad (5)$$

4. perfect gas equation:

$$B(T) = N_e + N_{\mathrm{O}_2} + N_{\mathrm{N}_2} + \sum N_i + \sum N_i^+ \quad i \to \text{Cu, Cr, N, O}; \qquad (6)$$

5. the ratio between the number of nitrogen and oxygen particles in the air:

$$\begin{aligned}(N_{\mathrm{N}} + 2N_{\mathrm{N}_2} + N_{\mathrm{N}^+}) &= \\ = 3.72(N_{\mathrm{O}} + 2N_{\mathrm{O}_2} + N_{\mathrm{O}^+}).\end{aligned} \tag{7}$$

To solve this set of equations, it was necessary to define additionally the radial profile of the ratio between the concentrations of different metals atoms $N_{Cu}$ and $N_{Cr}$. With this aim, the intensities of spectral lines Cu I 515.3 nm and Cr I 540.9 nm in the plasma radiation were experimentally measured in the assumption of Boltzmann population distribution of copper and chromium atom levels (see Fig.2).

## 3. Results and Discussions

Calculated plasma composition in assumption of LTE was used to obtain the radial profiles of copper $X_{Cu}$ and chromium $X_{Cr}$ vapour contents in mid-section of arc discharge channel. These contents are defined as follows:

$$X_{\mathrm{Cu}}, \% = (N_{\mathrm{Cu}} + N_{\mathrm{Cu}^+}) \cdot 100 / \sum_j N_j, \tag{8}$$

$$X_{\mathrm{Cr}}, \% = (N_{\mathrm{Cr}} + N_{\mathrm{Cr}^+}) \cdot 100 / \sum_j N_j. \tag{9}$$

The total concentration of particles in plasma $\sum_j N_j$ without electron density is used in Eq.8-9.

In Fig.3 the radial distributions of copper and chromium vapours contents in plasma of electric arc discharge in air between composite Cu–Cr electrodes at current 3.5 A are shown. It must be noted, that in spite the fact, that chromium in the electrode material plays role as refractory component, the content of both metals in plasma is comparable. Moreover, it must be stressed, that electrical conductivity in plasma of arc discharge channel was provided just mainly by thermal ionization of chromium component (see Fig.4). This phenomenon can be explained due to the relatively low ionization energy of chromium atom (6.76 eV) in comparison with copper atom (7.74 eV).

The qualitatively and quantitatively behavior of metal vapours contents in mid-section of arc discharge channel does not reflect in fact the real distribution of metal concentration near electrodes working surface. Nevertheless, the suggested in this study techniques can be used as indirect way with aim of study of composition material erosion properties.

To clarify the role of air atmosphere in such erosion behavior of both metal component in electrode composition it seems reasonable to carry out similar study of plasma parameters of arc discharge in argon flow at the next stage of investigation.

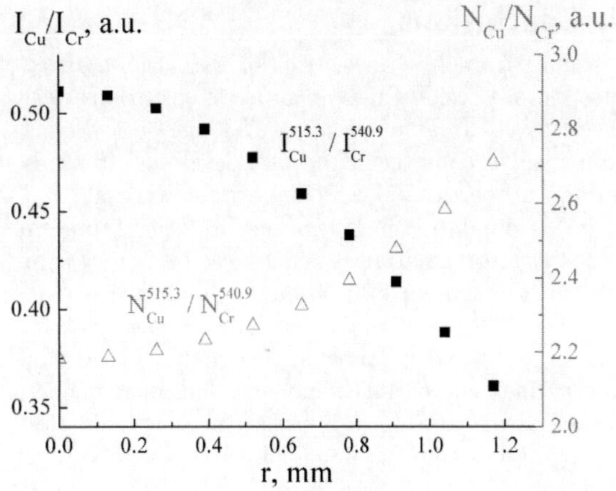


*Figure 2. The ratio of the radiation intensity of the spectral lines Cu I 515.3 nm and Cr I 540.9 nm (■). The ratio of concentrations of copper and chromium atoms (△).*

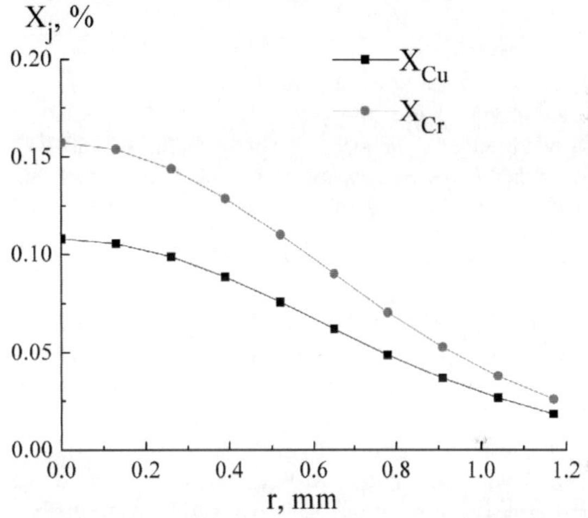


*Figure 3. The radial distributions of copper and chromium vapours contents in plasma of electric arc discharge between composite Cu–Cr electrodes at current 3.5 A.*

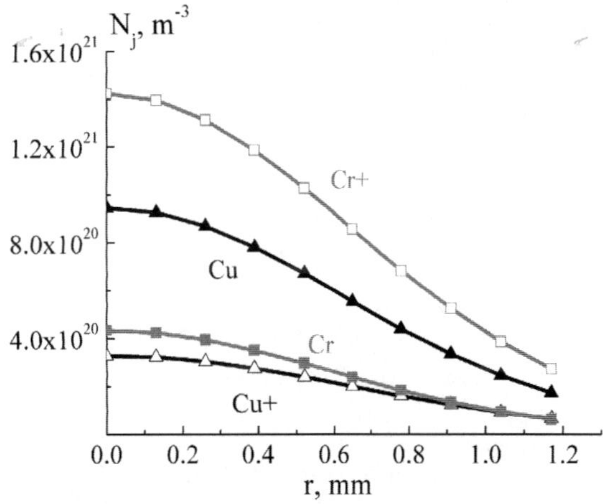


*Figure 4. The radial distributions of Cu, Cr atoms and Cu$^+$, Cr$^+$ ions concentrations.*

## 4. Conclusions

Plasma parameters of electric arc discharge in air atmosphere between Cu–Cr composite electrodes were investigated. In assumptions of LTE the optical emission spectroscopy techniques are developed to study erosion properties of such composite materials.

It was found, that in spite of the fact, that chromium in the electrode material plays role as refractory component, the content of both metals in plasma is comparable. The electrical conductivity in plasma of arc discharge channel was provided just mainly by thermal ionization of chromium component. This phenomenon can be explained due to the relatively low ionization energy of chromium atom in comparison with copper atom.

### Acknowledgements

This work was supported by joined project in the frame of research and technology collaboration between Ukraine and Czech Republic No.7AMB197UA.

### References

[1] P. G. Slade. *Electrical contacts, principles and applications.* CRC Press, 2014.

[2] A. Papillon, J. Missiaen, J. Chaix, S. Roure, and H. Schellekens. Sintering mechanisms of Cu-Cr metallic composites. *Int. Journal of Refractory Metals and Hard Materials*, 65:9–13, 2016. doi:10.1016/j.ijrmhm.2016.11.010.

[3] Z. Wang, W. Yan, Y. Jiang, Y. Li, J. Liu, W. Wang, Y. Geng, J. Wang, Z. Liu, and L. Sun. The effect and dynamic behavior of particles in high-current vacuum arc interruptions. *Journal of Physics D: Applied Physics*, 52(7):1–20, 2018. doi:10.1088/1361-6463/aaf557.

[4] R. Methling, S. Gorchakov, M. Lisnyak, S. Franke, A. Khakpour, S. Popov, A. Batrakov, D. Uhrlandt, and K. Weltmann. Spectroscopic investigation of a Cu-Cr vacuum arc. *IEEE Transactions on plasma science*, 43(8):2303–2309, 2015. doi:10.1109/TPS.2015.2443856.

[5] R. Semenyshyn, A. Veklich, I. Babich, and V. Boretskij. Spectroscopy peculiarities of thermal plasma of electric arc discharge between electrodes with Zn admixtures. *Advances in Space Research*, 54:1235–1241, 2014. doi:10.1016/j.asr.2013.11.042.

[6] K. Bockasten. Transformation of observed radiances into radial distribution of the emission of a plasma. *Journ. of Opt. Soc. of Am*, 9:943–947, 1961. doi:10.1364/JOSA.51.000943.

[7] A. Veklich, A. Lebid, and T. Tmenova. Spectroscopic data of W I, Mo I and Cr I spectral lines: selection and analysis. *J. Astrophys. Astr.*, 36(4):589–604, 2015. doi:10.1007/s12036-015-9342-0.

[8] J. J. Lowke. Simple theory of free-burning arcs. *J. Phys. D: Appl. Phys.*, 12:1873–1886, 1979. doi:10.1088/0022-3727/12/11/016.

[9] V. Boretskij, Y. Cressault, P. Teulet, and V. A. Plasma of electric arc discharge incarbon dioxide with copper vapours. *XIX th Symposium on Physics of Switching Arc (FSO 2011), Brno, Czech Republic*, pages 121–124, 5-9 September 2011.

[10] A. Veklich, S. Fesenko, V. Boretskij, Y. Cressault, A. Gleizes, P. Teulet, Y. Bondarenko, and L. Kryachko. Thermal plasma of electric arc discharge in air between composite Cu-C electrodes. *Problems of atomic science and technology. Series: Plasma Physics*, 6:226–229, 2014.

[11] A. Murmantsev, A. Veklich, V. Boretskij, S. Fesenko, and M. Kleshych. Thermal plasma of electric arc discharge between evaporated electrodes. *Plasma Physics and Technologies*, 5(2):87, 2018.

doi:10.14311/ppt.2019.1.31
*Plasma Physics and Technology* **6**(1):31–34, 2019


# IMPACT OF STARK SHIFTS ON THE RADIATION COOLING OF Cu-DOMINATED PLASMAS

Pierre Corfdir*, Gabriel Lantz, Markus Abplanalp, Thierry Delachaux, Frank Kassubek, Patrick Sütterlin, Matthias Bator

*ABB Corporate Research, Segelhofstrasse 1K, 5405 Baden-Dättwil, Switzerland*

* pierre.corfdir@ch.abb.com

**Abstract.** We study the impact of Stark line shifts reported recently for Cu I transitions on the radiative cooling of Cu-dominated plasmas. The observed detuning in absorption between the hot core and cold shell of the arc leads to a reduction in radiation reabsorption compared to the case where Stark line shifts are neglected. Using a modeling based on a phenomenological treatment of the Stark line shift, we show that this reduction is below 2%.

**Keywords:** Cu plasmas, Stark effect, radiation cooling, radiative transfer equation, spectroscopy.

## 1. Introduction

A proper modeling of current interruption in low-voltage breakers requires an accurate treatment of the radiation of Cu vapors produced by the evaporation of the contacts of the devices [1]. In that purpose, knowledge of the oscillator strengths and Stark broadening parameters of the optical transitions of the atoms and ions constituting the plasma is necessary. While Stark line shifts are usually neglected when computing the absorption of Cu plasmas [2], we have recently reported that Cu I $3d^94s5s$ transitions exhibit a Stark shift of about 0.1 nm with increasing temperatures from 5000 to 10000 K [3].

In this paper, we investigate the impact of the reported Stark line shifts on the radiation cooling of Cu-dominated electric arcs. While shifts of about 0.1 nm are observed with increasing plasma temperature from 5000 to 10000 K, the wavelength of Cu I $3d^94s5s$ remain essentially constant for higher temperatures. Using these experimental results, we compute the radiation intensity by a Cu arc with and without Stark line shift and show that these shifts have only a minor impact on the radiation cooling of the arc.

## 2. Experimental Methods

All experiments have been performed on arcs forming between a 5 mm diameter cylindrical and a ring (inner diameter of 20 mm) Cu electrodes separated by 58 mm. The interface of the cylindrical electrode with air is limited to 17 mm$^2$ using a Macor cap. We measured the arc current and the voltage difference between the electrodes using closed loop Hall and differential probes, respectively. Using an interleaved Buck converter with three parallel lines, we could generate a current profile consisting of a linear increase from 0 to 2 kA in 1 ms, followed by a 19 ms long plateau with a constant current of 2 kA, and a linear decrease to zero over 30 ms. Light emission from the arc discharge was detected with a high-speed camera with frame rate and exposure time of 6400 frames per second and 1 $\mu$s, respectively. For spectroscopy measurements, we collected the light emitted by a spot centered on the $z$–axis 3.5 mm away from the cathode surface and analyzed it with a 500 mm focal length spectrometer equipped with a UV–sensitized charge couple device cooled down with a Peltier module.

## 3. Results and Discussion

High-speed camera images taken at 20 and 38ms are shown in Figs. 1(a) and 1(b), respectively. The position of the Macor cap surrounding the Cu anode is indicated by white dashed lines in Fig. 1(a). The arc current for these two images is about 2 and 0.82 kA, respectively. Profiles corresponding to the fourth root of the intensity ($I$) along the $z = 3.5$ mm axis are shown in Fig. 1(c). Note that we plot $I^{1/4}$ rather than $I$ as the former quantity can be used to estimate the shape of the arc temperature profile [3]. A decrease in current leads both to a slight decrease in the arc peak emission intensity and to a reduction in the width of the intensity profile along the $y$–axis. As shown in Fig. 1(c), the width of the $I^{1/4}$ profile decreases from 9.0 to 7.5 mm with increasing time from 20 to 38 ms.

Figure 1(d) displays spectra collected at different times from a spot centered in $y = 0$ and $z = 3.5$ mm. With increasing time, one can notice significant changes to the spectra. First, lines at about 505.2 and 506.6 nm associated with Cu II states and visible between 20 and 30 ms cannot be detected for longer times. This decrease in the intensity of ion lines with decreasing current suggests a decrease in the arc temperature. Second, the shape of Cu I lines also varies during the experiment. In particular, lines at 515.32 and 521.82 nm and associated with Cu I $3d^{10}4d$ $^2$D 3/2 and $3d^{10}4d$ $^2$D 5/2 transitions, respectively, exhibit not only strong variations in their intensity, but also

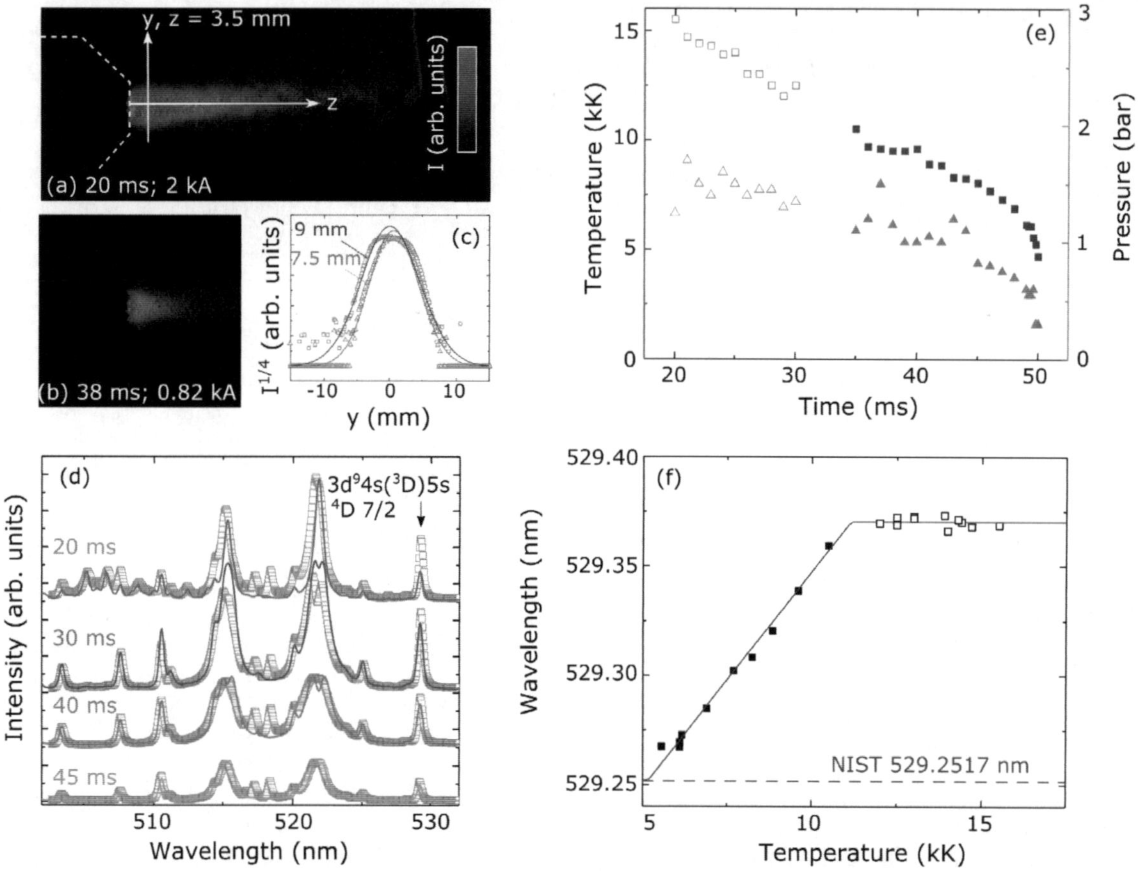


*Figure 1. (a,b) High-speed camera images taken at (a) 20 and (b) 38 ms after ignition and with an arc current of 2 and 0.82 kA, respectively. The intensity is color coded according to the linear scale shown in panel (a). The dashed line show the Macor cap surrounding the Cu cylindrical electrode. The y and z axis are also shown.(c) Fourth root of the intensity profile along the y-axis at 20 (squares) and 38 ms (triangles). The blue and red lines are the result of Gaussian fits for the profiles at 20 and 38 ms, respectively. The number on the left show the width of each profile. (d) Spectra detected at 20, 30, 40 and 45 ms from the spot at the intersection of the y and z axes in (a). The spectra have been shifted vertically for clarity. The solid lines show the result of the lineshape fit discussed in the text. (e) Time-dependence of the arc peak temperature and pressure. The solid and open symbols are the result of a lineshape fit of the Cu I lines only and of both the Cu II and Cu I lines, respectively. (f) Temperature-dependence of the peak wavelength of the $3d^9 4s(^3D)5s\ ^4D\ 7/2$ line highlighted by an arrow in (d). The dashed line shows the wavelength reported in the NIST database for this transition. The solid lines are guides-to-the-eye.*

in their broadening, indicating that both the pressure and the temperature of the arc vary with decreasing arc current.

We described in Ref. [3] a method to extract the time-dependence of the arc peak temperature and pressure from the analysis of high-speed camera images combined with a lineshape fit of the spectra. This method requires several assumptions, that we list in the following: (i) the arc exhibits a radial symmetry, (ii) the width of the fourth-root of the intensity profiles along the $y$-axis in Fig. 1(a) is equal to $0.72\,w$, with $w$ the width of the arc temperature profile [3], and (iii) the ratio between $w$ and the width of the arc pressure profiles is equal to 3 [4]. In Ref. [3], we applied this method in the range of temperature where only Cu I lines are visible. This range corresponds to times larger than 35 ms in Fig. 1(c). As shown in Fig. 1(d) for spectra taken at 40 and 45 ms, the lineshape of the detected spectra is well reproduced. The time-dependence of the plasma peak temperature and pressure deduced from the lineshape fit is shown in Fig. 1(e). For times longer than 35 ms, i.e., for arc currents below 1 kA, the plasma temperature and pressure at the core of the arc are below 10500 K and 1.5 bar, respectively.

The method detailed in Refs. [2–4] is not limited to Cu I plasmas, and we apply it here to times between 20 and 30 ms when both Cu I and Cu II transitions are detected [(Fig. 1(d))]. The best fits obtained for the 20 and 30 ms spectra are shown in Fig. 1(d). In view of the assumptions made above, the agreement between simulations and experiments is good. In particular, we reproduce the correct intensity ratio between Cu II and Cu I lines. The arc core temperature and pressure

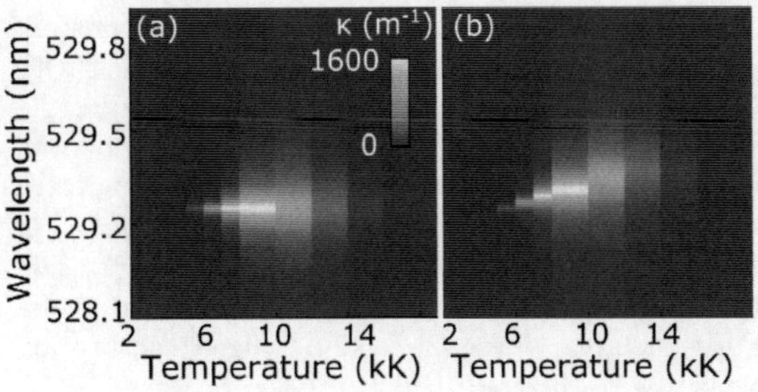


*Figure 2. Absorption spectra computed for a 1 bar Cu plasma as a function of temperature in the vicinity of the $3d^9 4s(^3D)5s\ ^4D\ 7/2$ line, (a) without and (b) with the temperature-dependent Stark line shift shown in Fig. 1(f). The absorption coefficients κ are color-coded according to the linear scale shown in (a).*

obtained from the fit of the spectra taken between 20 and 30 ms are shown in Fig. 1(e). In particular, 20 ms after ignition and for a current of 2 kA, the temperature at the core of the arc exceeds 15000 K.

In addition to the variations in the intensity and broadening of the transitions clearly seen in Fig. 1(d), we could also observe with decreasing current variations in the wavelength of specific lines. Such shifts in the peak wavelength have for instance been reported in Ref. [3] for the $3d^9 4s(^3\mathrm{D})5s\ ^4\mathrm{D}\ 7/2$ transition at 529.3 nm. Figure 1(f) shows the peak wavelength of this line as a function of the arc core temperature shown in Fig. 1(e). On the one hand, for temperatures between 5000 and 10000,K, the peak wavelength of the transition increases linearly with increasing temperature. On the other hand, for temperatures ranging between 12000 and 16000 K, the wavelength of the $3d^9 4s(^3\mathrm{D})5s\ ^4\mathrm{D}\ 7/2$ remains essentially constant.

Due to a lack of available data, Stark shifts are usually neglected when computing the absorption spectra of Cu plasmas. Accordingly, it is not clear how large is the impact of the Stark line shifts reported in Fig. 1(f) on the cooling of the arc by radiation. In particular, one can anticipate the fact that spectral shifts in absorption between the core and the shell of the arc result in reduced reabsorption phenomena and, therefore, in an increased efficiency of the arc cooling by radiation. To estimate how large is the impact of the measured Stark line shifts on the cooling of the arc, we compute in the following the intensity of the $3d^9 4s(^3\mathrm{D})5s\ ^4\mathrm{D}\ 7/2$ transition at about 529.3 nm. We perform this calculation with as well as without the temperature-dependent Stark line shift given in Fig. 1(e).

Figure 2(a) shows the temperature dependence of the absorption spectrum computed for a Cu vapor with a pressure of 1 bar. The Stark line shift was neglected when computing these spectra. In Fig. 2(b), the computed absorption spectra are in addition shifted spectrally according to the results shown in Fig. 1(f). Figure 2(b) thus shows the temperature dependence of the absorption spectrum of a 1 bar Cu vapor with the Stark line shift taken into account.

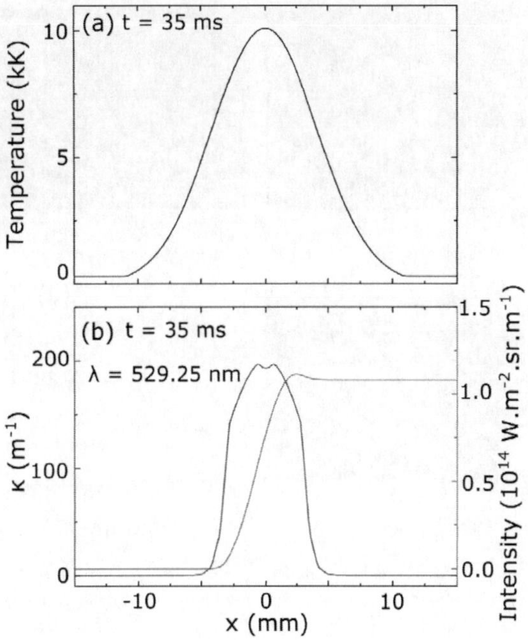


*Figure 3. (a) Temperature, (b) absorption coefficient (blue) and intensity at 529.25 nm (red) profiles for a 1 bar Cu plasma. The profiles have been computed using the peak temperature and pressure shown in Fig. 1(e) for a time $t = 35$ ms, a temperature profile width $w = 8.1$ mm and using the absorption spectra in Fig. 2(a). The intensity profile in (b) has been computed for a ray propagating from negative to positive values of $x$.*

Figure 3(a) displays the arc temperature profile along the line of sight using the temperature profile widths and peak values deduced for $t = 35$ ms from the high-speed camera images and the spectra. Figure 3(b) shows the evolution of the absorption coefficient of the vapor at $\lambda = 829.25$ nm, assuming that the vapor exhibits a Cu fraction of 1 and a pressure of 1 bar. In these simulations, we have used the absorption spectra without Stark line shift shown in Fig. 2(a). In addition, we set $\kappa = 0$ for temperatures below the boiling temperature of Cu. With the temperature and absorption profiles in Figs. 3(a) and 3(b), respectively, it is possible using the radiative heat transfer equation to compute the evolution of the intensity of a ray with $\lambda = 829.25$ nm that propagates along the line-of-sight. The radiative heat transfer equation reads $\nabla I(x, \lambda) = \kappa(I(x, \lambda) - B(\lambda, T))$, with $B(\lambda, T)$ the Planck function. The one-dimensional profiles in Fig. 3 are discretized and the radiative transfer equation has been solved using simple numerical techniques. The results of the line-of-sight integration is shown in Fig. 3(b). We assume that the intensity obtained in $x = 15$ mm in Fig. 3(b) is equal to the intensity of the detected signal $I_{\mathrm{det}}$. Computing $I_{\mathrm{det}}$ for the whole 528 to 530 nm wavelength range, it is possible to

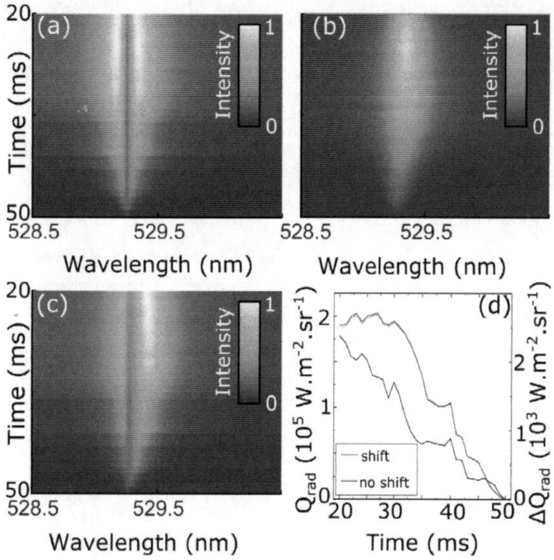


*Figure 4. Time-dependence of the spectra (a) computed for a 1 bar Cu plasma, neglecting the temperature-dependent Stark line shifts in Fig. 1(f). The intensities are color-coded according to the linear scale on the right. The measured time-dependent spectra are shown in (b). (c) Same as (a), this time accounting for the temperature-dependent Stark line shift in Fig. 1(f). (d) Radiation intensity $Q_{rad}$ spectrally integrated between 528.5 and 530.5 nm in $x = 15$ mm without (red) and accounting for (blue) the Stark line shift of the $3d^9 4s(^3D)5s\ ^4D\ 7/2$ transition. The intensity difference $\Delta Q_{rad}$ between the cases with and without Stark line shift is shown by the black line.*

deduce the exact lineshape of the $3d^9 4s(^3\mathrm{D})5s\ ^4\mathrm{D}\ 7/2$ transition. The result of our calculations are shown in Fig. 4(a).

In agreement with the experimental observations shown in Fig. 4(b), the line intensity remains essentially constant between 20 and 35 ms, and decreases for longer time. The constant intensity of the detected signal for times between 20 and 35 ms is related to the combination of the decrease in temperature from 15500 to 10500 K and the increase in peak absorption coefficient from 40 to above $10^3\ \mathrm{m}^{-1}$. For times longer than 35 ms, the detected line not only weakens, but also narrows. The latter observation arises from a reduction in Stark broadening that result from the decrease in pressure displayed in Fig. 1(e). Note that the pronounced absorption dip in the spectra in Fig. 4(a) is smeared out by the response of our spectrometer and is therefore not observed experimentally. Also, while there is a clear blueshift in the peak wavelength of the measured $3d^9 4s(^3\mathrm{D})5s\ ^4\mathrm{D}\ 7/2$ line with increasing time [Fig. 4(b)], the central wavelength of the computed transition in Fig. 4(a) remains constant.

We apply the procedure described above to compute the time-dependence of the lineshape of the $3d^9 4s(^3\mathrm{D})5s\ ^4\mathrm{D}\ 7/2$ transition using the shifted absorption spectra in Fig. 2(b). As shown in Fig. 4(c), we reproduce the experimental observation of a constant peak wavelength for times below 35 ms, together with the blueshift of the transition for longer times. As for the spectra without Stark line shift, a pronounced reabsorption dip is observed in the simulated spectra shown in Fig. 4(d). We emphasize, however, that in contrast to the peak wavelength, the spectral position of this dip remains constant between 20 and 50 ms. Consequently, it is mostly related to light reabsorption occurring in the cold shell of the arc.

To quantify the impact of the observed Stark line shifts on the heat transfer by radiation, we integrate spectrally the spectra in Figs. 4(a) and 4(c). The time-dependence of the spectrally integrated intensity obtained this way is plotted in Fig. 4(d). For all times, the intensity of the detected signal is the strongest when accounting for the Stark line shift. This finding is in agreement with the expected increase in cooling by radiation in the presence of Stark line shifts. This intensity increase is, however, rather small. As shown in Fig. 4(d), for times between 20 and 35 ms, i.e., when the impact of the Stark line shift is the largest [Fig. 4], the intensity increase is below 2%.

## 4. Conclusions

In conclusions, the wavelength of Cu I $3d^9 4s5s$ transitions in Cu-dominated plasmas increases nearly linearly between 5000 and 10000 K as a result of the Stark effect and remains essentially constant at higher temperatures. The spectral detuning in absorption between the hot core and cold shell of the arc results in a reduction in radiation cooling of the plasma. We have shown that this reduction is, however, below 2%. The Stark shift of Cu transitions can thus be neglected when simulating the dynamics of arcs in low-voltage switching applications.

## References


[1] H. Nordborg and A. A. Iordanidis. Self-consistent radiation based modelling of electric arcs: I. efficient radiation approximations. *Journal of Physics D: Applied Physics*, 41(13):135205, 2008. doi:10.1088/0022-3727/41/13/135205.

[2] S. Franke, R. Methling, D. Uhrlandt, R. Bianchetti, R. Gati, and M. Schwinne. Temperature determination in copper-dominated free-burning arcs. *Journal of Physics D: Applied Physics*, 47(1):015202, 2013. doi:10.1088/0022-3727/47/1/015202.

[3] P. Corfdir, G. Lantz, M. Abplanalp, T. Delachaux, F. Kassubek, P. SÃijtterlin, and M. Bator. Stark shift measurement as a temperature diagnostic of cu-dominated thermal plasmas. *Journal of Physics D: Applied Physics*, in press, 2019. doi:10.1088/1361-6463/ab188e.

[4] M. Abplanalp, K. Menzel, T. Delachaux, R.-P. SÃijtterlin, and F. Kassubek. Optical investigation of constricted vacuum arcs. In *27th International Symposium on Discharges and Electrical Insulation in Vacuum (ISDEIV)*, volume 1, pages 1–4, 2016. doi:10.1109/DEIV.2016.7748731.

doi:10.14311/ppt.2019.1.35
*Plasma Physics and Technology* **6**(1):35–38, 2019 

# 3D CFD ARC FAULT SIMULATION IN GAS-INSULATED SWITCHGEARS

F. REICHERT*, A. PETCHANKA

*Siemens AG, Nonnendammallee 104, 13629 Berlin, Germany*

* frank.reichert@siemens.com

**Abstract.** Arc fault processes can lead to strong damages in gas–insulated switchgears and have to be considered in the development process. In order to reduce test costs, the development of overpressure protection systems can be supported by CFD arc fault simulations. The paper deals with the modelling and simulation of arc fault processes in gas–insulated switchgears. The developed simulation tool takes into account a three–dimensional arc model and the opening of a rupture disc during the arc fault process. The influence of different insulating media as e.g. $SF_6$, Air and $CO_2$ on the arc fault process has been investigated. The simulation model has been validated by measured signals for pressure build-up and arc voltage.

**Keywords:** gas-insulated switchgear, internal arc, CFD arc fault simulation, modelling, validation.

## 1. Introduction

Internal arcs in gas–insulated switchgears can lead to strong damages of the switchgears and the surrounding volumes e.g. a substation. According to the standard [1], gas–insulated switchgears must be verified in type tests with respect to internal arcs. The effects of the respective arc fault process are the pressure increase of the insulating gas and the possible burn–through of the enclosure. In this context the effects of the arc fault process shall be limited to the appearance of a hole or a tear in the enclosure as it can be seen from the upper picture in Figure 1. Any kind of fragmentation is not allowed, see the lower picture in Figure 1.

*Figure 1. Visualization of different effects of the arc fault process.*

The pressure build–up during the arc fault process can be controlled by the number, location, cross section, and response pressure of rupture discs. The technical design of such overpressure protection system can be supported by arc fault simulations.

There are many simulation approaches in the literature to simulate the arc fault process. The most simple approach considers the involved compartments by their effective volumes and describes the pressure increase in dependence on the thermal energy $Q$ as a part of the electrical energy $W_{\mathrm{el}}$ according to eq. (1), see e.g. [2]

$$Q = k_{\mathrm{p}} \cdot W_{\mathrm{el}}. \tag{1}$$

In eq. (1) $k_{\mathrm{p}}$ corresponds to the heat transfer coefficient determining the fraction of the electrical arc energy, which directly results in a pressure rise. Such approach is based on assumptions for heat transfer coefficient and arc voltage as a part of the electrical energy.

In the present study, the developed CFD arc fault simulation tool that is based on ANSYS® Fluent software doesn't require an input for the heat transfer coefficient and the arc voltage, as the arc fault process is simulated using a three–dimensional arc model including a sub model describing the radiative transfer inside the arc and from the arc to the surrounding gas.

In Section 2 the basic aspects of the modelling are explained. The validation of the developed simulation tool is illustrated in Section 3 and the influence of the insulating medium on the arc fault process in Section 4. Finally, the conclusions are given in Section 5.

## 2. Modelling

### 2.1. Sum current

According to [3], it is possible to estimate the arc power $P_{\mathrm{arc}}$ in a three–phase system by eq. (2)

$$P_{\mathrm{arc}}(t) = I_{\mathrm{sum}}(t) \cdot U_{\mathrm{arc}}. \tag{2}$$

In eq. (2), $U_{arc}$ is the mean value of arc voltage and $I_{sum}(t)$ is the so-called sum current which corresponds to the sum of the three phase currents at any time $t$, see eq. (3)

$$I_{sum}(t) = 0.5 \cdot (|i_{L1}(t)| + |i_{L2}(t)| + |i_{L3}(t)|). \quad (3)$$

Figure 2 shows an example for the three-phase currents and the resulting sum current calculated with eq. (3) over time.

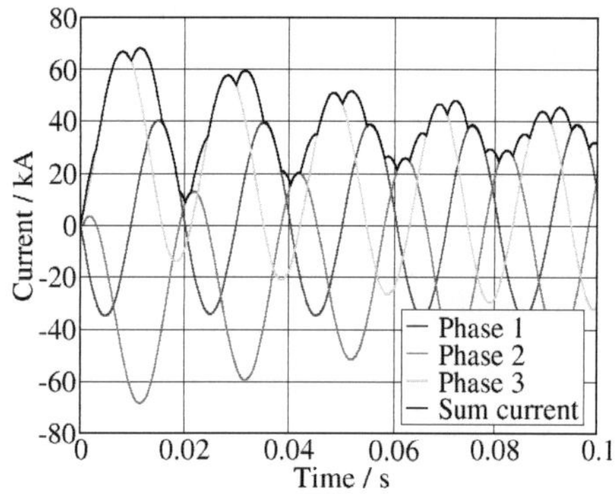


*Figure 2. Three-phase currents and sum current.*

## 2.2. Arc fault model

The implementation used for the arc fault in different insulating media is illustrated in Figure 3.

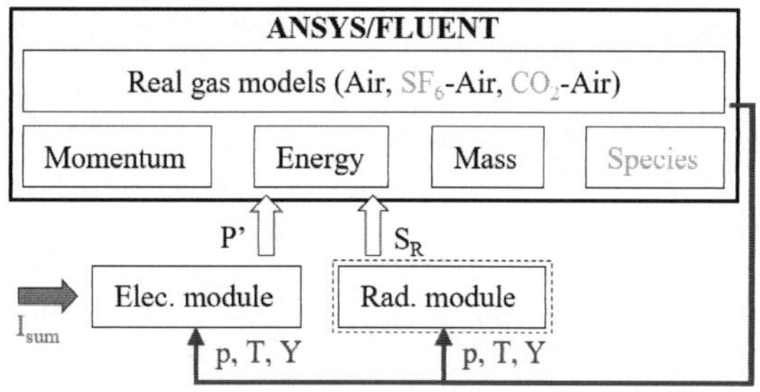


*Figure 3. Link of conservation equations and sub models via source terms and scalar quantities.*

The radiative transfer inside the arc fault and into its environment is reflected by the source term $S_R$ in the energy equation. This source term is calculated in the radiation module using the Discrete Ordinate Model.

The source term $P'$ in the energy equation for the consideration of the Joule effect is calculated in the electric module. This source term is evaluated with a varying arc radius but with the assumption of a constant current density. The arc radius variation is controlled by the following formula

$$R_{arc} = \sqrt{\frac{I_{sum}(t)}{\pi J_{char}}}. \quad (4)$$

In eq. (4), $J_{char}$ is a characteristic current density, which has been derived from measurements. The assumption taken allows a more stable CFD arc simulation in 3D and bigger time step values.

To simulate the plasma flow, thermodynamic and transport properties of the insulating media must be provided to the ANSYS® Fluent solver. This is realized by real gas models for pure Air and the mixtures $SF_6$ - Air and $CO_2$ - Air, based on the assumption of local thermodynamic equilibrium (LTE). The scalar quantities pressure $p$, temperature $T$ and mass fraction $Y$ ($SF_6$ or $CO_2$) in case of $SF_6$ - Air or $CO_2$ - Air mixtures represent the input variables to the radiation module and the electric module.

## 2.3. Operation of rupture disc

As mentioned in Section 1 the pressure build–up during the arc fault process can be controlled by the opening of a rupture disc. The rupture disc opens at a pre-defined response pressure $p_{open}$. To consider this operation a macro has been developed. At the end of each time step the surface averaged pressure value $p_{sim}$ is estimated at the rupture disc wall and compared with $p_{open}$. If $p_{sim}$ exceeds $p_{open}$ the rupture disc wall is removed and the vessel volume is connected to the surroundings.

# 3. Validation

The 3D CFD arc fault simulation tool has been initially validated using a test arrangement investigated in [4]. The corresponding measurements have been performed in a closed test vessel filled with Air, $CO_2$ or $SF_6$ at different filling pressure. Figure 4 shows the current curve used in the simulations and the comparison of measured and simulated pressure within the test vessel for a filling pressure of 1 bar (absolute value).

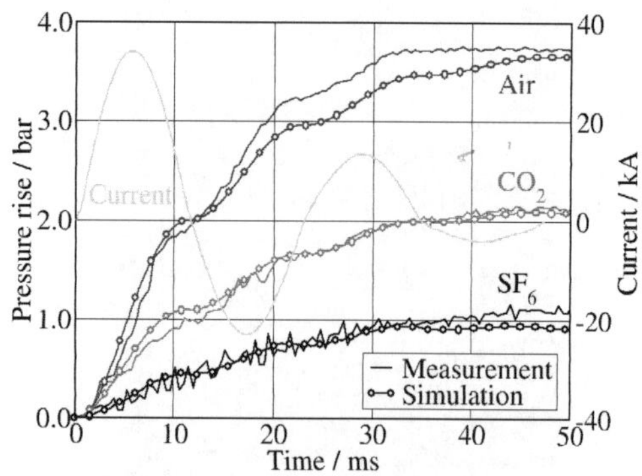


*Figure 4. Pressure rise for different insulating media.*

In addition, the 3D CFD arc fault simulation tool has been validated using a type test of a gas-insulated switchgear filled with Air. Figure 5 shows the discretized vessel without the surrounding solution domain. Using $I_{sum}(t)$ depicted in Figure 2 the arc

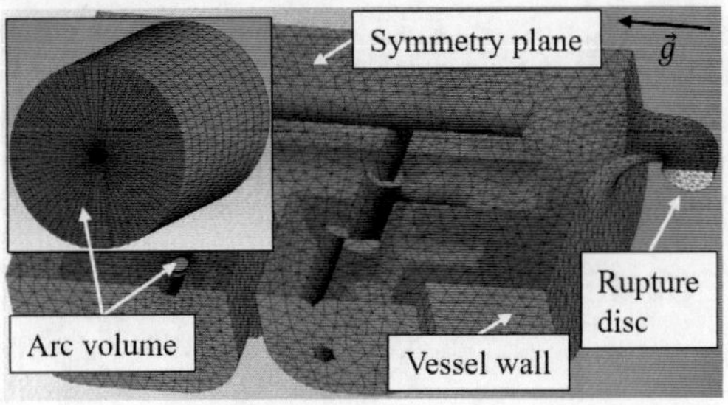


*Figure 5. Half model of the regarded vessel with discretized arc volume.*

fault process has been simulated for an arcing time of 0.5 s including the opening of the rupture disc and the following outflow of hot Air to the surroundings. For validation purposes concerning the type test, Figure 6 shows the comparison of measured and simulated arc voltage. In contrast to the oscillating arc voltages from the measurement the simulated arc voltage is more or less constant following the input by the sum current. The comparison with respect to the pressure build–up within the gas–insulated switchgear is illustrated in Figure 10.

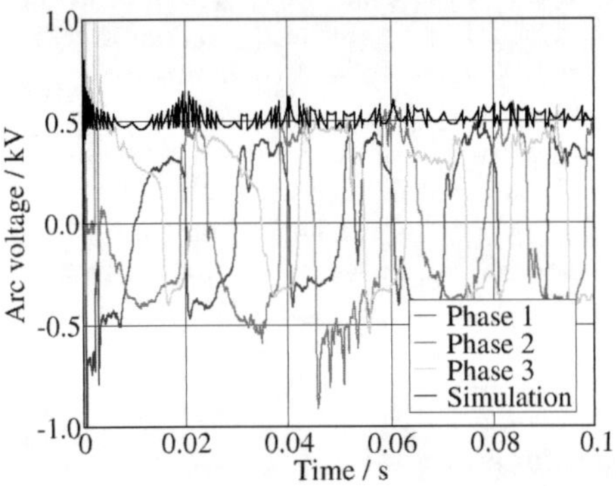


*Figure 6. Comparison of measured and simulated arc voltage.*

Despite the simplifications in arc modelling, the comparison according to Figure 4, 6 and 9 demonstrates a good agreement between measurement and simulation validating the 3D CFD arc fault simulation tool.

## 4. Influence of insulating medium

The illustrated CFD arc fault simulation tool allows to investigate the influence of different insulating media on the arc fault process.

Initially this influence should be demonstrated using the example of the arc fault process in the closed vessel from [4]. As it can be seen from Figure 5, the highest pressure build-up occurs in case of Air followed by $CO_2$ and $SF_6$. This behaviour is typical in case of a closed vessel and primarily a result of the arc heat capacity which can be calculated by the following formula

$$C_{\mathrm{arc}}\left(t'\right)=\frac{\int_0^{t'} P_{\mathrm{arc}}\left(t\right)\mathrm{d}t}{\Delta T'}. \tag{5}$$

$C_{\mathrm{arc}}$ is a function of transferred arc power $P_{\mathrm{arc}}$ and resulting temperature change $\Delta T' = T_{\mathrm{arc}}\left(t'\right) - T_{\mathrm{init}}$. The comparison of the arc heat capacity for the three regarded insulating media is depicted in Figure 7. As it can be seen from Figure 7, Air features the smallest arc heat capacity leading to the highest pressure build–up depicted in Figure 4. Due to the significantly larger arc heat capacity of $SF_6$, the corresponding pressure build–up is considerably lower. $CO_2$ is in between.

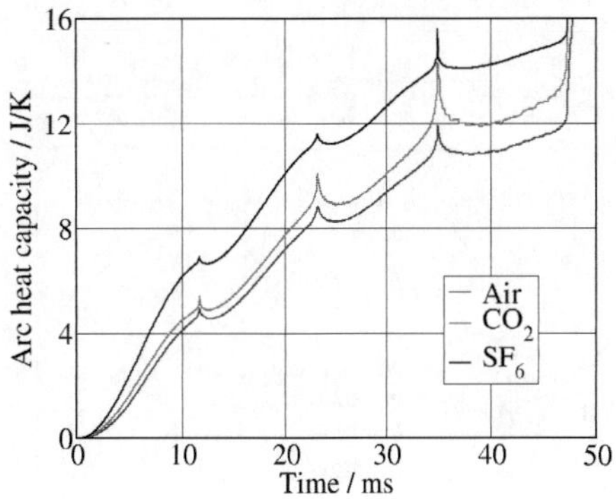


*Figure 7. Comparison of simulated arc heat capacities.*

According to eq. (5), the arc heat capacity depends on the transferred arc power and consequently on the arc voltage. Figure 8 shows the simulated arc voltages for the three regarded insulating media in comparison. The tooth–saw shape of the arc voltages is a result of the utilization of eq. (4) to control the arc radius variation in the discretized arc volume (see Figure 5). According to Figure 8, $CO_2$ features the highest arc voltage followed by Air and $SF_6$. This behaviour corresponds to the measurement results from [4].

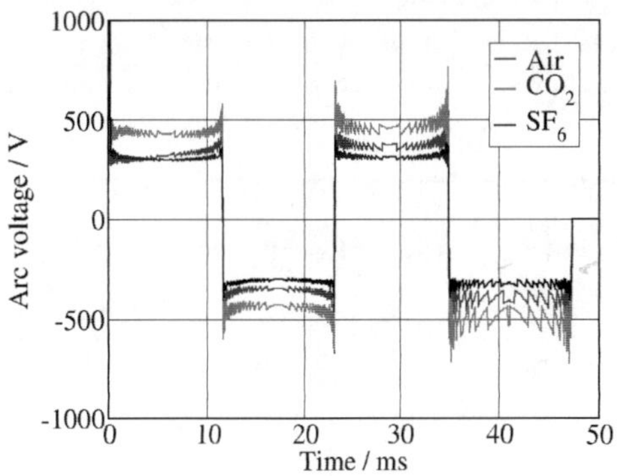


*Figure 8. Comparison of simulated arc voltages.*

The influence of the insulating media on the arc fault process should be explained in more detail using the type test highlighted in the previous section. For this purpose, the arc fault process has been additionally simulated with $CO_2$ and $SF_6$ under the same conditions for e.g. current input, filling pressure and response pressure of rupture disc. The operation of the rupture disc and the following outflow of hot gas to the surroundings influences the arc fault process significantly and changes the behaviour concerning the pressure build–up within the gas–insulated switchgear, see Figure 9. As long as the rupture disc is closed, the

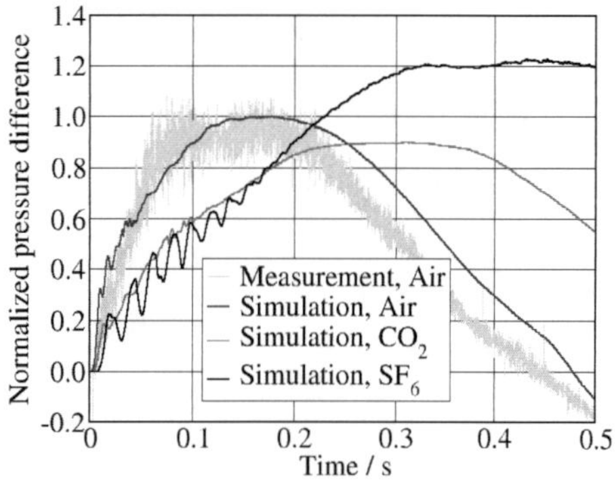


*Figure 9. Pressure build-up in the gas-insulated switchgear (normalized to the maximum value in the simulation with Air).*

arc fault process is mainly influenced by the arc heat capacity of the respective gas and the arc voltage as in the process in the closed vessel from [4]. After the opening of the rupture disc, the outflow of the hot gas is mainly influenced by the speed of sound, which depends inversely proportional on the gas density. As the initial pressure is the same, the gas density differs very strongly in the compared simulation variants. In case of $SF_6$, the highest gas density exists and the hot gas flows out with the lowest speed of sound, see Figure 10. The highest values for the speed of sound

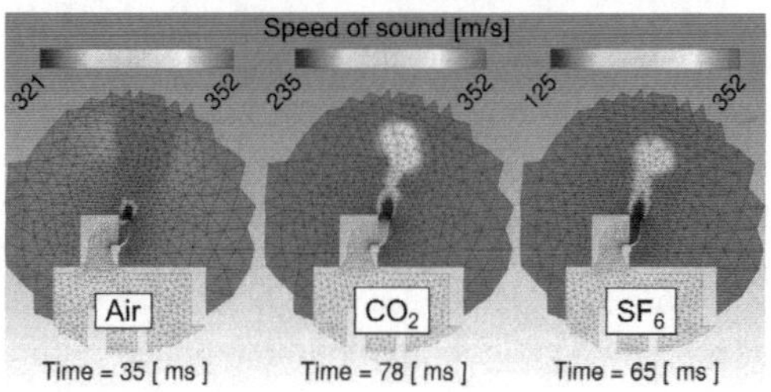


*Figure 10. Contour plot of speed of sound in the symmetry plane about 5 ms after opening of rupture disc.*

occur in case of Air. $CO_2$ lies in between. Regarding the described differences in arc heat capacity, arc voltage and speed of sound, the pressure curves in Figure 9 can be explained. In case of Air, the arc fault process is connected with the steepest pressure rise but by the enforced outflow of hot gas, the maximum of pressure is reached very fast followed by a strong drop of pressure. In case of $SF_6$, the pressure rises much slower compared to Air but the outflow of the hot gas is affected by the low speed of sound leading to a higher maximum of pressure. The lowest values for the maximum pressure are reached in case of $CO_2$.

## 5. Conclusions

The present paper focuses on the basics of a 3D CFD arc fault simulation tool which can be used in the development of overpressure protection systems for gas–insulated switchgears.

The illustrated tool has been validated by measured signals for pressure build–up and arc voltage and allows to investigate the influence of different insulating media on the arc fault process. In this paper, the three insulating media Air, $CO_2$ and $SF_6$ have been considered.

It has been clarified, that in case of a closed vessel the highest pressure rise occurs in Air followed by $CO_2$ and $SF_6$. This behaviour results mainly from the corresponding arc heat capacities.

In case of a gas–insulated switchgear including a overpressure protection system, the pressure build–up within the vessel is also mainly influenced by the arc heat capacity of the respective gas, as long as the rupture disc is closed. After the opening of the rupture disc, the outflow of the hot gas is mainly influenced by the speed of sound of the gas and hence by the density of the insulatig medium within the vessel. In the present paper, the same initial pressure within the vessel has been considered in the 3D CFD arc fault simulation for all regarded insulating media. Thus the highest pressure build–up has been simulated for $SF_6$ followed by Air and $CO_2$.

### Acknowledgements

The authors are grateful to Pierre Freton, Jean-Jacques Gonzalez, Yann Cressault and Philippe Teulet from the University Paul Sabatier in Toulouse for their support and participation in the developments of numerous submodels and for providing of thermodynamic, transport and radiation properties.

### References

[1] *High-voltage switchgear and controlgear - Part 203: Gas-insulated metal-enclosed switchgear for rated voltages above 52 kV.* IEC 62271-203, 2011.

[2] N. Uzelac et al. *Tools for the simulation of the effects of the internal arc in transmission and distribution switchgear.* Cigre, Technical Brochures, 602, WG A3.24, 2014.

[3] X. Zhang. *Modellierung der Auswirkungen von Störlichtbögen in elektrischen Anlagen.* Postdoctoral thesis. RWTH Aachen, 2002.

[4] S. Wetzeler et al. Influence of insulating gas on pressure rise in electrical installations due to internal arcs. *20th International Conference on Gas Discharges and their Applications*, pages 442–445, 2014.

doi:10.14311/ppt.2019.1.39
*Plasma Physics and Technology* **6**(1):39–42, 2019 

# BREAKDOWN IN $CO_2/O_2$ AND $CO_2/O_2/$ $C_2F_4$ MIXTURES AT ELEVATED TEMPERATURES IN THE RANGE 1000–4000 K

M. Seeger[a,*], T. Votteler[a], S. Pancheshnyi[b], J. Carstensen[a], A. Garyfallos[a], M. Schwinne[a]

[a] *ABB Switzerland Ltd., Corporate Research, Segelhofstrasse 1K, CH-5405 Baden-Dättwil, Switzerland*
[b] *ABB Switzerland Ltd., Business Unit Transformers, 8048 Zürich-Altstetten, Switzerland*
[*] martin.seeger@ch.abb.com

**Abstract.** $CO_2$ is a promising gas for replacement of $SF_6$ in high voltage circuit breakers. The electric breakdown in $CO_2$ and mixtures with $C_2F_4$ from PTFE nozzles is, however, still not fully understood. To understand the electric breakdown in $CO_2$ and mixtures with $C_2F_4$ from PTFE nozzles fundamental parameters like the temperature and pressure dependence of critical electric fields are needed. Data on critical fields is usually available based on simulations only and experimental validation is lacking so far. Our present contribution aims to close this gap and presents experiments where the breakdown fields in uniform and weakly non-uniform electric fields are determined at various temperatures and pressures. The gas temperatures were estimated from measurements with pyrometers. The results are compared to theoretical predictions.

**Keywords:** electric breakdown, $CO_2$, $C_2F_4$, dielectric recovery.

## 1. Introduction

Circuit breakers (CB) are important components in electrical power transmission and distribution networks. In gas circuit breakers, as used in high voltage transmission, mostly $SF_6$ is used today. Due to the high global warming potential of $SF_6$, search for alternative gases is ongoing since more than two decades and has intensified in the last decade. The most suitable alternative gases are identified to be $CO_2$ or mixtures of $CO_2$ with strongly attaching fluorinated gases [1, 2]. Since $CO_2$ is the base gas for all these new solutions it is very important to understand its properties under conditions of switching and insulation sufficiently well. For the dielectric interruption after arcing the dielectric properties of the hot gases need to be known. In the present contribution we present some results, where we investigated the breakdown fields in hot $CO_2$ and mixtures with PTFE ($C_2F_4$) vapor from the nozzles in uniform electric fields. Usually the breakdown in uniform fields occurs at streamer inception which is fulfilled slightly above the critical electric field, i.e. the zero crossing of the effective ionization coefficient, if voltage application time is sufficiently long to allow for spark transition. Therefore, measurement of the breakdown field in uniform or close to uniform electric fields may give an approximation for the critical field. At elevated gas temperatures, where precise measurements of the effective ionization coefficient are nearly impossible, such measurements give useful data to validate theoretical data based on solution of the Boltzmann equation. Section 2 describes the experimental set-ups, section 3 gives results and discussions. Conclusions are given in section 4.

## 2. Experimental setup

For the experiments a similar set-up as described in [3] was used. The test vessel was filled with $CO_2/O_2$ (90/10% mixture, molar fractions given throughout the paper) at 1, 2 and 4 bar. An arc was ignited between two contacts in a steel tube and the resulting hot gas was exhausted through a hole into a free gas space. This is denoted as hot gas generator in the following, see figure 1. The arcing gap was 35 mm, the relative contact speed about $3\,\mathrm{m{\cdot}s^{-1}}$ and the diameter of the plug contact 17 mm. The 50 Hz test current was varied between 4 . . . 5 kA peak. Arcing times were in the range of 25 . . . 28 ms, typically. The electrode with the hole on top of the hot gas generator was on ground potential. The diameter of the plate electrode was 150 mm and the hole had a diameter of 28 mm. Above the ground plate a parallel plate of same diameter was positioned, where HV was applied. The electrode gap was 38 mm in set-up 1. In set-up 2 a mesh grid was introduced 18 mm above the ground plate and electrically connected to it, resulting in an electrode gap of 20 mm. The hot gas generator was used with and without PTFE tube (see figure 1), which allowed for having $C_2F_4$ vapor or not in the hot gas stream.

The electrode arrangement was placed into a commercial gas insulated switchgear (GIS) vessel with a flange, equipped with a quartz window, allowing for optical observation of the discharges by high speed camera (Vision Research, Phantom V7.3, exposure time 20 . . . 80 µs). This window was also used to measure the infrared light in the gap with a fast infrared two color pyrometer (Kleiber, custom made, 990–1040 nm and 1100–1800 nm) in combination with a single-color pyrometer (Kleiber 740-LO, 2000–2200 nm). The time resolution of these pyrometers was less than 100 µs. The spot size for the pyrometer focus was about 2 mm in diameter and was adjusted to the mid gap position. Since in $CO_2$ and with PTFE nozzles as well we observed soot particles, the pyrometers measure the temperature of the soot particles. Assuming that these soot particles are in thermal equilibrium with the gas allows deducing the gas temperature from the pyrometer measurement. The test voltage was produced by a pre-charged capacitor (LC-circuit), which was switched onto the test gap by a triggered spark gap through a bushing in the GIS vessel. The test voltage had a rise time of 30 µs and a peak of 100 . . . 250 kV, depending on fill pressure and was switched onto the test gap at various instants during or after the sinusoidal high current half wave. In this way the breakdown voltage at different temperatures could be determined.

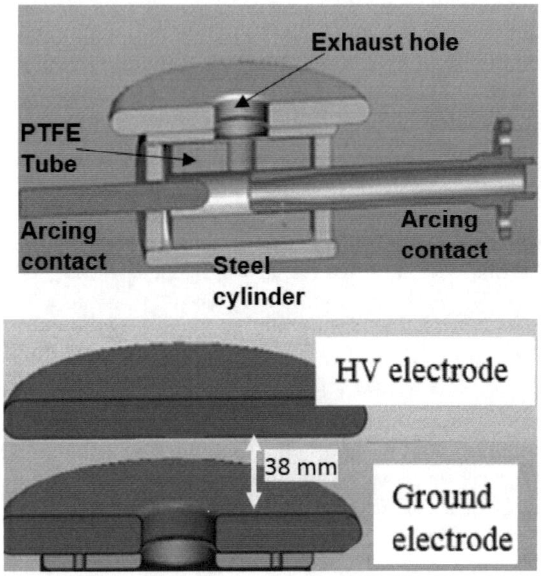


*Figure 1. Hot gas generator and exhaust hole in ground electrode (top), electrode arrangement of set-up 1 (bottom).*

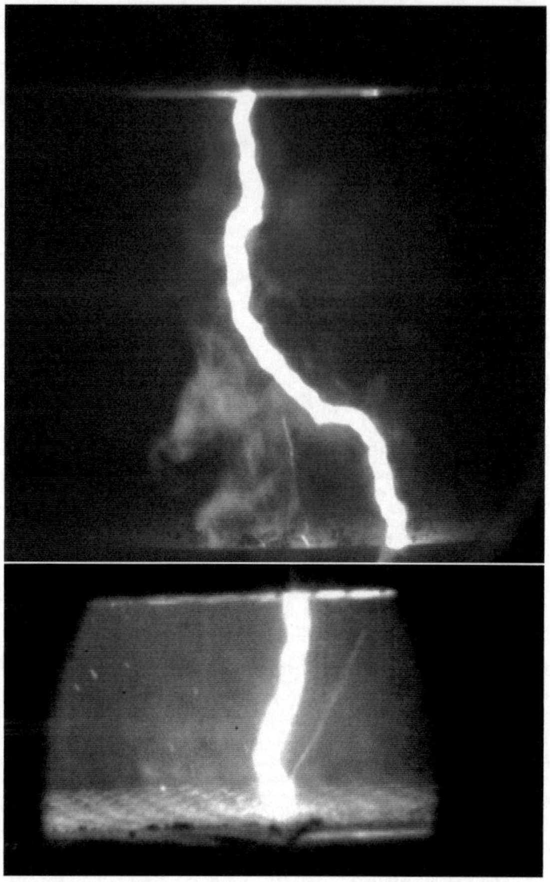

*Figure 2. Breakdowns at 1 bar in set-up 1 (top) and set-up 2 (bottom).*

## 3. Results

Typical hot gas distributions in the test gap and breakdown paths in both experiments are shown in figure 2. Usually in set-up 2, due to the mesh a more uniform light distribution of the hot gas was observed in the video frames, indicating possibly a more uniform temperature distribution. The breakdowns in set-up 1 happened sometimes at the hole edge. Set-up 2 was used for deducing the influence of the hole on the breakdown voltages.

From the pyrometer signals the temperature was deduced under assumption of same emissivity in all three bands. The resulting dependencies of breakdown voltages on temperatures for $CO_2/O_2$ and $CO_2/O_2/C_2F_4$ mixtures are shown in figure 3 and figure 4, respectively. The error bars show the uncertainties of the temperature evaluations. The figures show also predictions based on the critical electric field in hot $CO_2/O_2$ and $CO_2/O_2/C_2F_4$ mixtures. These predictions were obtained with an in-house code calculating the LTE particle composition and using Bolsig+ [4] for the electron energy distribution function with the cross sections retrieved from LXCAT [5].

The figures show that the experimental values exceed the predictions. This discrepancy can be partially understood by the higher field needed for streamer inception. This is shown for set-up 1 at 1 bar by the green curve. Additionally, the formative time lag plays a role, i.e. the time for streamer to spark transition, leading to higher breakdown voltages than only the critical field. Note that a formative time lag of a few microseconds translates into a voltage of several kV at the given steepness of the TRV (e.g. about 3 kV/µs at 1 bar). There are also variations in the temperature distributions which will lead to variations of breakdown voltages. Taking this into account, a reasonable agreement between calculated voltages from the critical fields and measured breakdown voltages can be stated, especially for the predicted trends of

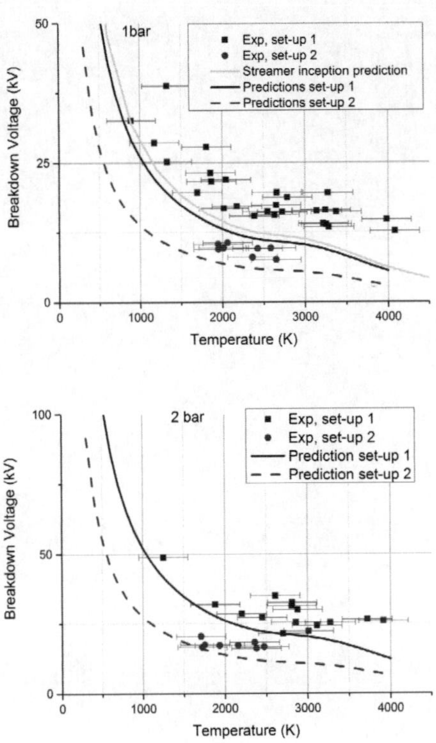


*Figure 3. Breakdown voltages in $CO_2/O_2$ mixture vs temperature at 1 and 2 bar.*

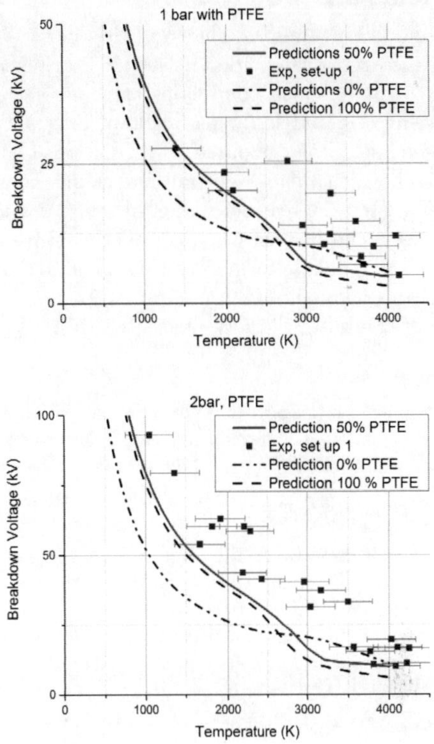


*Figure 4. Breakdown voltages in $CO_2/O_2/$ $C_2F_4$ mixture vs temperature at 1 and 2 bar.*

the temperature dependencies. Only at higher temperatures, above 3000 K and $CO_2/O_2$ mixtures in the measurements there is less decrease of the breakdown voltage with increasing temperature compared to the predictions in figure 3. Comparing the breakdown voltages for set-up 1 and 2 we note that they scale with the gap distance, i.e. breakdown fields in both set-ups are the same. This indicates that influences due to field enhancement at the exhaust hole and temperature variations in both set-ups are within the uncertainties. For $CO_2/O_2/C_2F_4$ mixtures in figure 4 the $C_2F_4$ content was not determined. Therefore, the predictions are shown for various $C_2F_4$ contents (0, 50 and 100%). A reasonable agreement is seen with 50% curve (blue). Since the difference between 50% and 100% content is small, we can deduce that the PTFE content in the tests was probably about 50% or higher. This seems to be plausible since the arc is burning in PTFE vapor in the hot gas generator, i.e. there should be a large $C_2F_4$ content. Interestingly the agreement for the $C_2F_4$ mixtures is also satisfactory in the range above 3000 K. A similar agreement is seen at 4 bar for $CO_2/O_2$ and $CO_2/O_2/$ $C_2F_4$ mixtures (not shown here).

The calculated critical electric fields can be plotted in particle density reduced form. This is shown in figure 5. These values can be compared to literature values, indicated by the symbols [6, 7]. A reasonable agreement of our calculations with the literature data is seen. Note that the literature values are for pure $CO_2$. For $CO_2/O_2$ mixture the critical fields are slightly higher than those for $CO_2$ below 3000 K, this was also observed in [7]. At 1 bar and temperatures above 3000 K we see a discrepancy between our predictions and those from literature. This needs further investigations. For $C_2F_4$ mixtures significantly higher critical fields can be seen below 2500 K compared to $CO_2$ and $CO_2/O_2$ mixtures.

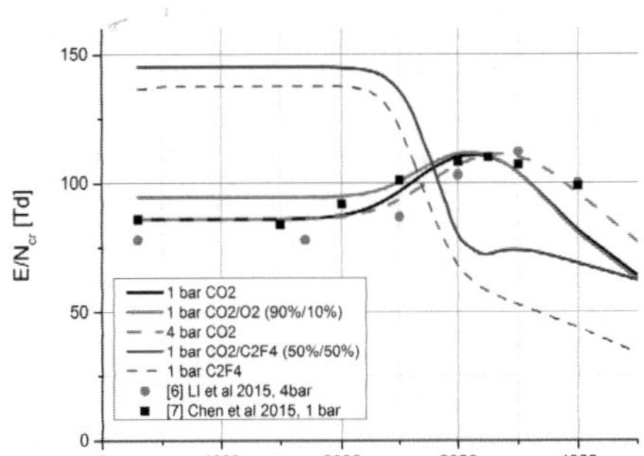


*Figure 5. Predicted particle density reduced critical fields normalized to standard conditions of 300 K and 1 bar for $CO_2/O_2$, $CO_2$ and $CO_2/O_2/$ $C_2F_4$ mixtures. Literature values for $CO_2$ are given by the symbols.*

## 4. Conclusions

The breakdown in uniform electric fields in $CO_2$ at temperatures in the range 1000 ... 4000 K is presented. Breakdown under such conditions was not experimentally investigated so far to our knowledge. The measurements can be compared to predictions of the critical field, which are a lower limit for the breakdown in uniform gaps. A reasonable agreement with the measurements can be seen when considering streamer inception and possible formative time lags. By this we validate, within the given uncertainties, not only our predictions but also those from literature.

### References

[1] P. C. Stoller, M. Seeger, A. A. Iordanidis, and G. V. Naidis. $CO_2$ as an arc interruption medium in gas circuit breakers. *IEEE Transactions on Plasma Science*, 41(8):2359–2369, 2013. doi:10.1109/TPS.2013.2259183.

[2] M. Seeger, R. Smeets, J. Yan, H. Ito, M. Claessens, E. Dullni, L. Falkingham, C. Franck, F. Gentil, W. Hartmann, et al. Recent trends in development of high voltage circuit breakers with $SF_6$ alternative gases. *Plasma Physics and Technology Journal*, 4(1), 2017. doi:10.14311/ppt.2017.1.8.

[3] P. Simka, M. Seeger, T. Votteler, and M. Schwinne. Experimental investigation of the dielectric strength of hot $SF_6$. In *2012 IEEE 10th International Conference on the Properties and Applications of Dielectric Materials*, pages 1–4. IEEE, 2012. doi:10.1109/ICPADM.2012.6319001.

[4] G. Hagelaar. Brief documentation of BOLSIG+ version 03/2016. *Laboratoire Plasma et Conversion dEnergie (LAPLACE), Universit Paul Sabatier*, 118, 2016.

[5] S. Pancheshnyi, S. Biagi, M. Bordage, G. Hagelaar, W. Morgan, A. Phelps, and L. Pitchford. The LXCAT project: Electron scattering cross sections and swarm parameters for low temperature plasma modeling. *Chemical Physics*, 398:148–153, 2012. doi:10.1016/j.chemphys.2011.04.020.

[6] X. Li, X. Guo, H. Zhao, S. Jia, and A. B. Murphy. Prediction of the critical reduced electric field strength for carbon dioxide and its mixtures with copper vapor from boltzmann analysis for a gas temperature range of 300 K to 4000 K at 0.4 MPa. *Journal of Applied Physics*, 117(14):143302, 2015. doi:10.1063/1.4917210.

[7] Z. Chen, C. Niu, H. Zhang, H. Sun, Y. Wu, F. Yang, M. Rong, and Z. Xu. Investigation on the reduced critical breakdown field of hot $CO_2$ gas and $CO_2$-based mixtures. In *2015 3rd International Conference on Electric Power Equipment–Switching Technology (ICEPE-ST)*, pages 36–39. IEEE, 2015. doi:10.1109/ICEPE-ST.2015.7368390.

doi:10.14311/ppt.2019.1.43
*Plasma Physics and Technology* **6**(1):43–46, 2019 

# FORMATION OF THE UPSTREAM REGION IN HV GAS BLAST INTERRUPTERS WITH SYNCHRONOUS GAS INJECTION

N.K. Kurakina[a, b, *], V.Ya. Frolov[a], E.N. Tonkonogov[a]

[a] *Peter the Great Saint-Petersburg Polytechnic University, Polytechnicheskaya 29, St.-Petersburg 195251, Russia*

[b] *Institute for Electrophysics and Electric Power of Russian Academy of Sciences, Dvortsovaya nab., 18, St.-Petersburg, 191186, Russia*

[*] nkuriee@gmail.com

**Abstract.** The problem of a synchronous gas injection at the vicinity of current zero for high voltage gas blast circuit breakers (GCB) is discussed. The basic aim of the investigation is to improve the efficiency of arc quenching in GCB by increasing interruption ability without changes in drive parameters. The paper reviews the upstream region influence on the interruption capability, gas injection - mass flow interactions at the region between electrical contact and nozzle throat of GCB. Numerical simulation is used to calculate gas–dynamics parameters at the upstream region of mono–flow GCB. The mass discharged as a result of the synchronous gas injection can rapidly modify both the density and pressure at the nozzle throat, regulate flow velocity in the research region. The selected method of the gas injection effects the breaking performance in the arc quenching device strongly. Some calculation results are presented.

**Keywords:** high–voltage gas blast interrupters, air blast circut breakers, interruption capability, arc quenching, synchronous gas injection, gas–dynamics, numerical simulation, OpenFOAM.

## 1. Introduction

The power growth in electrical networks imposes the continued efforts to improve the performance of the circuit breakers. According to the Kioto protocol, agreed in 1997, the use of sulfur hexafluoride in electric power transmission and distribution systems must to be minimized despite its high efficiency. The search of alternative gases of natural origin and the appliance of dry air, $CO_2$, and $N_2$ still remain an urgent problem [1, 2]. It is therefore useful to develop of GCB which are advanced to a high performance level. Today, the drive mechanics has a particular impact on the efficiency of the synchronous GCB. The high proportion failures (52–72 % of the overall failure rates [3]) are attributed to the drivers. Furthermore, the complicated drives construction contradicts the modern requirements of resource-intensiveness and the high criteria for operation according to the international standards [4].

The present study is aimed at increasing the interruption capability of the mono–flow gas arc quenching device without a complication of the drive constructions with an alternative to $SF_6$ medium. An example of such medium is dry air. It is proposed to influence the switching arc parameters by a short–time gas pulse of high pressure at the vicinity of current zero. The main reason of electrical breakdown post current zero is a very hot remnant of the arc channel between the contacts. The intensification of the cold gas flow removes the surplus of energy of the decaying arc column as well as the complexity of the pressure, mass, and velocity interchanges (due to the impulse gas injection) increases the turbulent mixing to enable a successful arc quenching.

This article presents the model validation, the gas dynamics parameters (mass flow rate, pressure and etc) in the upstream and nozzle throat regions for cold and hot models obtained by means of the free open-source package OpenFOAM for the numerical simulation.

## 2. Model device and methods

The model of the mono–flow gas arc quenching device (figure 1) was chosen by the authors due to the accessibility of experimental data, necessary for the conclusions in this work and the subsequent experimental studies.

Reference [5] identified which design parameters were dominant in establishing the optimum rate of rise of recovery voltage performance ($RRRV$). According to these results, the nozzle size and shape of the model device (figure 1) were chosen on the assumption $L/D$=1, where $L$ is the distance between the fixed contact and the critical section of the nozzle, and $D$ is the diameter of the nozzle. The working gas (air) of pressure $p_0$ is fed into the upstream area through the inlet. The gas flow interaction builds the upstream region.

There is a general relationship in the switchgear literature [6, 7] that the interruption capability of GCB increases significantly with gas pressure growth and reduces with increasing current. The logarithmic relationship is shown in figure 2 of the parameter $RV^*$=$(\mathrm{d}U/\mathrm{d}t)/p_{cr}$, where $p_{cr}$=$p/2$, versus the rate of

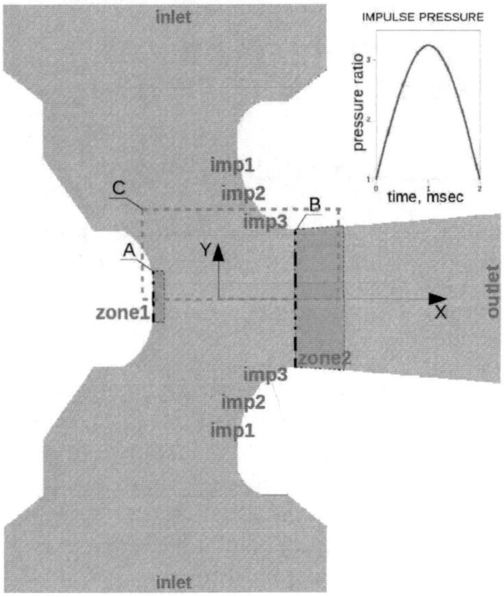


*Figure 1. The model of the mono–flow gas blast arc quenching device.*

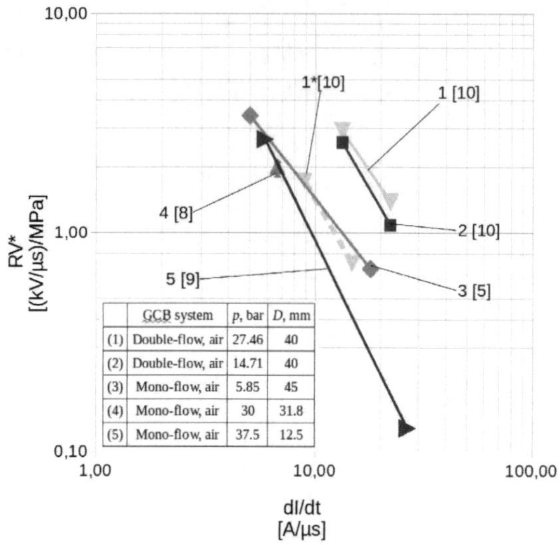


*Figure 2. Normalized parameter $RV^*$ as a function of $dI/dt$.*

change of the current $\mathrm{d}I/dt$ for air GCB. The data was obtained in experiments by the authors in Ref. [5, 8–10] at various values of upstream pressures and nozzles design constructions.

The analysis of the curves in figure 2 shows the general trend of interaction ability for air GCB. Yellow (1) and blue (2) curves were plotted for double flow arc quenching systems [10]. Their interrupting ability is by a factor of 1.5 higher [6, 7] than that of mono–flow devices, see curves (3), (4), (5). Yellow dashed line (1*) denotes the "actual" location of the curve (1) after dividing by the over–coefficient 1.5. The arching time in the experiment [10] was 1.5 ms. This means that the active arc–gas interaction after current zero took not more than 1 ms. The comparison of mono–flow systems and a synchronous commutation shows that a successful commutation can be reached by using a short-time gas pulse of high pressure with a duration of about 1.5 – 2 ms in the interrupting ability of the GCB.

It is suggested, therefore, to deliver a short–time powerful gas impulse of pressure $p$ at the vicinity of current zero by means of an injector (imp1, imp2 or imp3) with a diameter $d$ (here $D/d$=4.5). The gas pulse shape and duration (taken as 2 ms as a first approximation) are presented in figure 1. The zone 1 is a region of stagnation. It appears with the increase in pressure and the reduction of velocity near the contact. The most probable position of the arc remnant destruction is marked as zone 2. Zones 1 and 2 are the regions of our special interest. The injection pressure $p$ was chosen from the relationship $p/p_0$=3 [8].

The gas–dynamics parameters (pressure rate, flow velocity and mass flux) are simulated with the free open–source package OpenFOAM by adopting a two–dimensional model and a continuous medium approximation. The conservation laws of mass, momentum and energy complemented by the state equation serve as a basis for the description of the compressed gas flow with initial supersonic speed in similar systems. The method was described and verified in [11, 12] of the experimental stand [13, 14], where the calculated results agreed well with experimental data and dependences.

## 3. Cold gas flow phenomena with synchronous gas injection

The interplay of the gas–dynamics parameters has the major contribution to the interruption ability of GCB. The knowledge of the cold gas flow phenomena helps to understand better the nature of the arc–gas correlations and to choose the optimum position for the injection, which is important since it governs the efficiency of the arc quenching.

Figure 3 presents a comparative analysis of the normalized mass flux $Q/Q_0$ versus time for zone 1 of the cold model in relation to the basic gas flow value with different injection sites (imp1/imp2/imp3). The initial values are the following: $p/p_0$=3, $D/d$=4.5, $L/D$=1.

The plots in figure 3 show that the best injection site is imp3 for the model of mono–flow GCB due to the maximum time–mean value of the mass flux on the considered region.

Figure 4 presents calculation results of the system pressure ($p/p_0$) with a synchronous injection at the site imp3 along the chamber axis for various times from the pulse beginning. The normalized distance from the contact (A) is defined as $X=x/D$, where $x$ is

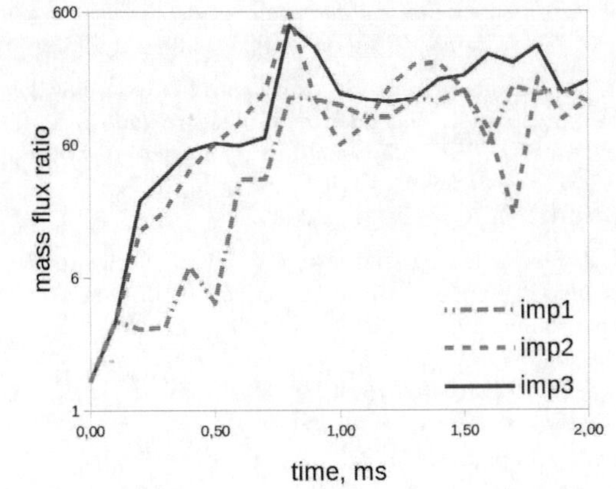


*Figure 3. Mass flux ratio $Q/Q_0$ in zone 1 for different injectors sites (imp1, imp2, imp3).*

the distance along the axis OX of the nozzle channel, and $D$ is the nozzle diameter.

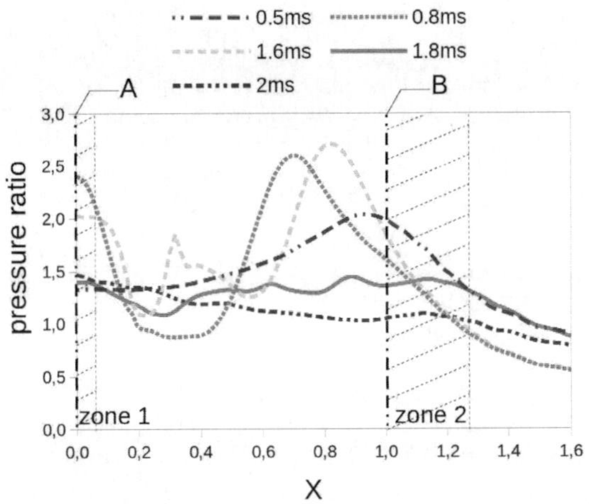


*Figure 4. Pressure ratio along OX for various times from injection beginning at imp3.*

The injection of high pressure gas begins at time $t$ in the vicinity of current zero. Here, the zero value is conditionally taken (see figure 4). The gas pulse is supplied to the computational domain when the basic stationary flow is reached. A sharp pressure jump occurs in front of the nozzle throat during the period between 0.5 and 0.8 ms (blue-red curves). Afterwards, the reflected shock wave begins to propagate to zone 1 near the contact and is reflected on it, intensely affecting the stagnation points. The pressure variation is determined by the interaction of the waves and their successive reflection from the fixed elements of the arc quenching chamber in the time interval (0.8–1.6) ms (red-yellow curves). At $t$=2 ms (green-black curves), unsteady decaying waves are observed and the pressure value reduces to the level $p_0$, determined by the basic gas flow.

As a result, the directional synchronous gas injection controls the density and pressure differential of the mass discharged. The pressure differential regulates the gas velocity. The maximum effect of a turbulent mixing is achieved. According to the calculation results, the gas pulse intensifies and prolongs mass flow fluctuations for more than 100 µs. It is instrumental in the recovery performance.

## 4. Interactions between the switching arc and the synchronous gas injection

To study how the decaying plasma column influences the upstream region in the mono–flow air arc quenching device with a synchronous gas injection, a "hot" simulation is used. For this purpose, a region of high temperature, small density, initial pressure and velocity was set at the position of the arc column. These parameters have been calculated in a previous simulation.

It is known that the arc diameter is proportional to the square root of the current magnitude. The diameter of the hot arc remnant $d_a$ at current zero does not exceed 1 mm for the rate of change of current $dI/dt$=24 A/µs [6]. The plasma temperature varies from 2000 K on the thermal boundary layer up to 10000 K on the axis [6]. In the simulation model, the bulk temperature of the arc and the diameter were taken respectively $T$=5000 K and $d_a$=4 mm near the contact (A) as a first approximation, taking into account the initial interaction between the injection and the thermal boundary layer at 400–600 µs before current zero (see below).

Calculation results of the deformation of the thermal boundary layer for the region C (see figure 1) are presented in figure 5. The temperature distribution with a contour line of $T$=2000 K defines the thermal boundary layer in the model. To trace the dynamics of the injection influence on the thermal boundary layer, the curves are plotted for various times from the beginning of the injection: $t_0$=0 ms – start of the injection (yellow line), $t_1$=0.35 ms – the initial time of active interaction between the front of the gas pulse and the arc boundary layer (blue line), $t_2$=$t_1$+100 µs (red line), $t_3$=$t_2$+50 µs (black line).

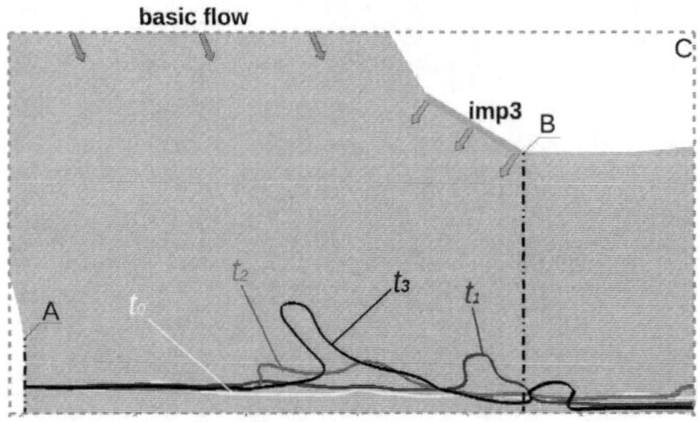


*Figure 5. Thermal boundary layer deformation for various times from injection beginning.*

Figure 5 confirms that the synchronous gas injection has a deforming effect on the boundary layer for the "hot" model simulation. The general deformation is defined by the mass connecting to the area. A boundary layer displacement and an increase of the interaction cross-section are observed, providing a turbulent mixing at the upstream region. This would cause a development of the interruption ability and efficiency of the arc quenching.

## 5. Conclusions

The influence of the synchronous gas injection on the gas–dynamics parameters in the HV mono–flow GCB are investigated by cold and hot models simulations. The synchronous gas injection changes the initial gas–dynamics characteristics (density, pressure, velocity, and mass flow rate), deforms the arc thermal boundary layer, provides the turbulent mixing in the upstream region at the vicinity of current zero. All of these points are essential for the improvement of the interruption performance of the GCB. The numerical simulation confirms the efficient contribution of the synchronous gas injection to increase the interruption ability of high voltage gas blast circuit breakers.

### References

[1] H. Hama, S. Okabe, R. Buhler, D. Gautschi, A. Girodet, M. Hering, D. Imamovic, K. Juhre, Y. Kieffel, J. Kindersberger, W. Koltunowicz, J. Lopes-Roldan, S. Neuhold, C. Neumann, R. Pietsch, U. Riechert, T. Rokunohe, U. Schichler, and T. Yasuoka. Dry air, $N_2$, $CO_2$ and $N_2/SF_6$ mixtures for gas-insulated systems. *Electra*, page 71, Aug 2018.

[2] M. Seeger, R. Smeets, J. Yan, H. Ito, M. Claessens, E. Dullni, L. Falkingham, C. Franck, F. Gentils, W. Hartmann, Y. Kieffel, S. Jia, G. Jones, J. Mantilla, S. Pawar, M. Rabie, P. Robin-Jouan, H. Schellekens, J. Spencer, T. Uchii, X. Li, and S. Yanabu. Recent trends in development of high voltage circuit breakers with $SF_6$ alternative gas. *Plasma Physics and Technology*, 4(1):8–12, 2017. doi:10.14311/ppt.2017.1.8.

[3] R. Thomas. Controlled switching of high voltage $SF_6$ circuit breakers for fault interruption, 2004. Thesis for the degree of licentiate of engineering. http://webfiles.portal.chalmers.se/et/Lic/ThomasRichardLic.pdf.

[4] International standard IEC 62271-100. High-voltage switchgear and controlgear. Part 100. Alternating current circuit-breakers, 2012.

[5] A. Briggs and L. King. Factors affecting thermal breakdown in gas blast interrupters. *IEE Conf. Dev in Design and Performance, London*, pages 52–55, Nov 1979.

[6] K. Ragaller. *Current Interruption in High Voltage Networks*. Moscow, Energoizdat, 1981.

[7] E. Tonkonogov. *HV AC Circuit Breakers*. Saint Petersburg, Polytechnic University Press, 2015.

[8] A. Briggs. The contribution of axial a radial components of gas velocity to the thermal interruption capability of a gas blast circuit breaker. *Proceedings of the 7th International Conference on Gas Discharges and their Applications, UK, London*, pages 28–31, 1982.

[9] G. Frind and J. Rich. Recovery speed of axial flow gas blast interrupter: dependence on pressure and di/ dt for air and $SF_6$. *IEEE Transactions on Power Apparatus and Systems*, PAS 93(5):1675–1684, Sep 1974. doi:10.1109/TPAS.1974.293900.

[10] T. Morita, M. Iwashita, and Y. Nitta. A theoretical analysis of dynamic arcs and test results of model synchronous air blast circuit breakers. *IEEE Transactions on Power Apparatus and Systems*, PAS-97(3):940–949, May 1978. doi:10.1109/TPAS.1978.354567.

[11] N. Kurakina, M. Pinchuk, A. Budin, A. Smirnovsky, and V. Frolov. Numerical simulation using openfoam of gas-dynamics in the discharge chamber with a movable electrode. *St. Petersburg Polytechnic University Journal of Engineering Science and Technology*, 24(2):69–81, 2018. doi:10.18721/JEST.240206.

[12] N. Kurakina, M. Pinchuk, A. Budin, and A. Smirnovsky. Analysis of the gas dynamics in the discharge chamber with a movable electrode. *Journal of Physics: Conference Series*, 1135:012094, Dec 2018. doi:10.1088/1742-6596/1135/1/012094.

[13] A. Budin, M. Pinchuk, V. Leontev, A. Leks, N. Kurakina, A. Kiselev, J. Simakova, and V. Frolov. Experimental stand for investigations of insulator degradation and electrode erosion in high-current breaker. *Plasma Physics and Technology*, 4(2):120–123, 2017. doi:10.14311/ppt.2017.2.120.

[14] A. Budin, M. Pinchuk, V. Kuznetsov, V. Leontev, and N. Kurakina. An experimental setup for investigation of arc and erosion processes in high-voltage high-current breakers. *Instruments and Experimental Techniques*, 60(6):837–842, Nov 2017. doi:10.1134/S0020441217060033.

doi:10.14311/ppt.2019.1.47
*Plasma Physics and Technology* **6**(1):47–50, 2019 

# ESTIMATION OF THE INTENSIVELY BLASTED ELECTRIC ARC MODEL SENSITIVITY TO SELECTED VARIABLES

J. Senk*, I. Laznickova, I. Jakubova, O. Coufal

*Faculty of Electrical Engineering and Communication, BUT, Technicka 12, 616 00 Brno, Czech Republic*

* senk@feec.vutbr.cz

**Abstract.** Results of measurements carried out on the fabricated experimental modular-type arc heater serve as input data for the designed simplified model of the intensively blasted electric arc burning in argon inside the cylindrical arc heater's anode channel. The axial dependence of the arc temperature and radius is expressed using the exponent, the current density on the cathode tip and the arc temperature at the end of the near-cathode boundary layer. These quantities form the vector of state variables that is sought to minimize the value of the objective function expressing the deviations between measured and computed values. On a typical example, the paper demonstrates the sensitivity of the modelling to individual state variables.

**Keywords:** electric arc, arc heater, argon, modeling.

## 1. Introduction

In the experimental modular-type arc heater, which has been designed and fabricated at the Department of Electrical Power Engineering of the Faculty of Electrical Engineering and Communication, Brno University of Technology for both educational and research purposes, the electric arc is burning in the cylindrical anode channel with water-cooled walls and is stabilized by intensive flow of argon (see Fig. 1 of [1]). For each experiment, a set of measured values is recorded, including the arc current, the total voltage, the gas flow rate, and power loss in individual mutually electrically and thermally isolated segments of the arc heater (cathode - subscript cat, input part of the anode channel - subscript in, the anode channel itself - subscript ch, the grounded anode - subscript an).

A simple model of the intensively blasted electric arc inside the anode channel has been designed to describe the behavior of the arc and to reveal mutual relationships between main arc heater's operational quantities and arc's properties in a simple and clear form suitable for educational purposes. The model is based on the mass and energy conservation laws and Ohm's law and includes several simplifying presumptions (e.g. LTE, cylindrical symmetry, rectangular arc temperature profile, etc.) [2, 3]. Measured values together with transport and thermodynamic properties of argon [4–6] serve as input data of the mathematical model.

## 2. The Vector of State Variables and the Objective Function

The behavior of the arc in the anode channel is characterized by the axial development of the arc temperature $T_A(z)$ and arc radius $r_A(z)$. A rectangular temperature profile across the anode channel cross-section is assumed, with the arc zone $T_A(r,z) = T_A(z)$ surrounded by cold argon. The dependence of the arc radius $r_A(z)$ on the axial coordinate $z$ is defined by the exponent $n_r$ and by the cathode spot radius $r_0$ corresponding to the current density on the cathode tip $j_0$ as follows [7]

$$r_A(z) = r_0 \left[1 + \left(\frac{z}{r_0}\right)^{(1/n_r)}\right]. \quad (1)$$

The behavior of the arc in the close vicinity of the cathode is characterized by the width of the near-cathode boundary layer $s$ and by the temperature $T_A(s)$ there [2].

These three quantities $[n_r; j_0; T_A(s)]$ form the vector of state variables that is sought for to minimize the value of the objective function $\Delta_{tot}$. The objective function $\Delta_{tot}$ is given by the deviations between the measured and computed values of the arc power loss in individual segments of the device and the voltage drop along the arc

$$\Delta_{tot} = \sqrt{\Delta_{in}^2 + \Delta_{ch}^2 + \Delta_U^2}. \quad (2)$$

The first term $\Delta_{in}$ describes the relative deviation between the measured (subscript m) and calculated (subscript c) values of the arc power loss in the input part of the arc heater, and similarly, the second term $\Delta_{ch}$ does for the main part of the anode channel

$$\Delta_{in} = \frac{P_{loss\_in\_c}}{P_{loss\_in\_m}} - 1,\ \Delta_{ch} = \frac{P_{loss\_ch\_c}}{P_{loss\_ch\_m}} - 1. \quad (3)$$

The measured power loss is determined from the cooling water flow rate in the corresponding segment and the increase of its temperature. The computed arc power loss is calculated using the theoretically determined values of the net emission coefficient of argon which is used as the working gas in the discussed experiments [1, 6, 8].

The third term $\Delta_U$ expresses the relative deviation between the measured and calculated values of the arc voltage drop. As neither the arc voltage drop over the whole arc $U_A(z_L)$ nor the power consumed for an arc-anode attachment can be measured separately, this last term compares the arc power $U_A(z_L)I_m$ decreased by the measured power loss of the cathode $P_{loss\_cat\_m}$ and the computed power loss of the arc-anode attachment $P_{loss\_as\_c}$ versus the total input power of the device $U_m I_m$ as follows

$$\Delta_U = \frac{U_{A\_c}(z_L)I_m + P_{loss\_cat\_m} + P_{loss\_as\_c}}{U_m I_m} - 1. \tag{4}$$

More details on the model and its gradual refinement can be found in [1–3]. Here, only the selected formulas and their short explanation is repeated to make the following discussion easy to understand.

## 3. Results and Discussion

The paper focuses on the sensitivity of the objective function to the state variables. The investigation aims at making the search of the optimum state variables' vector as effective as possible. Although some preliminary considerations on sensitivity of the model to individual state variables can be derived, qualitative knowledge about how significantly the sensitivity is changing in what a range must be obtained using concrete experimental data.

In the following, the dependence of the objective function as a whole and of its components on the state variables' vector $[n_r; j_0; T_A(s)]$ is shown for a chosen experiment. The experimental arrangement of the arc heater is the following: the anode channel radius $r_c = 8$ mm, the anode channel length $z_L = 109$ mm (the input part 22 mm, the main anode channel 60 mm, the grounded anode 27 mm). The arc was burning in argon with flow rate of 22.5 g/s. The arc current was set to 122.0 A and the total voltage drop over the arc heater was 111.8 V. The total input power was 13.64 kW and the total measured power loss was 3.23 kW. The chosen experiment corresponds to the middle slightly increasing region of the arc-voltage characteristic of the device where the arc exhibits a good stability.

The computation was made for $n_r$ around the value of 3 given by other authors [7], with a step of 0.1 which can be decreased, especially near the minimum of the objective function. For each $n_r$, the calculation was made in a mesh of $j_0$ (up to $4 \times 10^8\,\mathrm{Am^{-2}}$, step $0.01 \times 10^8\,\mathrm{Am^{-2}}$) and $T_A(s)$ (between 6000 K up to approx. 18000 K, step 10 K) and the minimum was found for this $n_r$. The absolute minimum value of the objective function $\Delta_{tot\_min} = 0.03705$ has been reached for $n_{r\_min} = 2.45$, $j_{0\_min} = 2.93 \times 10^8\,\mathrm{Am^{-2}}$ and $T_A(s)_{min} = 13400\,\mathrm{K}$.

Figure 1 shows the dependence of the computed minimum values of the objective function $\Delta_{tot}$ and the corresponding values of $j_0$, the cathode boundary layer width $s$, the arc temperature $T_A(s)$ and the electric field intensity $E(s)$, both at the end of the boundary layer, and the arc temperature $T_A(z_1)$ at the first computational step $z_1 = 1$ mm from the cathode tip, on the exponent $n_r$. Mutual correspondence of temperatures, near-cathode boundary layer width $s$ and $E(s)$ was discussed in previous works [3] and will be mentioned later. Obviously, in the investigated case, the found minimum values of $\Delta_{tot}$ first quickly decrease with increasing $n_r$ and after reaching the global minimum value $\Delta_{tot\_min} = 3.705\%$ for $n_{r\_min} = 2.45$ they start to increase slowly.

Figure 2 shows again the computed minimum values of the objective function $\Delta_{tot}$ but in dependence on other state variables ($j_0$, $T_A(s)$) and altogether with the other two state variables. While a pronounced global minimum for exponent $n_{r\_min} = 2.45$ can be seen in Fig. 1, in Fig. 2 the computed minimums of the objective function exhibit significant changes for low and high values of both $j_0$ and $T_A(s)$, but preserve only slightly changing values in a rather wide middle region. This observation confirms the leading role of the exponent $n_r$.

As explained in our previous work [3] also some other points of view should be taken into account to obtain physically acceptable results. More detailed analysis can show that for some obtained minimums of the objective function, the computed arc temperature $T_A(z)$ tends to oscillate, especially near the beginning. Another problem to be avoided is an extremely high electric field intensity at the end of the cathode boundary layer and consequently an extremely thin boundary layer. Requirements of fast but smooth and non-oscillating increase of the arc temperature near the cathode and reasonable values of $s$ and $E(s)$ play a role of complementary limiting conditions of the model. Their fulfillment can be ensured by a fine tuning of the state variables in a close vicinity of the global minimum. For this task, a deeper insight into the behavior of individual components of the objective function is useful.

In the following, results are computed for a constant exponent $n_r$ and are given in spatial diagrams and their projection to the $[T_A(s)\,j_0]$ plane to demonstrate sensitivity of the objective function and its components to the current density at the cathode tip $j_0$ and to the arc temperature $T_A(s)$ at the end of the cathode boundary layer. The chosen exponent $n_{r\_min} = 2.45$ corresponds to the global minimum of the objective function $\Delta_{tot\_min}$ for the investigated experiment.

In spatial diagrams, rather sharp shapes in Fig. 3 (a,b) clearly indicate that the first two components of the objective function, which express fidelity of the computed arc power loss distribution in the input and main parts of the anode channel in comparison with the measured power loss, are highly sensitive especially to the current density $j_0$ at the cathode tip. As far as the arc temperature $T_A(s)$ at the end of the

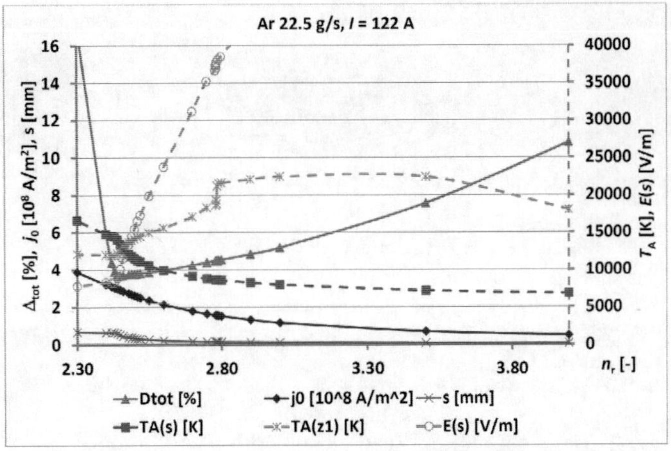


*Figure 1. Minimum values of the objective function $\Delta_{\text{tot}}$ together with the corresponding values of the current density at the cathode tip $j_0$, the boundary layer width $s$, the arc temperature $T_A(s)$ and the electric field intensity $E(s)$, both at the end of the boundary layer, and the arc temperature $T_A(z_1)$ at the first step $z_1 = 1$ mm vs. the exponent $n_r$, for the investigated experiment.*

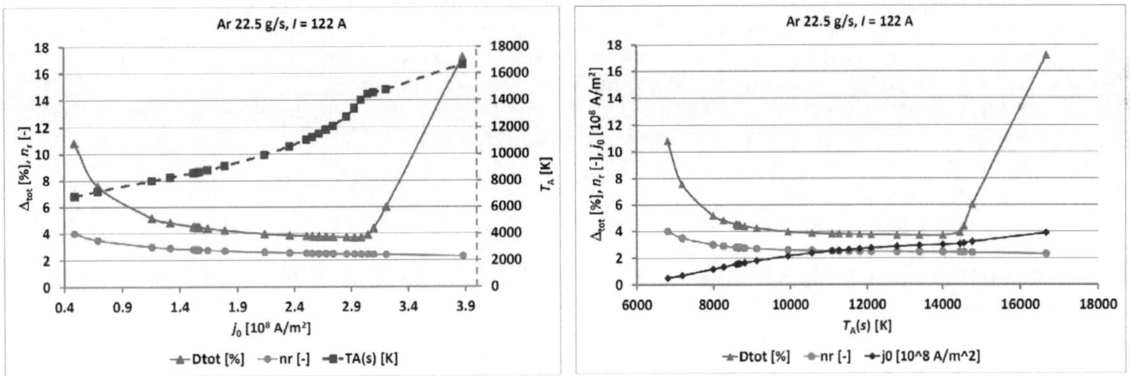


*Figure 2. Minimum values of the objective function $\Delta_{\text{tot}}$ together with the corresponding values of the exponent $n_r$ and the arc temperature $T_A(s)$ at the end of the boundary layer vs. the current density at the cathode tip $j_0$ (left), and together with the corresponding values of the exponent $n_r$ and the current density at the cathode tip $j_0$ vs. the arc temperature $T_A(s)$ at the end of the boundary layer (right), for the investigated experiment.*

cathode boundary layer is concerned, obviously $\Delta_{\text{in}}$ exhibits the highest sensitivity to this state variable; this observation is in accordance with preliminary expectations. Both $\Delta_{\text{in}}$ and $\Delta_{\text{ch}}$ change their sign in the tested range. In other words, their minimum absolute value is zero. This fact is stressed in Fig. 3 (d, e).

The spatial diagram of $\Delta_U$ in Fig. 3 (c) is given in fine resolution (values between 2% and 4%) and is much flatter. Finally, the spatial diagram of $\Delta_{\text{tot}}$ in Fig. 3 (f) is steep and its lowest values correspond to the intersection of $\Delta_{\text{in}} = 0$ and $\Delta_{\text{ch}} = 0$ lines (Fig. 3 d,e). It can be concluded that the behavior of $\Delta_{\text{in}}$, $\Delta_{\text{ch}}$ is crucial for reaching the minimum of the objective function while the third component $\Delta_U$ has a predominant influence on the reached minimum value of $\Delta_{\text{tot\_min}}$.

As can be seen from Fig. 1, in the global minimum of $\Delta_{\text{tot\_min}}$ the temperature $T_A(s)$ is higher than the temperature $T_A(z_1)$. Obviously, the requirement of the increasing arc temperature $T_A(z)$ near the arc beginning is not met in the found global minimum of the objective function in this case. Solving this problem only by a shift of $T_A(s)$ to lower values while keeping the found $n_r$ unchanged would significantly rise $\Delta_{\text{tot}}$, especially due to $\Delta_{\text{in}}$. Fortunately, close above $n_{\text{r\_min}} = 2.45$ the curves $T_A(s, n_r)$ and $T_A(z_1, n_r)$ intersect, thus a slight shift above this intersection, to a neighboring minimum with higher $n_r$ can ensure the required relation between $T_A(s)$ and $T_A(z_1)$. Of course, the shift out of the global minimum must lead to a worse value of the objective function, but the increase of $\Delta_{\text{tot}}$ is of the order of low tenths of percent and is acceptable. On the other hand, a steep increase of $E(s)$ over physically reasonable values does not allow to retreat too far from the global minimum (Fig. 1).

In the discussed case, the global minimum of $\Delta_{\text{tot\_min}}$=3.705% is found for the following values: $n_{\text{r\_min}} = 2.45$, $j_{0\_\text{min}} = 2.93 \times 10^8$ Am$^{-2}$, $T_A(s)_{\text{min}} =$

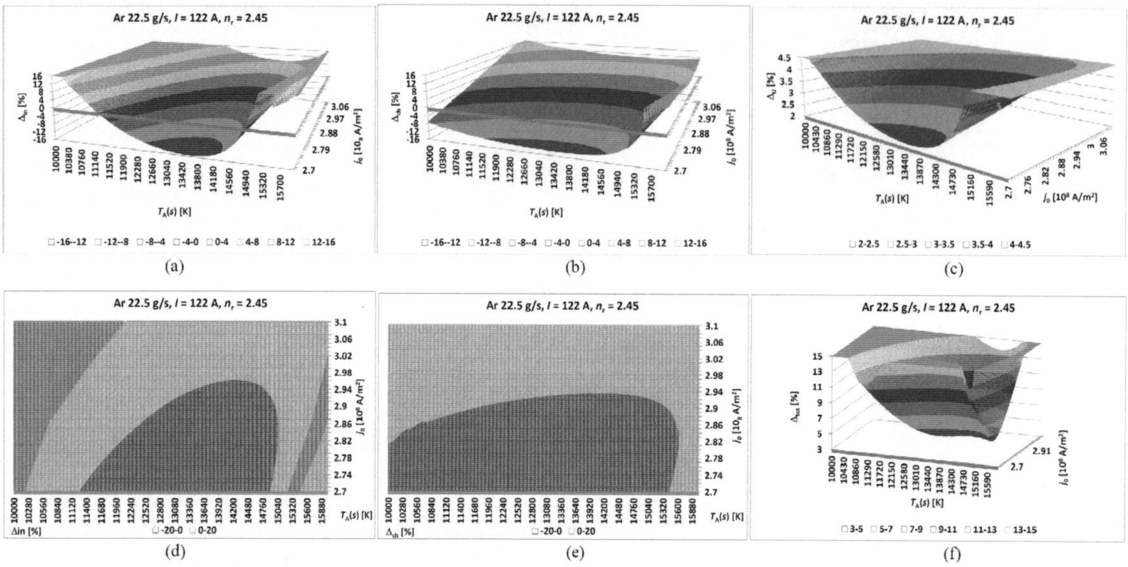


*Figure 3. The computed values for the investigated experiment and exponent $n_r = 2.45$ corresponding to the obtained global minimum of the objective function: individual components of the objective function $\Delta_{in}$ (a, d), $\Delta_{ch}$ (b, e) and $\Delta_U$ (c), the objective function $\Delta_{tot}$ as a whole (f), vs. the current density $j_0$ (in the range of $(2.7\text{–}3.1)\times10^8$ $Am^{-2}$) and the arc temperature $T_A(s)$ (in the range (10–16) kK).*

13400 K. The optimum solution for which the smooth increase of the arc temperature near the beginning and a reasonable value of $E(s)$ are ensured while only negligible deterioration of $\Delta_{tot\_opt} = 3.765\%$ is obtained is the following: $n_{r\_opt} = 2.49, j_{0\_opt} = 2.68 \times 10^8\,\mathrm{Am^{-2}}, T_A(s)_{opt} = 11810\,\mathrm{K}$.

## 4. Conclusions and Further Work

The paper deals with the sensitivity of the model of intensively blasted electric arc burning in argon inside the arc heater's anode channel to exponent $n_r$ and the current density $j_0$ at the cathode tip, which define the arc column development, and the arc temperature $T_A(s)$ at the end of the near-cathode boundary layer. Such a vector of these three quantities is sought when the objective function $\Delta_{tot}$ reaches its minimum. Fine tuning is then needed to ensure physically reasonable dependences of the arc temperature and electric field intensity near the arc beginning. Based on typical experimental results, the paper tries to demonstrate quantitatively how significantly changes of $n_r$, $j_0$ and $T_A(s)$ affect the objective function and its components $\Delta_{in}$, $\Delta_{ch}$, and $\Delta_U$. Just the analysis of individual components $\Delta_{in}$, $\Delta_{ch}$, and $\Delta_U$ of the objective function elucidates the behavior of the model as a whole and confirms the correctness and effectiveness of the computation procedure. Utilization of the model with numerous sets of different experimental results is intended to confirm or enhance its applicability.

### Acknowledgements

This research work has been carried out in the Centre for Research and Utilization of Renewable Energy (CVVOZE). Authors gratefully acknowledge financial support from the Ministry of Education, Youth and Sports of the Czech Republic under OP VVV Programme (project No. CZ.02.1.01/0.0/0.0/16013/0001638 CVVOZE Power Laboratories-Modernization of Research Infrastructure).

### References

[1] J. Senk, I. Laznickova, and I. Jakubova. Estimation of electric arc power loss using net emission coefficient of argon. In *Proceedings of EPE 2018*. Brno Univ. of Technology, 2018. doi:10.1109/EPE.2018.8396015.

[2] J. Senk, I. Jakubova, and I. Laznickova. Analysis of intensively blasted electric arc burning in the arc heater's anode channel. *Acta Polytechnica*, 56(5):395–401, 2016. doi:10.14311/AP.2016.56.0395.

[3] J. Senk, I. Jakubova, and I. Laznickova. Updated version of the simplified model of intensively blasted electric arc. *Acta Polytechnica*, 58(4):264–270, 2018. doi:10.14311/AP.2018.58.0264.

[4] A. Farmer and G. Haddad. *Material Functions and Equilibrium Composition of argon and Nitrogen Plasma*. Csiro Division of Applied Physics, 1989.

[5] O. Coufal. Composition and thermodynamic properties of thermal plasma up to 50 kk. *J Phys D Appl. Phys*, 40(11):3371–3385, 2007. doi:10.1088/0022-3727/40/11/018.

[6] P. Kloc. Computed values of net emission coefficient of argon. Private communication, 2018.

[7] H. Schlichting. *Grenzschicht Theorie*. Verlag G. Braun, Karlsruhe, Germany, 1965.

[8] J. Lowke. Predictions of arc temperature profiles using approximate emission coefficients for radiation losses. *J Quant Spectrosc Radiat Transfer*, 14:111–122, 1974. doi:10.1016/0022-4073(74)90004-1.

doi:10.14311/ppt.2019.1.51
*Plasma Physics and Technology* **6**(1):51–55, 2019 

# MODELLING AND EXPERIMENTAL INVESTIGATIONS OF DC ELECTRIC ARCS IN ARGON AND CARBON DIOXIDE

C. Mohsni[a,b,*], M. Baeva[a], St. Franke[a], S. Gortschakow[a], D. Gonzalez[a], Z. Araoud[b], K. Charrada[b]

[a] *Leibniz Institute for Plasma Science and Technology, Felix-Hausdorff-Str. 2, 17489 Greifswald, Germany*
[b] *National School of Engineering of Monastir, 5000 Rue Ibn Jazzar, Monastir 5035, Tunisia*
[*] mohsni_chayma@hotmail.fr

**Abstract.** In this work an arc model is employed along with electric and spectroscopic measurements to study DC electric arcs in Ar and $CO_2$. The model is aimed at describing the arc and the electrodes. Simulation and experimental results are shown for currents between 150 A and 210 A.

**Keywords:** thermal plasma, electric arcs, plasma-electrode interaction, plasma diagnostics.

## 1. Introduction

Experimental and simulation studies of electric arcs, which are used in industrial applications, are important for a deeper understanding of the arc process. Some gases such as $SF_6$, used in circuit breakers, have a negative impact on the environment. Therefore, alternatives are considered and one potential candidate is carbon dioxide ($CO_2$) due to its electrical and chemical properties.

The modelling of electric arcs, which combines the equilibrium description of the bulk plasma and a non-equilibrium treatment of the near-electrode regions [1, 2], has been considered as an alternative to the fully equilibrium approach applied over the course of many years [3, 4] and the fully non-equilibrium approaches suggested more recently in [5, 6]. The combined modelling of the plasma bulk and the cathode boundary layer in DC arcs with a refractory cathode allows accounting for the significant amount of electric power deposited in the thin boundary layer adjacent to the cathode, in particular for currents below 200 A, and to reduce the discrepancy to the measured arc voltage, i.e. to improve the predictive capability of the model. This modelling approach can be, in general, more easily applied to molecular gases, which are used in industrial devices, e.g. circuit breakers.

In this work, DC arcs in argon and $CO_2$ were studied. A Local Thermodynamic Equilibrium (LTE) arc model in $CO_2$ was established while the model for argon is aimed at coupling the thermionic cathode, the LTE bulk plasma and the anode. To assess the model accuracy, electrical and spectroscopic measurements in various gases were carried out.

## 2. Model description

The development of the combined arc model includes two steps: the establishment of the LTE model of the arc column and the non-equilibrium model of the near-electrode regions. The LTE model is based on a magneto-hydrodynamic approach which involves the Navier-Stokes equations for conservation of mass, momentum and energy. The electrodes are included in the model and coupled to the plasma through a non-equilibrium boundary layer. It is assumed that the gas flow is laminar and the plasma behaves like a fluid which state is characterized by LTE. A stationary solution is sought for the equations

$$\nabla \cdot (\rho \mathbf{u}) = 0 \tag{1}$$

$$\rho(\mathbf{u} \cdot \nabla)\mathbf{u} = \nabla \cdot (-p\hat{I} + \hat{\tau}) + \mathbf{F_L} \tag{2}$$

$$\rho C_p(\mathbf{u} \cdot \nabla T) + \nabla \cdot \mathbf{q} = Q_\mathrm{J} - Q_\mathrm{rad} \tag{3}$$

where $\rho$ is the mass density, $\mathbf{u}$ is the flow velocity, $p$ is the pressure, $\hat{\tau}$ is the stress tensor, $\hat{I}$ is the identity matrix, and $\mathbf{F_L} = \mathbf{j} \times \mathbf{B}$ is the Lorentz force per unit volume. In Eq. (3), $C_p$ is the specific heat at constant pressure, $T$ is the gas temperature, $\mathbf{q} = -\kappa\nabla T$ is the heat flux due to thermal conduction with $\kappa$ being the thermal conductivity. The gas is heated by Joule heating $Q_\mathrm{J} = \mathbf{j} \cdot \mathbf{E}$, where $j$ is the electric current density, and cooled by radiation $Q_\mathrm{rad}$. The Maxwell equations and the Ohm's law are used to obtain the self-induced magnetic field $\mathbf{B}$ and the electric potential $\varphi$

$$\nabla \times \mathbf{B} = \mu_0 \mathbf{j}; \quad \mathbf{B} = \nabla \times \mathbf{A} \tag{4}$$

$$\nabla \cdot \mathbf{j} = 0; \quad \mathbf{j} = \sigma \mathbf{E} = -\sigma \nabla \varphi \tag{5}$$

The electric field $\mathbf{E}$ and the magnetic field $\mathbf{B}$ couple the electromagnetic part of the model to the fluid through the Lorentz force per unit volume and the Joule heating. In Eq. (4) and (5), $\mathbf{A}$ is the vector potential and $\sigma$ is the electrical conductivity. The transport properties were provided by A. B. Murphy as calculated according to [7, 8] and applying a C-C collision integral from [9].

The heat conduction and current continuity equations are solved in the electrodes:

$$\nabla \cdot (k_\mathrm{s} \nabla T) + \mathbf{j} \cdot \mathbf{E} = 0 \tag{6}$$

$$\nabla \cdot \mathbf{j} = 0\,; \quad \mathbf{j} = \sigma_s \mathbf{E} = -\sigma_s \nabla\varphi \tag{7}$$

The electrical conductivity $\sigma_s$ and the thermal conductivity $k_s$ of the solid materials are taken as a function of the temperature $T$. The cathode boundary layer includes the region of space-charge adjacent to the cathode surface (the sheath) and the ionization layer.

The energy balance in the boundary layer [10, 11] accounts for 1) the heat flux leaving the boundary layer and going to the cathode, 2) the electric power density deposited in the boundary layer, 3) the power brought by the electrons emitted from the cathode, and 4) the heat flux entering the boundary layer from the LTE plasma. The energy balance and the total current density, which is the sum of the density of the thermionic emission current ($j_{em}$), the ion current density ($j_{ion}$), and the current density transported by counter-diffusing electrons ($j_{be}$), determine the heat flux $q_w$ and the current density $j_w$ on the cathode surface as functions of the voltage drop $U_{tot}$ in the boundary layer and the surface temperature $T_w$:

$$q_w(T_w, U_{tot}) = j_w U_{tot} - \frac{j_w}{e}(A_w - \Delta A_w + 3.2kT_e) \tag{8}$$

$$j_w\,(T_w, U_{tot}) = j_{em} + j_{ion} - j_{be} \tag{9}$$

Here, $A_w$ is the work function of the electrode material, $\Delta A_w$ is the Shottky correction, $T_e$ is the electron temperature obtained from the boundary layer model. The voltage drop in the near-cathode layer includes the voltage drop in the space-charge sheath ($U_d$) and that in the ionization layer ($U_i$), i.e. $U_{tot} = U_d + U_i$. A similar approach has been applied in [12].

Losses due to black-body radiation of the hot body are considered as $q_{rad} = \epsilon\sigma_{SB}T_w^4$. The resultant heat flux $q_c = q_w(T_w, U_{tot}) - q_{rad}$, and the current density $j_w(T_w, U_{tot})$ serve as boundary conditions of equations (6) and (7). The cathode boundary layer is solved prior to the arc simulation for a series of values of the temperature on the cathode boundary $T_w$ and the voltage drop $U_{tot}$. The results are provided as lookup tables and are used during the iterative solution of the combined model.

The consideration of the plasma-anode boundary layer is simpler [13]. The expression for the heat flux as in many LTE models [3, 4] is used. The normal component of the energy flux to the surface $q_a$, which is added to the conductive one, is given by

$$q_a = -\epsilon\sigma_{SB}T_w^4 + |j_e|\,A_{an} \tag{10}$$

Here, $\epsilon$ is the emissivity of the surface, $\sigma_{SB}$ is the Stefan-Boltzmann constant, $j_e$ is the electron component of the current density and $A_{an}$ is the work function of the anode material. The term $|j_e|A_{an}$ accounts for heating of the anode due to condensation of electrons. For pure tungsten, the work function and the Richardson constant are 4.55 eV and $0.602 \cdot 10^6$ Am$^{-2}$K$^{-2}$, respectively. The corresponding values for graphite are 4.6 eV and $0.600 \cdot 10^6$ Am$^{-2}$K$^{-2}$. The model is set up on the commercial computational

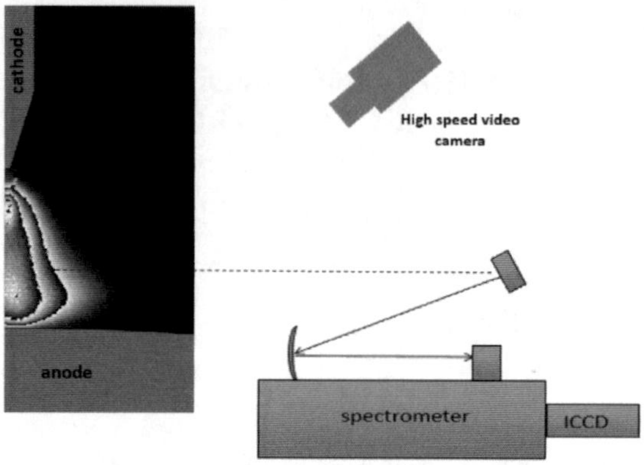


*Figure 1. A view of the computational domain and the OES arrangement used for investigation of the arc.*

platform COMSOL Multiphysics® using the interfaces Laminar flow, Electric current, Magnetic field and Heat transfer. The computational domain includes both electrodes and the inter-electrode space in a radial extend of 30 mm. A part of this domain is shown in Figure 1. Temperature of 300 K, electric and magnetic insulation are set on the external boundaries. The anode is grounded. The total arc voltage drop in the boundary layer is adjusted until the integral of the current density along the cathode surface equals the desired arc current.

## 3. Experimental setup

A view of the spectroscopic equipment and the computational domain is shown in Figure 1. The setup is tested for arcs in argon and $CO_2$. It comprises a spherically shaped anode, made of graphite and a cylindrical rod with a conical shape made of pure tungsten as cathode. The distance between the cathode tip and the anode is varied between 2 and 10 mm.

The image of the arc is taken by a high-speed video camera of the type IDT Motion Pro Y6. The arrangement is placed inside a chamber which is filled with gas up to the desired pressure. The latter is controlled by a manometer. The arc is ignited making use of a spark coil arrangement. The equipment for optical emission spectroscopy (OES) consists of a spectrograph Acton SP2500 with a focal length of 0.5 m and an intensified CCD camera Princeton Instruments PI-MAX4. Obtained images contain spectral and spatial side-on information about the arc. The arc is imaged by a spherical mirror on the entrance slit of the spectrometer with a width of 30 µm. Absolute calibration is performed by a tungsten strip lamp placed at the position of the arc.

The radiation of the Ar II line at 480 nm is spectrally resolved using a grating with 1800 lines/mm. The radiation of the O I line at 777 nm is resolved using a grating with 150 lines/mm. An integration over the wavelength range is performed and Abel inversion is done in order to get the radial dependence

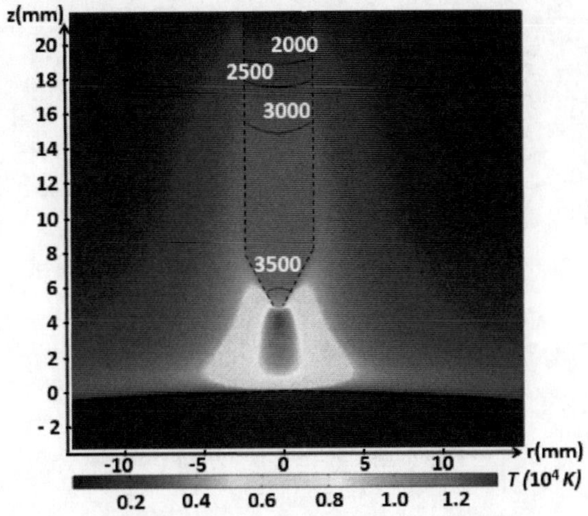


*Figure 2. Two-dimensional distribution of the plasma temperature and the temperature in the electrodes obtained with the LTE model for an arc in $CO_2$ at current of 200 A.*

of the emission coefficient. The number densities of the upper level corresponding to the transition of the Ar II 480 nm, O I 777 lines are derived assuming LTE. The population density is related to the plasma temperature. The temperature is obtained from a pre-calculated equilibrium composition of argon and carbon dioxide.

## 4. Results and discussion

Calculations with the LTE arc model were performed for the arrangement shown in Figure 1 with $CO_2$ at atmospheric pressure as working gas. Figure 2 shows the distribution of the temperature in the plasma and the electrodes for an arc current of 200 A. Since a matching of the boundary layer with the LTE plasma is not considered, the temperature in the arc seems to

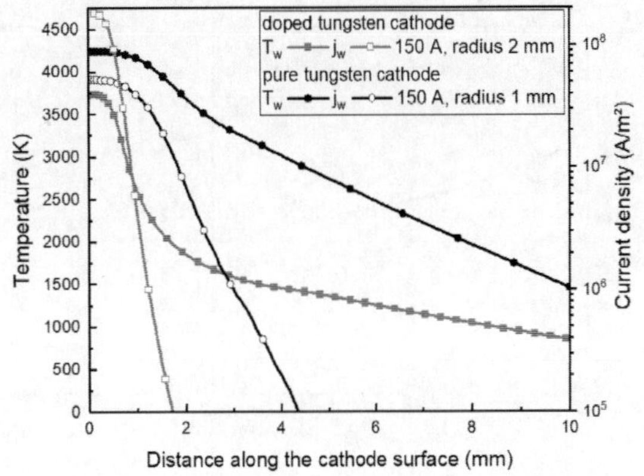


*Figure 4. Temperature and normal current density along cathode surface for doped and pure tungsten cathodes for an arc current of 150 A.*

be underestimated as the experimental results show in Figure 3.

The combined model has been applied first for argon as working gas to obtain the plasma parameters for arc currents of 150 A and 210 A. Figure 4 shows the temperature on the cathode surface and the current density for an arc current of 150 A for two cathode configurations: a) doped tungsten cathode (Ce-W) with a radius of 2 mm, and b) pure tungsten cathode with a radius of 1 mm. The profile of the temperature is more broadened and the arc attachment area is larger for the thinner electrode.

Melting of the electrodes and metal vapour are not considered in the model, which leads to temperatures higher than the melting point of tungsten. The normal current density at the cathode tip is 10% higher in the case of doped tungsten instead of pure tungsten. Figure 5 presents the components of the heat fluxes from the boundary layer to the cathode along the cathode surface for an arc current of 150 A. The

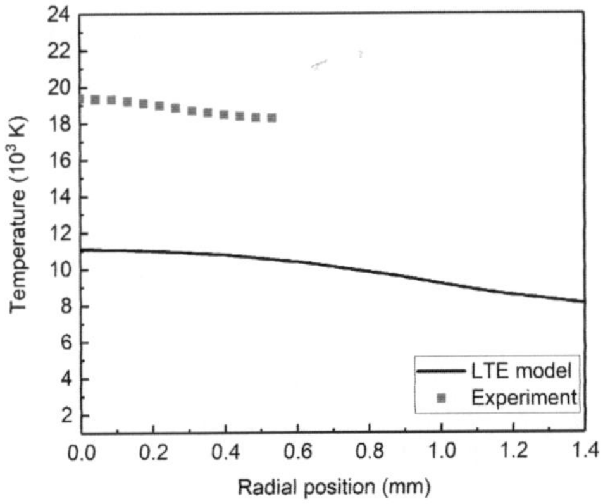


*Figure 3. Radial plasma temperature obtained in the LTE model for an arc in $CO_2$ and the plasma temperature from OES for an arc current of 200 A at axial position 1 mm away from the cathode tip.*

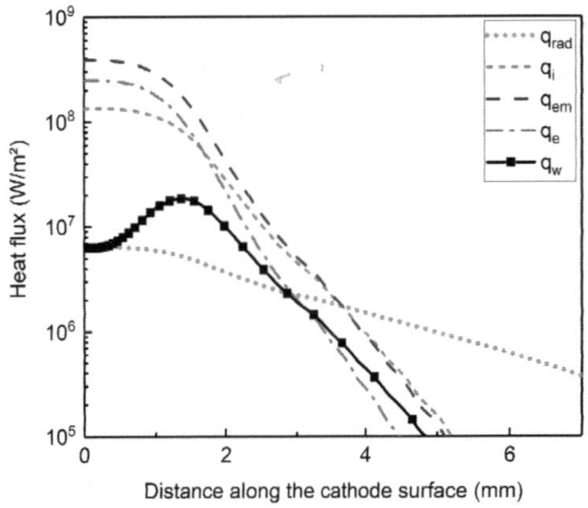


*Figure 5. Total heat flux from the plasma to the cathode surface, its components and the contribution of the black body radiation along the cathode for an arc current of 150 A.*

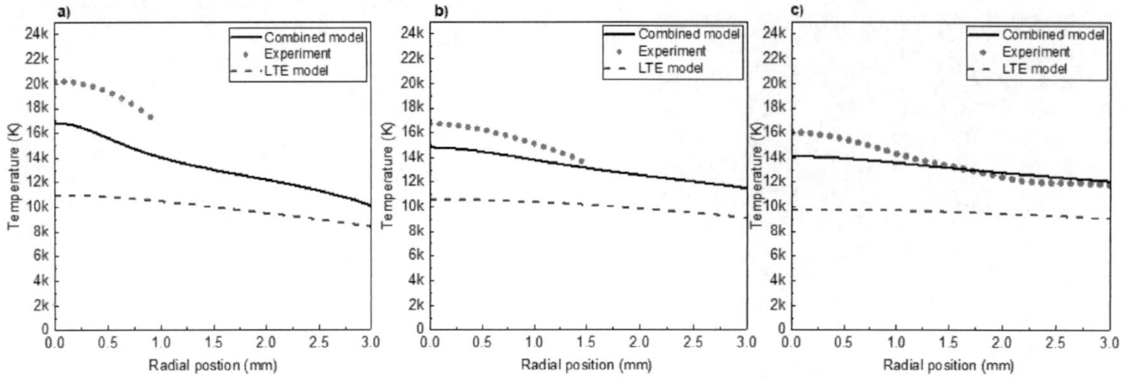


*Figure 6. Temperature predicted by the combined model, LTE model and plasma temperature from optical emission spectroscopy at axial position 1mm a), 2.5 mm b) and 4 mm c) below the cathode. Arc current is 210 A.*

solution shows that the dominant contribution at the cathode tip is that of the emitted electrons $q_{\mathrm{em}}$, which exceeds those of the back diffusion electrons $q_{\mathrm{e}}$ and ions $q_{\mathrm{i}}$ contributions. From the position of 2 mm on, the main contributions are those of ions and emitted electrons. The resulting flux $q_{\mathrm{w}}$ around the cathode tip is lower than the other components, but still exceeds the contribution of the black body radiation $q_{\mathrm{rad}}$.

The temperature predicted by the combined model in comparison with the LTE model and the values obtained from OES are shown in Figure 6. The results are presented for an arc current of 210 A and an arc length of 5 mm at three axial positions: 1 mm, 2.5 mm, and 4 mm away from the cathode tip. One can see that the radial distribution obtained by means of the combined model results in a more focused arc and in higher plasma temperatures near the cathode tip. This modelling approach provides a better agreement with the data obtained by emission spectroscopy.

## 5. Conclusions

The LTE model of the arc plasma is coupled with a model of the cathode boundary layer and is applied to an atmospheric pressure arc in argon. The consideration of the cathode boundary layer allows a more realistic description of the heat and current transfer in the electrodes and the plasma. Therefore, a similar approach has to be applied for $CO_2$. The model results are compared with experimental values and reasonable agreement is obtained.

### Acknowledgements

Financial support of the German Academic Exchange Service (DAAD) is highly acknowledged (Grant No. 91644228).

### References

[1] A. Shirvan et al. Effect of cathode model on arc attachment for short high-intensity arc on a refractory cathode. *J. Phys. D: Appl. Phys.*, 49(48):5201, 2016. doi:10.1088/0022-3727/49/48/485201/meta.

[2] M. Lisnyak et al. Numerical modelling of high-pressure arc discharges: matching the LTE arc core with the electrodes. *J. Phys. D: Appl. Phys.*, 50(31):5203, 2017. doi:10.1088/1361-6463/aa76d3/meta.

[3] J. J. Lowke et al. A simplified unified theory of arcs and their electrodes. *J. Phys. D: Appl. Phys.*, 30(14):2033–2042, 1997. doi:10.1088/0022-3727/30/14/011.

[4] L. Sansonnens et al. Prediction of properties of free burning arcs including effects of ambipolar diffusion. *J. Phys. D: Appl. Phys.*, 33(2):148–157, 2000. doi:10.1088/0022-3727/33/2/309.

[5] M. Baeva et al. Two-temperature chemically non-equilibrium modelling of transferred arcs. *Plasma Sources Science and Technology*, 21(5):5027, 2012. doi:10.1088/0963-0252/21/5/055027.

[6] M. Baeva et al. Novel non-equilibrium modelling of a DC electric arc in argon. *J. Phys. D: Appl. Phys.*, 49(24):5205, 2016. doi:10.1088/0022-3727/49/24/245205/meta.

[7] A. B. Murphy. Transport coefficients of air, argon-air, nitrogen-air, and oxygen-air plasmas. *Plasma Chem. Plasma Proc.*, 15(2):279, 1995. doi:10.1007/BF01459700.

[8] A. B. Murphy and C. J. Arundell. Transport-Coefficients of Argon, Nitrogen, Oxygen, Argon-Nitrogen, and Argon-Oxygen Plasmas. *Plasma Chem. Plasma Proc.*, 14(4):451–490, 1994. doi:10.1007/BF01570207.

[9] J. R. Stallcop at al. Potential energies and collision integrals for interactions of carbon and nitrogen atom. *J. Thermophys. Heat Transf.*, 14:480–488, 2000. doi:10.2514/2.6570.

[10] M. S. Benilov and A. Marotta. A Model of the Cathode Region of Atmospheric-Pressure Arcs. *J. Phys. D: Appl. Phys.*, 28(9):1869–1882, 1995. doi:10.1088/0022-3727/28/9/015/meta.

[11] M. S. Benilov and M. D. Cunha. Heating of refractory cathodes by high-pressure arc plasmas: I. *J. Phys. D: Appl. Phys.*, 35(14):1736–1750, 2002. doi:10.1088/0022-3727/35/14/314.

[12] S. Lichtenberg et al. The plasma boundary layer of HID-cathodes: modelling and numerical results. *J. Phys. D: Appl. Phys.*, 38(17):3112–3127, 2005. doi:10.1088/0022-3727/38/17/S13.

[13] J. Heberlein et al. The anode region of electric arcs: a survey. *J. Phys. D: Appl. Phys.*, 43(2):3001, 2010. doi:10.1088/0022-3727/43/2/023001/meta.

doi:10.14311/ppt.2019.1.56
*Plasma Physics and Technology* **6**(1):56–59, 2019


# COMPARATIVE STUDIES OF LOW-INTENSITY SHORT-LENGTH ARCS

M. Baeva[a,*], E. Siewert[b], D. Uhrlandt[a]

[a] *Leibniz Institute for Plasma Science and Technology, Felix-Hausdorff-Strasse 2, 17489 Greifswald, Germany*
[b] *Linde AG, Geschäftsbereich Linde Gas, Carl-von-Linde-Strasse 25, 85716 Unterschleissheim, Germany*
[*] baeva@inp-greifswald.de

**Abstract.** We present results obtained by two non-equilibrium modelling approaches and experiments on low-intensity short-length arcs in argon at atmospheric pressure. The first one considers a quasi-neutral arc column combined with boundary conditions on the electrodes based on the energy balance in the space-charge sheaths. The second approach applies a unified description over the entire gap and solves the Poisson equation for the self-consistent electric field. The experiments provide the arc voltage.

**Keywords:** non-equilibrium, unified, arc plasma, modelling, experiment.

## 1. Introduction

Electric arcs of short length operated at a current level of several Amperes are not well understood, neither from an experimental nor from a modelling point of view. Most of the reported measurements on electric arcs were made for arc lengths of 3 mm and more. Low-current electric arcs depart presumably from Local Thermodynamic Equilibrium (LTE) across the entire inter-electrode-region so that a non-equilibrium description is required. A self-consistent non-equilibrium modelling of tungsten-inert gas arcs of lengths down to 2 mm accomplished by experimental work have been reported in a recent authors' publication [1]. Some modelling studies on electric arcs of lengths 1 mm and below have been published in [2, 3]. These studies have shown the transition from glow to arc in argon at atmospheric pressure. However, the heating of the tungsten cathode resulting from the arc-cathode interaction has not been done in a self-consistent manner, since a prescribed value of the cathode fall equal to the energy of ionization has been used. In recent publications [4, 5] by two of the authors of the present work, a unified one-dimensional and a self-consistent modelling of microdischarges and microarcs (inter-electrode distance less than 1 mm) has been presented and the properties of the microarc plasma have been analyzed. The results are found to be in a fairy qualitative agreement with experimental studies reported by Park et al [6]. Detailed experimental studies on microarcs, which gain importance in micro-welding and manufacturing, have not been reported yet.

In this work, we present the first results of comparative studies on short and low current arcs as shown by the organisation chart in Fig. 1. Two non-equilibrium modelling approaches and electrical measurements are employed. The paper is organized as follows. A brief description of the modelling approaches and the experiments is given in Section 2. Results and discussion are given in Section 3. Concluding remarks are summarized in Section 4.

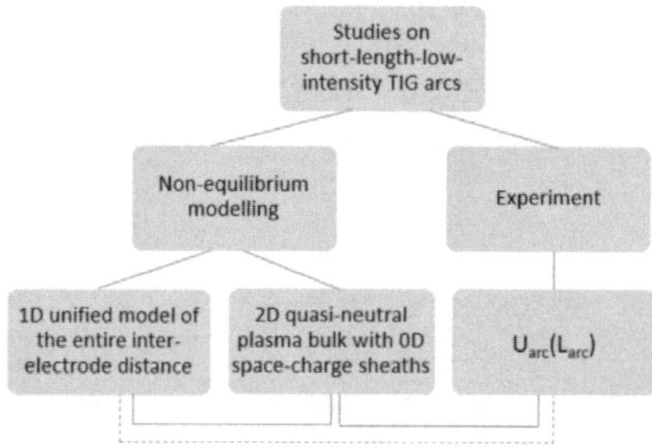


*Figure 1. Organisation chart of the study.*

## 2. Methodology

The studies comprise a non-equilibrium and self-consistent modelling of the arc plasma and the electrodes, and electrical measurements in a tungsten-inert gas configuration. The cathode is made of ceriated tungsten (WC20) and the anode is a water-cooled plate made of copper. The arc is burning in argon at atmospheric pressure. The inter-electrode distance, $L_{\mathrm{arc}}$, is varied both in the experiment and in the models. The cathode is considered in the models with a length of 20 mm.

### 2.1. Modelling approaches

Two modelling approaches are employed in the present study (see Fig. 1). They are referred to in the following as the 2D-fully-non-equlibrium model (2D-FNEM) and the 1D-unified-non-equilibrium-model (1D-UNEM). The detailed descriptions of the models are provided in [4, 5, 7], respectively. Here, we outline briefly their common and specific features. Both models describe the plasma as a fluid that contains electrons and neutral and charged species of argon. The electrons are assumed to obey a Maxwellian velocity distribution and are characterized by the electron temperature $T_e$, while the heavy particles are characterized by the common temperature $T$. Both models avoid assumptions of thermal and chemical equilibrium. Therefore, individual temperatures ($T_e \neq T_h$) are related to the electrons and the heavy particles, and the plasma composition results from plasma-chemical reactions. Both models imply the bidirectional interaction between the arc plasma and the electrodes, accounting for fluxes between the plasma and the electrodes in both directions. In particular, the cathode is of central importance in tungsten-inert gas arcs, since the arc plasma is sustained by the thermionic emission of electrons from the cathode.

The 2D-FNEM is based on the magneto-hydrodynamic description of the axially symmetric quasi-neutral plasma bulk. Hence, the Navier-Stokes equations of conservation of mass, momentum, energy of electrons and heavy species, and the conservation of ion species are solved along with the equation of current continuity, the Maxwell equations for the magnetic vector potential, the generalized Ohm's law, and the Dalton's law. The model further solves the heat and current transfer in the solid bodies of the electrodes. The 2D-FNEM employs the Stefan-Maxwell equations and the effective binary diffusion approach in a multi-component plasma, which is composed of electrons, singly charged argon ions and ground state atoms, to build a general diffusion formulation. This formulation gives rise to the generalized Ohm's law, in which the electric current density contains a drift component (proportional to the electric field) and a diffusion component (accounting for gradients of the mass fractions of species, the pressure, the temperatures). The space-charge sheath with a spatial extent of the order of the Debye length adjacent to the electrodes are not spatially resolved. Instead, their properties are accounted for in the boundary conditions on the boundary between the plasma and the electrodes. That is, the integral energy balance of electron energy and the integral balance of total energy in the space-charge sheath are important since they provide comprehensive boundary conditions.

The 1D-UNEM is aimed at solving the equations of particle and energy conservation for electrons and heavy species, the Poisson equation for the electric potential, the equation of heat-transfer in the cathode, and an electric circuit equation. The entire inter-electrode distance is spatially resolved, i.e. the quasi-neutral plasma bulk and the regions of space-charge are considered together. The 1D-UNEM employs the drift-diffusion approximation for the electron particle and energy flux and allows us therefore to account for the drift and diffusion components of the electric current density. Since low current arcs are of interest, the ionization degree of the generated plasma is assumed to be low enough so that the magnetic field and magnetohydrodynamic effects are ignored. A convective motion of the plasma and a self-induced magnetic field are not included in the model. The plasma bulk and the near-electrode regions are characterized by very different scales. Therefore, the spatial resolution in the unified model sets an important limit concerning the computational overhead even in 2D treatment. The unified model is so far realized in 1D and its applicability is therefore restricted to plane-parallel configurations which length is much shorter than their radial extent.

### 2.2. Electrical measurements

The detailed description of the experimental setup and the evaluation of the results covering the electrical measurements can be found in [1]. The measurements are aimed at providing the arc voltage $U_{arc}$ for various arc lengths $L_{arc}$ in a tungsten-inert gas configuration. The cathode is a cylindrical rod with a radius of 2 mm, and a conical tip with an angle of 30° and a plateau radius of 0.15 mm. An arc current of 5.3 A is considered in the present study. The arc voltage is measured for arc lengths starting from 2 mm in 0.025 mm steps until a short circuit occurs. The zero-value of $L_{arc}$ is determined by the position of the short circuit.

## 3. Results and discussion

The 2D-FNEM is applicable to an arbitrary axially symmetric tungsten-inert gas configuration. In order to enable a direct comparison with the experiment, the experimental setup with a conical tip of the cathode is reproduced in the model. The 1D-UNEM is inapplicable to this configuration due to the small radius of the cathode tip. Nevertheless, a comparison of the results between the 1D-UNEM and the 2D-FNEM can be done for a configuration with a cylindrical cathode rod with a radius $R$=2 mm and $L_{arc} \leq 1$ mm, for which the one-dimensional treatment is still justified. Experimental data for a configuration with a flat cathode is still not available. Therefore, the comparisons model-to-experiment and model-to-model as indicated in Fig. 1 can be done. The objective includes two important tasks that are new. First, the comparison 2D-FNEM to experiment will prove the predictive capability of the 2D-FNEM at very short arc lengths and low currents. Second, the comparison 1D-UNEM to 2D-FNEM will prove the accuracy level of the modelling approach based on the unresolved space-charge sheath and its average properties at the conditions under consideration.

### 3.1. 2D-FNEM and experiment

Figure 2 shows an image of the arc plasma and the calculated two-dimensional distribution of the temperature for an inter-electrode distance of 0.4 mm. The CCD image illustrates the plasma cloud around the cathode, which tip end is hidden by the plasma but still visible. The calculated distribution is taken in

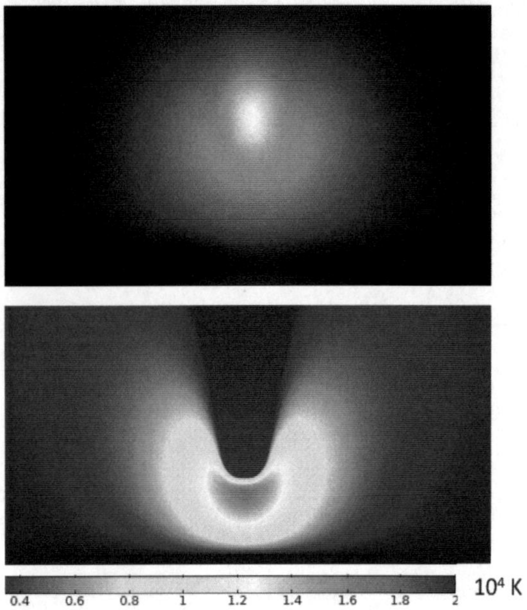


*Figure 2. CCD-image of the arc and the distribution of the gas temperature of the arc at $L_{arc}$=0.4 mm from the 2D-FNEM.*

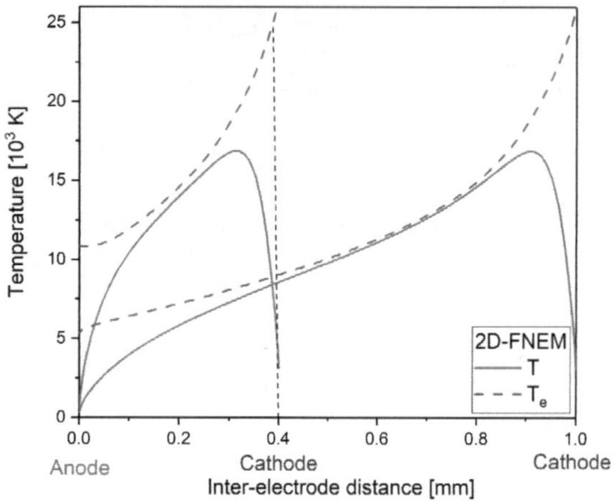


*Figure 3. Temperatures of electrons and heavy particles at an electric current of 5.3 A from the 2D-FNEM for arc lengths of 0.4 and 1 mm.*

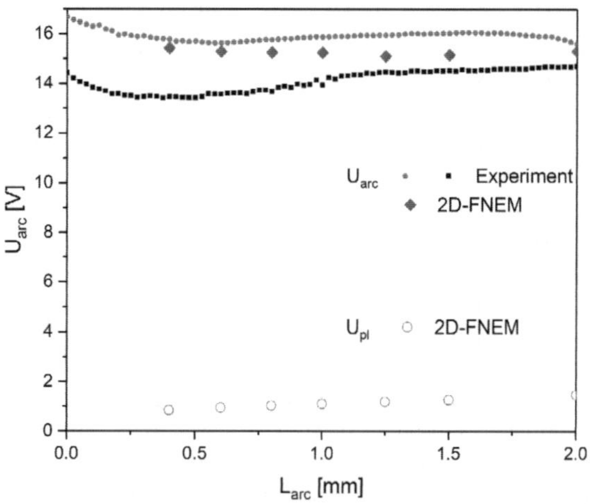


*Figure 4. The arc voltage as a function of the arc length at an electric current of 5.3 A from the experiment and the 2D-FNEM.*

the plane through the axis of symmetry and shows the end of the cathode tip. At the conditions under consideration ($L_{arc}$=0.4 mm, $I_{arc}$=5.3 A), the arc shape differs from the well known bell shape of high-current arcs. The magneto-hydrodynamic simulation shows that flow jet generated by the Lorentz force is of minor importance. The arc plasma spreads therefore in radial direction.

The short-length low-current arcs are in non-equilibrium (Fig. 3). The thermal non-equilibrium is stronger pronounced in the regions in front of the electrodes than in the middle of the plasma column. For shorter arc lengths, $T_e$ remains higher close to the anode, while close to the cathode it approaches almost the same value of about 25000 K. The departure between $T$ and $T_e$ is less pronounced in the middle of the arc for a longer $L_{arc}$.

Figure 4 presents the arc voltage $U_{arc}$ obtained in the experiment and the calculated one for arc lengths below 2 mm. The experimental values are obtained in two series and show a difference between 1 and 2 V due to the less stability of the arc at low current density. The values at $L_{arc} \to 0$ indicate the total voltage drop in the space-charge sheaths. The calculated arc voltage values lie well within the experimental span. The arc voltage in the model is the resultant of the voltage drops in the space-charge sheaths, the arc column ($U_{pl}$) and the solid bodies. Since the voltage drop in solids is in the order of 0.1 V and $U_{pl}$ is below 2 V in the range of arc lengths considered, the arc voltage contains most the voltage drop in the space-charge sheaths, in particular in the cathode sheath of the tungsten-inert gas arc. The good agreement between the experimental and the modelling results is not trivial. We have shown in [1] that the 2D-FNEM is capable of predicting the arc plasma properties and in particular the arc voltage over the range of arc currents (5–200) A and arc lengths (2–20) mm. For these $L_{arc}$ values, the axial extent of the arc is much longer than the spatial extent of the space-charge sheaths, which are unresolved in the model but are considered as zero-dimensional. This inequality becomes less justified for arc lengths in the sub-millimeter range. Below, we consider the spatial structure of short-length arcs.

## 3.2. 1D-UNEM and 2D-FNEM

Figure 5 shows the spatial distribution of the electron and ion density, and the electric potential obtained by the 1D-UNEM for the shortest arc length considered

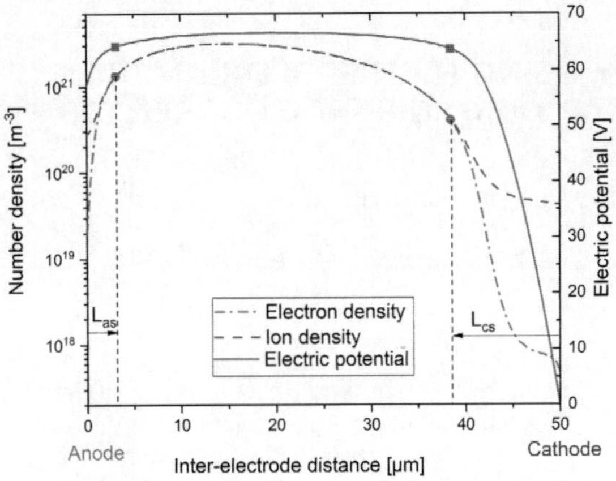


*Figure 5. The spatial distribution of the electron and ion density and the electric potential from the 1D-UNEM at an electric current of 5.3 A.*

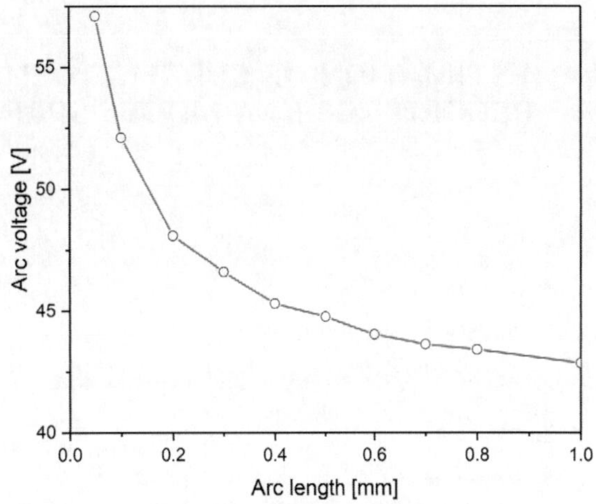


*Figure 6. The arc length-voltage characteristics from the 1D-UNEM at an electric current of 5.3 A.*

in the study ($L_{arc}$=0.05 mm) and an arc current of 5.3 A. The edge of the regions of quasi-neutrality and the space-charge sheaths adjacent to the electrodes are indicated by the symbols, which correspond to the position, where the ion density amounts 1% of the difference between the ion and electron density. The distances $L_{as} \simeq 3$ µm and $L_{cs} \simeq 13$ µm are then the estimates of the lengths of, respectively, the anode and the cathode space-charge sheath. With the decrease of the arc length, the length of the arc column tends to vanish and the anode and the cathode sheaths merge. The 2D-FNEM becomes inapplicable at such conditions. From Figure 5, one can obtain the voltage drops in the anode and the cathode sheaths. The voltage drop in the anode sheath is negative and amounts about -7.1 V while the voltage drop in the cathode sheath is about 64.4 V and exceeds the total arc voltage of about 57 V.

The arc voltage decreases with the increase of the arc length as shown in Fig. 6. The increase of the arc voltage at vanishing arc length is also observed experimentally and predicted by the 2D-FNEM (Fig. 4). Results from the 2D-FNEM for arc lengths below 2 mm and a cylindrically shaped cathode, which can be directly compared with those of the 1D-UNEM, are still forthcoming. For an arc length of 5 mm, the 2D-FNEM predicts an arc voltage of 27.6 V.

## 4. Conclusions

The comparative studies presented here show that the 2D-FNEM is a powerful tool for modelling of short-length low-current tungsten-inert gas arcs. The modelling results agree well with electric measurements. The 1D-UNEM delivers important knowledge about the spatial structure of the arc plasma and the plasma parameters in the space-charge sheaths. It can be applied in cases of very short arc lengths, for which the 2D-FNEM becomes inapplicable. Future works will focus on the analysis of the average plasma properties in the sheaths employed in the 2D-FNEM and spatially resolved ones in the 1D-UNEM.

## Acknowledgements

The work of M. Baeva is funded by the Deutsche Forschungsgemeinschaft (DFG, German Research Foundation) - Project number 390828847.

## References

[1] E. Siewert, M. Baeva, and D. Uhrlandt. The electric field and voltage of dc tungsten-inert gas arcs and their role in the bidirectional plasma-electrode interaction. *J. Phys. D: Appl. Phys. (under review)*, 2019.

[2] S. I. Eliseev, A. A. Kudryavtsev, H. Liu, Z. Ning, and A. S. Chirtsov. Transition from glow microdischarge to arc discharge with thermionic cathode in argon at atmospheric pressure. *IEEE Trans. Plasma Sci.*, 44:2536–2544, 2016. doi:10.1109/TPS.2016.2557587.

[3] A. I. Saifutdinov, I. I. Fairushin, and N. Kashapov. Analysis of various scenarios of the behavior of voltage-current characteristics of direct-current microdischarge at atmospheric pressure. *JETP Lett.*, 104:180–185, 2016. doi:10.1134/S0021364016150145.

[4] M. Baeva, D. Loffhagen, M. M. Becker, and D. Uhrlandt. Fluid modelling of DC argon microplasmas: effects of the electron transport description. *Plasma Chem. Plasma Process.*, 39:949–968, 2019. doi:10.1007/s11090-019-09994-5.

[5] M. Baeva, D. Loffhagen, and D. Uhrlandt. Unified non-equilibrium modelling of tungsten-inert gas microarcs in atmospheric pressure argon. *Plasma Chem. Plasma Process.*, 2019.

[6] M. Park, Y. Hirata, and T. Urabe. Development of controlled micro-discharge at the atmospheric pressure. *Weld. World*, 58:47–54, 2014. doi:10.1007/s40194-013-0092-9.

[7] M. Baeva, M. S. Benilov, N. A. Almeida, and D. Uhrlandt. Novel non-equilibrium modelling of a DC electric arc in argon. *J. Phys. D: Appl. Phys.*, 49:245205, 2016. doi:10.1088/0022-3727/49/24/245205/meta.

doi:10.14311/ppt.2019.1.60
*Plasma Physics and Technology* **6**(1):60–64, 2019 

# ESTIMATION OF CURRENT DENSITY USING HIGH-SPEED-CAMERA RECORDINGS IN A MODEL SPARK GAP DURING SURGE CURRENTS

T. H. Kopp*, E. Peters, M. Kurrat

*Institute for High Voltage Technology and Electrical Power Systems, Technische Universität Braunschweig, Braunschweig, Germany*

* t.kopp@tu-braunschweig.de

**Abstract.** For experimental investigations of short time plasma in spark gaps, as used in surge protective devices, high-speed camera recordings are used frequently. The analysis of these recordings provides further details regarding the plasma state and distribution. These deduced details are used to assist research and development processes.

To increase the benefit of high-speed camera recordings an empirical model is presented to improve the picture analysis. In this model the recorded radiation intensity is empirically related to the current density within a spark gap. Therefore a specially adapted model spark gap was developed and tested. In this model spark gap areas with homogenous current densities occur. These current densities are determined in the experimental setup through current measurements with separated electrodes. Additionally, the relative radiation intensity between the electrodes is identified using high-speed camera recordings. An empirical correlation between these two measurement values was found and is discussed. It confirmed that the determined correlation improve the mostly intuitive interpretation of high speed camera recordings in spark gaps.

**Keywords:** spark gap, plasma, surge protective device, high speed camera, current density, narrow gap.

## 1. Introduction

Surge protective devices, SPD class 1, are required within power grids to ensure the normal operation of the electrical installation [1]. With these devices surges are driven directly to the equipotential bonding bypassing the electrical installation. In order to investigate the conductive behavior of SPD's during operation, they are tested using current generators, which deliver standardized 8/20 µs pulse currents according to international standards [2].

Spark gaps are used as class 1 SPD's. Within spark gaps an electric arc forms the bypassing path during the protecting operation [3, 4]. During this process the main aspects are as follows: Firstly, SPD's have to create a highly conductive path to divert the surge current to the equipotential bonding. Thus, the protection of the connected electric devices is guaranteed. Secondly, the conductivity of the plasma has to be reduced by several orders of magnitude after the surge, in order to prevent the subsequent fault current fed by the power network [4]. These requirements are opposing each other and pose a challenge to undertake.

The behavior of the spark gap depends on the plasma properties and it's propagation within the plasma chamber of the model spark gap. To improve the protective function this behavior has to be known for the development process [2].

Hence information regarding the plasma distribution is necessary to investigate the switching behavior of spark gaps. In present research of SPD's, high speed camera recordings can be used to visualize the distribution and support investigations in modern research and development projects [5–9].

In actual fundamental research projects information regarding radiation and temperature [10], pressure [11], and electrical conductivity [12] are given for the transient time domain in spark gaps. These data are not yet measured simultaneously and therefore cannot verify a complete plasma model. However, during the development processes of spark gaps, empiric models are often sufficient to estimate the plasma state and to identify the changes caused by development steps of the SPD design.

The method presented in this publication shall provide an additional tool for analysis and evaluation of high speed camera recordings in spark gaps. As a result the current density can be deduced quickly from the high-speed camera recording of a spark gap. It is the aim to support the development processes of spark gaps. Therefore, it is sufficient to provide a basic empirical relation between the recordings and the current density. However the presented method is applicable for spark gaps due to their behavior being comparable to a black or grey radiator. For line radiators the application is not recommended.

The method, section 2, is divided into four subsections: description of the measurement setup and of the model spark gap, an introduction into the theoretic approach and finally the data acquisition.

To assess the applicability of the method, measurement results are given in section 3. Here, the possibilities of this method are presented while discussing the change of the current density in the plasma during a surge current.

In section 4 the functionality of this method will be discussed and an outlook for further investigations will be given.

## 2. Method

A measurement setup was built to investigate the current density in relation to high speed camera recordings of a model spark gap as shown in Fig. 1. The current source of this measurement circuit is an impulse generator generating an unipolar surge current up to 20 kA, with a rise time of 8 µs and a time to half-value of 20 µs, according to the international test standard [2].

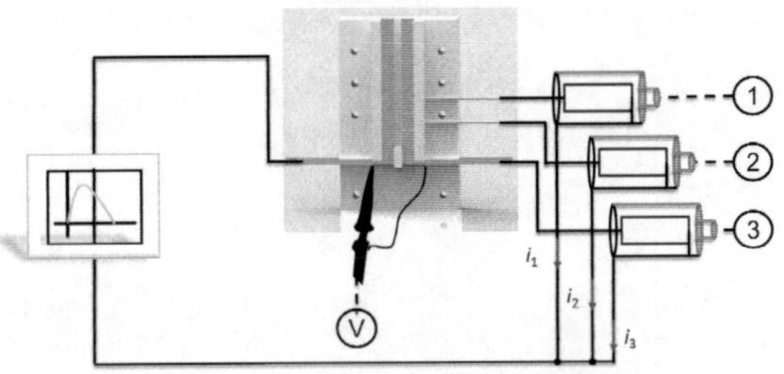


*Figure 1. Measurement setup.*

The measurement system provides a 14-bit amplitude resolution and a sampling rate of 100 MS/s [9, 11]. In this setup the surge current is separated inside the spark gap through different electrodes, therefore the partial currents $i_1$ to $i_3$ are measured separately with three coaxial shunts of the same type. The shunts have a resistance of 1 mΩ and show low inductive behavior. The voltage at the connector of the spark gap $u_{\mathrm{arc}}$ is measured with a high voltage probe (Type PMK 3.6 kV).

A model spark gap is used as device under test in this measurement setup [5], Fig. 1. The spark gap consists of one electrode (anode) and three separated electrodes C (cathode), they are connected with the measurement setup via copper bus bars, Fig. 2. The distance of the anode and cathode is 5 mm and the separated cathodes are isolated with a 1 mm thick polymer layer.

The plasma chamber P is surrounded with evaporating polymer and has a height of 80 mm, a width of 5 mm and a depth of 1 mm. The chosen polymer is poly-oxy-methylene (POM). To obtain optical access to the plasma chamber in the front of the spark gap a cover of poly-methyl-methacrylat (PMMA) is used.

The plasma was ignited using a spark discharge device (SDD) which is connected in series with a varistor to ensure a response with occurring high voltages.

The plasma radiance in the plasma chamber was recorded with a high speed camera of the type Motion Pro X4 from the manufacturer Redlake with a CMOS-sensor. The sensor has a resolution of 8 bit with an pixelarea of 16 µm x 16 µm. During all measurements

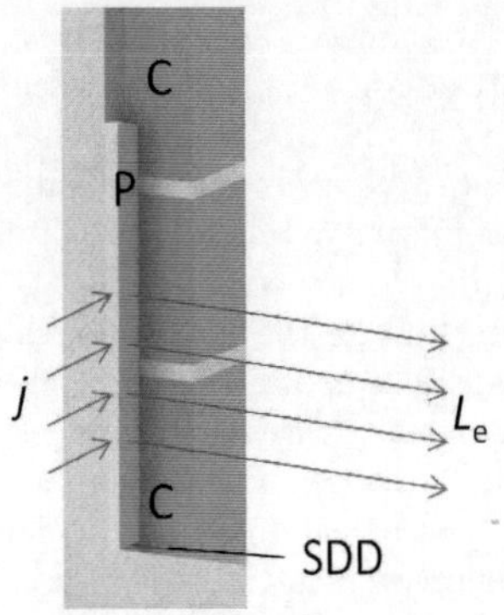


*Figure 2. Model spark gap with separated cathode.*

an integration time of 1 µs was applied. The frame rate for the recording was set to 94.500 fps. Different settings for aperture and neutral density filters were used for different recorded surge currents amplitudes, as considered in the analysis.

The plasma distribution between the electrodes depends on the surge current amplitude and the geometrie model spark gap. Within the following introduction into the theoretical approach firstly the relation between electrical conductivity and radiance has to be determined. This dependency could be deduced applying suitable material properties [13]. In order to do so, the temperature must be derived from its spectral radiance. Recent investigations with spark gaps have shown that, due to the high plasma pressure, the plasma has a low optical depth and therefore the radiance assume to be same as for a grey body radiator [10, 14], Fig. 3. In the respective spectra it should be noted that PMMA has a cut off wavelength by approximately 375 nm.

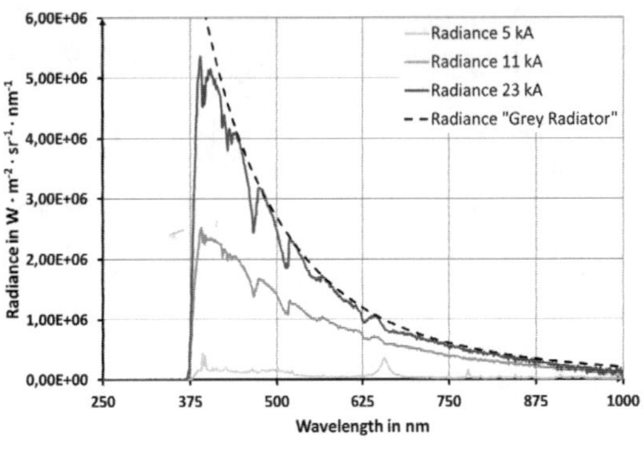


*Figure 3. Spectral Radiance acc. to [14].*

Thus, the temperature can be determined by comparing the measured radiance to the theoretical spectral radiance of a grey radiator with a suitable emission coefficient, which is between 0.8 and 1 [14]. Furthermore, the PMMA cover, the quantum efficiency of the sensor and the adjusted filters have to be considered. While using a high speed camera as radiance sensor, the recieved spectral radiance is converted to the recieved radiance. This is done by integrating the spectral radiance over the wavelength range of the sensitivity of the camera. With this radiance the temperature can be estimated and subsequently the electrical conductivity. However, a necessary requirement to compute the conductivity is the information of plasma pressure for the spark gap, as given in [11]. For spark gaps it is known that the pressure rises linearly with increasing current density [11]. Combining radiance and pressure the relation to the electrical conductivity can be deduced in more detail using the material properties [13], Fig. 4.

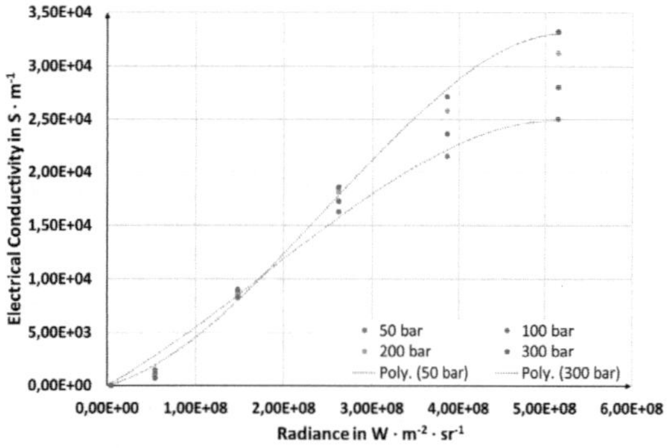


*Figure 4. : Relation between electrical conductivity and radiance.*

With the deduced conductivity and the geometry of the spark gap, the current density can be determined. However a measured arc column voltage $u_{\mathrm{arc}}$ (under consideration of the induced voltage [15]) and the voltage of the electrode sheath $u_{\mathrm{es}}$, is also necessary for the implementation in a simplified approach (1):

$$j = \sigma(p, T) \cdot \frac{u_{\mathrm{arc}} - u_{\mathrm{es}}}{l_{\mathrm{arc}}} \tag{1}$$

In this brief theoretical approach the dependence between current density (1) and radiation Fig. 3 (hence temperature) was shown. Due to the presented complex correlation between this quantities only an empirical approach will be used in the next chapter "Results". There, with the high speed camera recordings the correlation from the radiance to the current density is derived for different surge current amplitudes (5, 10 and 20 kA). For each current, three measurement series are done. Within each series ten surge currents was applied to an unused spark gap.

To develop an empirical model the current in the lower section of the spark gap (shunt 3, Fig. 1) was correlated with the mean value of the radiance of this area recorded by the high speed camera during its integration periode. This area has a width of 5 mm and a height of 10 mm. It is optically recorded from a distance of approx. 1.5 m on 12 × 24 pixel. The lower section has the advantage that information regarding the pressure was recorded in [11]. Furthermore, the current density can be assumed in this area as homogeneously distributed [5].

The mean radiance is considered relative to the maximum recorded brightness (255 bit number) with a transmission factor of $12.5 \cdot 10^{-6}$ % (aperture 32 and neutral density filters, ND 3.0 and ND 0.6).

## 3. Results

With the camera recordings and the data of the measured current with the shunt, the correlation between current density and relative radiance is determined. The relative radiance is the quotient of the mean radiance in the lower section to the maximum recordable brightness. The mean radiation was calculated by averaging the pixel values in the area of 12 × 24 pixel. Regarding the current densities and the radiance of every recording an empirical correlation can be found, Fig. 5.

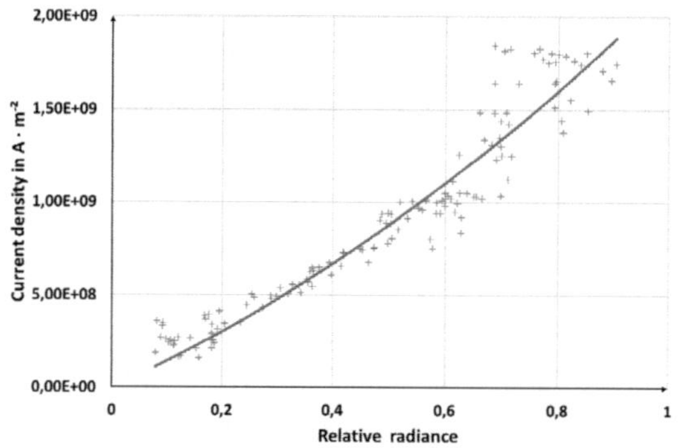


*Figure 5. Correlation Current density and relative radiance.*

Within this data plot 5, 10, and 20 kA measurements are considered. It is shown that with higher radiance the current density increases as well. This correlation can be described approximately using the square formula (2) with the current density $j$ and the relative radiance $L_{\mathrm{r}}$:

$$j = 8 \cdot 10^{8} \cdot L_{\mathrm{r}}^{2} + 1.36 \cdot 10^{9} \cdot L_{\mathrm{r}} \tag{2}$$

Formula (2) has a determination-coefficient of 0.913 to the recorded data. The relation between current density and radiance can be explained according to the shown theoretical approach (chapter 2). Exemplary this relation has been verified with measurement data of a reference spark gap which is discussed in [14]. For this reference geometry the relation of the radiance recorded with optical emission spectroscopy (OES) as published in [10] to the measured current density is evaluated, Fig. 6. To increase the significance of the presented model, the reference spark gap was used. This improving is obtained due to same material (POM) and a slightly different geometry of the reference spark gap.

Within these OES-data the optical transmission of the real optical path is considered. As a result the reference spark gap shows qualitatively similar results

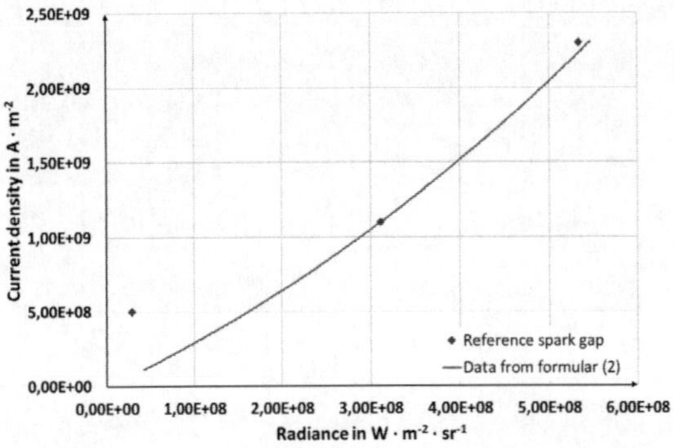


*Figure 6. Correlation Current density and radiance for reference spark gap.*

as the measurements with the high speed camera system. Therefore, the method to link the irradiance on the camera chip with the current density is considered applicable. The deviation of the curve at low current density is due to the spectral behavior observed for the 5 kA measurement (Fig. 3), whose spectrum is more like a line than a Planck radiator. Therefore the application of the method appears to have a bigger uncertainty for surge currents of lower amplitude.

Within the recorded data, at an occurring relative radiance over 0.6, the scattering is increasing (Fig. 5). In theory [14] the temperature determination based on the grey body model also shows differences according to a model based on transport properties. The method could be more accurate if the transport properties are taken into account. However, the aim to provide an easily and quickly applicable empirical model with a comprehensible theoretical behavior is achieved. The grey body radiation is sufficient due to a function uncertainty of 30%. For more accurate investigations the transport properties have to be considered.

## 4. Conclusions and Outlook

In this publication a method for current density estimation in spark gaps is presented. It is shown that the radiance during measurements with different surge currents can be empirically related to the current density. Furthermore, the empirically described relation shows also comparable patterns with the spectral data of the reference spark gap.

The approach was tested in this contribution with two different designs of model spark gaps. Yet, the empirical function still has to be verified with other types of spark gaps. This method can preferably be applied when the radiative characteristics of the plasma are like those of a grey or black body radiator. Due to the behavior of the plasma as a grey body radiator this method should be applicable also for spark gaps with other plasma geometrial arrangements. However, in further contributions the boundaries of the method has to be investigated to give a statement of the applicability under different circumstances.

## Acknowledgements

The authors gratefully thank the company Phoenix Contact GmbH and Co. KG for the financial and technical support. Additionally we thank F. Fitzel M. Sc. for the help during the execution of the experiments and the discussions during the writing of his bachelor thesis. In particular we specially thank Dr.-Ing. T. Runge for his efforts and valuable discussion.

## References

[1] *Electrical installations of buildungs – Part 5-53: Selection and erection of electrical equipment – Isolation, switching and control.* Geneva, Switzerland, 2002. IEC 60364-5-53.

[2] *Surge protective devices connected to low-voltage power systems – Requirements and test methods.* Geneva, Switzerland, 2011. IEC 61643-11.

[3] L. Hüttner, L. Jurcacko, F. Valent, A. Ehrhardt, S. Schreiterand, and M. Rock. Fundamental analysis of encapsulation of a low-voltage spark gap with deion chamber. In $XVIX^{th}$ *Symposium of Physics of Switching Arc*, Nove Mesto na Morave, Czech Republic, 2011.

[4] G. Finis, R. Durth, C. Depping, and M. Wetter. Laboratory for the qualification-testing of spds combining highest power performance parameters with unique fine adjustment possibilities. In *International Conference on Lightning Protection (ICLP)*, Shanghai, China, 2014. doi:10.1109/ICLP.2014.6973145.

[5] B. Schottel. *Plasmavorgänge im engen Spalt beim Ableiten hoher Stoßströme bis 25 kA.* PhD thesis, Technische Universität Braunschweig, 2014.

[6] A. Ehrhardt and S. Beier. New deion chamber for encapsulated switchgear. In $27^{th}$ *International Conference on Electric Contacts*, Dresden, Germany, 2014.

[7] S. Ait-Amar, G. Serrie, J. B. Docourneau, and M. Abplanalp. Arc extinguishing method of spd type 1. In $29^{th}$ *International Conference on Lightning Protection*, Uppsala, Schweden, 2008.

[8] B. Schottel, T. H. Kopp, T. Runge, and M. Kurrat. Experimental investigation of short-time plasma propagation recorded by high speed camera. In *International Conference on Gas Discharges and their Applications*, Orleans, France, 2014.

[9] T. Kopp, T. Runge, and M. Kurrat. Analysis of arc behavior in a model spark gap after surge currents. *IEEE Transactions on Components, Packaging and Manufacturing Technology*, 8(6):958–965, 2018. doi:10.1109/TCPMT.2018.2811940.

[10] T. Runge, S. Franke, S. Gortschakow, and M. Kurrat. Optical investigations on plasma temperature estimation in a model spark gap for surge currents. *Plasma Physics and Technology*, 4(2):108–111, 2017. doi:10.14311/ppt.2017.2.108.

[11] T. Runge, T. Kopp, and M. Kurrat. Novel approach for maximum plasma pressure estimation at surge current based on experimental investigations. *IEEE Transactions on Plasma Science*, 46(8):2935–2941, 2018. doi:10.1109/TPS.2018.2850364.

[12] T. Runge, T. Kopp, and M. Kurrat. Experimental investigations on electrical plasma conductivity in a model spark gap for surge currents. *Plasma Physics and Technology*, 4(1):24–27, 2017. doi:10.14311/ppt.2017.1.24.

[13] A. D'Angola, G. Colonna, C. Gorse, and M. Capitelli. Thermodynamic and transport properties in equilibrium air plasmas in a wide pressure and temperature range. *The European Physical Journal*, 46(1):129–150, 2008. doi:10.1140/epjd/e2007-00305-4.

[14] T. Runge. *Plasmaeigenschaten in Funkenstrecken unter Stoßstrombelastung.* PhD thesis, Technische Universität Braunschweig, 2018.

[15] O. Kerfin, T. H. Kopp, and M. Kurrat. Parasitic magnetic coupling in voltage measurement setups for impulse current tests. In *International Symposium on Electromagnetic Compatibility EMC Europe*, 2018. doi:10.1109/EMCEurope.2018.8485117.

doi:10.14311/ppt.2019.1.65
*Plasma Physics and Technology* **6**(1):65–68, 2019 

# 3D ANALYSIS OF LOW-VOLTAGE GAS-FILLED DC SWITCH USING SIMPLIFIED ARC MODEL

S. Gortschakow[a,*], D. Gonzalez[a], S. Yu[b], F. Werner[b]

[a] *Leibniz institute for plasma science and technology (INP), Felix-Hausdorff-Strasse 2, 17489 Greifswald, Germany*

[b] *TDK Electronics, Gradestrasse 48-50, 12347 Berlin, Germany*

[*] sergey.gortschakow@inp-greifswald.de

**Abstract.** Electro-magnetic simulations have been used for the visualization of distribution of Lorentz force acting on a DC switching arc in low-voltage contactor. A simplified plasma model (black-box model) was applied for the description of arc conductivity. Arc geometry was gained from the high-speed camera images. Influence of arc position, arc current and of external magnetic field has been studied. Results have been compared with optical observations of the arc dynamics.

**Keywords:** black-box model, Lorentz force, arc dynamics, 3D simulations.

## 1. Introduction

3D numerical simulations have a great potential for optimization of switching devices. The main drawback of such simulations is the complexity of the arc model, which must be applied in order to get the realistic results. Consideration of arc dynamics, changes in plasma composition, high temperature and pressure gradients requires enormous mathematical efforts and immense computing power. However, some questions of the design optimization, such as positioning of additional elements for arc control, influence of the electrode shape on the resulting distribution of electromagnetic fields can be roughly solved applying suitable black-box arc models.

Increasing use of pure electric as well as of hybrid cars promotes the research on DC switching devices and contactors with enhanced current interruption capability. Recent studies of DC switching devices in the gas mixtures [1], have shown that hydrogen mixtures, especially those with nitrogen exhibit a higher electric strength as well as a shorter arcing time comparing to air. Additional application of magnetic arc blowing causes further reduction of arcing time. According to the results of [1], the mixture with proportion of 80% $H_2$ and 20% $N_2$ is the most promising for DC applications. In other publications, like e.g. [2] the influence of working pressure and comparison between several gases ($H_2$, $N_2$, He and air) has been discussed putting again the mixture $H_2/N_2$ as a favourite due to significantly faster arc interruption (up to 7 time comparing to He). The arc duration was inversely proportional to working pressure. In [3, 4] the advantages of magnetic field blowing for DC air circuit breakers and relays for operating voltages below 1000 V.

Present contribution focusses on numerical simulations of DC switching arc in low-voltage contactor. Electro-magnetic simulations have been used for the visualization of distribution of Lorentz force in order to understand the general arc dynamics. Arc geometry was gained from the high-speed camera images. The influence of external magnets and of the current density was studied using a simple arc model. Qualitative comparison with corresponding experiments has been performed for model validation. The results are in good agreement and confirm the ability of used method.

## 2. Arc diagnostics

Numerical analysis has been performed for a model arc chamber presented schematically in Figure 1. The electrode system consists of double bridge contactor with two fixed contacts and moving bridge made of copper. Maximum electrode distance is set between and 3 mm. The chamber is made of plexiglass in order to get an optical access to the arc keeping all mechanical parts from original DC contactor. The chamber is filled with mixture $H_2/N_2$ in 80%/20% proportion to a total pressure between 1 and 7 bar.

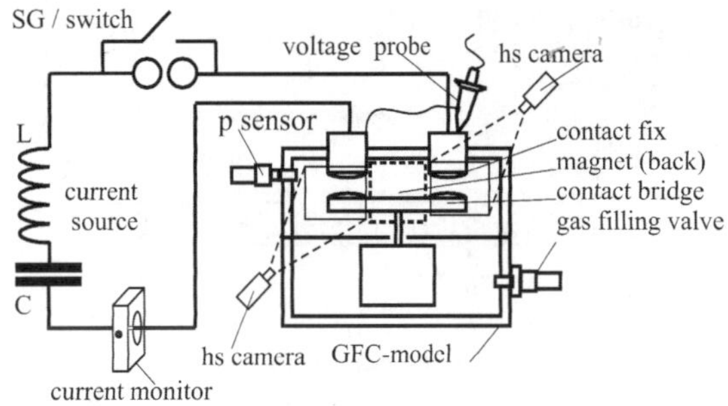


*Figure 1. Schematic view of experimental setup and used diagnostics.*

The diagnostics comprises electric measurements (Tektronix TCP202 Hall current probe, Tektronix 6015A voltage probe), pressure measurements (piezoelectric sensor PCB-105C), optical observation of arc dynamics by a high-speed camera (RedLake Motion Pro Y6) with frame rate of 4000 fps and acquisition time of 1 µs as well as acquisition of spatially integrated arc spectra using compact spectrometer (Avantes VIS-spectrometer, AvaSpec-ULS2048).

Detailed results of experimental diagnostics are presented elsewhere [5]. In the present paper mainly the results of arc dynamics observations are utilized.

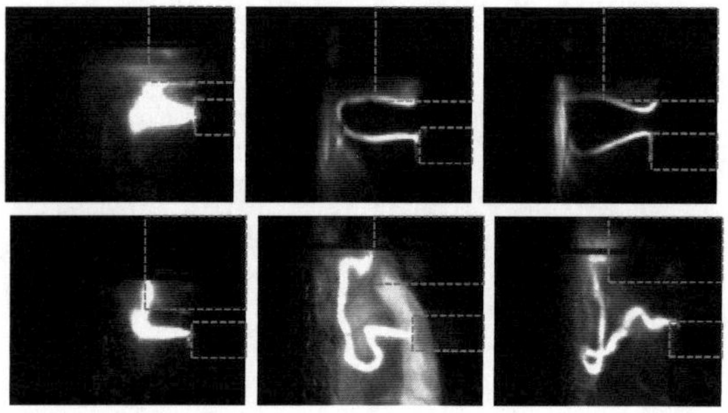

*Figure 2. Arc images: upper row - without external magnets, lower row - with magnets. Red lines show schematically positions of bridge and left electrode.*

Images of the switching arc for two variants without and with external magnets are shown in Fig. 2 for the current of 80 A. The arc is elongated for both geometries and in addition twisted when the permanent magnets have been used. Notice, that according to results of [6] the influence of magnetic fields causes not only the arc elongation through Lorentz force but also an increase of thermal losses due to created turbulent cooling. This leads to the reduction of arc temperature and consequently of plasma conductivity and, hence to more effecting switching-off process. Another positive effect of the arc column twisting is that the arc is kept away from the wall. This leads to the reduction of the wall erosion.

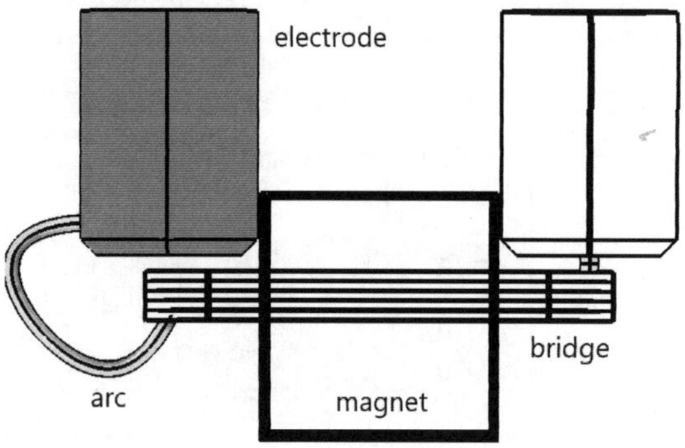


*Figure 3. Model geometry of arc chamber.*

## 3. Description of the model

Figure 3 presents the model geometry used in simulations. It consists from two cylindrical electrodes, bridge electrode, permanent magnets and arc plasma. The distance between the electrodes is fixed to 2 mm.

The system of Maxwell equations has been used to model the electro-magnetic field distribution. Taking into account the assumptions for the arc plasma (corresponding conductivity model) the following equation system was used for the determination of the electric potential $V$, magnetic vector potential $\mathbf{A}$ and magnetic field strength $B$:

$$\nabla \cdot (\sigma \nabla V) = 0 \tag{1}$$

$$\nabla^2 \mathbf{A} = \mu \sigma \nabla V \tag{2}$$

$$\mathbf{B} = \nabla \times \mathbf{A} + \mathbf{B}_r \tag{3}$$

where $\sigma$ is the electric conductivity, $\mu$ is permeability of the medium and $\mathbf{B}_r$ denotes the magnetic flux provided by external magnetic elements. Finally, the electromagnetic force $\mathbf{F}_\mathrm{L}$(Lorentz force) which is responsible for arc plasma movement is determined by following relation

$$\mathbf{F}_\mathrm{L} = \mathbf{j} \times \mathbf{B} \tag{4}$$

with the current density

$$\mathbf{j} = -\sigma \nabla V \tag{5}$$

The system was complemented by appropriate boundary conditions, i.e. ground potential at one electrode, fixed current at the second cylindrical electrode as well as zero current density and magnetic insulation at the outer boundaries. The model was developed using the commercial code COMSOL Multiphysics Version 5.4. The combination of modules „Electric currents" and „Magnetic fields" has been used to describe the DC arc.

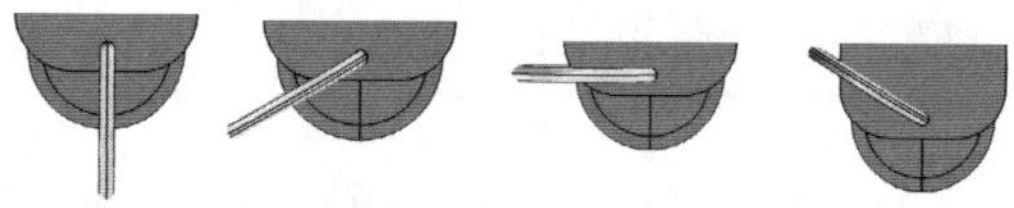

*Figure 4. Chosen arc positions in the electrode system, rotation by 0, 60, 90 and 120 degree. Bottom view from bridge electrode.*

### 3.1. Arc plasma model

A simplified plasma model (black-box model) has been used for the investigations. The shape of the arc was chosen based on experimental observations. The initial stage of arc development during the contact separation corresponds to an cylindrical and axial symmetric plasma. Then, due to the action of Lorentz force the arc elongation begins (cf. images in the first column in Fig. 2). This shape was chosen for numerical analysis. The arc column is assumed to has a form of sleeve with homogeneous conductivity distribution in axial and radial directions direction. Since the arc plasma shows high dynamics, several positions (Fig. 4 have been analysed.

The black-box model has several parameters: arc diameter, arc shape (length) and the conductivity. The arc length was fixed from the images of the arc, as mentioned above. The arc diameter was estimated also from images and assumed to be of about 1 mm. As the experimental results [5] clarify, the arc spectrum includes intensive lines of copper. It is well known that the presence of only few percent admixture of metal vapour in the plasma conductivity becomes close to that of pure metal plasma. Furthermore, the temperature of metal affected plasma is in the range 8000 – 12000 K. Taking into account those aspects the plasma conductivity was estimated to be around $10^4$ S/m.

## 4. Results and discussion

### 4.1. Influence of external magnets

Figure 5 shows the comparison of simulation results for the cases without and with external magnets. The colour scale gives the information about the magnitude of Lorentz force, while the arrows give the force direction. The length of the arcs is proportional to the magnitude of $F_L$.

In the case without external magnets the arc always moves straight to the chamber wall. The same tendency was observed in the experiments (cf. Fig.2 upper row) where the arc reaches the wall after certain time. Such a behaviour finally leads to wall erosion and contamination of the walls with metal vapour. In term of optimization of switching behaviour, the walls should be made of gassing polymers in order to hold the arc away from the wall and to keep the surface clean. Application of external magnets leads to several effects in the arc dynamics. Firstly, the magnitude of Lorentz force increases by about factor of two for the positions 0 and 60 degree and by about factor of three for 90 and 120 degree which are close to the magnet. The second, more important impact, is the change of force distribution along the arc and of its direction. The sections of the arc column experience different direction and magnitude. This causes more efficient increase of the arc length and additional twisting of the column (cf. Fig. 2 lower row). Thus, the arc is kept away from the walls and has longer length providing more efficient arc cooling, shorter arcing time and, therefore, more effective current termination. The results for the arc positions 90 and 120 degree detect also the potential danger for these arc positions. Hot plasma is pushed toward the bridge electrode and, hence, can comes close to the central part of the device, which is not favour position for the arc burning. Fortunately, the experiments [5] have clarified, that such a scenario occurs quite seldom. Nevertheless, one has to keep this possibility in mind and to provide corresponding countermeasures.

### 4.2. Influence of arc current

Influence of current magnitude on the distribution of electromagnetic force is shown in Fig. 6. An increase of the arc current causes as expected to a proportional growth of the Lorentz force magnitude. Also, the distribution of force magnitude and direction along the arc length change. When the current increases, the self-induced magnetic field start to compete with external magnetic field. Finally, at very high currents the self-induced field becomes dominant and the spatial distribution and the direction of Lorentz force will be close to that presented in upper row of Fig. 5. The case with 400 A is still far away from this situation. However, the change of force direction is clearly visible. The results show the advantages of enhanced current in case of switching-off process: the arc will be more elongated and stronger twisted. In addition, the electromagnetic force for positions 90 and 120 degree is directed away from the bridge electrode.

## 5. Summary

The present contribution discusses the influence of additional arc control elements in model arc chamber. The shape of the arc plasma was adopted from experimental observations by high-speed camera. Distribution of electromagnetic fields and forces was calculated for various arc positions. It was found that external permanent magnets causes the twisting and elongation of the arc plasma and, therefore, positively influence the switching-off process. However, when the arc current increases the magnetic filed of the arc becomes stronger and the arc is less influenced by external magnets. This means that for higher current some additional arc extinguishing elements, like e.g. splitter plates or gassing polymers have to be used. The results of simulations have been compared with experimental observations of arc dynamics. A good qualitative agreement was found which gives the validation of applied approach.

### References

[1] Y. Shiba et al. Study of DC Circuit breaker of $H_2$-$N_2$ Gas Mixture for High Voltage. *Electrical Engineering in Japan*, 174(2):1407–1413, 2011. doi:10.1002/eej.21042.

[2] K. Sawa et al. Influence of sealed gas and its pressure on arc discharge in electromagnetic contactor. *IEEE Holm Conference on Electrical Contacts*, pages 236–241, 2017. doi:10.1109/HOLM.2017.8088093.

[3] Breaking performance of a circuit breaker influenced by a permanent magnetic field at dc voltages up to 450 v. *Proceedings of the ICEC-ICREPE*, pages 236–241, 2012.

[4] D. Volm and F. Winkler. Development of a compact relay for high voltage switching of up to 1000 v and 40 a. *International Conference on Electrical Contacts*, 27:144–148, 2014.

[5] D. Gonzalez, S. Gortschakow, S. Yu, and F. Werner. Investigation of dc switching arc characteristics in hydrogen containing gas mixtures. *acceptred for publication in Plasma Physics and Technology*, 2019.

[6] J. Shea, E. Heckman, and J. Suarez Guevara. DC Arc Properties in a DC Magnetic Fields. *64th IEEE Holm Conference on Electrical Contacts/29th International Conference on Electrical Contacts*, 64(1):195–202, 2017.

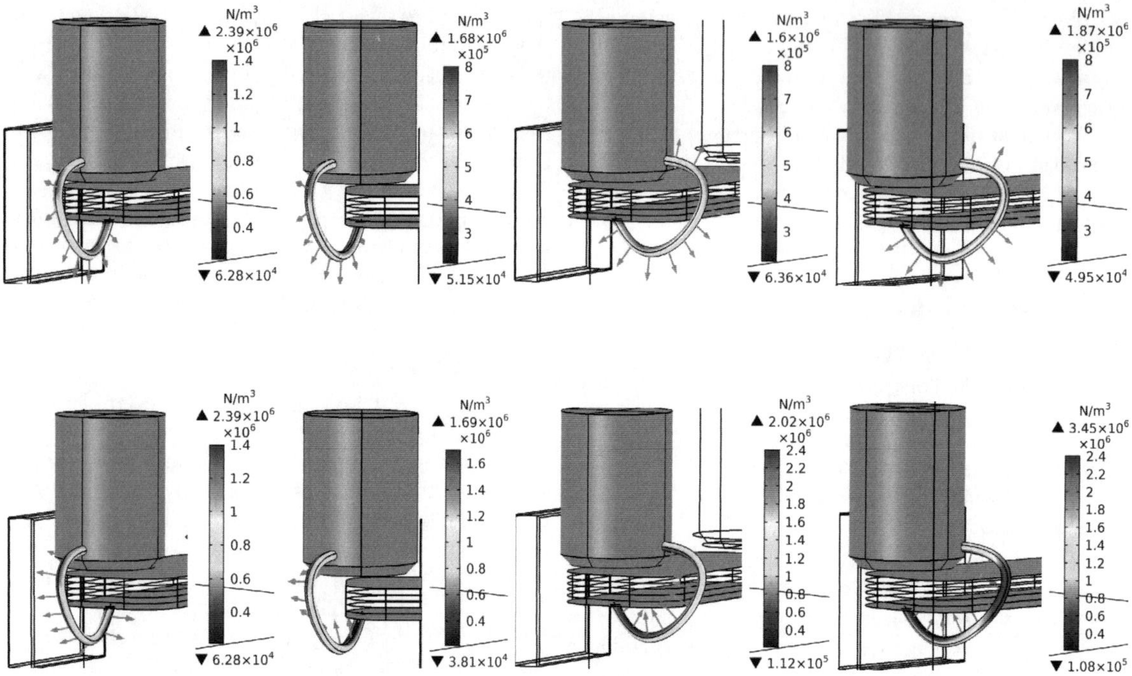


*Figure 5. Lorentz force distribution in case without (upper row) and with (lower row) external magnets. Arc positions (from left to right) 0, 60, 90 and 120 degree.*

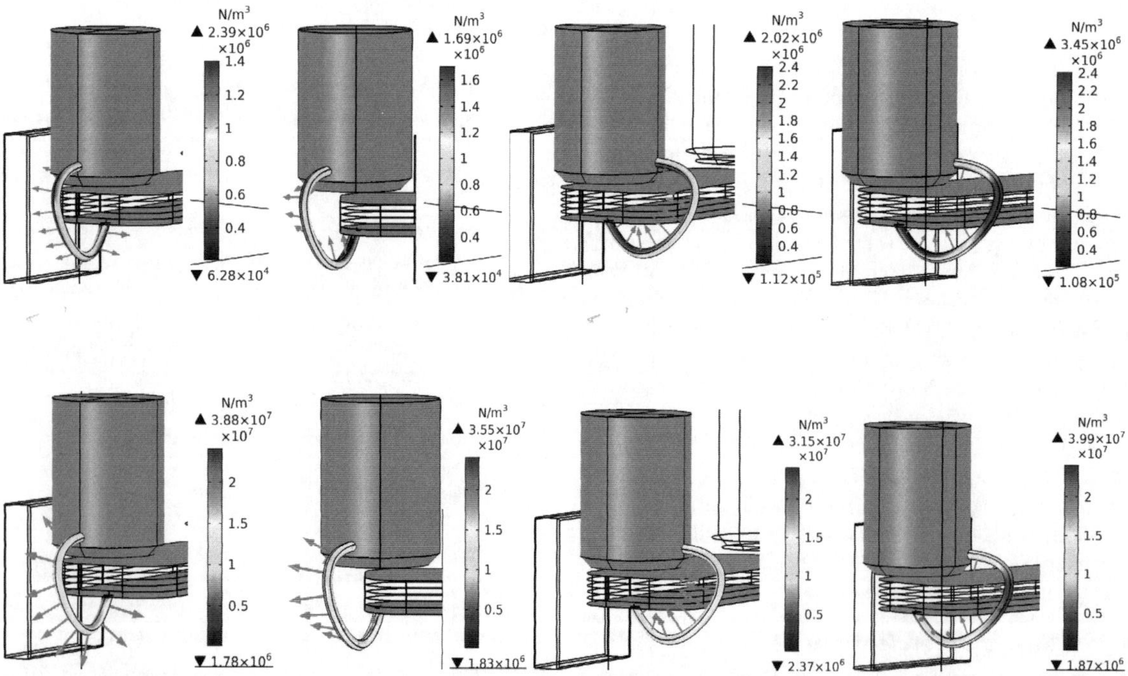


*Figure 6. Lorentz force distribution in case with 80 A (upper row) and 400 A (lower row) current. Arc positions (from left to right) 0, 60, 90 and 120 degree.*

doi:10.14311/ppt.2019.1.69
*Plasma Physics and Technology* **6**(1):69–72, 2019 

# INVESTIGATION OF THE ARC CHARACTERISTICS OF SWITCHING DC ARCS ON HYDROGEN CONTAINING GAS MIXTURES

D. Gonzalez[a,*], S. Gortschakow[a], S. Yu[b], F. Werner[b]

[a] *INP Greifswald, Felix-Hausdorff-Str. 2, 17489 Greifswald, Germany*
[b] *TDK Electronics, Gradestraße 50, 12347 Berlin, Germany*
[*] diego.gonzalez@inp-greifswald.de

**Abstract.** The characteristics of switching DC current arcs in hydrogen containing gas mixtures under pressure were investigated using a model chamber. The switching device consists of an electro-mechanic double breaker unit with copper contacts. High-speed imaging and spectroscopy were used to observe and to characterize the switching arc. The experiments indicate how the dynamic interaction of an external magnetic field with a high-pressure discharge causes an elongation and twisting of the arc-channel and consequently a voltage increase. Comparative measurements with and without external magnetic field indicate a strong influence of this factor. Cu and N lines together with strongly broadened H lines were observed.

**Keywords:** arc behaviour, external magnetic fields, gas filled contactors, hydrogen, overpressure.

## 1. Introduction

The increasing use of hybrid cars and more recently the growing market of pure electric cars with batteries of also increasing capacity requires the use of switching devices, contactors, of also increased current interrupting capability.

Recent results about DC- switching devices under $H_2/N_2$ like by Shiba, Kaneko and co-workers [1], which studied the breakdown characteristics of the $H_2/N_2$ gas mixture as well as its influence on the interrupting performance of relays, have shown that compared to air $H_2/N_2$ at 1 bar possess a higher electric strength and permits to obtain shorter arcing times. Also the combination with magnetic arc blowing allows for further reduction of arcing time. According to their results the mixture proportion of 80% for $H_2$ and 20% for $N_2$ appeared as the most practical. Sawa, Suzuki and co-workers [2] presented results comparing the influence of gas ($H_2$, $N_2$, He and air) and working pressures (0.5 bar to 4 bar) on the arc voltage and current characteristics as well as arcing time for 100 V and 30 A DC relays. These authors found that the arcing time was shorter using $H_2$ followed by $N_2$, air and He, which shows up to 7 time larger arcing times than $H_2$. Also the arc duration was inversely proportional to working pressure. This effect was more clear by arc currents over 20 A and independently of filling gas. Regarding contact resistance, its increase by switching on air was of about factor two compared to the other gases, which is a consequence of oxidation processes on air. Schrank, Gerdinand and co-workers [3] presented results regarding the advantages of magnetic field blowing for DC air circuit breakers (operating voltages around 330 V DC). Also results of Volm [4] indicate improvements of arc switching performance for 1000 V and 40 A air relays by means of magnetic blowing combined with outgassing from polymer components.

The main interest of our investigation was to observe and subsequently to characterize the arc inside gas filled contactors (GFC). To avoid the formation of explosive atmospheres containing hydrogen are GFC's manufactured gas-tight, making difficult the optical access to the arc chamber, which is made of non transparent materials like Aluminum-oxide ceramics and furthermore enclosed in polymer casings. For this reason we decide instead of modify a real contactor, to recreate the geometry of its chamber using a transparent material like polymethyl methacrylate (PMMA) but keeping the mechanics and other components of the real device. The constructed GFC-model allows for optical observation of the arcing volume, making therefore possible to observe the arc dynamics, to collect its radiation and to measure the dynamic pressure change during arcing.

## 2. Experimental methods

### 2.1. Experimental setup

An outline of the experimental setup and of the GFC-model is presented in Fig. 1. It consists of a double bridge contactor with a 12 V coil. The fixed contacts and moving bridge are made of copper. The PMMA chamber has an internal volume of around 20 cm$^3$. The filling gas pressure can be varied from 1 ... 7 bar. Although different gas mixtures can be used, our investigation focused on $H_2/N_2$ in a 80%/20% mixture (purity of 5.0).

### 2.2. Current delivery

For the delivering of the arc current a CLR-circuit with 10 CL-groups in a transmission-line like configuration was used [5]. The resulting rectangular-shaped

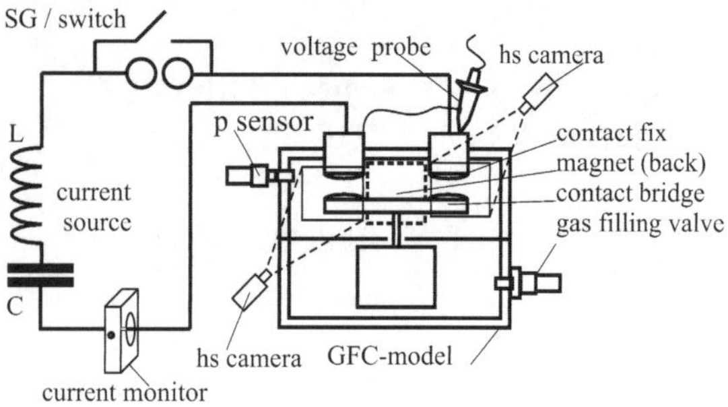


*Figure 1. Experimental setup for the investigation of switching arcs in DC gas filled contactors.*

current-pulse has a duration of 28 ms. The charging voltage for the different experiments series was varied between 500 and up to 1000 V. The amplitude of the current was of 80 and up to 160 A respectively. Due to the relative high effective inductivity of the circuit (64 mH) the delivered currents show a near impressed nature.

### 2.3. Methods

The arc current and voltage signals were measured using a Pearson 1423 current monitor and a Tektronix 6015A voltage probe respectively. The current of the contactor's coil was detected using a Tektronix TCP202 Hall current probe. For the dynamic measurement of pressure during arcing a Piezosensor model PCB-105C was used. The electrical signals were recorded with a Tektronix oscilloscope model DPO 4054. The arc images were recorded using two high-speed cameras of the series Y4 and Y6 from Integrated Design Tools. The arc spectrum was determined with a compact spectrometer with a temporal resolution of 1 – 2 ms (Avantes VIS-spectrometer, AvaSpec-ULS2048, wavelength 300 to 1000 nm). The fiber optic of the spectrometer was covered by a cosinus difussor to reduce light intensity. It was oriented over the same line of view of the cameras. The spectrometer was calibrated by means of a tungsten strip lamp (OSRAM Wi 17/G) in units of spectral radiance. Metal Interference Filters (MIF) for the $H_\alpha$ line at 656 nm and for the Cu-line at 525 nm with a full width at half maximum - FWHM of about 10 nm were placed before the lens of the high-speed cameras.

Values between 4 and up to 7 bar were used for filling pressure during a given experiment series. Before filling the chamber was evacuated down to 150 mbar by a membrane pump. Different experiments to determine the arc dynamics, pressure change during switching and the influence of filling pressure on arcing time and arc appearance were carried out.

## 3. Results and discussion

### 3.1. Characteristics of the switching arc

The characteristics of the switching arc are shown in Fig. 2. The images of the high speed camera indicate how the arc will be elongated and twisted, which explains the observed strong voltage fluctuations. According to results of Shea [6] and Rieder [7] the influence of magnetic fields should cause, besides the elongation of the arc through Lorentz-forces, also an increase of thermal losses because of the created turbulent cooling. This leads to a reduction of arc temperature and consequently of plasma conductivity, hence the additional increase on arc voltage.

### 3.2. Influence of external magnetic field on arc-behavior and voltage characteristics

Fig. 3 shows the typical characteristic of the arc current and voltage signals as well as dynamic pressure during a switching operation with the GFC-model. The considered time for the definition of arcing time and arc-extinction time are also depicted in Fig. 3. The arc-extinction time represents the interval where the arc is effectively elongated as a result of the combined effects of Lorentz-forces and turbulent convective gas flow inside the arc chamber. The dynamic characteristic is noticed on the abrupt changes of voltage value, depending on the effective change of plasma conductivity and arc-length. Also the observed pressure increase corresponds to the beginning of the arc-extinction time indicating the start of the noticeable influence of magnetic field and filling pressure on the thermal characteristic of the arc plasma. The arc-extinction time ends logically when the current is effectively driven to zero. After the effective switching, the pressure decays slowly because of the thermal delay of the heated filling gas. Also the rest voltage of the current source capacitors stays on the GFC-model contacs a longer time.

The recordings of Fig. 2 and Fig. 3 regarding arc-voltage indicate a strong fluctuating behaviour, where its value can change very quickly from some hundreds to thousand of volts. This is due to the combined effects of arc-elongation and resulting turbulent deflection through the interaction of arc-current and external magnetic field. To estimate this influence we choose to use the equivalent arc-voltage vs. current characteristic, plotting the temporally associated values of current and arc-voltage from several measurements (Fig. 4). Comparing the considered cases it is evident that a external magnetic field causes an additional voltage increase and therefore a more efficient current limiting. Although the depicted linear regression curves of the gathered data in Fig. 4 show a relative small difference, it is important to remark that with external magnetic fields the fluctuations are stronger, both in frequency and magnitude and in particular when the current droops below 20 A.

### 3.3. Influence of filling pressure on arcing time

According to results of [2] the filling pressure influences considerably the arc extinction characteristics. The results of our experiments regarding this factor are

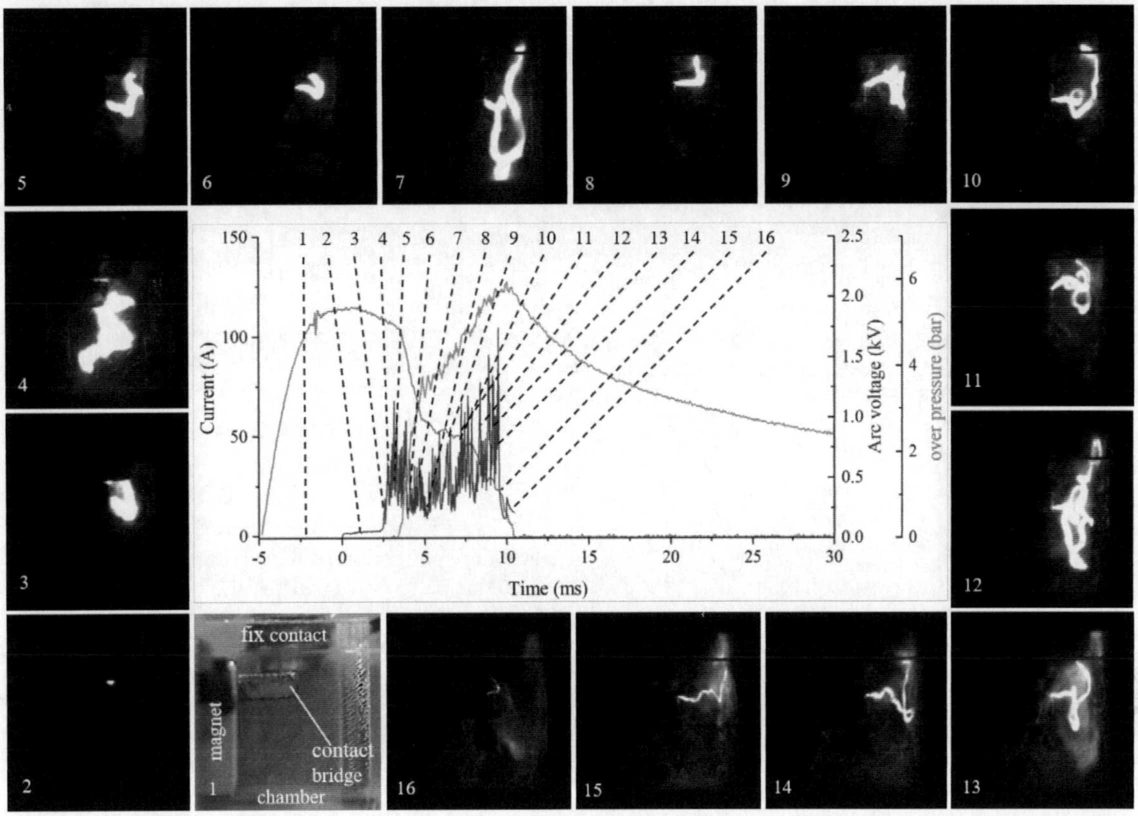


*Figure 2. High-speed images of the switching arc, right side of the GFC-model, and corresponding time sequence on the arc-voltage, current and pressure waves. Frame rate of Y6 camera: 4000 fps; Exposure time: 1 µs*

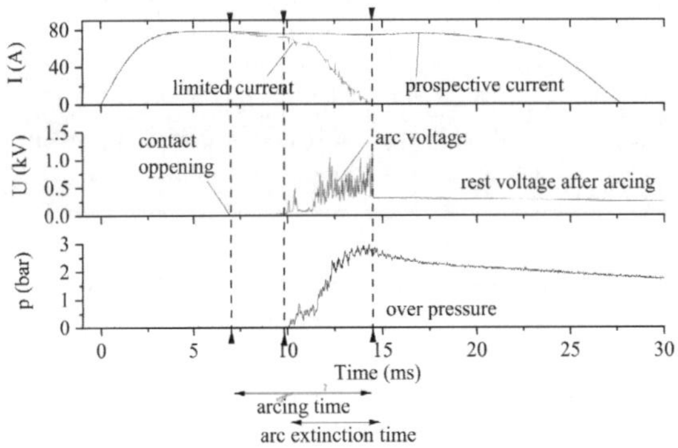


*Figure 3. current, arc-voltage and over-pressure of a typical switching operation of the GFC-model.*

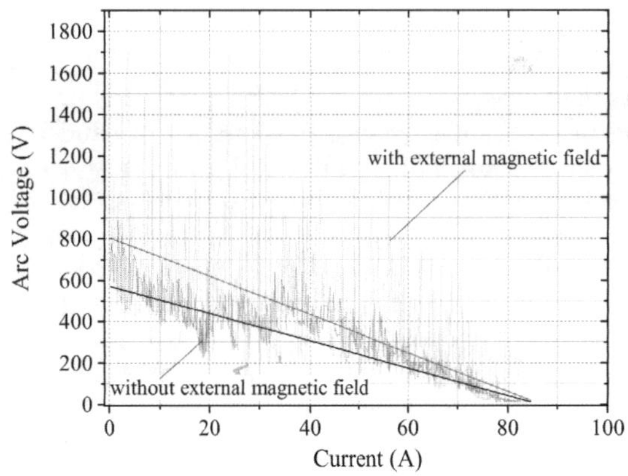


*Figure 4. Influence of the external magnetic field on the arc-voltage vs. current characteristic for the GFC-model.*

depicted in Fig. 5. Although arcing time and arc extinction time appear to be independent on the filling pressure, Fig. 5 shows a slight reduction of the data scattering, in particular by 7 bar the arc-extinction time is notably reduced compared to that of pressures between 4 and 6 bar. Although the considered pressure interval is small, this result points to the influence of filling gas and pressure on the switching behavior, which depends rather on the burning conditions for the arc than on the switching device mechanics. The registered times for the 80 A are nevertheless around 2.5 times higher than those reported by [2]. This is most probably due to the differences between used current-sources, since for [2] the voltage was of 100 V and as it appears, the circuit was of purely ohmic nature. The tendency of reduction of arc extinction time as a function of filling pressure reported there was not clearly observed during our measurements, which is probably due to some kind of saturation

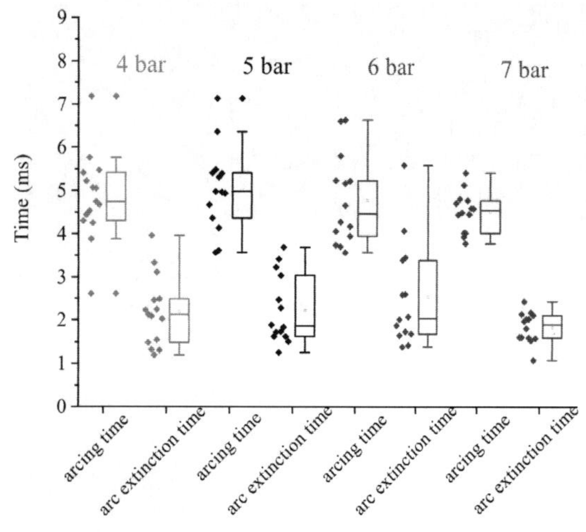


*Figure 5. Influence of the filling pressure on the total arcing time and an the total arc extinction time for the GFC-model.*

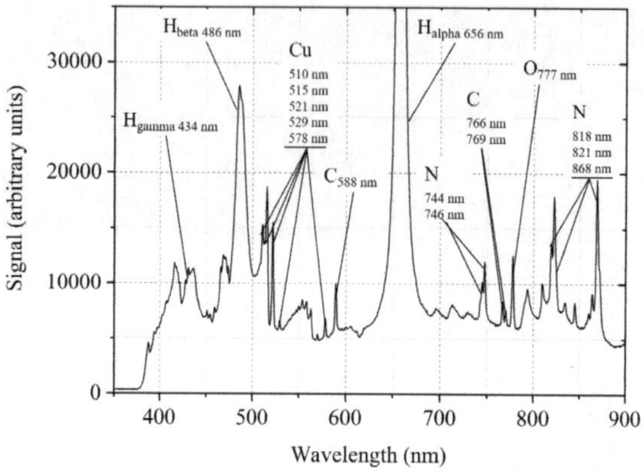


*Figure 6. Emision spectrum of the switching arc inside of the GFC-model.*

effect because of the geometric characteristics and components of the GFC-model.

### 3.4. Characteristics of the arc spectrum

In a first attempt to characterize the thermal properties of the arc plasma inside the GFC-model, qualitative spectroscopic measurements were carried out. The results are shown in Fig. 6. One of the difficulties presented here was the unstable behavior of the arc inside the chamber (see Fig. 2), preventing a spatially resolved measurement of its spectrum. It was nevertheless characterized by a strong broadened $H_\alpha$ line. Further measurements overriding the intensity of the $H_\alpha$ permit to identify other species like $H_\beta$ and $H_\gamma$ lines as well as copper and nitrogen lines. Also carbon and oxygen lines could be identified, which should result because of arc ablative effects on polymer parts of the mechanics and on the PMMA chamber. Although it is not yet possible to quantify the plasma characteristics from the obtained spectrum, the absence of ion-lines is an indication that the arc temperature is probably under 12000 K.

## 4. Summary

The characteristics of DC switching arcs in gas filled contactors was investigated with a experimental model using a PMMA chamber. The results indicate how the arc will be elongated and twisted because of the combined effects of external magnetic field and turbulent cooling in the high pressure chamber. The consequence is a remarkable increase of arc-voltage though accompanied by strong fluctuations. The final effect is however an effective improvement of the current limiting performance of the device. External magnetic fields allow also to reduce the effective switching time and therefore the arcing stress on the contactor. An increase on filling pressure appears to influence the resulting arc-extinction time scattering. Its mean value was though similar within the considered pressure interval between 4 and 7 bar. The emission spectrum shows predominantly atomic lines. In particular a strong broadened $H_\alpha$ dominates the collected radiation intensity. Also copper and nitrogen lines as well as carbon and oxygen lines, probably because of arc ablative processes on polymer components and on the arc chamber (PMMA), were observed. Further experiments regarding the thermal characteristics of the arc plasma are in preparation.

## References

[1] Y. Shiba et al. Study of DC Circuit breaker of $H_2$-$N_2$ Gas Mixture for High Voltage. *Electrical Engineering in Japan.*, 174(2):1407–1413, 2011. doi:10.1002/eej.21042.

[2] K. Sawa et al. Influence of sealed gas and its pressure on arc discharge in electromagnetic contactor. *IEEE Holm Conference on Electrical Contacts*, pages 236–241, 2017. doi:10.1109/HOLM.2017.8088093.

[3] T. Schrank et al. Breaking performance of a circuit breaker influenced by a permanent magnetic field at DC voltages up to 450 V. *Proceedings of the ICEC-ICREPE*, pages 236–241, 2012. doi:10.1049/cp.2012.0618.

[4] D. Volm and F. Winkler. Development of a compact relay for high voltage switching of up to 1000 V and 40 A. *International Conference on Electrical Contacts*, 27:144–148, 2014.

[5] A. Khakpour et al. Impact of different vacuum interrupter properties on high-current anode phenomena. *IEEE Transactions on Plasma Science*, 44:3337–3345, 2018. doi:10.1109/TPS.2016.2625862.

[6] J. Shea, E. Heckman, and J. Suarez Guevara. DC Arc Properties in a DC Magnetic Fields. *64th IEEE Holm Conference on Electrical Contacts/29th International Conference on Electrical Contacts*, 64(1):195–202, 2017. doi:10.1109/HOLM.2018.8611725.

[7] W. Rieder and A. Eidinger. Das Verhalten des Lichtbogens im transversalen Magnetfeld (Magnetische Blasung). *Archiv für Elektrotechnik*, 43(2):94–114, 1957. doi:10.1007/BF01479337.

doi:10.14311/ppt.2019.1.73
*Plasma Physics and Technology* **6**(1):73–77, 2019 

# INTERRUPTION OF MEDIUM–VOLTAGE DIRECT–CURRENTS BY SEPARATION OF CONTACT ELEMENTS IN MINERAL OIL USING AN ULTRA FAST ELECTRO-MAGNETIC ACTUATOR

S. Jugelt*, C. Leu

*Research Unit High-Voltage Technologies, Technische Universität Ilmenau, Gustav-Kirchhoff-Straße 1, 98693, Ilmenau, Germany*

* stefan.jugelt@tu-ilmenau.de

**Abstract.** The increasing usage of medium-voltage direct-current in upcoming electrical energy grid topologies requests novel solutions for MVDC switching. The interruption of direct-currents is accomplished by enforcing a current zero crossing by adequate means and preventing reignition due to the recovering dc voltage. This paper evaluates the rapid separation of the contact elements in mineral oil leading to a liquid flow around the contact elements and the switching arc. The energy turnover of dielectric liquids interacting with an electric arc is considerably higher leading to heavily increased arc voltages compared to dielectric gases. This paper confirms results of earlier publications and carries them further towards a possible usage in an MVDC switching or protection device. Thus a contact arrangement surrounded by mineral oil in combination with an ultra fast electro-magnetic actuator is introduced and performed measurements are discussed.

**Keywords:** dc interruption, bulk oil, dc switching arc; mvdc; mvdc protection device.

## 1. Introduction

The availability of high voltage power electronics leads to new concepts of HVDC and MVDC electrical energy grid topologies these days. But the usage of high and medium voltage direct currents imposes new difficulties regarding current interruption. With direct currents a current zero crossing does not naturally appear hence AC switchgear technologies are no capable of high and medium voltage current interruption. To overcome the lack of a current zero crossing three main principles are known [1]. Divergent oscillation (1) and current injection (2) utilize synthetic resonant networks in parallel to an ultra fast disconnecter (1) or main circuit breaker (2). The third principle, counter voltage (3), requires an arc (or semiconductor circuits) to build up a voltage drop along the current path higher than the driving voltage thereby forcing the current to zero. In this paper an approach to build up a high arc voltage conducting a contact separation in mineral oil as proposed in [2] is shown. The mechanical force and contact velocity needed is provided by an ultra fast electro-magnetic actuator.

## 2. Test and measurement setup

### 2.1. Test setup

To accomplish a fast contact separation in mineral oil a contact configuration within an extinguishing chamber as shown in figure 1 is used. This extinguishing chamber (1) is filled with mineral oil of type Shell Diala S4 ZX-I. The contact separation is accomplished by pushing (3) on the moving contact (4) by the electro-magnetic actuator (2) (see section 2.2).

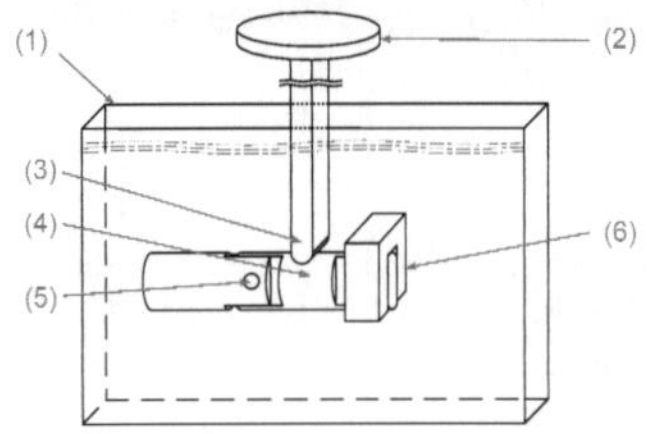


*Figure 1. Contact configuration in the mineral oil filled extinguishing chamber with electro-magnetic actuator.*

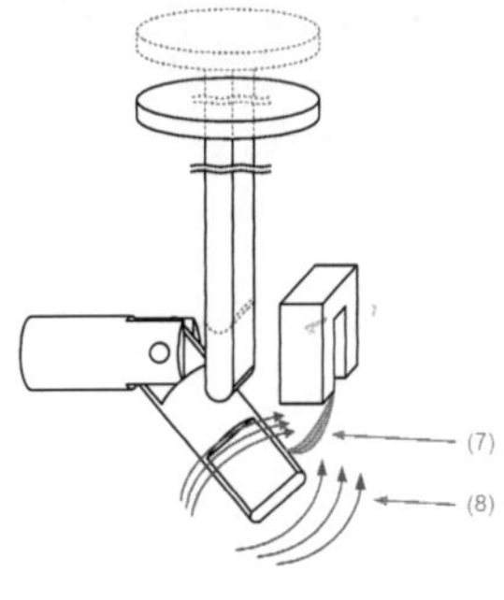


*Figure 2. Electric arc (7) and liquid flow (8) during movement.*

The moving contact (4) pivots around the pivot point (5) from 0° (contact closed) to −90°. Physical contact separation from the fixed contact (6) does occur 7 mm in to the mechanical movement of the moving contact (4) and pivoting ends 49 mm into the movement of the electro-magnetic actuator. After contact separation an electric arc (7) originates. Due to pivoting this contact (4) a liquid flow (8) of mineral oil on the tip of the contact (4) does occur, see figure 2, draining power from the electric arc.

### 2.2. Electro-magnetic actuator

It is known from [2] that contact separation or arc elongation in mineral oil needs to be very rapid in order to generate high arc voltages. Previously propellant powered drives were used to achieve these high velocities. Disadvantageous to propellent drives are there properties to be single use and there low velocity after ignition. Hence an ultra fast electro-magnetic actuator driven by a capacitor discharge is used here. These types of drives are also proposed for usage in ultra fast vacuum circuit-breakers [3].

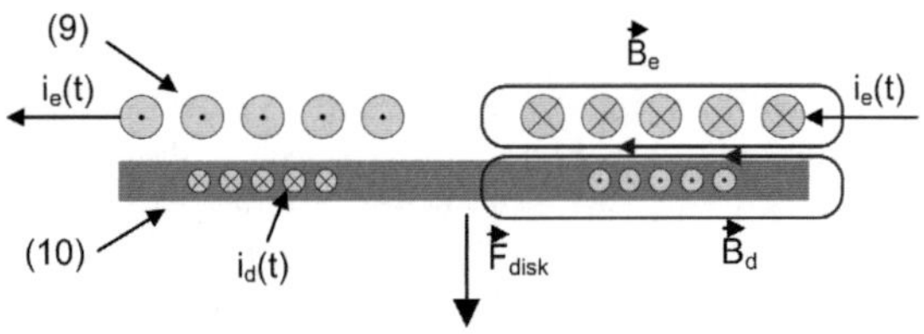


*Figure 3. Schematic illustration of Thomson coil actuator.*

A schematic illustration of such a ultra fast electro-magnetic actuator is given in figure 3. For the purpose of the conducted experiments only unifilar windings were used for the flat coil (9), in this shape often called THOMSON coil. The current $i_e(t)$ (1) induces a voltage into the aluminium disk (10) resulting in eddy currents opposing the magnetic field of the Thomson coil (9).

$$L_{B_e} \cdot \frac{di_e(t)}{dt} - \frac{dM_{B_e,B_d}(x(t)) \cdot i_d(t)}{dt} = \int C_0 \cdot i_e(t)\,dt - L_{par} \cdot \frac{di_e(t)}{dt} - R_{par} \cdot i_e(t) \quad (1)$$

($x(t)$ – Displacement of the disk (10))

And the resulting force accelerating disk (10) is proportionate

$$\vec{F}_{disk} = \iint \vec{J}_d(x(t))d\vec{A}_{disk} \cdot \left(\vec{d}(t) \times \mu \frac{\partial}{\partial \vec{x}} \iint \vec{J}_e(x(t))d\vec{A}_{wire}\right) \quad (2)$$

$$F_{disk} \sim i_e(t) \cdot i_d(t). \quad (3)$$

So with variation of load voltage $U_{TD}$ and/or capacitance $C_0$ the speed of contact separation can be controlled.

### 2.3. Electrical test circuit

The conducted measurements were carried out using the electrical test circuit shown in figure 4.

A low inductance resistor damped capacitor discharge provides the energy against which the arc voltage is build up. This capacitor discharge is initiated

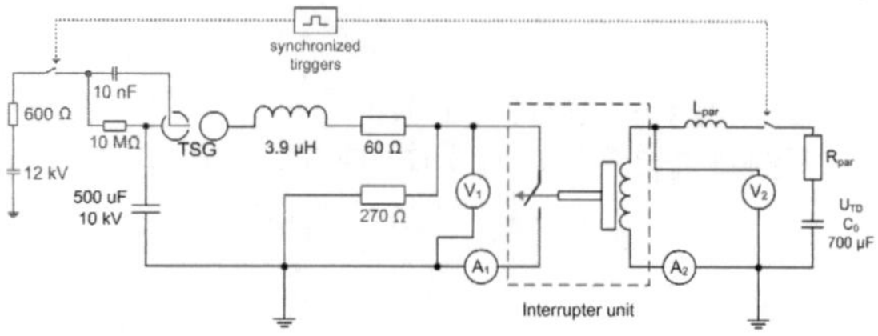


*Figure 4. Electrical test circuit as used in measurements ($\tau_{discharge} = 30\,ms$)*
*left: low inductance medium voltage dc circuit*
*right: electrical circuit of the electro-magnetic actuator.*

by a spark gap (TSG) triggered by a controlled voltage exaltation on the third (pin) electrode. This voltage exaltation is synchronized to the excitation of the electro-magnetic actuator by timer/counter blocks of a micro controller. Additionally the velocity of the electro-magnetic actuator is measured in direction of the movement using a laser triangulation sensor (40 kHz, measuring range: 50 mm, resolution: 50 µm).

## 3. Theoretical considerations on the decomposition of mineral oil

Electric arcs in dielectric liquids generate a gaseous phase around them [4]. As for highly molecular liquids such as the used Shell Diala S4 ZX-I due to the high arc temperature dissociation leading up to the atomic level does occur. Following diffusion into colder areas of the gaseous phase partial recombination does occur releasing part of the bond-dissociation energy heating up the gaseous phase. Furthermore the chemical dissociation reaction can be described by

$$C_{18}H_{38} \rightarrow 10H_2 + 5C_2H_2 + 2CH_4 + 6C \quad (4)$$

where only the shorted possible alkane ($C_n$ $H_{2n+2}$) [5] is taken into consideration because of lower bond-dissociation energy. Which is defined by

$$\begin{aligned} H-H &\quad 103.4\,\text{kcal/mole} && (5)\\ C-C &\quad 58.6\,\text{kcal/mole} && (6)\\ C \equiv C &\quad 123\,\text{kcal/mole} && (7)\\ C-H &\quad 87.3\,\text{kcal/mole} && (8)\\ C_{gaseous} \rightarrow C_{solid} &\quad 124.3\,\text{kcal/mole} && (9) \end{aligned}$$

as per [6]. So to dissolve a hydrocarbon molecule 4314 kcal/mole of energy is needed. Likewise 3967 kcal/mole do get converted during recombination. So for the chemical reaction

$$347\,\text{kcal/mole} \quad (10)$$

of energy needs to be supplied. With a molecular weight of 254 g/mole ($C_{18}$ $H_{38}$) the chemical reaction energy is

$$E_{decomp} = 1.37\,\text{kcal/g} \approx 5.7\,\text{kJ/g}. \quad (11)$$

Additionally for the electric arc to sustain within the gaseous phase energy is needed to heat up the products of decomposition. For heated highly pressurized gases of this matter the specific heat is dominated by the concentration of $H_2$ [7]. Based on the enthalpy of high temperature pressurized $H_2$ which according to [8] can be estimated to $31\,\mathrm{kJ/g_{H_2}}$ and the mass faction of $H_2$ in $C_{18}\,H_{38}$ of $0.18\ /\mathrm{g_{C_{18}H_{38}}}$ an additional energy of

$$\mathrm{E_{heat}} \approx 5.5\,\mathrm{kJ/g} \tag{12}$$

is required to heat up the products of decomposition by the electric arc.

## 4. Measurements

With the contact configuration described in section 2.1 measurements were carry out varying the velocity of the electro-magnetic actuator hence varying the speed of contact separation.The results can be grouped into two voltage/current-time characteristics.

For the purpose of this paper the first voltage/current-time characteristic is treated as a successful current interruption. An example of a measured successful current interruption is given in figure 5.

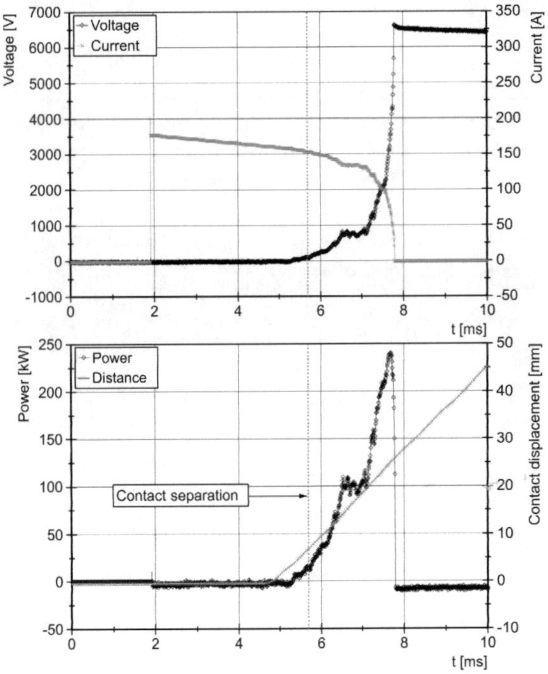


*Figure 5. Top: arc voltage and current for a current interruption at $v_{contact} = 9.0\,m/s$;*
*bottom: power output of the electric arc and measured displacement of the moving contact.*

After contact separation at $\sim 5.8$ ms the arc voltage increases nearly exponentially until it reaches the capacitor discharge voltage. Contrary to that the current decreases exponentially following the voltage rise until it reaches zero and current interruption is accomplished within 1.8 ms. With the applied velocity of the moving contact of 9.0 m/s the displacement at current interruption is just 18 mm hence the contact is not fully pivoted in this case. Due to the continuing pivot movement and the very high dielectric strength of the mineral oil during all measurements no rearcing process could be observed. For the case of the current interruption of figure 5 the power output of the electric arc reaches $240\,\mathrm{kW_{peak}}$ immediately before the current declines to zero, inducing 210 J of energy into the liquid.

Opposed to the voltage/current-time characteristic of figure 5, under the same constrains of capacitor load voltage, discharge current and trigger delay of the electro-magnetic actuator but lower velocity (7.2 m/s) an example of a non successful current interruption is given in figure 6.

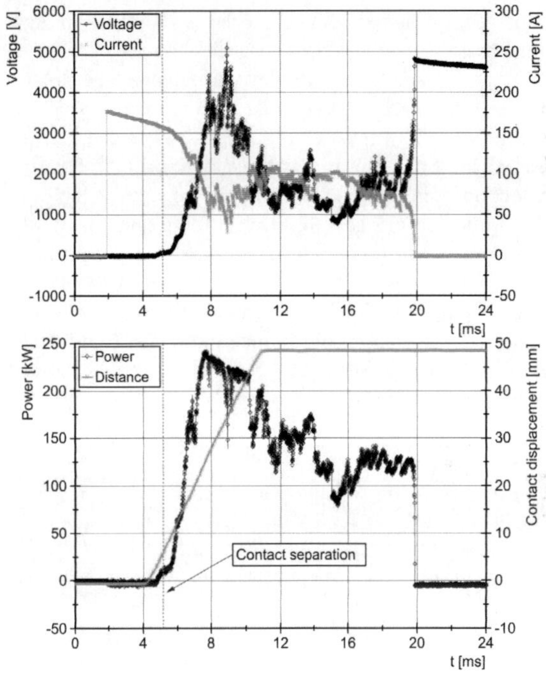


*Figure 6. Top: arc voltage and current for a current interruption at $v_{contact} = 7.2\,m/s$;*
*bottom: power output of the electric arc and measured displacement of the moving contact.*

After contact separation at 5.8 ms the arc voltage at first rises exponentially to but the current does not follow in the same manner as before. At 7.8 ms (moment of current interruption of fig. 5) the power output of the arc peaks at 245 kW but due to the lower velocity of the moving contact the induced energy at 1.8 ms after contact separation is already 320 J hence an increased volume of mineral oil is decomposed (cf. (11)). The major gaseous phase provides sufficient volume for the arc to further heat up the decomposition products (cf. (12)) and the increasing gas volume leads to an electric arc surrounded only by gaseous $H_2$. Current interruption is still accomplished

14.2 ms after contact separation due to the decreasing capacitor voltage and the continuing pivot movement.

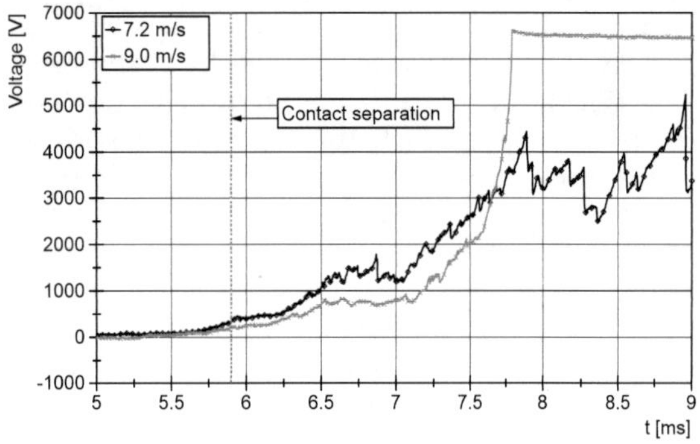


*Figure 7. Arc voltage immediately after contact separation.*

As seen in figure 7 the arc voltage increases higher immediately after contact separation in case of a lower velocity. This is due to the low turbulence flow of the mineral oil in this case. The electric arc is able to decompose more (stationary) liquid hence the higher arc voltage but the generated gas can not removed from the arcing zone sufficiently. For high velocities the liquid stream can remove more gas but arc voltage is limited by the distorted surface of the phase boundary.

To further verify this behaviour the accumulated energy of the electric arc induced into the liquid 1 ms after contact separation is evaluated for the measured velocities, see figure 8.

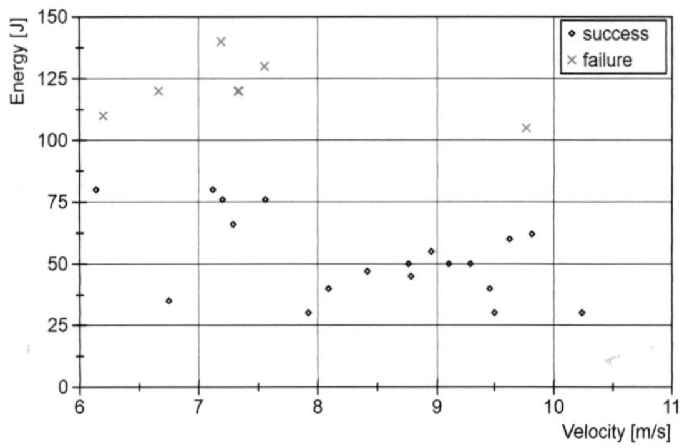


*Figure 8. Accumulated arc energy 1 ms after contact separation.*

The results reveal that a successful current interruption (as seen in figure 5) could only be accomplished when a low amount of energy is induced into the liquid by the electric arc within the first moments after contact separation. If however the energy turnover of the electric arc immediately after contact separation is rather high, current interruption will fail afterwards. As for dependency of the velocity, due to slight variations of the electro-magnetic actuator trigger delay, current interruptions for low ($\sim 6-8\,\mathrm{m/s}$) velocities are observed too, because voltage and current of the charged capacitor are decreased further leading to lower power output of the arc for a increased trigger delay. Futher it can be seen in figure 8 that the induced energy 1 ms after contact separation decreases with increasing velocity. It should be assumed that due to the turbulent flow for high velocities the electric arc is quenched more leading to a increased current density but less radiated heat hence lower energy induced into the liquid.

## 5. Conclusions

Based on the measurements shown before multiple conclusions regarding medium voltage direct current interruption utilizing the counter voltage principle can be drawn.

1. Ultra fast electro-magnetic actuators, which are of huge interest for UFDs and vacuum circuit breakers, are very well suited for medium voltage direct-current interruption within dielectric liquids like mineral oils.

2. Mechanical contact separation for MVDC interruption in dielectric liquids needs to be very fast to avoid excessive generation of gaseous decomposition products.

3. Utilizing appropriative velocities low current MVDC interruption can be accomplished with a contracted electric arc avoiding intermediate current rise and voltage drop within very short periods of time (for mechanical current interruption).

4. For high current interruption the previously shown principle of high speed contact separation may not be sufficient because of the power dissipation of the electric arc ($P \sim I^2$) hence a forced liquid flow is needed.

## References

[1] V. Lenz. DC current breaking solutions in HVDC applications. Master's thesis, High Voltage Laboratories, ETH Zurich, 2015.

[2] H. Ann. *Untersuchungen über die Erzeugung sehr hoher Lichtbogenspannungen unter Flüssigkeiten.* PhD thesis, TH Braunschweig, 1966.

[3] A. Bissal. *Modeling and Verification of Ultra-Fast Electro-Mechanical Actuators for HVDC Breakers.* PhD thesis, KTH, Electromagnetic Engineering, 2015. QC 20150422.

[4] F. Kesselring. Das Schaltproblem der Hochspannungstechnik. *Archiv für Elektrotechnik*, 35(3):155–184, Mar 1941.

[5] Shell Deutschland Oil GmbH. Sicherheitsdatenblatt Shell Diala S4 ZX-I. Technical report, 2018.

[6] R. Heizmann. *Abschaltversuche mit verschiedenen Löschflüssigkeiten und die Zusammensetzung der entstehenden Schaltergase.* PhD thesis, ETH Zurich, 1957.

[7] R. Kuthe. *Temperaturen und Zusammensetzung eines aus Kohlenstoff und Wasserstoff entstandenen hocherhitzten Gases in Abhängigkeit vom Druck.* PhD thesis, TH Braunschweig, 1961.

[8] F. Burhorn et al. Plasmazusammensetzung, Plasmadichte, Enthalpie und spezifische Wärme von Wasserstoff und Wasser bei 1,3,10 und 30 atm im Temperaturbereich zwischen 1000 und 30000 K. *Zeitschrift für Physikalische Chemie*, 215, 1960. doi:10.1515/zpch-1960-21528.

doi:10.14311/ppt.2019.1.78
*Plasma Physics and Technology* **6**(1):78–81, 2019 

# NUMERICAL INVESTIGATION, INCLUDING EXPERIMENTAL VALIDATION, OF AN AXIALLY BLOWN, STABLE ARC IN ARGON

B. Galletti[a,*], F. Kassubek[a], M. Buffoni[a], J. Carstensen[a], P. C. Stoller[b]

[a] *ABB Switzerland Ltd., Corporate Research, 5405 Baden-Daettwil, Switzerland*
[b] *ABB Switzerland Ltd., Technology Center - Interrupter Development, 5401 Zurich, Switzerland*
[*] bernardo.galletti@ch.abb.com

**Abstract.** In this work we present the outcome of a numerical validation study conducted with an arc model developed within a computational fluid dynamics (CFD) tool. The numerical investigations were aimed at reproducing the spatially resolved temperature data obtained with an experiment in which an axially symmetric argon arc was established in the observation region. The full absorption spectrum has been computed for argon and then compressed with minimum loss of information to a relatively small set of bands. The latter has been used for solving the radiative transfer equation in a computationally affordable, yet accurate way. The comparison between the arc temperature simulated with the reduced absorption data and the measured one is presented.

**Keywords:** high-voltage circuit breaker, radiation, optical diagnostics, arc simulation, validation.

## 1. Introduction

The behavior of the electric arc in high voltage gas circuit breakers is the result of the balance between the Joule heating due to the current flowing through it and several mechanisms of removal and redistribution of the energy, such as radiation, convection and turbulent mixing. Moreover, the arc interacts with the solid surfaces around it, generating ablation from the insulating nozzles and erosion from the contacts. As a result, the gas that enters in the arc has a multi-species composition, whose complexity is further enhanced by the dissociation and ionization processes caused by the strong ohmic power density. Developing a high-fidelity numerical description of these arc phenomena is challenging, not only because of their inherent complexity, but also because of the strong coupling between them. For this reason, we have conceived an experiment [1] for measuring the spatially resolved temperature of an electric arc in a part of which some of the physical aspects described above can be disregarded, thus allowing a better validation of the physical and numerical models of the remaining ones. In particular, since in this experiment the arc was blown along two opposite axial directions, the effects of ablation, erosion and turbulence were minimized in the measurement window, located around the stagnation point. Another important feature of the test setup was that it achieved the stabilization of a cylindrical arc column in the observation region, thanks to the combined effects of bi-directional axial blowing and expansion of the gas from rest. The latter condition was obtained by pressurizing the test setup, so that the arc was drawn in high-pressure argon that discharges to ambient with small flow velocities upstream from the arc. The spatially resolved arc temperature was obtained through the application of non-intrusive diagnostics techniques based on imaging interferometry [1]. The arc stability achieved in this experiment not only enhanced significantly the measurements accuracy and their repeatability, but brought benefits also on the numerical side. It has, in fact, enabled the use of two-dimensional axisymmetric models, which allows a higher numerical accuracy for the same computational cost as a full three-dimensional model. In view of the above, the main physical aspects that we have tried to represent numerically were the correct characterization of the gas properties and the radiation transport; the latter being the topic on which the present work is focused. The thermodynamics and transport properties of the gas in the conditions existing in a plasma arc have been investigated by many authors, either assuming local thermodynamic equilibrium or departures from these equilibrium conditions, e.g. [2]. The available studies concerning radiation are focused on both the calculation of the full absorption spectra and on simplified models that reduce the complexity of the problem to a level that make them suitable for engineering applications [3]. The physical model of the arc in the experiments used for the numerical validation is discussed in Section 2, while Section 3 presents the numerical methodology and the comparison between simulation and experimental results.

## 2. The physical model

The experimental reference for the numerical validation campaign was obtained with the test device sketched in Figure 1 of [1]. Our arc simulation tool can take into account nozzle ablation and contact erosion. However, they were excluded from present numerical investigation, since we have verified that they did not influence the arc behavior in the near-stagnation region, where the physical model is to be validated. In fact, the material generated by erosion and ablation of the surfaces exposed to the arc is conveyed downstream by the strong bidirectional axial flow. However, the ablation has an indirect impact on the flow at stagnation as it is responsible for a slight widening of the nozzle throat.

The physical model that describes the behavior of the plasma arc in this experiment has been derived under the assumption that the conditions for the chemical equilibrium (CE) and local thermodynamic equilibrium (LTE) are fulfilled. The governing equations of the arc model used in this work are:

$$\partial_t \rho + \nabla\cdot(\rho \mathbf{U}) = 0 \quad (1)$$

$$\rho D_t \mathbf{U} = -\nabla p + \nabla\cdot\boldsymbol{\tau} - \sigma\nabla\varphi\times(\nabla\times\mathbf{A}) \quad (2)$$

$$\rho D_t h^o = \partial_t p + \nabla\cdot(k\nabla T + \boldsymbol{\tau}\cdot\mathbf{U}) - \nabla\cdot\mathbf{q}_{\mathrm{rad}} + \sigma(\nabla\varphi)^2 \quad (3)$$

$$\nabla\cdot(\sigma\nabla\varphi) = 0 \quad (4)$$

$$\nabla^2\mathbf{A} = \mu_0\sigma\nabla\varphi \quad (5)$$

$$\hat{\mathbf{s}}\cdot\nabla I_\lambda = a_\lambda(I_{b\lambda} - I_\lambda) \quad (6)$$

where $D_t = \partial_t + \mathbf{U}\cdot\nabla$ denotes the material derivative.

Equations (1), (2) and (3) express the conservation of mass, momentum, and energy of the gas, respectively. In these equations, $\rho$ is the mass density, $\mathbf{U}$ is the velocity vector, $p$ is the pressure, $\boldsymbol{\tau}$ is the stress tensor, $h^o$ is the total enthalpy per unit mass, $k$ is the thermal conductivity, $T$ is the temperature, $\mathbf{q}_{\mathrm{rad}}$ is the radiative heat flux, $\varphi$ is the electric potential, $\mathbf{A}$ is the magnetic vector potential, $\sigma$ is the electrical conductivity, and $\mu_0$ is the magnetic constant.

Equations (4) and (5) represent current conservation and Ampere's law in the Coulomb gauge, respectively, under the magnetostatics approximation. The latter is valid when, as in the present case, the derivatives of electromagnetic fields can be neglected. The Lorentz force per unit volume $-\sigma\nabla\varphi\times(\nabla\times\mathbf{A})$ was numerically proven to have a negligible influence on the arc, which was expected as the arc was strongly axially blown and the currents were of the order of 100 A.

The model is completed by the quasi-steady radiative transfer equation (RTE) (6) for a non-scattering medium with absorption coefficient $a_\lambda$. The RTE governs the transport through the gas of the spectral radiative intensity $I_\lambda(\mathbf{r},\hat{\mathbf{s}})$, which is the amount of radiative power in the wavelength interval $[\lambda, \lambda + d\lambda]$ at position $\mathbf{r}$ in the direction $\hat{\mathbf{s}}$ per unit solid angle and unit area normal to $\hat{\mathbf{s}}$ [4]. The term $I_{b\lambda}(\mathbf{r})$ indicates the spectral radiative intensity for a blackbody. The RTE is bidirectionally coupled with the energy equation (3) through the divergence of the radiative heat flux $\mathbf{q}_{rad} = \int_0^\infty\int_{4\pi} I_\lambda\,\hat{\mathbf{s}}\,\mathrm{d}\Omega\,\mathrm{d}\lambda$ and through the absorption coefficient $a_\lambda(\mathbf{r})$, which in local thermodynamic equilibrium is a function only of pressure and temperature of the gas at the position $\mathbf{r}$.

With regard to turbulence, an axially blown arc induces turbulence through the flow in which it passes. The reason is the copresence of a radial gradient of the density across the arc with an axial gradient of the pressure due to the flow expansion. These two gradients result in a strong baroclinic production term of vorticity, $S_\omega = \nabla p \times \nabla\rho^{-1}$, which has its largest magnitude inside the arc mantle where $\nabla\rho$ is maximal, and is equal to zero at the stagnation point where $\nabla p$ is zero [5]. The arc can be then thought of as having a nearly irrotational core surrounded by a turbulent mixing zone. Since in our experiment the flow reaches the stagnation point almost undisturbed and $S_\omega = 0$ at stagnation, the flow can be considered laminar in the near-stagnation region where the arc model validation is performed. The turbulent mixing becomes more and more significant as the flow moves downstream along the two opposite axial directions. However, the throats of both the upper electrode and the nozzle are choked during the tests. Thus, the flow conditions around the stagnation point are unaffected by the downstream turbulent mixing. For these reasons, we have used the laminar flow model in our simulations. In fact, the validity of this approach has been confirmed by comparing the laminar flow quantities around stagnation with those computed with the $k\omega$-SST turbulence model. The differences were quite small if the Kato-Launder correction [6] was applied for suppressing the spurious generation of turbulent kinetic energy in the stagnation region.

The thermodynamics and transport properties of argon have been computed by an in-house Cantera based tool [7] that includes Debye-Hückel non-ideality corrections under the CE and LTE assumptions. These properties, functions of temperature and pressure, were stored as lookup tables to be used within the CFD solver. With the same in-house we have computed the absorption spectrum of argon using the databases [8, 9].

## 3. The numerical model

The simulations were performed with the computational domain depicted in Figure 1 along with the pressure sensors' locations, in the tank ($S_T$) and along the nozzle ($S_1, S_2, S_3$).

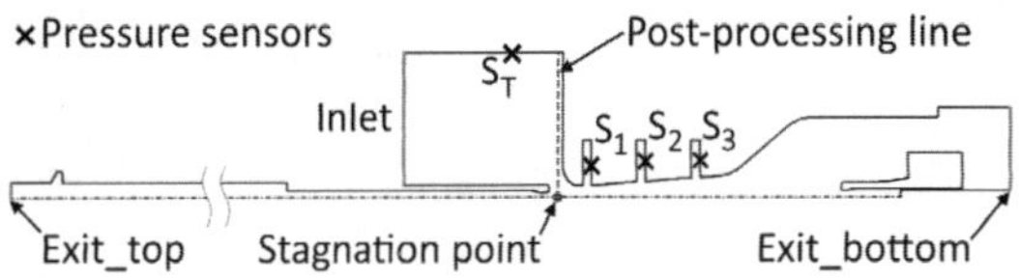


*Figure 1. 2D-axisymmetric computational domain with plug in full-open position.*

This is a two-dimensional axisymmetric domain which is derived with some modifications from the test device geometry used in [1]. In particular, the tank was not modeled in its entirety, in that its top boundary was shifted close to the nozzle and converted from wall into an inlet at which the measured pressure was prescribed. This sped up the simulations without affecting the accuracy, as we have verified by comparing the pressure distribution along the nozzle with that of the case in which the full tank discharge was simulated. At the bottom outlet the ambient pressure was prescribed, while at the top outlet we imposed the time-decaying pressure read from a sensor placed at the end of the hollow contact and immediately upstream from the system of hoses that redirect the flow to the bottom part of the pressurized volume [1]. The pressure was measured at three axial locations along the nozzle.

The physical model described above was numerically implemented in ANSYS Fluent 17.1. This commercial software solves for the flow equations with the finite-volume discretization method. It offers the user the freedom to implement a generic transport equation for the scalar $\phi$, of the following form $\partial_t(\rho\phi) + \nabla\cdot(\rho\mathbf{U}\phi) = \nabla\cdot(\Gamma\nabla\phi) + S_\phi$. The steady version of this transport equation with no convection term can be used for solving equations (4) and (5), by properly selecting the diffusivity coefficient $\Gamma$ and the the source term $S_\phi$. A structured mesh was used everywhere in the computational domain. The spatial resolution and the time stepping were selected in order to ensure a properly converged solution. The grid was refined, and the time steps reduced until a satisfactory grid convergence was achieved. The motion of the plug that draws the arc between the two hollow contacts was simulated as a moving porous region impermeable to the gas and having the same electrical and thermal conductivities as the plug. This numerical strategy allowed for no mesh alterations, preserving its high quality during the contact opening phase. The numerical integration of the RTE (6) in ANSYS Fluent was done with the gray-band discrete ordinates (GBDO) radiation model, due to the fine spectral and spatial resolution it allows. In this model, the RTE is solved for a finite number of discrete solid angles, each associated with a direction vector $\hat{\mathbf{s}}_j$, and for a finite number of bands $[\lambda_i, \lambda_{i+1}]$ in which the absorption spectrum is divided, with the assumption that the absorption coefficient is constant within each band. Under these conditions and assuming a refractive index of 1 for argon, the (6) and the divergence of the radiative heat flux can be written as:

$$\pi\left(\hat{\mathbf{s}}_j\cdot\nabla\mathcal{I}_{i,j} + a_i\mathcal{I}_{i,j}\right) = \sigma_{\mathrm{sb}}T^4 a_i\mathcal{E}_i \tag{7}$$

$$\nabla\cdot\mathbf{q}_{\mathrm{rad}} = \sum_{i=1}^{N_{\mathrm{bands}}}\left[4\sigma_{\mathrm{sb}}T^4 a_i\mathcal{E}_i - a_i\sum_{j=1}^{N_{\mathrm{beams}}}\mathcal{I}_{i,j}\Delta\Omega_j\right] \tag{8}$$

which are the equations of the GBDO model. In (7), $i$ runs from 1 to $N_{\mathrm{bands}}$ and $j$ from 1 to $N_{\mathrm{beams}}$. For the remaining notations, $I_{i,j}$ is the radiative intensity for the i-th band $[\lambda_i, \lambda_{i+1}]$ and in the beam of solid angle $\Delta\Omega_j$ along the $\hat{\mathbf{s}}_j$ direction, $\mathcal{I}_{i,j} = \int_{\lambda_i}^{\lambda_{i+1}} I_{i,j}d\lambda = I_{i,j}(\lambda_{i+1}-\lambda_i)$ is the band-integrated radiative intensity for the j-th beam, $a_i$ is the absorption coefficient in the i-th band, $\sigma_{\mathrm{sb}}$ is the Stefan-Boltzmann constant, $\mathcal{E}_i = f(\lambda_{i+1}T) - f(\lambda_i T)$ is the emissivity weighting factor, i.e. the fraction of the total energy emitted by a black body in the i-th band at the temperature $T$. The integral function $f(\lambda T)$ represents the fraction of the same energy in the wavelength interval from 0 to $\lambda$ and can be found tabulated (e.g. in [4]). In case of rotational symmetry, Fluent discretizes only half of the full angular space, by dividing it in four octants, each of which is in turn divided into $N_\theta \times N_\phi$ solid angles. We have retained the default division $N_\phi = N_\theta = 2$, which yields a total of $16\,N_{\mathrm{bands}}$ equations of type (7) that are to be solved at each cell of the computational domain. The linearity of the equations (7) and (8) with respect to the product $a_i\mathcal{E}_i$ implies that the solution of the radiation problem does not change if two or more bands (e.g. bands $[\lambda_p, \lambda_{p+1}]$ and $[\lambda_q, \lambda_{q+1}]$) in which the absorption coefficient value is the same (e.g. $a_p = a_q$) are grouped together, and if the new grouped band is given the emissivity weighting factor that is the sum of those of the original bands (e.g. $\mathcal{E}_p + \mathcal{E}_q$). ANSYS Fluent provides a macro for assigning directly the emissivity weighting factors of the bands, which makes it straightforward to use the band grouping technique described above. It should be noted that in this case the solution no longer depends on the wavelength limits of the grouped bands, as these are already included in the $\mathcal{E}_i$ jumps. This method was generalized to the full argon spectrum which contains about $10^6$ wavelengths, see Figure 2.

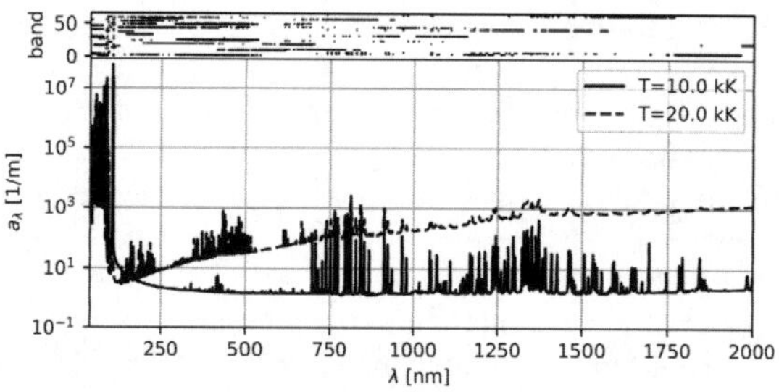


*Figure 2. Argon absorption spectrum for different temperatures at 10 bar. Top graph: attribution of the wavelength range to the different bands in the 64 band model.*

The idea was to group together those wavelengths in which the absorption coefficient had values comprised in a certain small range. However, a partition of a single spectrum in segments with similar $a_\lambda$ is unfortunately not sufficient; as seen in Figure 2, absorption spectra differ strongly depending on temperature. This is due to the different composition of the plasma at different temperatures; at low temperatures atomic lines dominate whereas at higher temperatures ion lines appear. We have implemented an algorithm that, for a given pressure, groups together bands $[\lambda_i, \lambda_{i+1}]$ whose $a_i(T)$ coefficients differ from each other only slightly over the full temperature range. The distance between these $a_i(T)$ is determined based on an appropriate functional norm. The details of this procedure will be illustrated in a separate article. The points in the upper part of Figure 2 show the results of this technique for the compression of the full spectrum into 64 grouped bands. A detailed analysis shows that indeed, spectral lines of the same origin (i.e. same species and spectroscopic term) are typically found in the same clustered band. The validation of the absorption data obtained with this procedure was accomplished by solving the RTE for a rotationally symmetric distribution of temperature $T(x, r, \theta) = T_{\exp}(r)$, where $T_{\exp}(r)$ was the radial distribution measured at stagnation at $p = 12.7$ bar in [1]. In Figure 3, $T_{\exp}(r)$ is depicted along with the comparison between the $\mathbf{q}_{\text{rad}}$ computed by integrating the RTE with a 64-band grouped absorption spectrum and with the full argon spectrum. In the latter case, the RTE was solved at 10 radial positions only.

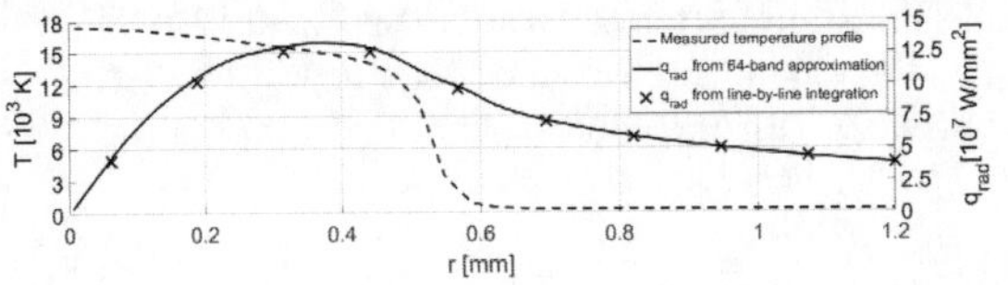


*Figure 3. Radiative heat flux computed with the full spectrum and with its 64-band approximation*

As it can be seen, the reduced spectrum with 64 grouped bands gives an accurate representation of the radiation transport with a maximum relative error on $\mathbf{q}_{\text{rad}}$ of 6% with respect to the full spectrum.

### 3.1. The results of the validation

The 64-band reduced absorption data was used in the numerical reproduction of the spatially resolved measurements of the arc temperature obtained in [1]. That experimental campaign consisted of several arcing operations (shots), each of which resulted in a slight change of the nozzle contour due to arc-induced ablation. The nozzle in the shot that we have selected for the numerical reproduction (shot #41) was not new, thus we increased its throat diameter of 0.5 mm, so that the simulated pressure matched the values measured at $S_1, S_2, S_3$. In this test a slowly decaying quasi-DC current was injected through the system that had a value of 158 A at the instant in which the spatially resolved measurements of the arc were acquired at stagnation. At the same instant the pressure in the tank pressure sensor reading was 11.8 bar, against an initial value of 12.8 bar. In Figure 4 the measured and simulated temperature profiles are represented along the radial post-processing line depicted in Figure 1.

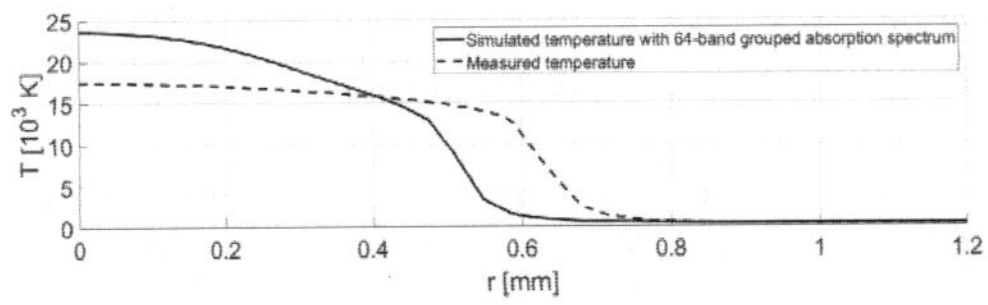


*Figure 4. Simulated vs. measured radial profile of the temperature at stagnation.*

Both profiles refer to the same axial position close to the stagnation point and to the same time. As can be seen, the numerical model underpredicts by 17% the measured arc radius. The predicted arc temperature at the axis is instead overestimated by 25%.

## 4. Conclusions

In this work, we have presented the results of the validation study conducted for a CFD-based arc model and aimed at reproducing the spatially resolved temperature measurements of a stabilized axially-blown arc. The test device was conceived so that ablation, erosion and turbulence could be neglected in the observation region, around its stagnation point, thus allowing a more precise validation of the radiation model, which was the main focus of this article. The full absorption spectrum has been computed for argon and then compressed to 64 clustered bands that were proven to approximate it very well. This reduced absorption spectrum was then used as an input for the gray-band discrete ordinates model of ANSYS Fluent to perform arc simulations in an accurate and computationally affordable way. Despite the high precision with which the radiation transport has been resolved numerically, the simulated arc temperature distribution exhibited a non-satisfactory agreement with the measured one. An additional research effort is thus needed on both the experimental and numerical side to further refine the diagnostics technique and to assess the effect of the implementation of non-equilibrium phenomena on the numerical representation of the arc behavior.

## References

[1] J. Carstensen et al. Measuring density profiles of electrons and heavy particles in a stable axially blown arc. *Phys. Rev. Appl.*, 8(2):024002, 2017. doi:10.1103/PhysRevApplied.8.024002.

[2] A. Gleizes et al. Topical review: Thermal plasma modelling. *J. Phys. D: Appl. Phys.*, 38:R153–R183, 2005. doi:10.1088/0022-3727/38/9/R01.

[3] L. Fagiano and R. Gati. On the order reduction of the radiative heat transfer model for the simulation of plasma arcs in switchgear devices. *JQSRT*, 169:58–78, 2016. doi:10.1016/j.jqsrt.2015.10.002.

[4] M. F. Modest. *Radiative Heat Transfer (Third Edition)*. Academic Press, 2013.

[5] L. Niemeyer et al. On the development of turbulence by the interaction of gas flows with plasmas. In *AIAA 9th Electr. Propul. Conf.*, pages 1–9, Bethesda, April 1972.

[6] M. Kato and B. E. Launder. The modelling of turbulent flow around stationary and vibrating square cylinders. In *Proc. 9th Symp. on Turbulent Shear Flows*, pages 10.4.1–10.4.6, Kyoto, August 1993.

[7] C. Doiron and K. Hencken. *APS Meeting Abstracts*, 3:3001, 2013.

[8] Kurucz database. http://kurucz.harvard.edu/.

[9] TopBase database. http://cdsweb.u-strasbg.fr/topbase/topbase.html.

doi:10.14311/ppt.2019.1.82
*Plasma Physics and Technology* **6**(1):82–86, 2019 

# ANALYSIS OF $C_2$ SWAN BANDS IN ABLATION-DOMINATED ARCS IN $CO_2$ ATMOSPHERE

R. Methling[a,*], St. Franke[a], N. Götte[b], S. Wetzeler[b], D. Uhrlandt[a]

[a] *Leibniz Institute for Plasma Science and Technology (INP), Felix-Hausdorff-Str. 2, 17489 Greifswald, Germany*
[b] *Institute for High Voltage Technology, RWTH Aachen University, Schinkelstrasse 2, 52056 Aachen, Germany*
[*] methling@inp-greifswald.de

**Abstract.** A model circuit breaker in a high-pressure chamber filled with $CO_2$ atmosphere is used to operate a wall-stabilized arc of several kilo-amperes between tungsten-copper electrodes surrounded by polytetrafluoroethylene nozzles. Optical emission spectroscopy (OES) is carried out via quartz plates inserted into the nozzles using a combination of an imaging spectrometer either with a high-speed video camera or with an ICCD camera. Depending on the nozzle geometry and the current, continuum from $C_2$ Swan bands was detected as absorption as well as emission pattern. After current zero, optical absorption spectroscopy (OAS) using a xenon flashlamp as broadband background radiator was applied. An absorption around 493 nm was detected and attributed to CuF molecules. The study proofs the existence of $C_2$ in the active phase and the formation of CuF near to current zero.

**Keywords:** ablation, switching arc, spectroscopy, PTFE, $CO_2$.

## 1. Introduction

Self-blast circuit breakers are based on the ablation of material from a nozzle producing a pressure build-up in a heating volume that is used for arc quenching around current zero (CZ). The CZ crossing and the period immediately after current interruption are of high importance for the interruption performance. In these time intervals several physical effects occur, such as flow reversal in the heating channel, transition from an ablation-controlled to an axially blown arc, the extinction of the arc and a continued evaporation of nozzle material after current zero due to the preceding thermal stress. The deeper understanding of the transient behavior and effects on the dielectric recovery are key issues for further development of high voltage switchgear, either based on sulfur hexa-fluoride ($SF_6$) or alternative gases, e.g. carbon dioxide or mixtures containing carbon dioxide.

The determination of dielectric properties, i.e. the plasma composition, pressure and temperature, as close as possible to current zero is mandatory for an improvement of interruption capability. Therefore, optical access to the arc plasma is needed but not available in commercial circuit breakers. Simplifications have to be allowed to obtain visual access at comparable plasma conditions. Model chambers are applied either with movable pin and tulip electrodes [1], with fixed electrodes ignited by explosion of thin wires [2], or with non-rotationally symmetric nozzles [3]. Recently, we reported on experiments with two pin electrodes of fixed distance placed in a single, long PTFE nozzle without a chamber [4]. The influence of ignition wire and surrounding gas (ambient air) were investigated as well as the PTFE ablation at low current phases, i.e. during the current increase, when the discharge is initiated, and during current decrease, when approaching CZ. Determination of plasma temperature profiles was limited to 0.4 ms before CZ.

In this work, a switching arc of a model circuit breaker similar to [5, 6] with optical access in a $CO_2$ atmosphere is investigated. The model circuit breaker consists of two nozzles surrounding the electrodes and forming a heating channel. Whereas in [6] the plasma in the heating channel was analyzed during the high-current phase by means of optical emission spectroscopy (OES), the observation position was shifted towards the electrodes by insertion of quartz windows into the nozzle wall in this study. Recently, the temporal evolution of plasma properties during the last millisecond before current zero was investigated for this geometry [7]. It was shown by means of OES that the gas flow changed from a blow-out of the nozzle, due to wall ablation into the heat volume (characterized by atomic F I and ionic C II lines), to a flow-in into the nozzle from the heating volume (characterized by C II and O I lines) at about 2 ms before CZ. While the C II extinguished about 100 µs before CZ, the oxygen triplet at 777 nm could be used for determination of temperature profiles until 10 µs before CZ. Arc constriction from diameters of several mm to less than 0.5 mm and a temperature decrease in the center of the arc from about 20 000 K at 300 µs to about 11 000 K at 10 µs before CZ were measured.

Although the OES of atomic and ionic lines has been established for plasma analysis, there are severe limitations. Usually, these lines have their maximum in the arc center and the temperature profiles are typically limited to minimum values of 8 000 – 10 000 K, due low emission intensities at lower currents and the cold gas flow close to CZ. Thus, spatial temperature profiles do not reach the close vicinity of the wall

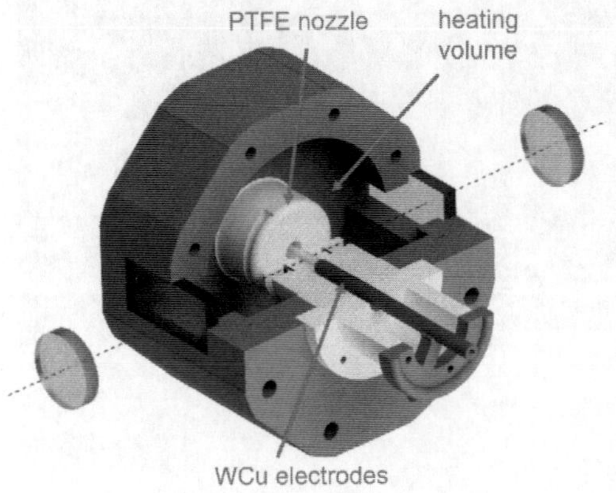


*Figure 1. Model circuit breaker geometry. The arc was operated between W-Cu electrodes surrounded by PTFE nozzles. Observation along the arc diameter was provided by vertical slits in the nozzles. Note that windows at both sides of the model chamber and the outer vessel allowed emission and absorption spectroscopy.*

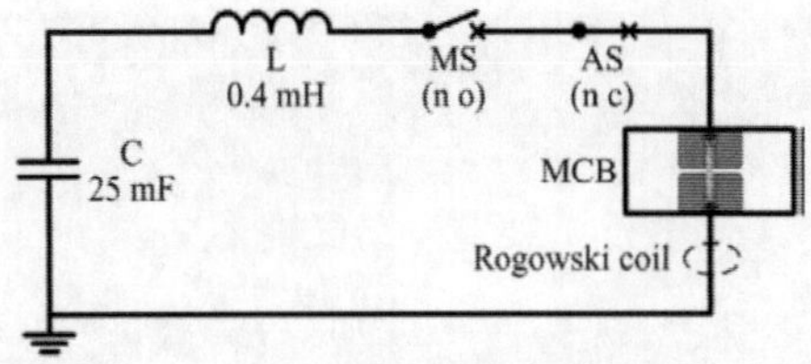


*Figure 2. Electric setup for generation of sine-like 50 Hz discharges, initiated and interrupted after one half-wave by the switches MS and AS, respectively.*

region that is of high interest due to importance of the nozzle ablation [8, 9].

However, at reduced temperatures the majority of atoms is in the ground state and could be analyzed by optical absorption spectroscopy (OAS) of resonant lines. Nevertheless, resonant lines of the relevant species are in the UV region (C, F, Cu) that is difficult to investigate under switching-relevant conditions and demands a lot of technical effort. Many other lines are very weak (O I at 630.03 nm). An alternative to the investigation of atomic and ionic lines is the spectroscopic investigation of molecules. Most molecules dissociate above critical temperatures of less than 10 000 K. They may provide insight into the region close to the wall at lower temperature.

## 2. Experimental Setup

A model circuit breaker in a high-pressure vessel was prepared. A sketch of the setup is shown in Figure 1. Two fixed W-Cu electrodes of 10 mm diameter were placed horizontally with a distance of 40 mm in a metal chamber. Both electrodes were surrounded by nozzles of PTFE doped with <0.5 wt% molybdenum disulfide ($MoS_2$). The distance between these two nozzles was about 4 mm forming a heating channel in the middle between both electrodes. The inner diameter of the nozzle was about 16 mm at the electrode side end of the nozzle and 12 mm at the nozzle exhaust. The transition from wide to narrow diameter was placed close to the electrode tip. In former experiments, the observation was carried out in the heating channel between the two nozzles [6]. The turbulent gas flow in the direction of observation induces several complication concerning the radial symmetry that is needed for plasma temperature determination. Therefore, a technique recently introduced for a single long PTFE nozzle [4] was adapted: Two vertical slits of 2 mm width, allowing access to the full nozzle diameter, were mortised directly into the nozzles. They were positioned about 9 mm away from both, the electrode tip and the nozzle exhaust. To seal the nozzle and to reduce the disturbing influence of the slit, 2 mm-thick quartz plates were applied. In combination with windows of both, model chamber and vessel, a free view through the nozzle has been enabled for observation from opposite directions as well as absorption measurements.

The electrical setup is shown in Figure 2. The LC circuit was discharged via the switch MS. A sine-like current waveform of about 50 Hz frequency and 5 kA peak current was applied for all discharges. After one half-wave the current was switched off by AS. Thin copper wires were used for initiation of the arc discharge. The pressure was measured in the heating chamber by a Kistler sensor of type 603A. A comparison with CFD simulated pressures allowed to estimate the pressure values inside the nozzles.

The spectroscopic setup consisted of an imaging spectrograph with 0.5 m focal length (Roper Acton SpectraPro SP2500i) equipped with either a high-speed video camera (Y4 series of Integrated Design Tools, 10bits monochrome) to record a series of images per shot or an intensified CCD camera (Princeton Instruments PI-MAX) to obtain higher sensitivity. The observation slit of the nozzle was imaged on the entrance slit of the spectrograph by a focusing mirror to observe the cross section of the arc in the direction perpendicular to the arc axis. Deflecting mirrors were used to guide the radiation. Since the sensitivity of non-intensified cameras is limited, the exposure times have to be relatively high.

For absorption measurements around CZ an intense xenon flashlamp (nearly rectangular pulse with 1 MW electric power input and 1 ms duration) was applied as an external broadband background radiator with a spectrum comparable to a thermal radiator with a temperature of about 14 000 K. A second high-speed video camera (Y6 series of Integrated Design Tools) was applied for the observation of the general behavior of the discharge.

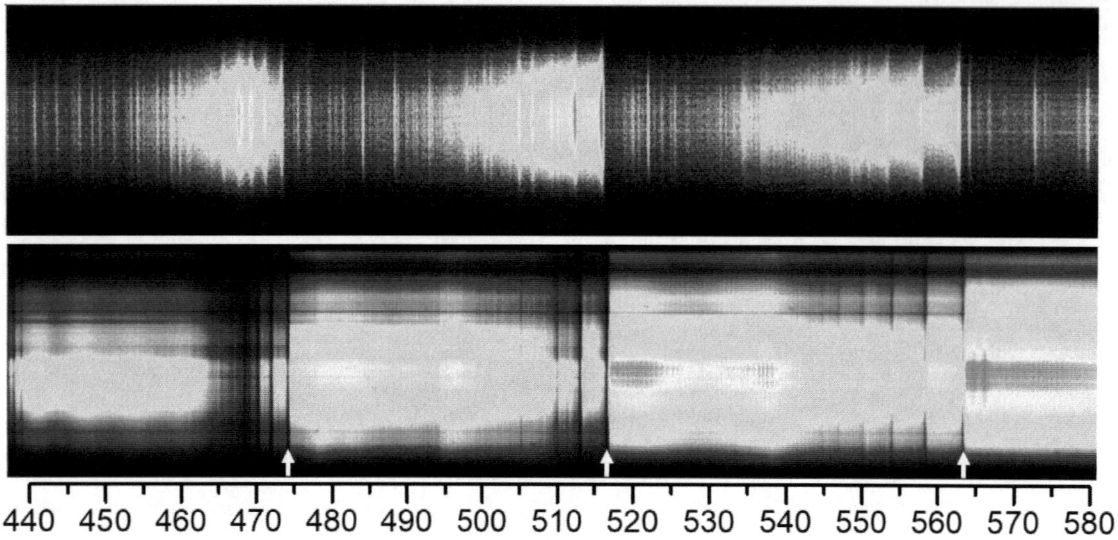


*Figure 3. Spectrum during high-current phase of the discharge dominated by molecular radiation of $C_2$ Swan bands. Depending on plasma conditions (arc current, nozzle geometry) the Swan bands could be observed as emission (top) or absorption pattern (bottom). The band head are labeled by arrows.*

## 3. Results and Discussion

### 3.1. Optical Emission Spectroscopy

Series of two-dimensional spectra containing spatial as well as spectral information were obtained by the video spectroscopy. The low-current phases after ignition and near CZ are dominated by line emission (Cu, C, O, and F) with highest intensities in the central part of the axis. Those lines could be utilized for an analysis of temperature profiles along most of the arc radius. In cases with significant nozzle ablation, usually fluorine atomic or carbon ionic lines were applied [1, 2, 4, 6]. In the phase of low currents, i.e. without sufficient nozzle ablation, oxygen atomic lines could be applied for the determination of temperature profiles, reaching down to few microseconds before current zero [7]. With increasing current of just 2.5 kA (1.5 ms), also continuum emission was observed as a pattern near to the nozzle wall [10]. It could be attributed to an emission of diatomic carbon molecules, i.e. the Swan band system arising from transitions between the $d^3\,\Pi_g$ and the $a^3\,\Pi_u$ electronic states of the $C_2$ molecules [10].

The Swan bands were found to occur under quite different conditions. The most pronounced emission spectra were obtained in the experiments with a single long PTFE nozzle [4]. In the example shown in the upper part of Figure 3, the complete spectrum at all radial positions (vertical axis) is dominated by the Swan bands (observed at currents above 10 kA). With the current setup in the model circuit breaker, however, the $C_2$ continuum was visible in emission only close to the nozzle wall. More typically, it appeared as an absorption pattern at moderate currents >4 kA, as shown in the lower part of Figure 3, taken 6.1 ms after ignition at (4.7 kA), i.e. shortly after the current maximum. Here, the hot plasma in the core served as an internal background radiator and the absorption close to the wall becomes visible. The band heads at 473.7 nm, 516.5 nm, and 563.6 nm are indicated by arrows (another one is at 438.2 nm). The spectra also contain several C II lines from carbon ions, e.g. at 564.06 nm, 564.81 nm, and 566.26 nm, mainly emitting in the center of the arc with higher temperatures. For the above described arcs with strong flow of ablated material towards the electrodes, no copper lines were visible. The F I lines were not in the chosen wavelength range.

The Swan bands have been widely investigated by laser-induced optical breakdown spectroscopy (LIBS) of graphite targets [11, 12]. For laser irradiation it was reported that the dominant mechanism for the production of $C_2$ molecules at low power is the collision of electrons with larger molecules like $C_3$, $C_4$ followed by photodefragmentation whereby one of the emitted products is an exited $C_2$ molecule. At higher laser irradiance, the Swan band emission is mainly caused by excitation resulting from electron-ion and ion-ion recombination [11]. The shape of continuum is influenced by pressure and temperature in the range of 5 000 – 8 000 K. Thus, an estimation of the vibrational temperature could be obtained by comparison of measured and simulated spectra [13, 14].

### 3.2. Optical Absorption Spectroscopy around Current Zero

Around current zero, neither emission nor absorption of Swan bands was observed, although weak O I line radiation was observed in the last 10 µs. As described above, most of the ablated wall material is blown towards the electrodes by the reversed gas flow from the heating channel. However, the spectral absorption was carried out systematically using a xenon flashlamp as an external wide-band background radiator (Figure) 4). Scanning through the wavelength ranges from about 400 nm to 800 nm, no hint on Swan bands or any absorbing lines could be detected around current zero, even with the application of the intensified camera with higher sensitivity and dynamic range. The only feature that was found was a multi-peak structure with a pronounced maximum around 493 nm and a weaker above 500 nm. An example for an absorption spectrum with increased spectral resolution (using a grating with 1800 instead of 150 lines per mm and an exposure time of 98 µs) is shown in the upper part of Figure 4 recorded 0.5 ms after current zero. The edges of the nozzle are indicated by a sharp transition from the bright stripe caused by illumination with the xenon lamp and the dark areas. The horizontal stripes are due to deposition on the quartz glass sealing the slits, e.g. by particles, and do not change from one video frame to the next. The occurrence of this structure was very reproducible and faded out within several hundreds of microseconds after current zero. Line absorption (Cu, O, C, F, H, or Xe) could be easily excluded. The only known absorber in that wavelength was the molecule of CuF.

Since no absorption spectra were available for CuF, a comparison was carried out with emission spectra of photofragments in a gas phase photochemical fragmentation process excited by 308 nm-laser radiation [15]. In the lower part of Figure 4 the emission spectrum from Cheon et al. [15] was added as an overlay (black curves) to the calculated absorption spectrum (red curve) for comparison. Taking into account the different experimental conditions and methods, a compelling agreement was found.

The occurrence of a considerable amount of CuF molecules was not expected for several reasons, especially regarding that no other molecules were observed around CZ. A possible explanation is as follows: Around CZ it is expected that convective fluxes are significantly reduced due to equalization of pressures. This allows diffusive expansion of copper from the hot electrode along the nozzle towards the slit position. At the same time there is still some release of fluorine from the PTFE nozzle wall. The chemical reaction of the fluorine and copper atoms forming CuF molecules could happen at the hot electrode surface followed by evaporation of molecules or in the gas phase with copper atoms evaporated from the electrode.

## 4. Summary and Conclusion

A wall-stabilized arc of several kilo-amperes between tungsten-copper electrodes surrounded by polytetrafluoroethylene nozzles was operated in a model circuit breaker in $CO_2$ atmosphere. Optical emission spectroscopy as well as absorption spectroscopy were carried out via quartz plates inserted into the nozzles.

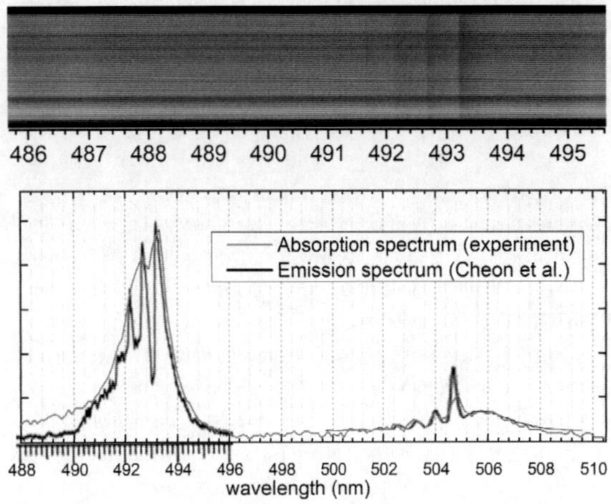


*Figure 4. Absorption spectrum around current zero and comparison with emission spectra of CuF molecules from [15].*

The results of spectroscopic analysis can be summarized as follows:

Line as well as continuum emission was observed. At the beginning of the discharge, i.e. when the current is rather low and no significant ablation of the PTFE nozzle takes place, only emission from ignition wire (Cu I) followed by the filling gas (C II, O I from $CO_2$) and the electrodes (Cu I from W-Cu) can be observed.

By increasing current with time, the wall ablation changes the behavior of the arc. The filling gas is blown out due to the pressure increase resulting from the ablated and dissociated PTFE wall material. Thus, line emission of fluorine atoms F I accompanies that of the carbon ions C II whereas O I and Cu I can not be observed any more. This state remains most of the discharge time, i.e. just down close to current zero. Nonetheless, in cases with strong wall ablation and pressure build-up, e.g. with long and narrow PTFE nozzles and rather high arc currents, Swan bands from $C_2$ molecules can be observed. Here, for the spectra three forms of appearance can be distinguished:

1. The Swan bands are observed mainly close to the nozzle wall in emission [10].

2. With stronger ablation of wall material, the Swan bands are observed as absorption pattern all over the nozzle cross section (lower part of Figure 3). The occurrence of rather intense C II lines at 564.1, 564.8, and 566.2 nm hints at relatively high temperatures at least in the center of the arc.

3. With even higher impact of the $C_2$ molecules, the Swan bands appear as pure emission pattern (upper part of Figure 3). Here, the C II ionic lines are less prominent (or not observed any more) hinting on lower arc temperatures than in the case above.

With decreasing current and wall ablation, the pressure inside the nozzle decreases, too. When it falls below the pressure inside the heating volume, the direction of the gas flow turns back, now flooding the nozzle with cold gas from the heating volume that is dominated by $CO_2$. Consequently, the spectra of $C_2$ and ablated wall material vanishes, and species from $CO_2$ appear (again, as in the starting phase). While the origin of C II lines remains uncertain, the flow reversal can be easily followed by the relation of line emission of F I to O I. Temperature profiles of the arc can be deduced down to few µs before current zero from the O I line emission.

Around current zero, optical absorption spectroscopy using a xenon flashlamp as an external broadband background radiator was carried out in addition to the OES. In the wavelength range under investigation (400-800 nm), only one absorption pattern around 500 nm was found, clearly belonging neither to any Swan band nor to lines of the expected elements C, O, F, and Cu. The absorption with a stronger maximum around 493 nm and a weaker around 506 nm could be attributed to CuF molecules by comparison with known CuF emission spectra. The occurrence of CuF absorption at the observation slit position around CZ is a little surprising since after the reversal of gas flow the nozzle is filled mainly by $CO_2$ and the electrodes as possible source of copper is some mm away from the observation slit in the nozzle. However, there must be a chemical reaction of fluorine atoms with copper from the electrodes either at the electrode surface that is hot and followed by evaporation of molecules, or directly in the gas phase with copper atoms evaporated from the electrode.

The study proofs the existence of $C_2$ in the active phase and the formation of CuF close to current zero. As a next steps, the absorption experiments should be repeated without Cu-containing electrodes. Furthermore, a quantification of the continuum analysis would be very interesting since it might reveal the less-known field of processes near to the nozzle wall as well as after current zero.

## Acknowledgements

The work was supported by grant UH 106/13-1 of the Deutsche Forschungsgemeinschaft.

## References

[1] R. Methling, S. Franke, D. Uhrlandt, S. Gorchakov, F. Reichert, and A. Petchanka. Spectroscopic study of arc temperature profiles of a switching-off process in a model chamber. *Plasma Phys. Technol.*, 2(2):163–166, 2015.

[2] R. Kozakov, M. Kettlitz, K.-D. Weltmann, A. Steffens, and C. M. Franck. Temperature profiles of an ablation controlled arc in PTFE: I. spectroscopic measurements. *J. Phys. D: Appl. Phys.*, 40(8):2499–2506, 2007. doi:10.1088/0022-3727/40/8/013.

[3] D. Eichhoff, P. Nikolic, and A. Schnettler. Basic study on the dynamic behaviour of asymmetrically blown arcs in non–rotationally symmetric nozzle arrangements. In *Proc. XXth Symposium on Physics of Switching Arc*, pages 126–129. Brno University of Technology, 2013.

[4] R. Methling, A. Khakpour, S. Wetzeler, and D. Uhrlandt. Investigation of an ablation–dominated arc in a model chamber by optical emission spectroscopy. *Plasma Phys. Technol.*, 4(2):153–156, 2017. doi:10.14311/ppt.2017.2.153.

[5] P. Nikolic, M. Seeger, and A. Schnettler. Investigations on the dielectric recovery of $SF_6$ and air in insulating nozzles. In *Proc. XXth Symposium on Physics of Switching Arc*, pages 265–268. Brno University of Technology, 2013.

[6] D. Eichhoff, A. Kurz, R. Kozakov, G. Gött, D. Uhrlandt, and A. Schnettler. Study of an ablation–dominated arc in a model circuit chamber. *J. Phys. D: Appl. Phys.*, 45:305204, 2012. doi:10.1088/0022-3727/45/30/305204.

[7] K.-D. Weltmann, R. Methling, N. Götte, S. Wetzeler, and D. Uhrlandt. Optical emission spectroscopy of ablation–dominated arcs during high–current phase and around current zero. In *Proc. XXIIth Int. Conf. on Gas Discharges and Their Applications*, volume 1, pages 83–86. Serbian Academy of Sciences and Arts, 2018.

[8] M. Seeger, J. Tepper, T. Christen, and J. Abrahamson. Experimental study on ptfe ablation in high voltage circuit-breakers. *J. Phys. D: Appl. Phys.*, 39(23):5016–5024, 2006. doi:10.1088/0022-3727/39/23/018.

[9] M. Seeger, L. Niemeyer, T. Christen, M. Schwinne, and R. Dommerque. An integral arc model for ablation controlled arcs based on CFD simulations. *J. Phys. D: Appl. Phys.*, 39(10):2180–2191, 2006. doi:10.1088/0022-3727/39/10/029.

[10] R. Methling, N. Götte, S. Wetzeler, D. Uhrlandt, and K.-D. Weltmann. Investigation of $C_2$ swan bands in optical emission and absorption spectroscopy of ablation–dominated arcs. In *Proc. XXIIth Int. Conf. on Gas Discharges and Their Applications*, volume 1, pages 227–230. Serbian Academy of Sciences and Arts, 2018.

[11] S. S. Harilal, R. C. Issac, C. V. Bindhu, V. P. N. Nampoori, and C. P. G. Vallabhan. Optical emission studies of $C_2$ species in laser–produced plasma from carbon. *J. Phys. D: Appl. Phys.*, 30:1703–1709, 1997. doi:10.1088/0022-3727/30/12/003.

[12] J. J. Camacho, L. Diaz, M. Santos, D. Reyman, and J. M. L. Poyato. Optical emission spectroscopic study of plasma plumes generated by IR $CO_2$ pulsed laser on carbon targets. *J. Phys. D: Appl. Phys.*, 41:105201, 2008. doi:10.1088/0022-3727/41/10/105201.

[13] H. S. Park, S. H. Nam, and S. M. Park. Optical emission studies of a plume produced by laser ablation of a graphite target in a nitrogen atmosphere. *Bull. Korean Chem. Soc.*, 25(5):620–626, 2004. doi:10.5012/bkcs.2004.25.5.620.

[14] C. G. Parigger, J. O. Hornkohl, and L. Nemes. Time-resolved spectroscopy diagnostic of laser-induced optical breakdown. *Int. J. Spectroscopy*, 2010:593820, 2010. doi:10.1155/2010/593820.

[15] J. Cheon, H.-K. Kang, and J. I. Zink. Spectroscopic identification of gas phase photofragments from coordination compound chemical vapor deposition precursors. *Coordination Chem. Rev.*, 200–202:1009–1032, 2000. doi:10.1016/S0010-8545(00)00325-8.

doi:10.14311/ppt.2019.1.87
*Plasma Physics and Technology* **6**(1):87–90, 2019


# OPTICAL EMISSION SPECTROSCOPY OF MAGNETHRON DISCHARGE Ar/Cu PLASMA

A. Murmantsev[a,*], A. Veklich[a], V. Boretskij[a], A. Shapovalov[b], A. Kalenyuk[c]

[a] *Taras Shevchenko National University of Kyiv, 64/13, Volodymyrska str., Kyiv, Ukraine*
[b] *V. N. Bakul Institute for Superhard Materials, National Academy of Sciences of Ukraine, Avtozavodska Str., 2, Kyiv 04074, Ukraine*
[c] *G. V. Kurdyumov Institute for Metal Physics, National Academy of Sciences of Ukraine, Academician Vernadsky Blvd., 36, Kyiv 03142, Ukraine*
[*] murmantsev.aleksandr@gmail.com

**Abstract.** Plasma parameters (excitation temperature and electron density) of pulsing magnetron discharge is studied by optical emission spectroscopy. Such discharges are usually used as effective sources in sputtering or deposition processes. Vapor admixtures in argon plasma define mainly the temperature and electron density in such discharges. This is the feature, which is typically takes place in plasma of discharge between contacts/electrodes in switching devices of electric technology circuits.

**Keywords:** magnetron, argon plasma, spectroscopy, Boltzmann plots.

## 1. Introduction

Magnetron sputtering is a plasma technology in which atoms of inert gas (usually argon) are ionized and accelerated as a result of the potential difference between a negatively biased (with negative potential) cathode and anode. The interaction of accelerated ions with the cathode surface causes sputtering of atoms which condense on the substrate and form a film [1, 2]. The condensation and growth of films often occur in conditions far from the thermodynamic equilibrium, due to the presence of kinetic constraints [2, 3]. Thus, the real control over the thermodynamic and kinetic factors of the deposition process provides an opportunity to control the growth of the films and allows to change the structural, optical, electrical and mechanical properties of the films [4]. One way of controlling the growth of a film is to heat the substrate during its deposition [2]. The deposition temperature affects the energy of the adatoms on the substrate surface, from which the condensate layer is formed [2]. This energy is critical for the activation of surface and mass diffusion [5, 6] and allows you to control the morphology of the deposited film. Another source of energy is the plasma particles that transfer energy and impulse to adatoms [5, 6] of the deposited film. At magnetron sputtering, the degree of ionization of plasma particles is relatively low [7], in addition, the quantity of ionization of sputtered cathode atoms is usually less than 1 % [7]. As a result, upon the magnetron deposition of films by direct current the plasma particles have relatively low energies.

In the mid-1990s Mozgrin et al. [8] and Bugaev et al. [9] demonstrated that the use of an ordinary magnetron voltage source in pulsed mode (pulse duration from 1 µs to 1 s and frequencies below 1 kHz) leads to an increase in pulse currents by two orders of magnitude as compared to the average currents in a conventional magnetron working at DC (DC magnetron) [8, 9]. The use of these high-pulse currents led, in turn, to the formation of a plasma with electron density of about $10^{18}\,\mathrm{m}^{-3}$ [8, 9], which significantly exceeds for DC magnetrons (the density of $10^{14} - 10^{16}\,\mathrm{m}^{-3}$) [10].

This study is focused on the discharge plasma parameters of magnetron voltage source in pulsed mode for material sputtering. Optical emission spectroscopy (OES) are used to obtain excitation temperature of Ar and Cu atoms, as well as, electron density in plasma. Such parameters at the next step of investigation should make it possible to optimise the deposition processes.

## 2. Experimental setup

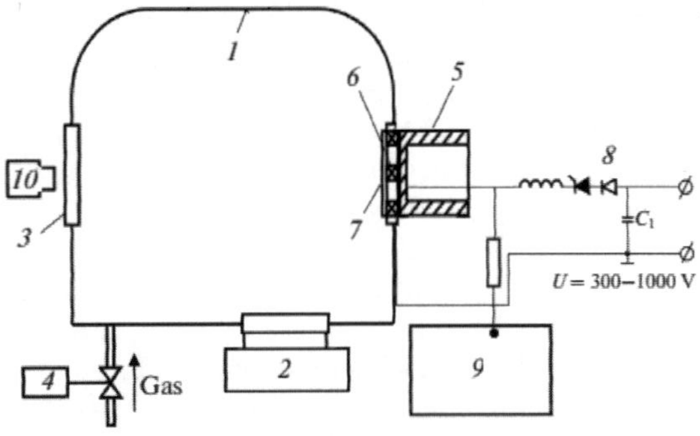


*Figure 1. Scheme of the experimental setup.*

Investigation are carried out using experimental setup shown in Fig. 1. Discharge was ignited in vacuum chamber (1). Initial vacuum (residual gas pressure of

0.4 Pa) is created by vacuum pump (2). Discharge parameters and ambient atmosphere were controlled by pulsed power source (8), magnetron (5), ring magnet (6) and gas flow meters (4) respectively. The sputtered target (7) is placed at the vacuum chamber to realise plasma treatment. The discharge power is controlled by two-channels oscilloscope (9). Plasma optical emission is registered through diagnostic windows (3) by Solar LS SDH-IV spectrometer (10) with a 4-position manually switchable diffraction gratings turret. The spectroscopy measurements are carried out at magnetron pulse voltage 550 V with pulse duration 100 µs and are performed in wide spectral range 444-908 nm [11]. Spectral sensitivity of this device is determined by tungsten band-lamp calibration and is taken into account in spectral data treatment (see Fig. 2).

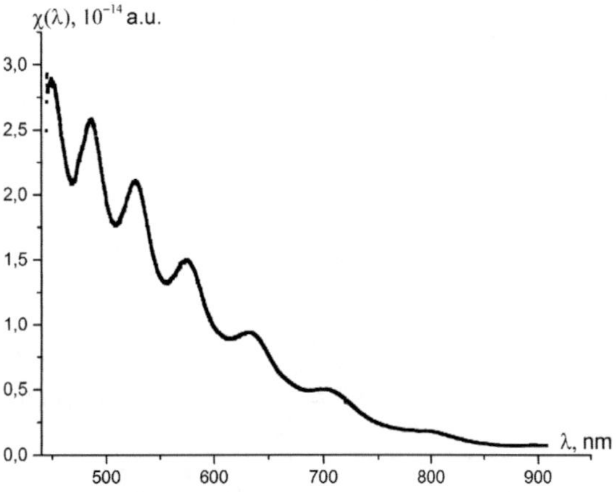


*Figure 2. Spectral sensitivity of spectrometer.*

## 3. Results and Discussions

It must be noted that spectral lines of different elements are observed during the spectroscopy measurements at the stage of vacuum chamber pumping and at the stage of invariable argon atmosphere pressure of 4 Pa. During the pumping process, just Cu I and $H_\alpha$ lines are mainly observed (see Fig. 3). The presence in spectra of such elements is due to the fact that they are presented in chamber volume as desorbed material from construction elements. At the stage of the invariable argon atmosphere Cu I are disappeared and Ar I with $H_\alpha$ spectral lines can be clearly registered (see Fig. 4).

Spectral lines profiles are fitted by Gaussian or Voigt function depending on the dominated broadening mechanism. The approximated experimental profiles of Cu I and Ar I lines are shown in Fig. 5 and 6. As one can see, there is the clearly observed additional noise signal of sinusoidal form in the spectrum. This signal must be withdrawn from consideration in measurement of spectral line intensity. For spectrum with Ar I lines this noise is not taken into account, since its effect was specially minimized and it can be neglected.

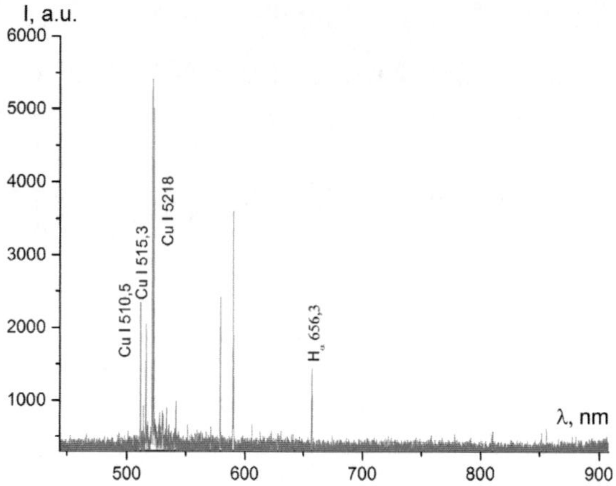


*Figure 3. The spectrum with marked Cu I and $H_\alpha$ spectral lines at the stage of vacuum chamber pumping.*

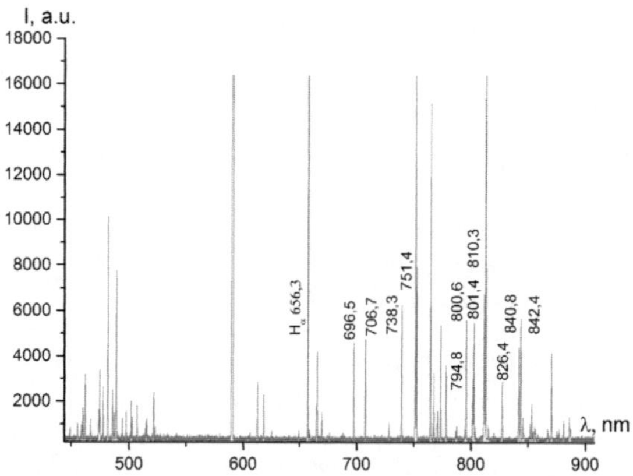


*Figure 4. The spectrum with marked Ar I and $H_\alpha$ spectral lines at the stage of invariable argon atmosphere.*

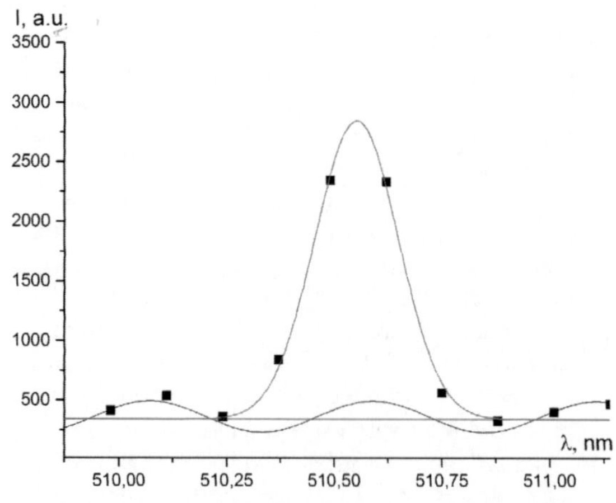


*Figure 5. The approximation of Cu I 510.5 nm spectral line profile by Gaussian.*

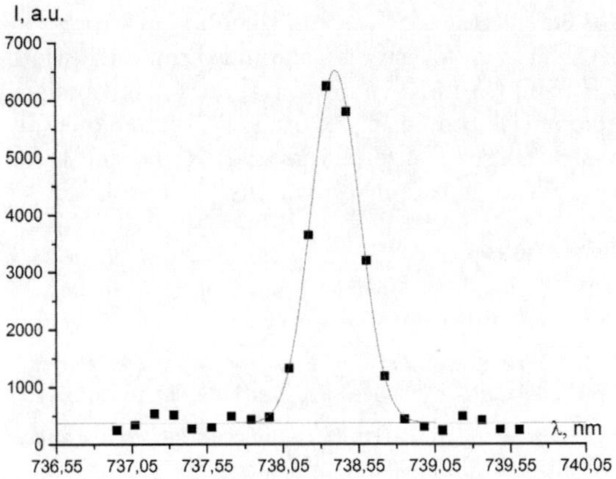


*Figure 6. The approximation of Ar I 738.3 nm spectral line profile by Gaussian.*

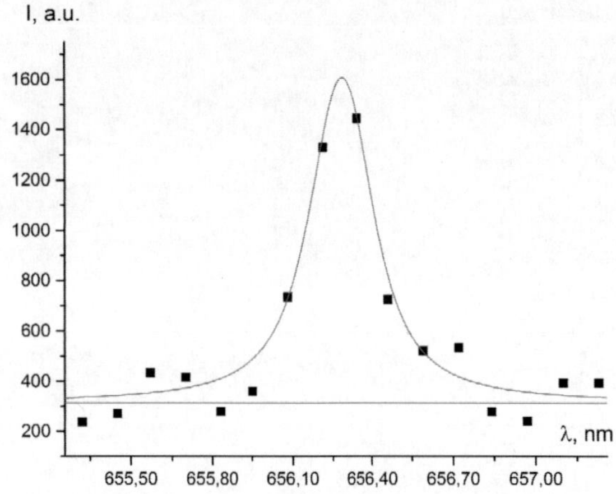


*Figure 8. The approximation of $H_\alpha$ 656.3 nm spectral line profile for electron density measurements.*

The electron density in plasma is obtained from half-width of Cu I 515.3 nm spectral line broadened by dominated mechanism of quadratic Stark effect [12]. The Lorentz function is typical approximation for this line, but the Voigt function is used in this study. The matter is that the spectrometer instrument function must be taken into account as Gaussian component of Voigt approximation profile (see Fig. 7). Just the Lorentz component of this profile is used to obtain the electron density.

Also for comparison, the electron density was calculated from the half-width of the $H_\alpha$ spectral line [13], which is attended in spectrum (see Fig. 3). The approximation of this line is shown in Fig. 8.

Electron density, obtained from half-width of Cu I 515.3 nm spectral line is equal $4.1 \times 10^{16}\,\mathrm{cm}^{-3}$, from $H_\alpha$ – $1.0 \times 10^{16}\,\mathrm{cm}^{-3}$.

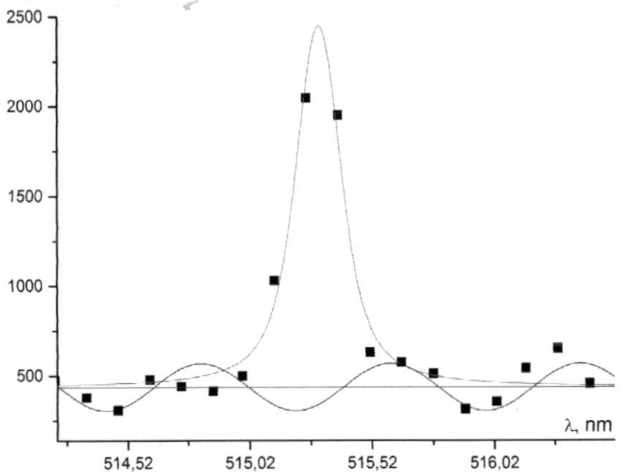


*Figure 7. The approximation of Cu I 515.3 nm spectral line profile for electron density measurements.*

Plasma excitation temperature is determined by the Boltzmann plot technique. In Fig. 9 and 10 typical Boltzmann plots for Cu I 510.5, 515.3 and 521.8 nm and Ar I 696.5, 706.7, 738.3, 751.4, 794.8, 800.6, 801.4, 810.3, 826.4, 840.8 and 842.4 nm are shown. Temperatures are equal $8600 \pm 1200$ K, obtained by Cu I spectral lines and $5400 \pm 1600$ K, obtained by Ar I spectral lines, accordingly.

As one can conclude, the determined temperatures are differ considerably. But it must be taken into account few points. Firstly, temperature obtained by Ar I spectral lines is determined with significant error. It can be explained that the registered argon lines have very low difference of the upper energy levels (about 0.3 eV). This difference should be as much as possible (at least 1 eV) for optimal application of this technique. Secondly, Cu I and Ar I spectral lines are registered at the different stages of experimental study as it is mentioned above. Thirdly, it is assumed that the plasma is homogeneous during investigation procedure. The last one point must be clarified in the future investigation.

One can turn a special attention that the electron density obtained in this investigation is exceed by more than some orders of magnitude in comparison with [8, 9]. But it must be stressed that results within frame of this work are determined from two spectral lines of different elements registered simultaneously in the same sample of spectra measurements.

So, the obtained results are presented in Tab.1.

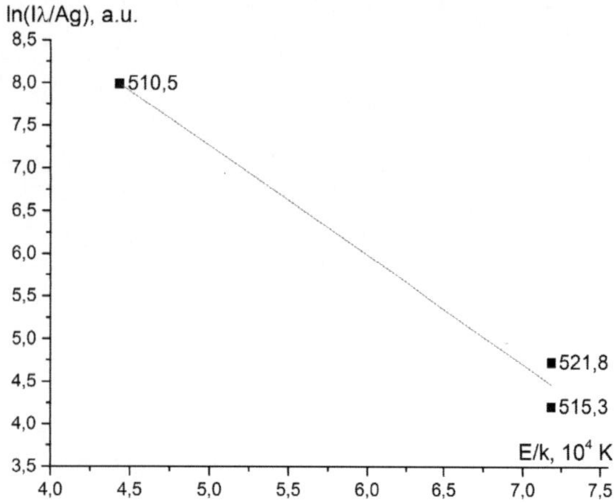


*Figure 9. Boltzmann plot on the base of Cu I spectral lines.*

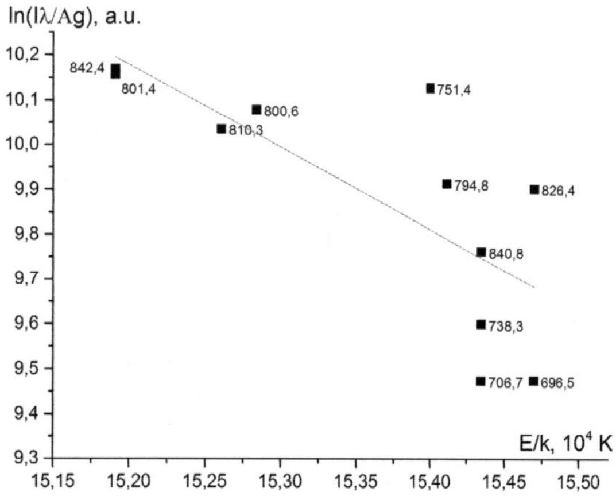


*Figure 10. Boltzmann plot on the base of Ar I spectral lines.*

| Parameter | Value |
|---|---|
| $T^{Cu}$ | $8600 \pm 1200$ K |
| $T^{Ar}$ | $5400 \pm 1600$ K |
| $N_e^{Cu}$ | $4.1 \times 10^{16}\,\mathrm{cm}^{-3}$ |
| $N_e^{H_\alpha}$ | $1.0 \times 10^{16}\,\mathrm{cm}^{-3}$ |

*Table 1. $T^{Cu}$ – temperature obtained by Cu I spectral lines, $T^{Ar}$ – temperature obtained by Ar I spectral lines, $N_e^{Cu}$ – electron density, obtained from half-width of Cu I 515.3 nm spectral line, $N_e^{H_\alpha}$ – electron density, obtained from half-width of $H_\alpha$ spectral line.*

## 4. Conclusions

Optical emission spectroscopy is suggested to obtain the discharge plasma parameters of magnetron voltage source in pulsed mode.

Excitation temperatures of Cu and Ar atomic levels population are determined from Boltzmann plots. The difference of these obtained temperatures can be explained by partial variation of plasma composition at different stages of vacuum chamber operation.

The electron density is determined from half-width of $H_\alpha$ and Cu I 515.3 nm spectral lines. The obtained value of electron density is original in frame of study of such magnetron discharge plasma and can be as object of the future investigations.

## References

[1] D. Mattox. *The Material Science of Thin Films.* Academic Press, San Diego, 1992.

[2] M. Ohring. *Handbook of Physical Vapor Deposition (PVD) Processing.* Noyes Publications, Westwood, 1998.

[3] T. Michely and J. Krug. *Islands Mounts and Atoms.* Springer, Berlin, 2004.

[4] O. Auciello and R. Kelly. *Ion Bombardment Modification of Surfaces: Fundamentals and Applications.* Elsevier Science Ltd, Amsterdam, 1984.

[5] P. Petrov, Barna, L. Hultman, and J. Greene. Microstructural evolution during film growth. *Journal of Vacuum Science & Technology A: Vacuum, Surfaces, and Films*, 21(5):S117–S128, 2003. doi:10.1116/1.1601610.

[6] P. Barna and M. Adamik. Fundamental structure forming phenomena of polycrystalline films and the structure zone models. *Thin solid films*, 317(1-2):27–33, 1998. doi:10.1016/S0040-6090(97)00503-8.

[7] C. Christou and Z. Barber. Ionization of sputtered material in a planar magnetron discharge. *Journal of Vacuum Science & Technology A: Vacuum, Surfaces, and Films*, 18(6):2897–2907, 2000. doi:10.1116/1.1312370.

[8] D. Mozgrin, I. Fetisov, and G. Khodachenko. Pulsed probe technique for determining the plasma parameters of a high-current low-pressure diffuse discharge. *Plasma Physics Reports*, 25(3):255–260, 1999.

[9] S. Bugaev, N. Koval, N. Sochugov, and A. Zakharov. *XVIIth International on Discharges and Electrical Insulation in Vacuum.* Noyes Publications, Westwood, 1996.

[10] B. Chapman. *Glow Discharge Processes.* John Wiley & Sons, 1981.

[11] A. Veklich, A. Lebid, T. Tmenova, V. Boretskij, Y. Cressault, F. Valensi, K. Lopatko, and Y. Aftandilyants. Plasma assisted generation of micro- and nanoparticles. *Plasma Physics and Technology*, 21:28–31, 2017. doi:10.14311/ppt.2017.1.28.

[12] R. Konjevi'c and N. Konjevi'c. Stark broadening and shift of neutral copper spectral lines. *Fizika*, 18:327–335, 1986.

[13] R. Venger, T. Tmenova, F. Valensi, A. Veklich, Y. Cressault, and V. Boretskij. Detailed investigation of the electric discharge plasma between copper electrodes immersed into water. *Atoms*, 40(5):1–17, 2017. doi:10.3390/atoms5040040.

doi:10.14311/ppt.2019.1.91
*Plasma Physics and Technology* **6**(1):91–94, 2019 

# PYROTECHNIC SWITCH WITH FUSE FUNCTION

A. Ehrhardt[a,*], S. Wolfram[b]

[a] *DEHN + SÖHNE GmbH + Co.KG., Hans-Dehn-Straße 1, 92318 Neumarkt, Germany*

[b] *Institute for Electric Power and Control Engineering, Technische Universität Ilmenau, Gustav-Kirchhoff-Str. 1, 98693 Ilmenau, Germany*

[*] arnd.ehrhardt@dehn.de

**Abstract.** The article will present the combination of a low-voltage fuse with a pyrotechnic switch in compact design. The limited switching capacity of the very simple pyrotechnic switch is increased considerably when combined with a fuse which has classical passive time/current characteristic and high switching capacity. FEM calculations will demonstrate the functional principle and the passive and active characteristics of functional models will be examined. Switching behaviour with different overcurrents in case of passive or active triggering will be discussed on the basis of measuring results. The behaviour of such a fuse with high impulse currents will also be presented.

**Keywords:** fuse, pyrotechnic switch, impulse current, numerical simulation, SPD.

## 1. Introduction

For more than 100 years, overcurrent protective devices in the form of fuses and switches have been successfully enhanced and adapted for the most diverse networks, consumers and protective objectives. The massive modification of the power supply networks which is currently underway also places new demands on these protective devices. As numerous power generators feed electricity into the grid, the flow of power in the distribution network is no longer just unidirectional, but frequently bidirectional. What is more, the source characteristics of numerous electricity generators and the new regulating mechanisms for controlling the power flow reduce the short-circuit currents in the grid. Both effects influence, among other things, the tripping conditions for overcurrent protective devices in the power grid. In order to keep up with requirements, additional compact and low-cost functions are needed alongside the optimum configuration and selection of devices.

Fuses have the advantage over switches that they are smaller while ensuring an extremely high short-circuit breaking capacity and that they have a very high current limitation and thus low total clearing integrals. In addition, they are inexpensive and do not exhaust ionised gases. The disadvantage, however, is that fuses only protect devices and systems once and that they respond slowly in case of small overcurrents which results in long melting times and also long arcing times over some half waves. In case of such small currents, the installation conditions and the ambient temperatures frequently have an impact on the disconnection behaviour. With the modification of networks described above, these disadvantages may come to the fore.

Switches, in contrast, have the advantage that the overcurrent performance is easily adjustable and that they can often be externally triggered. This external triggering by additional monitoring devices allows safe disconnection in a relatively short period of time, even in the event of faults which do not directly cause high short-circuit currents or in networks where the operating and short-circuit current are almost equal. In networks with active power regulation, however, the passive protection characteristic of protective devices is frequently only decisive if these triggering systems fail. As with high short-circuit currents, the one-time switching function in combination with external triggering is therefore sufficient in case of such faults.

For this reason, a combination of separate circuit breakers and backup fuses has become an established part of medium-voltage systems to get the benefit of the advantages of both devices [1]. Measures aimed at disconnecting small currents faster have been available for fuses for many years. To this end, e.g., exothermal masses are used which passively or actively heat the solders on fuse-elements or directly heat subareas [2]. These procedures can considerably reduce the natural melting time, however, disconnection within 10 ms, as is common with switches, cannot be achieved. It is also common knowledge that disconnection of fuses can be accelerated by a short-circuit enabling fast disconnection times, at least directly after the short-circuit has been generated [3]. This, however, requires conventional networks with a high short-circuit power. An overview of the different approaches can be found in [4].

The principle of the $I_S$ - Limiter, namely parallel connection of a pyrotechnic switch and a fuse, is known from the field of high-voltage [5]. This principle can also be used for low-voltage applications at a small design size. Similar devices have also been developed for the automotive sector and DC systems. However, these devices require a trigger signal and have neither a passive short-circuit performance nor, without additional programming, a defined time-current characteristic. The design targets for the switching device for low-voltage applications were:

- □ High passive short-circuit breaking capacity of a low-voltage fuse
- □ Time/current characteristic of a low-voltage fuse without triggering
- □ Size and costs similar to those of a fuse
- □ Additional feature: External triggering with a disconnection time < 30 ms
- □ High capacity to carry impulse current, particularly lightning currents
- □ Nominal current up to approx. 250 A
- □ Nominal voltage up to approx. 400 V AC

## 2. Set-up and components of the approach examined

The sample solution consists of components of an enclosed fuse, size 22x58. This size is borderline in every way with regard to the desired values, however, this approach allows a realistic assessment of the physical limits and the construction volume required for implementation.

Conventional compacted silica sand was chosen as a quenching medium. Two parallel copper fuse-elements with modulation, which were, as far as possible, installed straight through the housing, were used for the required high impulse current carrying capability. The nominal current of the two fuse-elements is 250 A. The fuse can carry lightning impulse currents with an amplitude of 25 kA (wave form 10/350 µs with $I^2t = 156\,\mathrm{kA^2s}$). Thus a configuration was selected by which the nominal current and impulse current are at the upper limit, resulting in maximum requirements on the switching capacity. Consequently, lower values can be used for the same size or the same values can be used for larger sizes.

A very simple pyrotechnic switch was integrated between two separated sand-filled parts of a fuse. The restricted sections of the fuse-element required for disconnecting the short-circuit current are distributed evenly between the two parts of the fuse-link. Figure 1 shows the basic test set-up with an electric match.

A hollow space a2 is located between the two separated sand-filled parts a1 of the fuse-link. The two parts a1 of the fuse-link filled with silica sand are separated from the hollow space a2 by a thin wall < 1 mm so that no arc extinguishing medium can enter the hollow space. A cutting blade c is situated in the hollow space between the two separated parts of the fuse-link nearby the fuse-elements b.

Consequently, only little force is required to actuate the cutting blade. The overall width of the hollow space is smaller than 5 mm. The two copper fuse-elements b routed directly through the two sand-filled parts of the fuse-link and also through the hollow space a2. The two fuse-elements b are connected with

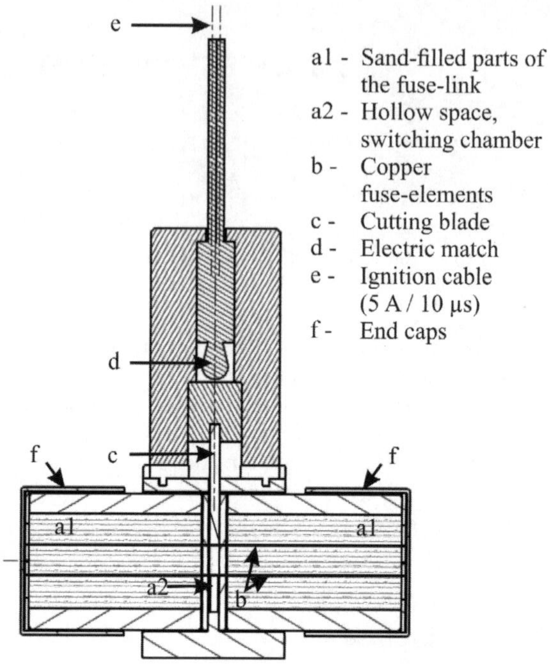


*Figure 1. Test set-up with actuator.*

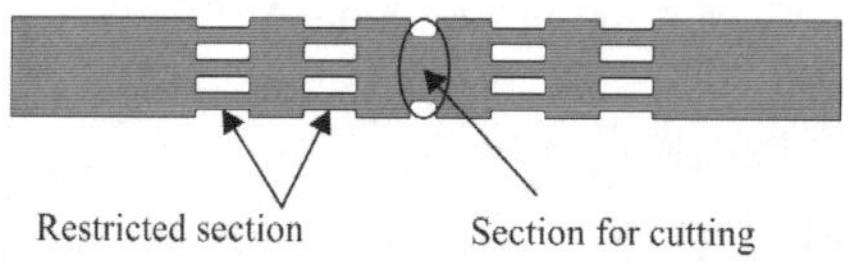


*Figure 2. Fuse-element with restricted sections and additional section for cutting.*

the end caps f. The fuse-elements have an additional restricted section in this hollow space which is dimensioned in such a way that it does not trip in case of the common tripping integral of the fuse when short-circuit currents are interrupted. Figure 2 shows a single fuse-element.

This additional restricted section does not melt over the entire time/current range as long as the fuse shows normal passive behaviour. It simply reduces the force required for the interruption of the fuse-elements b. The arrangement of the hollow space, the thin walls, the cutting blade, the additional restricted section and the electric match constituted a simple pyrotechnic switch. This pyrotechnical switch has a limited self-switching capacity. The electric match can be ignited with a short current flow (5 A, 10 µs). The cutting blade disconnects the fuse-elements. An arc can burn back the fuse-element up to the two parts of the fuse-link filled with silica sand where then the arc safely quenched at the latest to the next current zero.

## 3. Functional principle

### 3.1. Passive function

The time/current performance (heating until melting of one restricted section of the first fuse-element) of the sample arrangement was simulated based on FEM calculations of transient impulse currents up to the current which melts the fuse after 1 h and compared with a conventional fuse [6]. Figure 3 shows the temperature rise of the fuse-element when impulse current flows through it. The calculated time/current characteristics were compared with real currents in the range of about 600 A to 50 kA. The characteristics are illustrated in Figure 4 and show enough concordance.

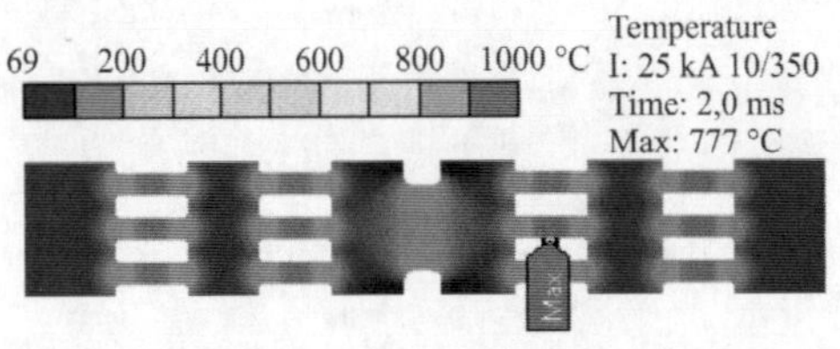


*Figure 3. Calculated maximum temperature at 25 kA, 10/350 µs impulse.*

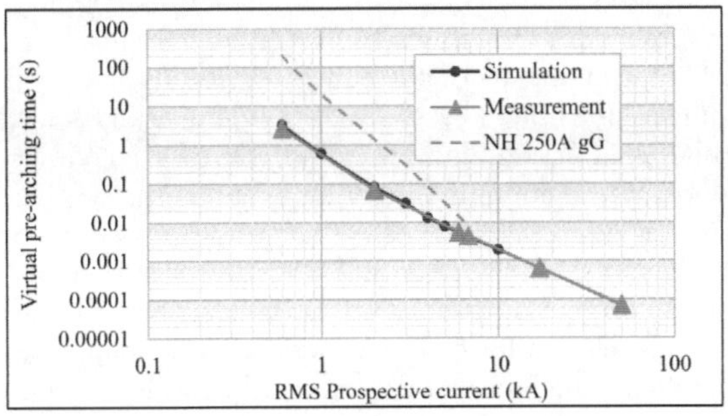


*Figure 4. Comparison of the measured with the calculated passive time/current characteristic.*

The passive switching performance (no triggering over the ignition cable) of the sample arrangement was tested at 440 V in the current range of 600 A to 50 kA. Figure 5 shows the current and voltage characteristic at 50 kA compared with the characteristic of a conventional fuse design.

## 3.2. Active function

In addition to the passive function, the function of the pyrotechnic switch was also tested under load (overcurrents). The ignition of the electric match and the current load of the test arrangment were managed via a time-dependent sequential control. In this context, the trigger time $T_{Delay}$ and the time until both fuse-elements are disconnected $T_{move}$ are relevant. The resulting delay $T_V$ defines the maximum current load which has to be mastered. Figure 6a shows the disconnection of a current of 600 A and Figure 6b the disconnection of a current of approx. 2 kA.

In principle, extremely high short-circuit currents are only switched passively. According to the passive time/current characteristic, currents > 3 kA are only interrupted passively. The pyrotechnic switch was modified for accelerated triggering for the test

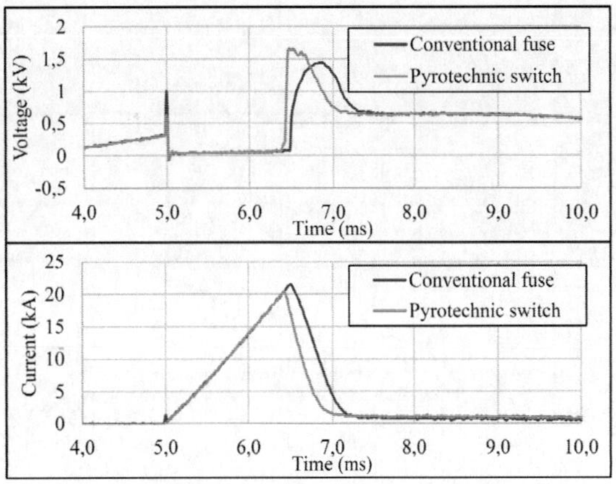


*Figure 5. Measured voltage/current characteristic at $I_p$ = 50 kA AC of the conventional fuse and the pyrotechnic switch.*

in Figure 6b. The test current in Figure 6b and the switching capacity of the pyrotechnic switch thus already exceed the maximum capacity of the switch. This high switching capacity of the simple pyrotechnic switch and the combination with the passive protection characteristic of the fuse thus allow safe and reliable series connection both in case of active and passive functions.

Figure 7 shows the passive time/current characteristic and the actual active characteristic. It becomes evident that the sample arrangement examined ensures safe disconnection at 440 V over the entire current range within a duration of < 30 ms. In case of a longer delay or non-triggering, the arrangement shows the time/current performance of a fuse with the relevant characteristic.

## 4. Condition after impulse current load

In addition to the time/current performance and the switching capacity, the condition of the fuse-element was assessed after a sequence of impulse current loads which correspond to a normative load of type 1 surge protective devices [7]. Figure 8 shows such a fuse-element. The restricted sections have changed colour, but this is uncritical. No partial melting resulting from excessive current densities or deformations of the fuse-element caused by electrodynamic forces were found.

After the switching loads, which were mastered with active and passive function, the impulse withstand voltage (insulation) was also tested by determining the impulse sparkover voltage of the sample set-up. It became evident that in case of active disconnection the impulse sparkover voltages were always > 6 kV independent of the load current. This is identical with the value achieved with short-circuit currents in case of passive function. Without active function, only electric strengths which are partly below 2 kV occurred when passively switching currents by means of fuse-

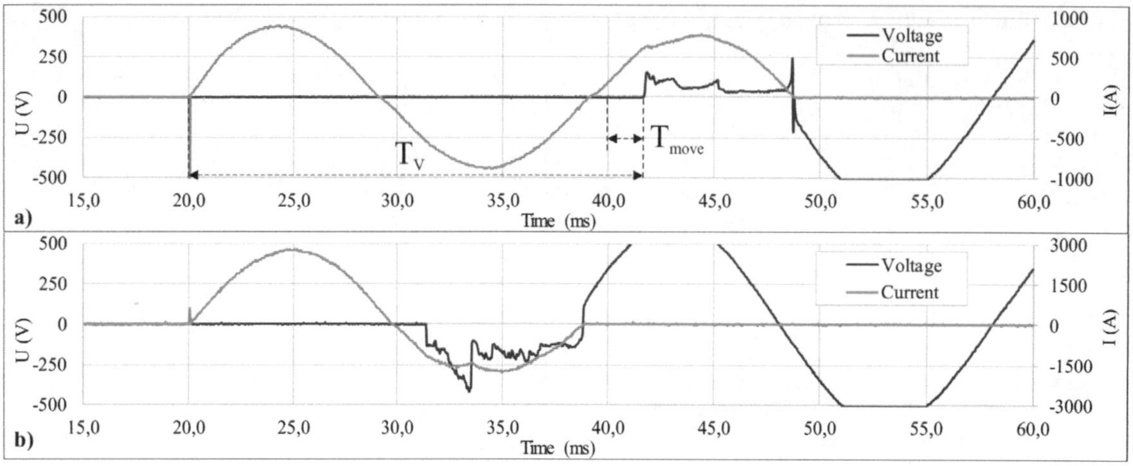


*Figure 6. Measured voltage/current characteristic in case of disconnection for: a)* $I_p = 600\,A$ $(U_C = 434\,V, \cos(\varphi) = 0.92)$; *b)* $I_p = 2\,kA$ $(U_C = 434\,V, \cos(\varphi) = 0.83)$.

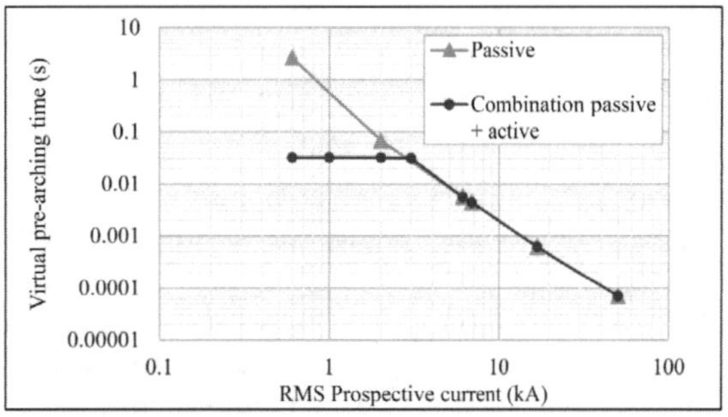


*Figure 7. Measured time/current characteristic of the pyrotechnic switch.*

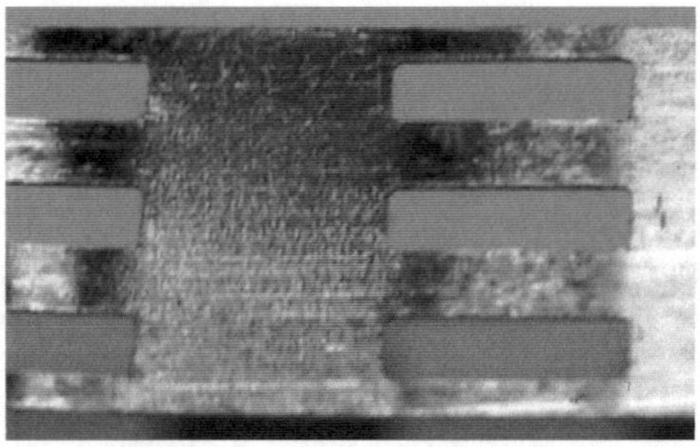

*Figure 8. Colour change of the restricted sections after normative load.*

elements in the seconds range (interruption of only one restricted section per fuse-element). The sample arrangement examined thus safely prevents repeated current flow through the fuse after disconnection even in case of higher voltages.

## 5. Summary

The sample arrangement presented in this paper is a space-saving and cost-effective compromise for networks, plants or equipment for which a one-time switching function is sufficient. It was demonstrated that high switching capacities and low total clearing integrals similar to fuses of comparable size can be achieved by means of additional triggering. The disconnection time in case of external triggering is the same as in case of conventional externally triggered switching devices. If triggering fails, the protective effect is equivalent to that of a normal passive fuse so that at least conventional overcurrent and short-circuit protection is ensured. Moreover, the sample arrangement can safely carry high transient impulse currents. Safe active disconnection is ensured independent of the current load, namely from zero current to the short-circuit current range, thus achieving a higher and at the same time defined electric strength compared to passive fuses.

## References


[1] IEC 616272-105 High-voltage switchgear and controlgear – Part 105: Alternating current switch-fuse combinations, March 2002.

[2] R. Ranjan. Current surge protector for power fuses. U.S. Patent 4807082(A), May 14, 1987.

[3] M. Koprivsek. Triggered fuse. In *10th ICEFA*, pages 69–78, September 2015.

[4] H. Bessei and F. Glinka. Smart fuses for smart grids: Considerations about the need, potential product features and feasibility. In *10th ICEFA*, pages 79–86, September 2015.

[5] K. H. Hartung and V. Schmidt. Limitation of short circuit current by an $I_S$ - Limiter. In *10th EPQU*, September 2009.

[6] D. Rochette, W. Bussière, and R. Touzani. Modelling of the pre-arcing period in hbc fuses including solid - liquid - vapour phase changes of the fuse element. In *64th IEEE Holm Conference on Electrical Contacts*, pages 87–93, September 2007. doi:10.1109/ICEFA.2007.4419972.

[7] IEC 61643-11 Low-voltage surge protective devices – Part 11: Surge protective devices connected to low-voltage power systems – Requirements and test methods, March 2011.

doi:10.14311/ppt.2019.1.95
*Plasma Physics and Technology* **6**(1):95–98, 2019 

# DIAGNOSTICS OF PLASMA PENCIL DISCHARGE FOR CHEMICAL ANALYSIS

P. Slavicek[a,*], A. Hrdlicka[b], M. Dvorakova[b], V. Kanicky[b]

[a] *Department of Physical Electronics, Faculty of Science, Masaryk University, Kotlarska 267/2, 61137 Brno, Czech Republic*

[b] *Department of Chemistry, Faculty of Science, Masaryk University , Kotlarska 2, 61137, Brno, Czech Republic*

[*] ps94@sci.muni.cz

**Abstract.** We present diagnostics of plasma pencil discharge as alternative excitation source for analytical chemistry. The plasma pencil is special type of rf plasma nozzle at atmospheric pressure. Through this nozzle flows working gas argon with aerosol. The aerosol sample introduction system employed a double pass Scott spray chamber with a pneumatic concentric nebulizer. The parameters of the plasma were calculated by optical emission spectroscopy.

**Keywords:** plasma pencil, atmospheric pressure plasma, OES, chemical analysis.

## 1. Introduction

Different plasma sources were used for chemical analysis. Inductively coupled plasma (ICP) and glow discharges [1] were used as very successful and versatile sources. Some alternatives, e.g. capacitively coupled plasma (CCP) [2, 3] or microwave-induced plasmas (MIPs) were studied for analytical purposes and various sub-types were constructed. A combination of a radio frequency (rf) plasma jet operating in the continuous mode and a laser-induced plasma was used for breakdown analysis of complex materials [4]. Plasma sources at low pressure were tested for chemical analysis too [5, 6].

Discharges at atmospheric pressure were used in most cases of measurements. Discharge sensitivity to water presence was limited stability and availability for chemical analysis. In case of ICP relatively high radio frequency (rf) power and high gas flow rates were used to maintain a big plasma, which is stable and resistant against water solutions.

Our study was aimed to rf plasma nozzle (plasma pencil) at atmospheric pressure as alternative excitation source for analytical chemistry. Plasma pencil had been a capacitively coupled plasma (CCP) radio frequency jet discharge, operated typically in argon, at atmospheric pressure [7]. It had been originally designed and used for surface treatment of both recent and ancient objects. The convenience for special polymer film deposition has also been proven [8]. It has been further tested as an alternative excitation source for elemental analysis of liquids [9]. The analytical profit of this plasma source has been shown during the determination of Li, Na, Mg, Ca, Cu and Zn in aqueous solutions as representatives of alkali, alkali-earth and transition metals [9–11]. In this paper, we present diagnostics of rf plasma nozzle at atmospheric pressure. The parameters of the plasma source (rotational and excitation temperature) were calculated by optical emission spectroscopy [12] in the spectral range 200 – 750 nm for different parameters of rf discharge (power 140 W and 160 W, working gas flow rate 4 slpm and 5 slpm, position of detector).

Parameters (input rf power, working gas flow rate) were selected so that discharge was stable and resistant against water aerosol. Rotational and excitation temperature were measured along the discharge tube because strong dependence of intensity of emission lines on position was observed during experiments with aqueous solutions of alkali, alkali-earth and transition metals [9].

Advantage of tested discharge had been lower operational costs compared with conventional inductively coupled plasma or microwave induced plasma excitation sources for chemical analysis.

## 2. Experimental

The plasma was boosted by a Cesar 136 rf generator (Advanced Energy, China) at 13.56 MHz frequency and the correct feedback was driven by a laboratory-made matching unit keeping the reflected power close to zero. The plasma was ignited with a spark coming from a hand Tesla transformer.The discharge is depicted in Fig. 1. It was enclosed in a 250 mm long quartz tube (i.d. 2 mm) with a fused inlet of the same diameter under normal direction for aerosol introduction. The position 6.5 cm corresponds to the ground electrode and the power electrode is on position 0 cm (Fig. 1). The main argon stream was dosed with a mass flow controller. The aerosol sample introduction system employed a peristaltic pump and a double pass Scott spray chamber with a pneumatic concentric nebulizer. The created aerosol was transported by the carrier gas perpendicularly into the main discharge argon stream. The carrier argon entering the nebulizer was controlled with a mechanical rotameter (0.3 slpm – standard litre per minute).

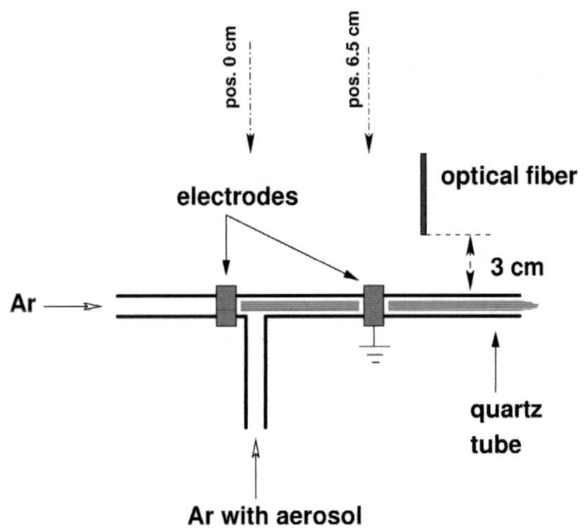


*Figure 1. Experimental arrangement of the discharge.*

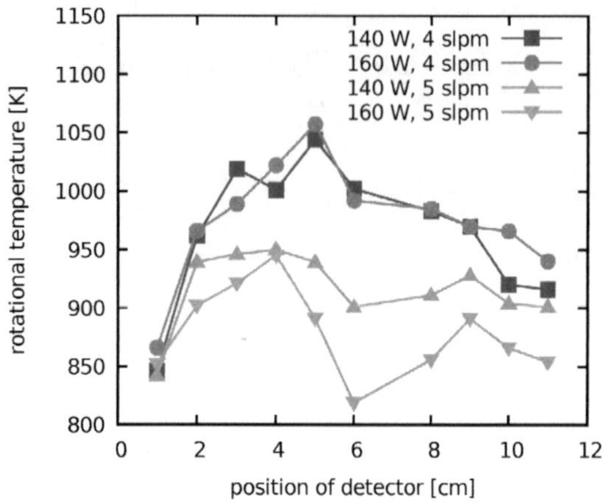


*Figure 2. Rotational temperature profile along the discharge axis without aerosol for different power and working gas flow rate.*

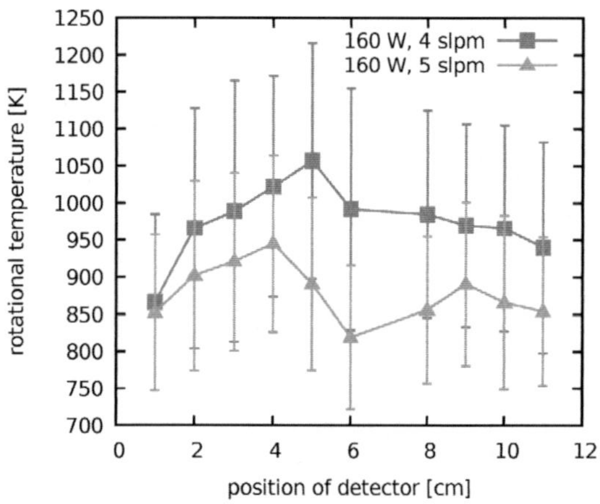


*Figure 3. Rotational temperature profile along the discharge axis without aerosol for power 160 W and different working gas flow rate with error bars.*

Plasma radiation was collected with a quartz optical fibre, fitted to a movable holder on a scaled rail for exact positioning with respect to the discharge tube, and delivered to Andor Shamrock 750 spectrometer (grating 2400 grooves $mm^{-1}$ ) equipped with iDus 416 CCD camera cooled by a Peltier cooler.

The rotational and excitation temperatures were calculated by optical emission spectroscopy for different input power of discharge, working gas flow rate and position along the discharge tube. These experiments were performed with and without aerosol. Pure deionised water was used for the experiment.

## 3. Results and discussion

Main task of our experiments was the optical emission diagnostics of the plasma. Rotational temperature from Boltzmann plots of rotational lines of OH spectra (Q1 branch in 0 – 0 band, transition $A^2\Sigma \rightarrow X^2\Pi$) and excitation temperature from Boltzmann plots of Ar lines (603.21, 667.73, 675.28, 687.13, 714.70 nm) were calculated for different parameters of rf discharge. Errors of rotational and excitation temperatures were calculated from linear regression of Boltzmann plots and were about 10 %. Identification of spectral lines and calculation of rotational and excitation temperatures were done using the Spectrum Analyzer software [13]. Rotational temperature can approximate temperature of heavy particles in discharge. Excitation temperature can approximate electron temperature. Temperature of plasma has a big impact on excitation and ionization of chemical elements.

Rotational temperatures from OH for different input power (140 W, 160 W) and different working gas flow rate (4 slpm, 5 slpm) without aerosol were measured first. Results of profiles along the discharge tube are depicted in Fig. 2. Errors of rotational temperature for input power 160 W and different working gas flow rate are presented in Fig. 3. The display of the error bars for all curves in Fig. 2 would be much cluttered. The different shapes of temperature dependencies Fig. 2, Fig. 3 for higher working gas flow rate were caused by cooling effect of gas flow.

Next measurements were done with same parameters of input power and working gas flow rate with aerosol. Results are presented in Fig. 4. Comparison of rotational temperatures for input power 140 W and 160 W for same working gas flow rate 4 slpm with and without aerosol is presented in Fig. 5. Rotational temperature has maximum at position about 5 cm. Discharge with aerosol has higher temperature than discharge without aerosol (Fig. 5). The temperature of discharge without aerosol can be changed by argon flow rate (Fig. 2). The temperature of discharge with aerosol is practical independent on input rf power and argon flow rate (Fig. 4).

Excitation temperatures from Ar lines for different input power (140 W, 160 W) and different working gas flow rate (4 slpm, 5 slpm) without aerosol were

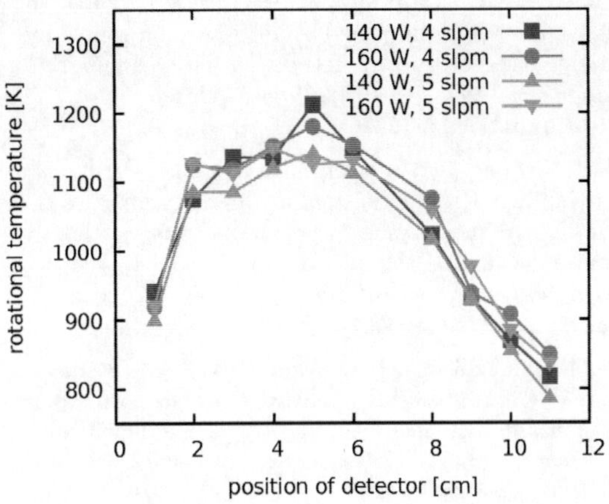


*Figure 4. Rotational temperature profile along the discharge axis with aerosol for different power and working gas flow rate.*

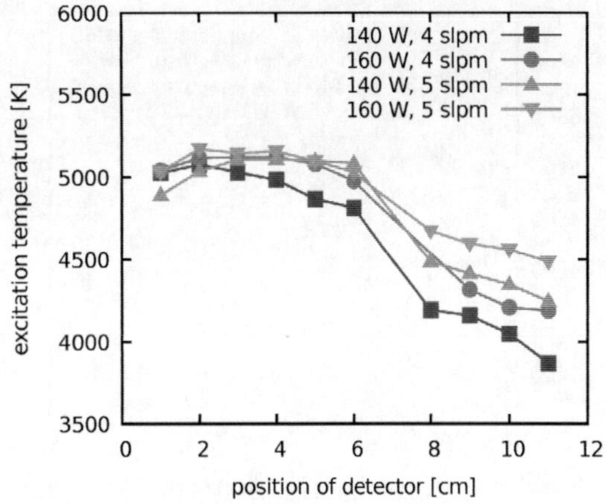


*Figure 6. Excitation temperature profile from the argon lines without aerosol for different power and working gas flow rate.*

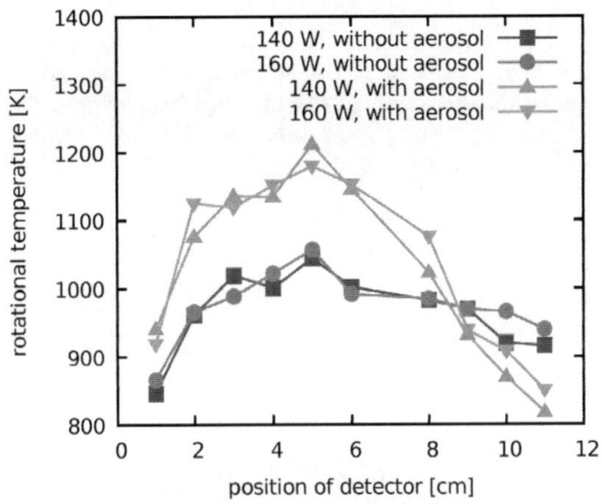


*Figure 5. Rotational temperature profile along the discharge axis with and without aerosol for different power for same working gas flow rate (4 slpm).*

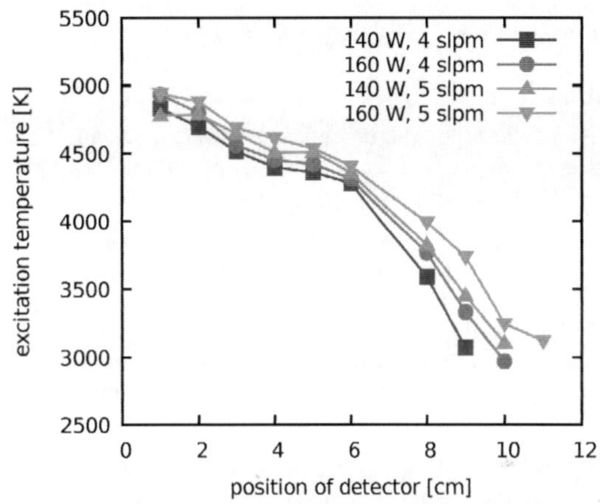


*Figure 7. Excitation temperature profile from the argon lines with aerosol for different power and working gas flow rate.*

measured. Results of profiles along the discharge tube are depicted in Fig. 6. Measurements were done with same parameters of input power and working gas flow rate with aerosol. Results are presented in Fig. 7. Comparison of excitation temperatures for input power 140 W and 160 W for same working gas flow rate 4 slpm is presented in Fig. 8.

Excitation temperature has not maximum as rotational temperature. Discharge with water aerosol has lower excitation temperature than discharge without aerosol (Fig. 8). Decline of the temperature along the discharge tube is significant in case of discharge with aerosol.

Significant contraction of plasma along the discharge tube was observed in case with aerosol during all measurement.

## 4. Conclusions

The plasma nozzle at atmospheric pressure presented an alternative excitation source for analytical chemistry. Diagnostics by optical emission spectroscopy of this plasma nozzle were presented in this contribution. Rotational and excitation temperature of discharge were calculated for different parameters (input rf power, gas flow rate, position along discharge tube) with and without aerosol. Discharge with aerosol has higher rotational and lower excitation temperature then in case without aerosol. Discharge parameters have influence on application of this capacitively coupled plasma source for chemical analysis.

### Acknowledgements

This research has been supported by the project CZ.1.05/2.1.00/03.0086 funded by European Regional Development Fund, by the project LO1411 (NPU I) funded by Ministry of Education Youth and Sports of Czech Republic

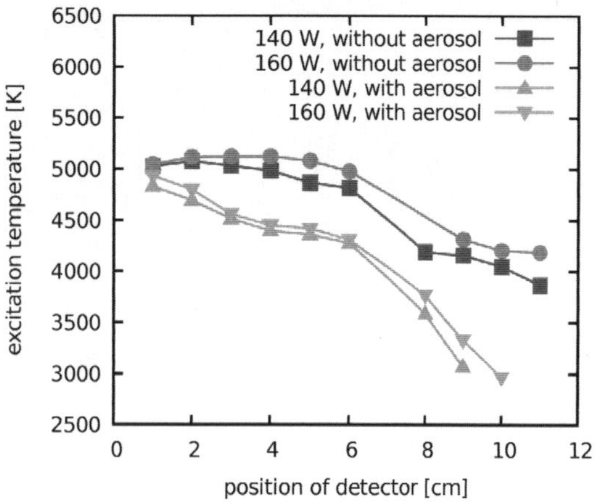


*Figure 8. Excitation temperature profile from the argon lines with and without aerosol for different power for same working gas flow rate (4 slpm).*

and by the project MUNI/A/1375/2018.

## References

[1] N. Bings, A. Bogaerts, and J. Broekaert. Atomic spectroscopy. *Analytical Chemistry*, 80:4317–4347, 2008. doi:10.1021/ac8006297.

[2] T. Frentiu, M. Ponta, M. Senila, A. Mihaltan, E. Darvasi, and M. Frentiu. Evaluation of figures of merit for zn determination in environmental and biological samples using EDL excited AFS in a new radiofrequency capacitively coupled plasma. *Journal of Analytical Atomic Spectrometry*, 25:739–742, 2010. doi:10.1039/b918872j.

[3] R. Gross, B. Platzer, E. Leitner, A. Schalk, H. Sinabel, H. Zach, and G. Knapp. Atomic emission gas chromatographic detection chemical and spectral interferences in the stabilized capacitive plasma (SCP). *Spectrochimica Acta Part B*, 47:95–106, 1992. doi:10.1016/0584-8547(92)80010-E.

[4] J. Pedarnig, J. Heitz, E. Ionita, G. Dinescu, B. Praher, and R. Viskup. Combination of RF-plasma jet and laser-induced plasma for breakdown spectroscopy analysis of complex materials. *Applied Surface Science*, 257:5452–5455, 2011. doi:10.1016/j.apsusc.2010.11.112.

[5] V. Mazakova, D. Trunec, and F. Krcma. Study of nitrogen flowing afterglow with mercury vapor injection. *The Journal of Chemical Physics*, 141:154307–1 – 154307–9, 2014. doi:10.1063/1.4898367.

[6] V. Mazakova, D. Trunec, Z. Navratil, J. Raud, and F. Krcma. Study of argon-oxygen flowing afterglow. *Plasma Sources Science and Technology*, 25:1–10, 2016. doi:10.1088/0963-0252/25/3/035008.

[7] J. Janca, M. Klima, P. Slavicek, and L. Zajickova. HF plasma pencil – new source for plasma surface processing. *Surface and Coatings Technology*, 116:547–551, 1999.

[8] P. Slavicek, V. Bursikova, A. Brablec, V. Kapicka, and M. Klima. Deposition of polymer films by rf discharge at atmospheric pressure. *Czechoslovak Journal of Physics*, 54(3 Supp):C586, 2004. doi:10.1007/BF03166455.

[9] L. Novosad, A. Hrdlicka, P. Slavicek, V. Otruba, and V. Kanicky. Plasma pencil as an excitation source for atomic emission spectrometry. *Journal of Analytical Atomic Spectrometry*, 27:0267–9477, 2012. doi:10.1039/c1ja10278h.

[10] L. Novosad, A. Hrdlicka, P. Slavicek, V. Otruba, and V. Kanicky. Possibilities and analytical properties of the radiofrequency plasma pencil operating in the continuous and in the pulsed mode. *Journal of Analytical Atomic Spectrometry*, 30:459–467, 2015. doi:10.1039/c4ja00402g.

[11] M. Dvorakova, A. Hrdlicka, P. Slavicek, V. Kanicky, and V. Otruba. Effects of easily ionisable elements on copper and zinc lines excited in a plasma pencil. *Journal of Analytical Atomic Spectrometry*, 31:2031–2036, 2016. doi:10.1039/c6ja00215c.

[12] R. Korneva, P. Sennikov, L. Shabarova, A. Shishkin, T. Drozdova, and S. Sintsov. Reduction of boron trichloride in atmospheric-pressure argon-hydrogen radiofrequency induction plasma. *High Energy Chemistry*, 53:246–253, 2019. doi:10.1134/S001814391903010X.

[13] Z. Navratil, D. Trunec, R. Smid, and L. Lazar. A software for optical emission spectroscopy - problem formulation and application to plasma diagnostics. *Czechoslovak Journal of Physics*, 56:B944–B951, 2006. doi:10.1007/s10582-006-0308-y.

doi:10.14311/ppt.2019.1.99
*Plasma Physics and Technology* **6**(1):99–102, 2019 

# SELF-CONSISTENT MODELING OF ELECTRODE MOTION IN A MODEL CIRCUIT BREAKER

R. Fuchs

*Institute for Energy Technology, HSR University of Applied Sciences Rapperswil, 8640 Rapperswil, Switzerland*

roman.fuchs@hsr.ch

**Abstract.** Numerical simulations of low-voltage circuit breakers require a coupled solution of gas flow, electromagnetism, electrical circuit, and other aspects. Including electrode motion is challenging because the computational grid is deformed and data is to be exchanged among dedicated solvers. A central issue is to keep them synchronized. This is addressed with a single framework that allows for a continuously morphing grid and accounting for the cumulative effects of mechanics, Lorentz force, and gas pressure. It is shown that gas pressure has negligible effect.

**Keywords:** circuit breaker, electrical arc simulation, self-consistent, rigid body motion.

## 1. Introduction

Low-voltage circuit breakers are designed to carry electrical current in normal operation as well as safely interrupt short-circuit currents in case of failures. In case of a fault, the contacts are opened by means of mechanical actuators and/or electromagnetic forces available due to fault current. This contact opening inevitably leads to an electrical arc. Depending on geometrical and other factors, the arc is ultimately driven into a deion chamber with splitter plates, in which it is elongated and cooled until extinction.

Contact opening can be a major design aspect in the current interruption process and numerical simulations should allow for including this effect appropriately and efficiently. In [1], the computational domain was deformed and remeshed to accommodate for the motion of the opening contact; however, the motion as well as current were not computed but followed input values obtained from experiment. Later, it was shown in [2, 3] that numerical simulations should include contact motion appropriately, especially if large contact gaps and contact motion are studied right in front of splitter plates. The adopted workflow consisted of creating a sequence of meshes that correspond to various contact positions. Contact position was calculated by a separate ordinary differential equation. While the meshes remained constant without any deformation, they are swapped frequently to follow contact position. However, the numerical framework involved three codes that must be kept synchronized and being a challenging task. [3] The same approach has been recently adopted in [4].

This contribution presents a self-consistent and seamless implementation of contact motion for a three dimensional model geometry of a low-voltage circuit breaker. The moving contact is assumed to be subject to a mechanical actuator represented as a torsional spring for simplicity, the Lorentz force that becomes important at larger current values, and we account for plasma pressure that acts on the moving contact surface. Additionally, we include an electrical circuit that defines electrical boundary conditions for the circuit breaker. In contrast to [1–4], our software framework allows for a continuous deformation and event-based remeshing. Data interpolation and exchange among dedicated solvers is inherently supported in a single user interface, that results in a uniform setup process. An earlier version of the same software framework was used in [5, 6] with an advanced plasma model; however, details on specifications of rigid body motion and electrical boundary conditions are not given.

## 2. Numerical Simulation Model

A model circuit breaker (see figure 1) is considered that consists of a fixed electrode, a rotationally opening bridge, and five U-shaped splitter plates enclosed in an insulating box with venting openings. The virtual geometry is designed to be of comparable size to a real device; for reference, the enclosure size is $47\,\mathrm{mm} \times 36\,\mathrm{mm} \times 24\,\mathrm{mm}$. The ambient gas is dry air at atmospheric pressure and gas flow is considered in a box of size $69\,\mathrm{mm} \times 49\,\mathrm{mm} \times 36\,\mathrm{mm}$ with pressure outlets in all directions and solid bodies located centrally. The moving electrode is 2.5 mm thick and the fixed electrode cross-section is $2\,\mathrm{mm} \times 5\,\mathrm{mm}$. Splitter plates are 1.5 mm thick and separated in $y$-direction by the same distance. A symmetry plane is applied at $z = 0$. The model circuit breaker is thought to be installed serially in an electrical circuit (see figure 2) that consists of a voltage source ($V_{\mathrm{rms}} = 230\,\mathrm{V}, 50\,\mathrm{Hz}$), a resistor $R = 2\,\mathrm{m}\Omega$, and an inductor $L = 25\,\mu\mathrm{H}$.

We solve the magnetohydrodynamic equations for a thermal plasma in local thermodynamic equilibrium (LTE), i.e., we consider the Navier-Stokes equations

$$\partial_t(\rho) + \nabla \cdot (\rho \mathbf{u}) = 0, \tag{1a}$$

$$\partial_t(\rho \mathbf{u}) + \nabla \cdot (\rho \mathbf{u} \otimes \mathbf{u}) = -\nabla p + \mathbf{f}_\mathrm{L}, \tag{1b}$$

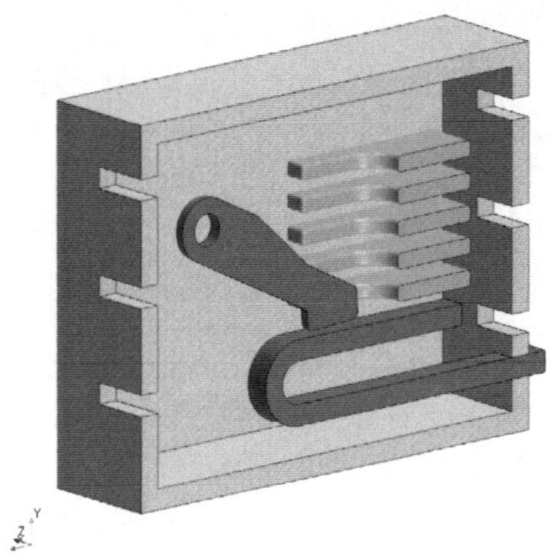

*Figure 1. Model geometry of a low-voltage circuit breaker, cut at symmetry plane $z = 0$.*

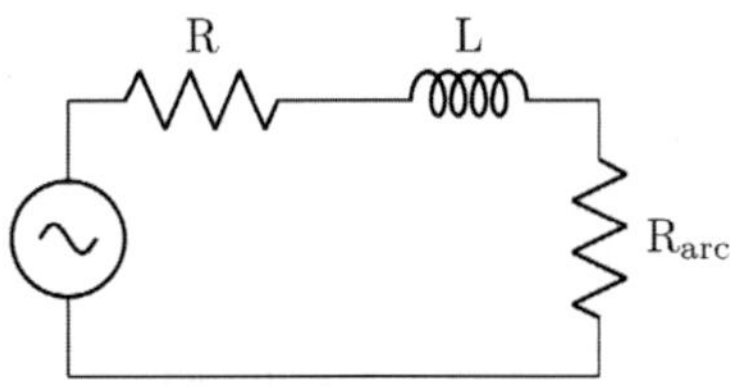


*Figure 2. Electrical circuit.*

$$\partial_t(\rho e_{\mathrm{tot}}) + \nabla \cdot ((\rho e_{\mathrm{tot}} + p)\mathbf{u}) = S_{\mathrm{ohm}} + \nabla \cdot (\lambda \nabla T) - \nabla \cdot \mathbf{q}_{\mathrm{rad}}, \quad (1c)$$

together with Maxwell's equations in low-frequency limit in $A$-$\phi$ formulation, i.e.,

$$\nabla \times \mu^{-1} \nabla \times \mathbf{A} = \mathbf{J}, \quad (1d)$$

$$\nabla \cdot \mathbf{J} = 0. \quad (1e)$$

In these equations, $\rho$ stands for gas density, $\mathbf{u}$ gas velocity, $p$ gas pressure, $e_{\mathrm{tot}}$ total energy, $T$ gas temperature, $\mathbf{q}_{\mathrm{rad}}$ radiative heat flux, $\mathbf{J}$ electric current density, $\mathbf{B}$ magnetic flux density, $\mu$ magnetic permeability, $\sigma$ electrical conductivity, and $\mathbf{A}$ magnetic vector potential. Moreover, we have by definition $\mathbf{B} = \nabla \times \mathbf{A}$ and $\mathbf{E} = -\nabla\phi$, and Ohm's law $\mathbf{J} = \sigma\mathbf{E}$, where $\mathbf{E}$ denotes electric field and $\phi$ is the electric potential. The gas flow is considered as laminar. The system of equations is coupled via the source terms for Lorentz force density $\mathbf{f}_{\mathrm{L}} = \mathbf{J} \times \mathbf{B}$ and Ohmic heating source density $S_{\mathrm{ohm}} = \mathbf{J} \cdot \mathbf{E}$. Radiative heat flux $\mathbf{q}_{\mathrm{rad}}$ is calculated using the Discrete Ordinate method (see, e.g., [7]) with six frequency bands defined by the ionization energies of nitrogen [8]. Band-averaged absorption coefficients were computed by Planck averaging from spectral data provided by [9]. Thermodynamic and transport properties of dry air plasma are taken from [10]. Material data for electrodes and splitter plates are those of copper and iron, respectively, with relative permeability $\mu_{\mathrm{Fe,rel}} = 1 \times 10^4$.

We account for voltage drop in arc roots on the anode and cathode following [8],

$$U_{a/c}(J) = \frac{a\tilde{J} + b\tilde{J}^d}{c + \tilde{J}^d}, \quad \tilde{J} = eJ, \quad (2)$$

with $a = 5 \times 10^8$, $c = 9 \times 10^{14}$, $d = 2$, $e = 5$, and $b = 5$ on the anode and $b = 10$ on the cathode. Comparing to [8], we use modified values for coefficients $a$, $c$, and $e$, so that maximum voltage drop on anode and cathode are reduced to 11.2 V and 14.7 V, respectively. Hence, the arc is able to run faster into the deion chamber since new arc spots are formed more easily on the splitter plates. The corresponding heat source is included at the plasma-solid interface. This difference in voltage drop modeling is assumed to be of minor importance towards the aim of this study, since we also neglect evaporation and erosion of any solid bodies.

The bridge is assumed to move as a rigid body with rotation axis parallel to $z$-axis. Total torque is computed by the contributions of a torsional spring, gas pressure of the plasma acting on the electrode surface, and Lorentz force density inside the electrode body:

$$\boldsymbol{\tau}_{\mathrm{tot}} = \boldsymbol{\tau}_{\mathrm{S}} + \boldsymbol{\tau}_{\mathrm{P}} + \boldsymbol{\tau}_{\mathrm{L}}, \quad \boldsymbol{\tau}_{\mathrm{S}} = -k(\alpha - \alpha_{\mathrm{r}})\hat{\mathbf{z}},$$
$$\boldsymbol{\tau}_{\mathrm{P}} = \int_A \mathbf{r} \times (-p\mathbf{n})\, \mathrm{d}A, \quad \boldsymbol{\tau}_{\mathrm{L}} = \int_V \mathbf{r} \times \mathbf{f}_{\mathrm{L}}\, \mathrm{d}V. \quad (3)$$

The torsional spring constant $k = 0.2\,\mathrm{N\,m\,rad^{-1}}$ and relaxation angle $\alpha_{\mathrm{r}} = 30\,\mathrm{deg}$ are defined such that the bridge would open to $\alpha_{\max} = 45\,\mathrm{deg}$ in 2.5 ms, at which it would be gradually slowed down before hitting the enclosure wall. The natural angular frequency of the torsional spring is $\omega_{\mathrm{n}} = \sqrt{k/I}$ where $I = 2.75 \times 10^{-7}\,\mathrm{kg\,m^2}$ denotes moment of inertia of the moving electrode. If torsional spring was the only acting force, the rotation angle of the moving electrode with respect to $z$-axis is given by

$$\alpha_{\mathrm{S}}(t) = \alpha_{\mathrm{r}}(1 - \cos(\omega_{\mathrm{n}} t)). \quad (4)$$

Initial position of the electrode bridge is defined with a gap of 2 mm to the fixed electrode. Arc ignition is modeled by a cylinder of radius 1 mm located centrally between the contacts with electrical conductivity set to $\sigma = 1 \times 10^4\,\mathrm{S\,m^{-1}}$ for 20 µs. For simplicity, voltage source is assumed to be at peak voltage.

The outlined model is implemented in the numerical framework of Simcenter STAR-CCM+ (v2019.1). The equations for gas flow, electric potential, and radiative heat transfer are solved with finite volume method, and the magnetic vector potential is formulated in finite elements because of the discontinuous magnetic permeability. Therefore, the computational domain is discretized with two meshes, a polyhedral grid for the FV solvers and a tetrahedral grid for the FE solver, with maximum cell sizes 0.5 mm and 2 mm, respectively. The meshes are refined at solid-fluid interfaces including prism layers. Timestep size is

1 µs. After each timestep, electric current density **J** is mapped to the tetrahedral grid, the magnetic vector potential is solved for, and magnetic flux density **B** is mapped back to the polyhedral grid. The domain can be remeshed whenever a specific criterion is met. Here, it is triggered if the upper electrode has moved 0.5 deg since the last remeshing event, or if the simulation had to be restarted. The rigid body dynamics and electrical circuit are also implemented in the user interface.

## 3. Results

Results are presented for time until $t = 400$ µs. Figure 3 shows arc voltage and current, as well as current through the lowest two splitter plates. We see that arc current rises almost linearly as defined by the electrical circuit. After ignition, arc voltage is at 26 V comparing well to the sum of voltage drop on anode and cathode, and increases gradually until $t = 200$ µs. This is explained by plasma cloud expanding from the ignition region into a volume bounded by the moving electrode tip, the fixed electrode and the lower two splitter plates. We note that a fraction of the current flows through the first splitter plate, and after $t = 200$ µs, we see that electrical current starts to flow through the second splitter plate. At the same time value, we also note that arc voltage increases more substantially. This is related to the plasma gas expanding upwards and arc root formation at second splitter plate. Since the focus of this work is on electrode motion modeling, we limit the discussion of plasma flow to this level as it becomes more complex.

Figure 4 shows total torque on the moving electrode and its contributing terms due to torsional spring, Lorentz force, and plasma pressure. We see that torsional spring and Lorentz force are dominant, while plasma pressure yields a small contribution at arc ignition and quickly becomes negligible. Torsional spring yields an almost constant torque because of mass inertia and therefore relatively slow angular motion in the time interval considered. As current increases, Lorentz force becomes gradually more important and its torque exceeds that of torsional spring after $t = 300$ µs.

To support the understanding of the data discussed above, the spatial distribution of plasma pressure and Lorentz force density at $t = 100$ µs are shown in figures 5 and 6, respectively. We see that the pressure wave due to arc ignition has already expanded from electrode tip towards splitter plates. Hence, torque due to plasma pressure is small. However, Lorentz force density increases gradually as total current increases and the self-induced magnetic field due to electric flux. Hence, their contribution to total torque becomes more important.

Figure 7 shows rotation angles due to total torque as obtained by the simulation and that of the analytical model in equation (4). We clearly see that torque due to Lorentz force results in faster electrode motion.

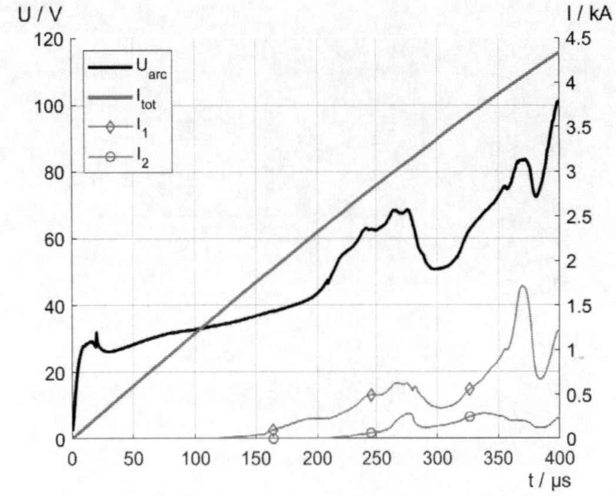


*Figure 3. Arc voltage and current in circuit breaker. The lines $I_1$ and $I_2$ show electrical current through the lowest splitter plates.*

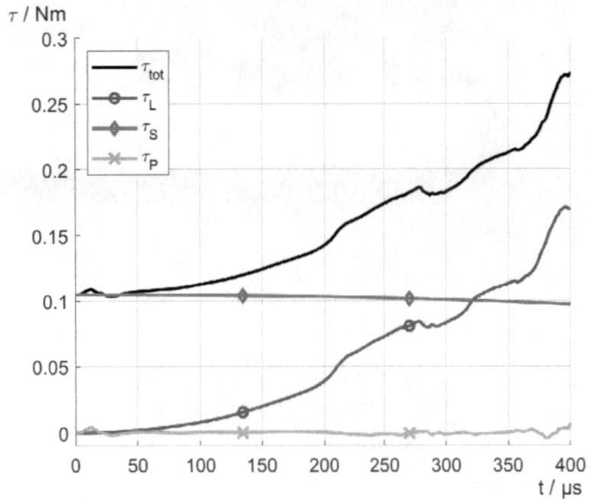


*Figure 4. Total torque on the moving electrode, with contributions from Lorentz force density, torsional spring, and plasma pressure.*

This effect becomes more pronounced with increasing time values.

## 4. Discussion and Conclusions

This contribution presents a self-consistent model for electrode motion in a virtual low-voltage circuit breaker. The results show that torque is mainly due to torsional spring and Lorentz force. Plasma pressure has negligible influence on rigid body dynamics because of its small torque value and its short interaction time. This does not mean that accurate plasma modeling is of minor importance; it is of uttermost importance because plasma gas flow and electromagnetic fields are directly affected. The mechanical actuator should be modeled appropriately because it affects rigid body dynamics and its interactions with arc plasma. Therefore, a self-consistent arc model should be used that allows to compute electrical current and voltage defined by the circuit elements, as well as to compute rigid body motion including the effects of mechanics and Lorentz force.

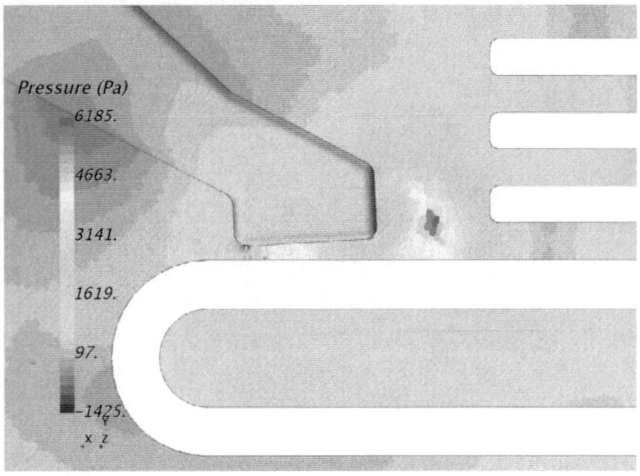


*Figure 5. Plasma pressure at* $t = 100\,\mu s$.

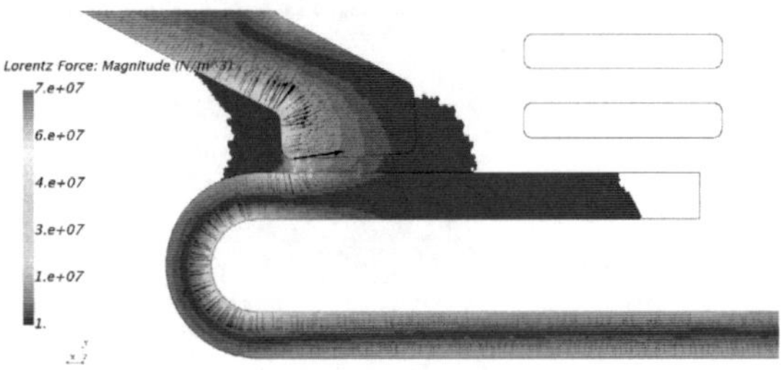


*Figure 6. Lorentz force density at* $t = 100\,\mu s$.

In contrast to previous work, the presented model is implemented with a continuously deforming mesh that is remeshed whenever needed. The user interface allows to define rigid body motions directly inside the framework without particular programming. It also inherently allows for data exchange among the solvers. As a consequence, the workflow for preparing simulations with complex body motions and interactions of the plasma with electrical circuit may be eased.

Future work will extend this study to longer simulation time, additional plasma physics including copper vapor due to electrode erosion and wall ablation, nonlinearity of $B$-$H$ curve of iron and further modeling techniques to ease user interaction with the software. Experiments would serve ideally to complement this study.

## Acknowledgements

This work was supported by Siemens PLM Software. The author thanks Dr Boris Klauderic and Dr Angelo Limone for many valuable discussions.

## References

[1] Y. Wu, M. Rong, Z. Sun, X. Wang, F. Yang, and X. Li. Numerical analysis of arc plasma behaviour during contact opening process in low-voltage switching device. *Journal of Physics D: Applied Physics*, 40(3):795–802, 2007. doi:10.1088/0022-3727/40/3/016.

[2] C. Rümpler, A. Zacharias, and H. Stammberger. Low-voltage circuit breaker arc simulation including contact arm motion. In F. Berger, editor, *The 27th International Conference on Electrical Contacts*, pages 290–294, 2014.

[3] C. Rümpler and V. Naranyanan. Arc Modeling Challenges. *Plasma Physics and Technology*, 2(3):261–270, 2015.

[4] D. Shin, J. W. McBride, and I. O. Golosnoy. Arc Modeling to Predict Arc Extinction in Low-Voltage Switching Devices. In *2018 IEEE Holm Conference on Electrical Contacts*, pages 222–228. IEEE, 14.10.2018 - 18.10.2018. doi:10.1109/HOLM.2018.8611712.

[5] Y. Enami and M. Sakata. Simulation of arc in molded-case circuit breaker with metal vapor and moving electrode. In *2013 2nd International Conference on Electric Power Equipment - Switching Technology (ICEPE-ST)*, pages 1–4. IEEE, 20.10.2013 - 23.10.2013. doi:10.1109/ICEPE-ST.2013.6804390.

[6] M. Sakata and Y. Enami. Arc Simulation Technology. *Fuji Electric Review*, 60(3):195–200, 2014.

[7] M. F. Modest. *Radiative heat transfer*. Academic Press, New York, 3. edition, 2013.

[8] C. Rümpler. *Lichtbogensimulation für Niederspannungsschaltgeräte*. Dissertation, TU Ilmenau, Ilmenau, 2009.

[9] P. Kloc, V. Aubrecht, M. Bartlova, and O. Coufal. Radiation transfer in air and air-Cu plasmas for two temperature profiles. *Journal of Physics D: Applied Physics*, 48(5):055208, 2015. doi:10.1088/0022-3727/48/5/055208.

[10] R. N. Gupta, K.-P. Lee, R. A. Thompson, and J. M. Yos. Calculations and curve fits of thermodynamic and transport properties for equilibrium air to 30000 K. N92-11285, NASA, 1991.

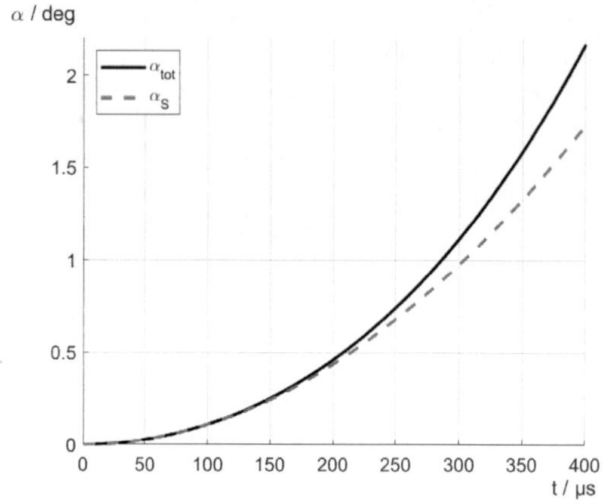


*Figure 7. Rotation angle due to total torque and compared to that due to torsional spring only.*

doi:10.14311/ppt.2019.1.103
*Plasma Physics and Technology* **6**(1):103–106, 2019 

# INFLUENCE OF VARYING GAPS BETWEEN TMF CONTACTS ON CONSTRICTED HIGH CURRENT VACUUM ARCS

B. Kuehn[a,*], B. Weber[a], D. Gentsch[b], M. Kurrat[a]

[a] *Institute for High Voltage Technology and Electrical Power Systems, Technische Universitaet Braunschweig Schleinitzstraße 23, 38106 Braunschweig, Germany*

[b] *ABB AG, Calor Emag Mittelspannungsprodukte, Oberhausener Str. 33, 40472 Ratingen, Germany*

[*] B.Kuehn@TU-Braunschweig.de

**Abstract.** The behavior of high current arcs in vacuum circuit breaker (VCB) is interesting for research and industrial development purpose which lead to further products. To improve the interruption capability of VCB, two approaches to control the arc have been proven successful. Applying transversal magnetic fields (TMF) on the arc is use for industrial VCB in medium voltage ranges. For greater gap distances the behavior of the arc is less thoroughly investigated. In this paper, the appearance of metal vapor arcs drawn by common TMF contacts in a vacuum-test-interrupter is investigated. An adapted drive mechanism enables to interrupt a fixed current with varying gaps from 5 to 25 mm and a constant opening time. Breaking operations with a 50 Hz current are observed with a high speed camera. With increasing gap distance a changed arc appearance can be observed. The goals of this work are to be understood as a feasibility study for optical evaluation methods for vacuum arcs under TMF.

**Keywords:** vacuum circuit breaker, VCB, TMF, constricted arcs, automatized evaluation.

## 1. Introduction

Vacuum circuit breakers are a main component of medium voltage grids and are used as power switches to ensure a safe operation under all switching cases according to the standard. The performance of VCB is dependent on different variables. One is the behavior during arcing. The aim is to distribute the converted interruption energy on the contact surfaces. Solutions are available by applying transversal (TMF) or axial magnetic fields (AMF) on the switching arc. In case of TMF the arc forms a constricted column, in dependence of the contact gap and the short circuit current, and starts moving over the contact surface. To investigate the exact behavior researcherss evaluate electrical and optical data during the arcing phase [1, 2]. The usage of electrical data has been established for years and the community is used to work with data of arc voltage and current. Some researchers measure the external magnetic field of the constricted vacuum arc to evaluate the rotation behavior, which also indicates the energy distribution on the surface [3]. In case of optical data the recording and the evaluation of data is not uniform. This makes comparability more difficult. First suggestions to improve comparability are given in [4]. In this paper the new approach based on [4] is used to evaluate recorded data during short circuit breaking operations of TMF contacts. Therefore an effective current of 20 kA (50 Hz) was interrupted with a vacuum test interrupter. In different test series the final gap distance during interruption is changed from 5 to 25 mm. The change of arc modes in the investigated window is well known and described by many researcherss [5]. The new evaluation method enables deeper insights and increases the available data. The focus of this paper is to show the possibilities of automatized data evaluation on the example of current density determination of constricted vacuum arcs under TMF.

## 2. Experimental Setup

A schematic diagram of the test setup is shown in Figure 1. Main part is a vacuum test interrupter which is evacuated by a two-level pump system. The measured pressure before and after each test series is below $4 \cdot 10^5$ Pa.

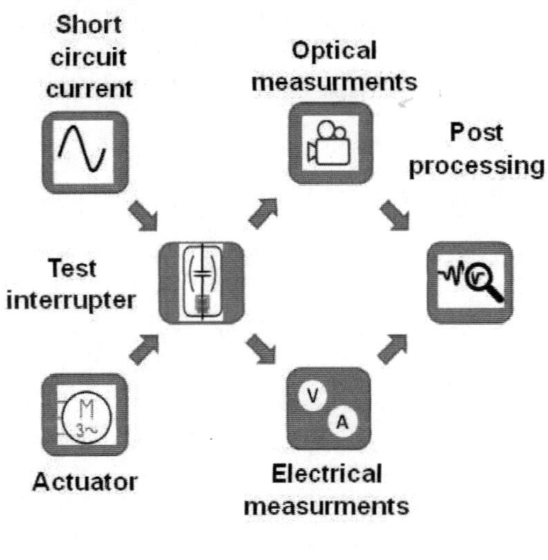


*Figure 1. Schematic of test setup in combination with recorded data during the test series.*

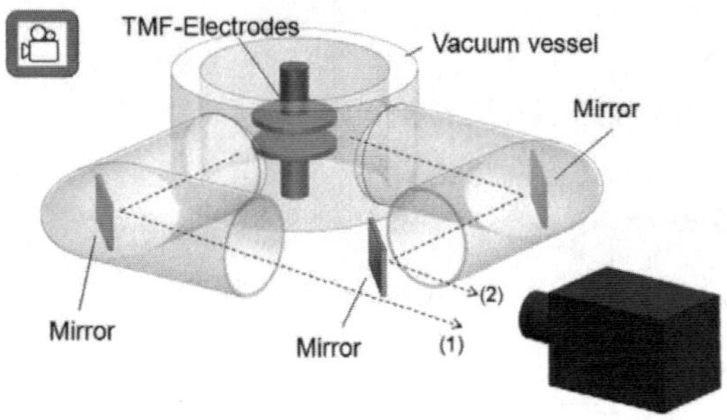


*Figure 2. Schematic of optical plasma recording.*

The used contacts are spiral type and supply a transversal magnetic field (TMF) on the arc. The test interrupter is driven by a special actuator based on a servo motor to enable adjustable switching gaps and velocities. The current used for the breaking operation is a 50 Hz sinusoidal half-wave generated by a capacitor bank and coordinated coils. Wolf [5] explains the setup of the supply circuit in detail. This circuit generates a 20 kA RMS current. A 20 mΩ shunt measures the current. For voltage measuring a voltage probe (Tektronix) over the gap is used. A linear potentiometer connected to the interruption rod measures the contact gap. The electrical signals are transformed into light signals and transmitted via fibre optical cable to a HBM Gen7t transient recorder. In addition to the electrical measuring, the test interrupter enables an optical measurement over two 90° angle arms equipped with a mirror system according to Figure 2. The optical set-up is used to observe the metal vapor plasma between the separating contact from two perspectives. Therefore, rectangular tubes are used in combination with a mirror system to enable the recording of two optical paths with one high-speed camera. The utilized high-speed camera is a Redlake MotionPro X4 with an aperture f/16 and a recording frequency of 50500 Hz. No additional filter is used. The high-speed camera integrates brightness

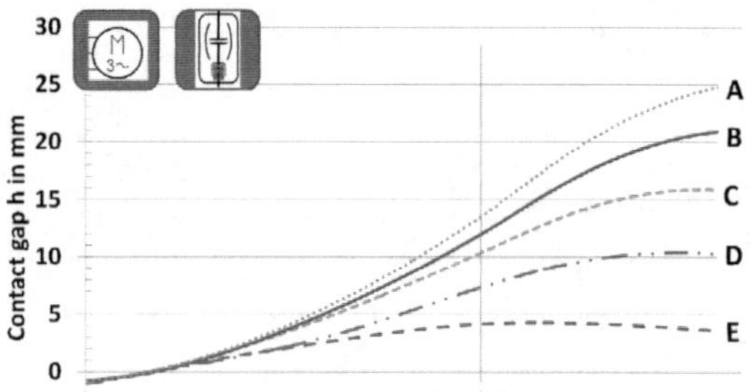


*Figure 3. Exemplary course of five different interrupting gaps used for the test series.*

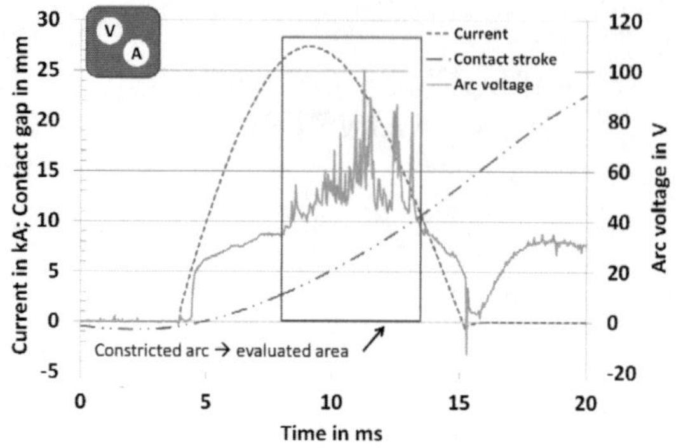


*Figure 4. Exemplary course of one interruption operation with current, gap and voltage measurement, used for the test series.*

levels within 1 µs. The camera sensor has 255 bit levels to differentiate brightness. The first 50 bit levels are presumed to be noise of background reflection and scattered light. In the test series five different contact gaps with a final gap from 5 to 25 mm are chosen. According to Figure 3 the contact spring is tightened over 3 mm to supply a contact force of 2 kN. The capacitor bank is switched onto the test interrupter at a time immediately after 0 ms. This leads to arcing times in a range of 9 ms for each interrupting test. All gaps are tested with five current applications. One representative breaking operation is displayed in Figure 4. The interrupter opens shortly after the zero crossing of the half wave current and an arc ignites. The arc voltage rises to ranges between 20 and 40 V. At one point the voltage starts to change its potential with spikes up to 100 V. In this area the plasma of the vacuum arc constricts to one defined arc column. With a falling current of the half wave the arc mode returns to a diffuse arc with a smooth voltage signal until the arc is extinguished at the zero crossing of the current. The behavior of the constricted arc column will be evaluated in the next steps.

## 3. Post-processing

For an automated analysis the images are simplified using a separation with an Otsu threshold and a morphological filtering. Weber [4] presents the exact processing of the images. The binarized images deliver different features of the recognized objects and enable a detailed evaluation of the recorded data. One representative image and its features is shown in Figure 5. From the recorded plasma image geometrical properties such as height, width and position on the contacts are extracted. The images and the electrical data are coordinated using the voltage step as a reference for the first plasma phenomena. To evaluate the plasma behaviour in dependece of the contact gaps over optical observation, the averaged high of the plasma in all five test series according to Figure 3 is considered. The size of objects is derived by calculating the length

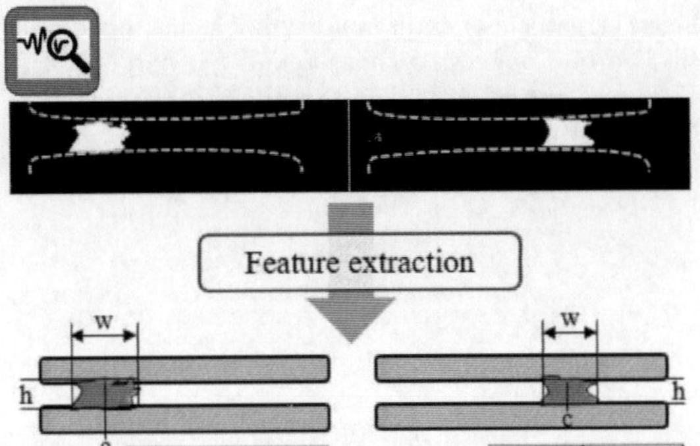


| | Side view | Front view | | Side view | Front view |
|---|---|---|---|---|---|
| Width w (px) | 40 | 37 | Center of mass c (px) | 58,6 | 149,4 |
| Height h (px) | 20 | 17 | Area (px) | 579 | 487 |

*Figure 5. Schematic for post-processing of optical recorded data.*

each pixel represents, which is done with the known diameter of the contacts. A reference picture is taken for this purpose. Subsequently the derived pixel to length ratio is then used for every calculation. With a 63 mm diameter contact, the ratio is 0.74 mm per pixel.

## 4. Results

The results of all evaluated optical data sets is shown in Figure 6. It depicts the height of a constricted plasma column, derived from the optically recorded data. The evaluation of the heigth also indicates the actual interruption gap. Thus, comparability between optical evaluation and results of the recorded electric measurement of switching gaps according to Figure 3 is expected. Due to the high noise, the presented data has been smoothed using a third degree polynomial function. The measured data of the 5 mm gap (E) could not be evaluated with this method, which is assumed to be due to the arc mode. During switching operation with a final gap of 5 mm no observable constricted arc is detectable which affords another data treatment. For the other gaps (A – C) a comparison between electrical data (Figure 3) and optical data Figure 6 enables evaluation of the functionality of the approach but also of its restrictions. The graph of (E) does not start at zero mm, which is caused by the pixel based evaluation. Due to the high luminescence in the small gap more pixels are illuminated which leads to a misinterpretation. During the interruption process the different trajectories show similar courses. After 10 ms the current flow goes to zero, the evaluated curves show a decreasing gap which is related to the luminescences in the test set-up. The 90° recording angle enables the calculation of an ellipsoid arc column. The averaged column area during the constriction mode (marked in Figure 4 over the rectangle) is calculated. A reduction of the effective arc area in dependence of the final contact gap was discovered. Combining the optical and electrical data, while assuming an equal current flow over the calculated arc column and measured current flow, a current density can be estimated. The resulting current densities are shown in Figure 7.

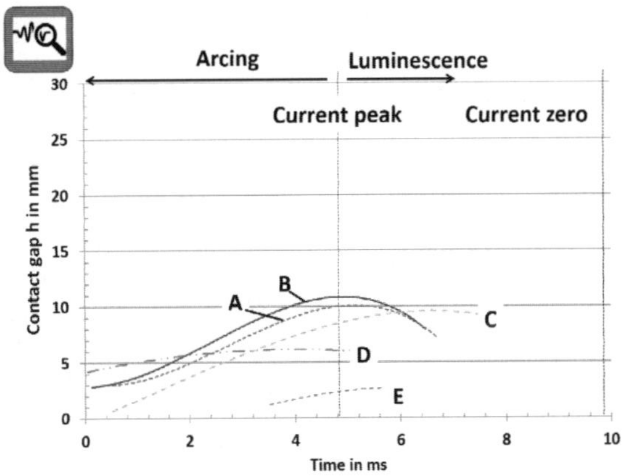


*Figure 6. Optical evaluated opening curves.*

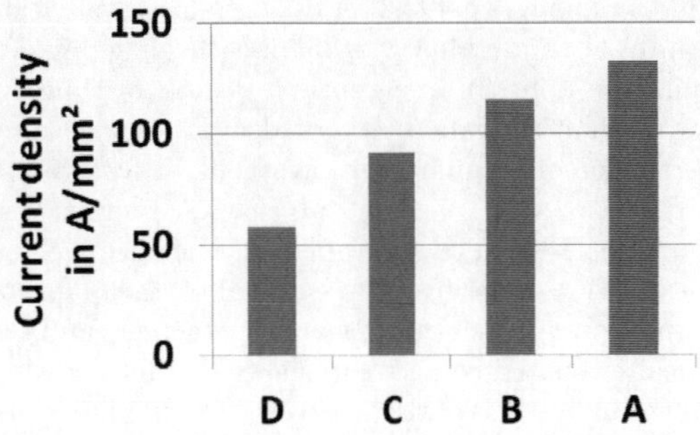


*Figure 7. Current densities of constricted plasma columns in dependence of the contact gap.*

In spite of the simplicy of the assumptions, the current densities are comparable to the ones reported in other works [6, 7]. The data show a connection between the current density and the final contact gap. The current density in the arc rises with the length of the final gap. This occurs due to the contraction of the plasma column which increases with the final gap. The reason for this constriction is assumed to be based on an interaction between the pinching force, the Lorentz force and the internal plasma pressure.

## 5. Discussion and Conclusion

The optical evaluation of vacuum arcs is an established tool to deduce arc properties. The variety of different cameras, optical observation methods and disfferent evaluation approaches complicate the comparison between the available results. In addition, the technology is becoming faster, which increases the number of images per test series. A manual evaluation requires the individual viewingof each image, which cannot be done in an acceptable period of time. A new method of evaluating recorded data concerning observation of vacuum arc is presented. It features an automatized tool based on LabView. A vacuum test interrupter is used to interrupt the same effective current of 20 kA under five different final contact gaps. A comparison between the measured contact gap and the optical evaluated contact gap shows the functionality and the limitations of the approach. The luminescence in the test setup could lead to a misinterpretation of data and shows that the optical investigation has to be combined with recorded electrical data to achieve a correct interpretation. Finally the areas of the different arcs are automatically determined to calculate the current densities of constricted arcs in dependence of the contact gap. The simple approach shows a rising current density with an increasing gap. The results of this work are to be understood as a feasibility study for optical evaluation methods and present a first step to automatically derive scientifically valuable data from optical observations of plasmas.

## References

[1] W. Shi, L. Wang, R. Lin, Y. Wang, M. J., and S. Jia. Experimental investigation of drawing vacuum arc under different TMF contacts in vacuum interrupter. *IEEE Trans. Plasma Sci.*, 47(4):1827–1834, 2019. doi:10.1109/TPS.2019.2903557.

[2] W. Haas and W. Hartmann. Investigation of arc roots of constricted high current vacuum arcs. *IEEE Trans. Plasma Sci.*, 27(4):954–960, 1999. doi:10.1109/27.782266.

[3] T. Rettenmaier, V. Hinrichsen, and E. Taylor. Investigations on arc movement in vacuum interrupters by arc rotation measurements with external magnetic field sensors. In *International Symposium on Discharges and Electrical Insulation in Vacuum (ISDEIV)*. IEEE, 2014. doi:10.1109/DEIV.2014.6961641.

[4] B. Weber, D. Gentsch, T. Pieniak, and M. Kurrat. Software-based processing of the radiation intensity distribution of high-current vacuum arcs between transversal magnetic field contacts. In *Proceedings of the 28th International Symposium on Discharges and Electrical Insulation in Vacuum*. IEEE, 2018. doi:10.1109/DEIV.2018.8537111.

[5] C. Wolf, M. Kurrat, M. Lindmayer, and D. Gentsch. Arcing behavior on different TMF contacts at high-current interrupting operations. *IEEE Trans. Plasma Sci.*, 39(6):1284–1290, 2011. doi:10.1109/TPS.2011.2135379.

[6] C. Wolf, M. Kurrat, M. Lindmayer, E.-D. Wilkening, and D. Gentsch. Optical investigations of high-current vacuum arc behavior on spiral-shaped and cup-shaped RMF-contacts. In *55th IEEE Holm Conference on Electrical Contacts*, pages 270–275. IEEE, 2009. doi:10.1109/HOLM.2009.5284391.

[7] T. Delachaux, O. Fritz, D. Gentsch, E. Schade, and D. L. Shmelev. Modeling and simulation results of a high current vacuum arc in a transverse magnetic field. In *23rd International Symposium on Discharges and Electrical Insulation in Vacuum*. IEEE, 2008. doi:10.1109/DEIV.2008.4676869.

doi:10.14311/ppt.2019.2.107
*Plasma Physics and Technology* **6**(2):107–110, 2019 

# INVESTIGATION OF THE METHOD OF DYNAMIC MICROWAVE POWER REDISTRIBUTION IN A RESONATOR-TYPE PLASMATRON

S. Madveika, S. Bordusau*, A. Zemlyakov, M. Lushakova

*Belarusian State University of Informatics and Radioelectronics, P.Brovki 6, 220013 Minsk, Belarus*

* bordusov@bsuir.by

**Abstract.** The investigation results of a dynamic microwave power $f_{\text{microwave}} = 2.45 \pm 0.05$ GHz redistribution in a 9000 cm$^3$ reaction-discharge chamber of a microwave resonator-type plasmatron are presented. In order to redistribute the microwave power, a rotating metallic four-blade L-form dissector placed above the reaction-discharge chamber was used. The microwave power in the local points at the axis of the chamber with plasma and without it was measured applying the "active probe" method. During the experiments the chamber contained silicon wafers. Periodical interchange of maximum and minimum microwave power values along the chamber axis was established experimentally. Note, when the dissector was rotating, the range of maximum and minimum "active probe" values dispersion decreased. It has been established that during the dissector rotation the microwave power in the local discharge areas changes with periodic repetition every quarter of revolution.

**Keywords:** microwave plasma, microwave power, redistribution, dissector.

## 1. Introduction

The development of microwave gas discharge devices for technological purposes is moving in the direction of developing microwave plasmatrons with big plasma volume. It is connected with the necessity to treat semiconductor wafers of 200 and 300 mm diameter and the need to treat groups of wafers of smaller diameter at the operations of plasma cleaning, substrate surface activation, radical and plasma etching, plasma-chemical deposition of thin films, etc. [1–8]. Such technological plasma-chemical installations are developed on the base of microwave cavity resonators with discharge vacuum chambers located in the center [9–12].

Reasons for using the resonator systems are the next: they provide a high level of power dissipation in plasma; resonator is a well matched connector of microwave power feeder with plasma because its impedance changes insignificantly with the change of plasma impedance [13]; resonator preserves sufficiently high Q-factor even with the presence of plasma in it [14].

It is known [2] that the rates of plasma processes (ionization, dissociation, etc.) in microwave discharge plasma significantly correlate with the distribution of electromagnetic field in plasma. The concentration of active plasma particles, in its turn, is connected with the rate of plasma treatment processes of materials [9]. Depending on the process active plasma particles may be ions, radicals, atoms or etc.

In [15] it was shown that microwave power and spatial electric conductivity distributions in the resonator-type plasmatron's discharge chamber is unsteady when the values of the specific power applied to the microwave discharge are in the range $W_{MW} = 0.06$–$0.08$ mW/cm$^3$. This is due to the formation of standing waves with a certain structure in the resonator [16].

There are some technical methods [2, 12, 17] allowing to change the microwave power distribution in a resonator. The simplest one is the use of a rotating metallic dissector. Such method of a dynamic microwave power redistribution does not require significant modification of the existing microwave plasmatrons' construction. As per today, there is no information in the scientific literature on the results of investigation of dissector's influence on the redistribution of microwave power inside and outside of the plasma volume in resonator-type plasmatrons. Thus, such research is of a scientific and practical interest as it may be useful for developing new microwave plasma systems or upgrading the existing ones.

## 2. Experimental apparatus and procedures

The research of the effect of a dynamic microwave power redistribution in a microwave discharge plasma was carried out with the help of a small-size resonator-type microwave plasma installation. The principle of operation of the vacuum-plasma system under study is based on the transmission of microwaves' energy from microwave magnetron M-105 ($f_{\text{microwave}} = 2.45 \pm 0.05$ GHz) into a vacuumed reaction-discharge chamber where a low-temperature gas microwave discharge is excited under the influence of the microwave electromagnetic field.

The discharge chamber is made of a quartz pipe, ⌀20 cm and 34 cm long, with a volume about 9000 cm$^3$, placed in the center of a $40 \times 40 \times 30$ cm$^3$ resonator (Fig. 1). The electromagnetic waves from the microwave magnetron are inlet into the resonator

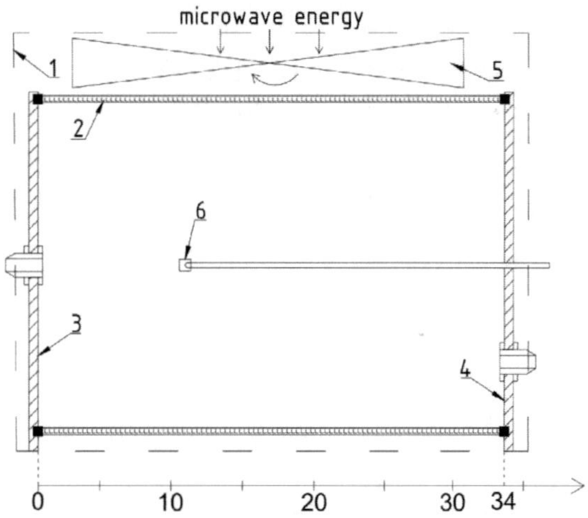


*Figure 1. Reaction-discharge system of the microwave plasma installation:*
*1 — resonator; 2 — discharge chamber; 3 — front cover; 4 — back wall; 5 — dissector; 6 — "active probe"*

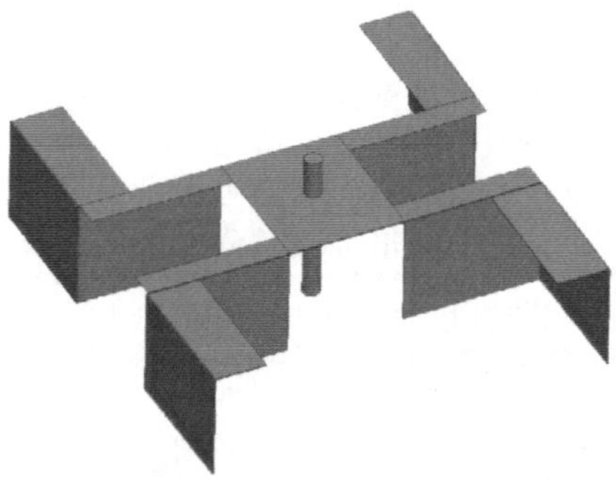

*Figure 2. L-form dissector*

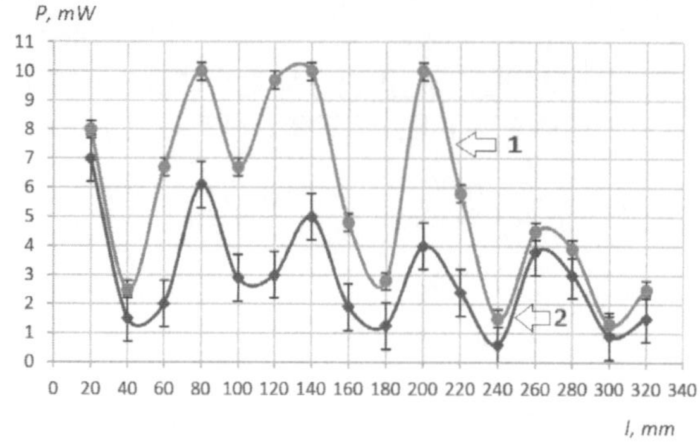


*Figure 3. Readings of the active probe along the axis of the discharge chamber without plasma:*
*1 — without a dissector; 2 — with a moving dissector*

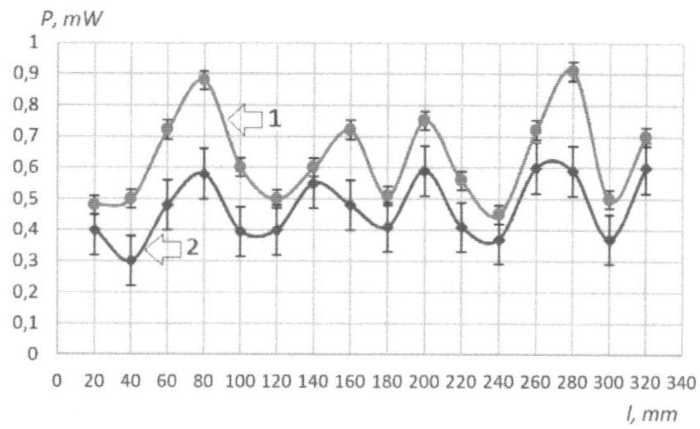


*Figure 4. Readings of the active probe along the axis of the discharge chamber with plasma:*
*1 — without a dissector; 2 — with a moving dissector*

through a rectangular hole in the resonator's top wall. An L-form metallic dissector is fixed on the resonator's top wall (Fig. 2). The dissector was placed in a position that allowed its blades to pass under the hole while rotating.

The microwave power in local points on the discharge chamber's axis with plasma and without plasma was measured using the "active probe" method [15]. The readings of the "active probe" were registered with a microwave power meter.

The character of plasma luminescence was controlled by registering the integral optical luminescence from the microwave discharge. The optical luminescence was sensed by an optical fiber, placed perpendicularly to the discharge chamber's axis, and was passed to the photoelectron multiplier. The photoelectron multiplier was connected to a digital oscilloscope.

During the experiments, there were 6 silicon wafers in the discharge chamber. The wafers of ø100 mm were evenly placed in two lines along the chamber, 3 pcs from each side of the probe. The distance between the wafer pairs was 60 mm.

$O_2$ was used as a plasma-forming gas. The operating pressure in the chamber was 130 Pa.

## 3. Results and discussion

Figure 3 shows the microwave power meter's readings when the "active probe" was moving along the axis of the discharge chamber without plasma. A point near the front chamber's cover was taken as the beginning of reading the distance along the discharge chamber axis.

It was established that when using a rotating metal dissector the range of microwave power values measured by the "active probe" along the axis of the discharge chamber decreases from 1.5–10 mW to 0.5–7 mW. This indicates a more uniform microwave power distribution in the discharge chamber volume.

Figure 4 shows the microwave power meter's readings for the case when the "active probe" was moved along the axis of the discharge chamber with plasma.

The presented data show a significant space irregularity decrease of the microwave power values in the discharge chamber's volume measured by the "active probe" when plasma is present and the dissector is moving. The maximum and minimum microwave

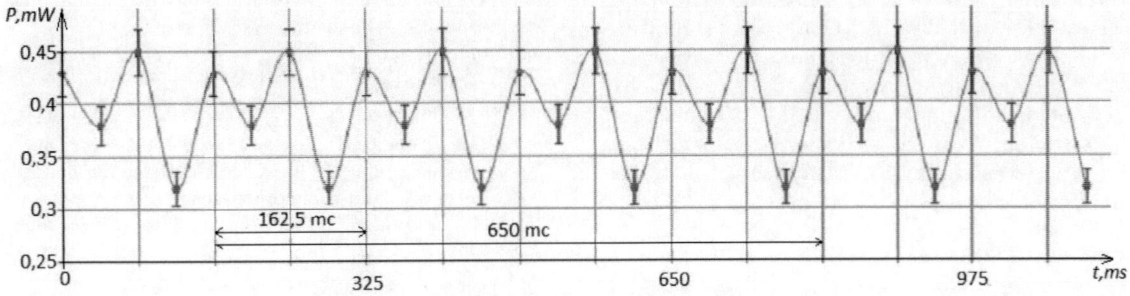


*Figure 5. Changing of the "active probe" readings in the local area of the microwave discharge during the rotation of the dissector*

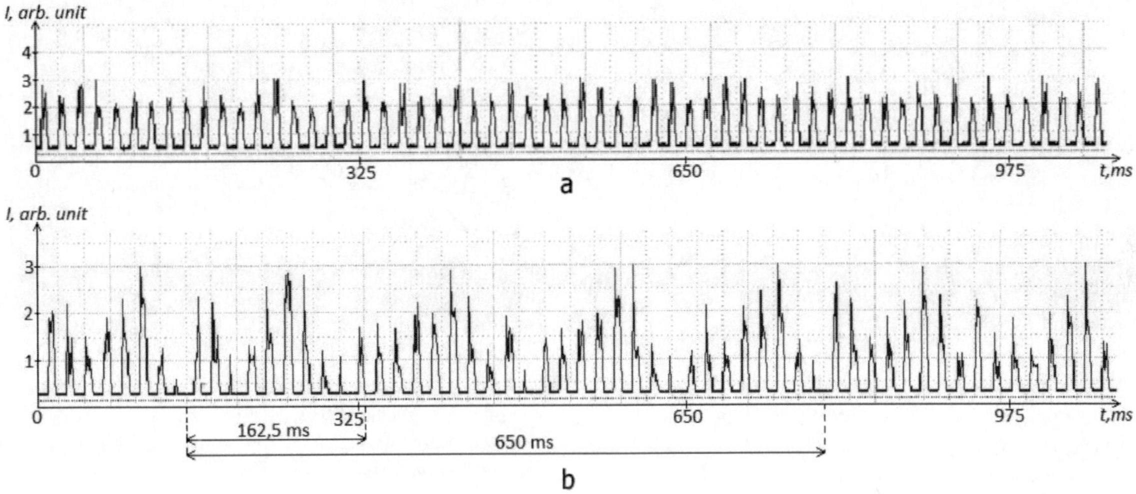


*Figure 6. The registered oscillograms of integrated optical plasma luminescence signals from the local region of the microwave discharge:*
*a — without a dissector; b — with a moving dissector*

power values in this case are 0.3 mW and 0.6 mW respectively.

The obtained results may be explained in the following way. When the dissector is moving, the conditions of excitation for various types of waves change depending on the blades position with respect to the hole of inletting the microwave power into the resonator. Because of this, the structure of the field in the resonator is permanently changing. It results in the decrease of irregularity of the electromagnetic power distribution in the microwave plasmatron's discharge chamber. The registered periodicity of maximum and minimum values in both series of experiments corresponds to about a half length of the electromagnetic wave ($f_{\text{microwave}} = 2.45 \pm 0.05\,\text{GHz}$) exciting the plasma.

It was found out that during the rotation of the dissector, the value of the microwave power in the local discharge areas changes with periodic repetition every quarter of its revolution (Fig. 5) (the time of a complete dissector's revolution is 650 msec).

Using a photoelectron multiplier, the changes of plasma luminescence intensity from different discharge chamber areas was studied. As an example, Figure 6 shows oscillograms of integral optical plasma luminescence signals from a local area when the dissector is absent (a) and in case it moves (b).

As it can be seen from Figure 6, the intensity of the plasma luminescence from the local region varies with periodic repetition every quarter of a turn of the dissector. This correlates with the data presented in Figure 5.

The obtained results indicate that during the dissector rotation a spatial redistribution of the microwave power in the volume of the discharge chamber takes place. This contributes to a decrease in the spatial inhomogeneity of the plasma in the microwave discharge chamber of the resonator-type plasmatron.

## 4. Conclusions

The performed investigation showed that, when the dissector is rotating, the microwave power is redistributed within the volume of the chamber of the microwave resonator-type plasmatron. The maximum and minimum values of the "active probe's" readings along the discharge chamber's axis decrease. It has been established that, when the dissector is rotating, the microwave power in the local areas of the discharge changes periodically every quarter of its revolution.

The obtained results can be used in designing new and upgrading of existing technological equipment for microwave plasma-chemical treatment of materials and electronic devices.

## Acknowledgements

This research has been supported by the Belarusian State Program of Scientific Research "Physical materials science, new materials and technologies", assignment 3.5.02.

## References

[1] M. Moisan and J. Pelletier. *Microwave Excited Plasmas.* Amsterdam: Elsevier, 1992.

[2] S. V. Bordusov. *Microwave plasma technologies in the production of electronic devices.* Minsk: Bestprint, 2002.

[3] M. Moisan and J. Pelletier. *Physics of Collisional Plasma. Introduction to High-Frequency Discharges.* Dordrecht: Springer, 2012.

[4] C. Boisse-Laporte and J. Marec, editors. *Microwave discharges: fundamentals and applications Proc. 3rd Int. Workshop (Fontevraud, France, 20-25 April 1997).* France: EDP Sciences, 1998.

[5] A. Ohl, editor. *Microwave discharges: fundamentals and applications Proc. 5th Int. Workshop (Greifswald, Germany, 8-12 July 2003).* Greifswald: INP, 2003.

[6] M. Kando and M. Nagatsu, editors. *Microwave discharges: fundamentals and applications Proc. 7th Int. Workshop (Hamamatsu, Japan, 22-27 September 2009).* Japan, 2009.

[7] Y. A. Lebedev, editor. *Microwave discharges: fundamentals and applications Proc. 8th Int. Workshop (Zvenigorod, Russia, 10-14 September 2012).* Moscow: Yanus-K, 2012.

[8] Y. A. Lebedev. Microwave discharges at low pressures and peculiarities of the processes in strongly non-uniform plasma. *Plasma Sources Science and Technology*, 24(5):053001, 2015. doi:10.1088/0963-0252/24/5/053001.

[9] T. Nakano and S. Samukawa. The correlation between an electric field and the metastable chlorine ion density distributions in an ultrahigh-frequency plasma. *Japanese Journal of Applied Physics*, 37(5R):2686–2692, 1998. doi:10.1143/JJAP.37.2686.

[10] Plasmatic Systems. Plasma-preen plasma cleaning/etching system. URL: http://www.plasmapreen.com.

[11] Terra Universal. Plasma-preen cleaner/etcher (patent pending). URL: https://www.terrauniversal.com/cleaning-systems/plasma-preen-cleaners.php.

[12] A. I. Kuzmichev. Ion plasma sources based on a microwave oven. *Instruments and Experimental Techniques*, 37(05):176–180, 1994.

[13] V. M. Batenin, I. I. Klimovsky, G. V. Lysov, and V. N. Troizky. *Microwave Plasma Generators: Physics, Engineering, Applications.* Moscow: Energoatomizdat, 1988.

[14] S. I. Madveika and S. V. Bordusau. Theoretical analysis of low vacuum microwave discharge exciting and maintaining conditions in resonator type plasmatron. *Plasma Physics and Technology*, 2(2):155–158, 2015.

[15] S. V. Bordusau, S. I. Madveika, and A. P. Dostanko. Investigation of microwave energy distribution character in a resonator type plasmatron. *Plasma Physics and Technology*, 3(3):122–125, 2016.

[16] E. Okress. *Microwave power engineering.* Number 1 in Electrical science series. Academic Press, 1968.

[17] J. Doehler. Large area microwave plasma apparatus with adaptable applicator, April 3 2001. US Patent 6.209.482. URL: https://encrypted.google.com/patents/US6209482.

doi:10.14311/ppt.2019.2.111
Plasma Physics and Technology 6(2):111–114, 2019 

# INVESTIGATION OF CONVERSION PROCESS OF $SiCl_4$ + $CCl_4$ MIXTURE BY RF (40.68 MHZ) ARC DISCHARGE

R. Kornev[a,*], P. Sennikov[a], V. Nazarov[a,b], A. Kut'in[a], A. Plekhovich[a]

[a] G.G. Devyatykh Institute of Chemistry of High-Purity Substances of RAS, 49 Tropinin Str., Nizhny Novgorod, 603951, Russia

[b] Institute of Applied Physics of RAS, 46 Ulyanova Str., Nizhny Novgorod, 603950, Russia

[*] kornev@ihps.nnov.ru

**Abstract.** A contracted RF (40.68 MHz) arc discharge of atmospheric pressure, stabilized between two rod electrodes, was used to obtain trichlorosilane by the reaction of hydrogen reduction of silicon tetrachloride ($SiCl_4$). In model mixtures of macro-composition in the ratio $H_2/SiCl_4/CCl_4=10/1/1$, it was shown that C and SiC are the main solid-phase product which are deposited on the surface of electrodes in the form of dendrides. The temperature of the ends of the electrodes determined using emission thermometry is 1600 K. The thermodynamic analysis of $H_2$+$SiCl_4$+$CCl_4$ system confirms that the formation of C and SiC occurs in the temperature range of 1600 K. The deposition of solid-phase products occurs on the electrodes in the zone of high electric field strength.

**Keywords:** RF arc discharge, conversion process of $SiCl_4$+$CCl_4$, thermodynamic analysis.

## 1. Introduction

The study of the process of hydrogen reduction of $SiCl_4$ to $SiHCl_3$ is an urgent task since these compounds are the main by-product and the starting material, respectively, in the processes of obtaining silicon and silane [1]. In [2] a method of plasma-chemical synthesis of trichlorosilane under conditions of RF (40.68 MHz) arc discharge, providing a high (up to 60%) yield of trichlorosilane with low energy consumption, was proposed. It is known that, along with the main synthesis reaction, the reactions involving micro-impurities occur in the volume of gas discharge which leads to the change in their chemical form and concentration [3–6]. The content of impurities in silicon tetrachloride, that are formed in the process of obtaining polycrystalline silicon and silane, is at the purity level of the original trichlorosilane. A significant place among impurities is occupied by the impurities of chlorohydrocarbons and, in particular,$CCl_4$. In [7] it was shown that the content of chlorine hydrocarbon impurities in the products of hydrogen reduction of $SiCl_4$ in this type of discharge decreases from $10^{-4}$ to $10^{-7}$ mass %. Lowering the concentration of these micro-impurities at the stage of trichlorosilane synthesis will allow to obtain semiconductor silicon of higher purity in the future. It is necessary, nevertheless, to establish the chemical form of these impurities.

## 2. Description of RF plasmatron

RF plasmatron consists of silica tube with two coaxially arranged silicon electrodes. The distance between the electrodes is 10 mm. To initiate a discharge, it is necessary for the $E/P$ value be at least 26 V/(cmTorr) [8]. The measurement of the electric field was carried out using the indicator Ya6P-110. The maximum value of the electric field $E$ was $6 \times 10^3$ V/cm. When the pressure in the reactor is 760 Torr, the field strength is not sufficient for ionization of the inter-electrode space. The initiation of the discharge was carried out under reduced pressure of 100 Torr. In this case, the ratio $E/P = 60$ V/(cmTorr) which is quite enough for gas breakdown. The type of discharge is shown in Fig. 1a. The discharge glow occupies the inter-electrode area, envelops the electrodes and spreads in the volume of the reactor. When pressure increases up to atmospheric the contraction of the discharge occurs. The discharge at atmospheric pressure using hydrogen is shown in Fig. 1b. The discharge is an AC electric arc with a frequency of 40.68 MHz burning between the electrodes. When $SiCl_4$ entered the plasma, the discharge was transformed into a stable spherical formation of bright blue color shown in Fig. 1c. In this case there is a strong heating of the end parts of the electrodes in contact with the plasma. The electric current in the gas under these conditions is determined by the characteristics of the external circuit. The characteristic value of the total current determined by the method described in [9] was $I \approx 2$ A.

Estimation of the current can also be made using the current density of thermionic emission, and the temperature of the electrodes. For the used material of the electrodes (silicon) and the known area of the heated surface, the current density is determined by the formula:

$$j = A \times T^2 exp(-\frac{e\varphi}{kT}), \qquad (1)$$

where $A$ for silicon is in the range 50–120, $e\varphi$ - 3.6 eV [8].

In order to determine the temperature of the ends of the electrodes in the range of 350–950 nm, the

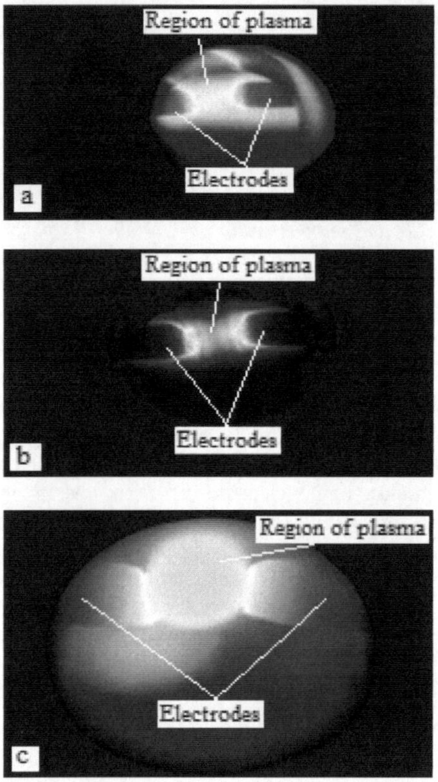


*Figure 1. View of gas discharge; a - P=100 Torr $H_2$, b - 760 Torr $H_2$, c - 760 Torr $H_2$+$SiCl_4$.*

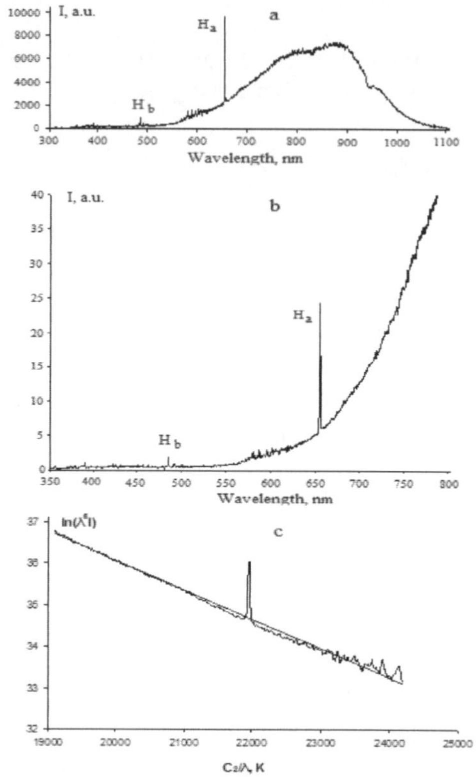


*Figure 2. Emission spectrum of the discharge in hydrogen before (a) and after (b) adjustments and in Wynn coordinates (c)*

emission spectrum of hydrogen, as well as chemically active plasma was investigated using the emission spectrometer HR4000CJ-UV-NIR. To determine body temperature with unknown values of emissivity, the emission spectrum was analyzed in the approximation of a gray radiator, the spectrum of which with accuracy to emissivity $\varepsilon$ corresponds to the Planck spectrum [10]:

$$I(\lambda,T) = \varepsilon(\lambda,T)\frac{2hc^2}{\lambda^5}\frac{1}{exp(\frac{hc}{\lambda k_b T}-1)}, \quad (2)$$

In the course of the analysis, the spectral range of measurements was chosen for which the dependence of the emissivity on the wavelength is weak compared to the curvature of the continuous spectrum. In the selected wavelength range, the spectrum constructed in Wynn coordinates $\ln(\lambda^5 I)$ - $C_2/\lambda$ is linear (the unit is neglected in the sum), and the proportionality coefficient is $1/T$:

$$\ln(\lambda^5 I) - \ln(\frac{\varepsilon C_1}{hc}) = -\frac{C_2}{\lambda}\frac{1}{T}, \quad (3)$$

where $C_1 = 2hc^2 = 37418\,\text{W}\cdot\mu\text{m}^4/\text{cm}^2$, $C_2 = hc/k_b = 1.4388\cdot10^4\,\mu\text{m}\cdot\text{K}$.

Fig. 2 shows the emission spectrum of the discharge in hydrogen before (a) and after (b) the correction which amends the spectral sensitivity of the measuring circuit. The same spectrum after adjustment, in Wynn coordinates is shown in Fig. 2c. The temperature of the ends of the electrodes determined in this way is 1600 K. It can be assumed that gas in the near-electrode region has the same temperature. Knowing the area of the heated surface of the electrode, estimated by the formula (1), the current is $I \approx 2.5$ A which by order of magnitude coincides with the value $I$ determined by the method [9].

## 3. Experimental

In order to determine the main products into which chlorine hydrocarbon impurities and in particular $CCl_4$ can transfer, the hydrogen reduction process was studied on the model mixture of macro-composition in the $H_2/SiCl_4/HCl_4$ ratio from 10/0.1/1 to 10/1/1. The total flow of plasma gases was 700 $\text{cm}^3$/min. Power input to the plasma discharge zone was calculated as difference in power of respectively reflected and incident waves measured by directional coupler with directive gain of 30 dB and an oscilloscope, and was equal to $500 \pm 30$ W. The total degree of chloride conversion and the presence of gas-phase reaction products were determined by gas chromatography with an accuracy of 1 mol %. X-ray analysis of the obtained solid-phase products was carried out on XRD-7000 diffractometer. Morphological studies were conducted using scanning electron microscopy methods.

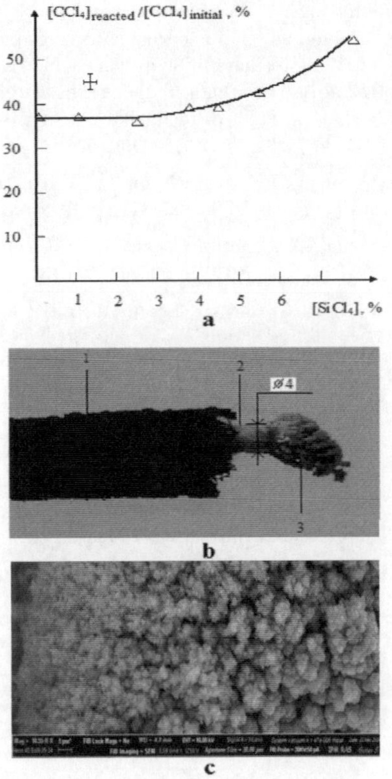


*Figure 3. a - Dependence of conversion degree of $CCl_4$ on the content of $SiCl_4$ in the initial mixture;b - View of electrode: 1 - nozzle, 2 - silicon electrode, 3 - deposited SiC and c - morphology of deposited SiC.*

The dependence of the degree of conversion of $CCl_4$ on the content of silicon tetrachloride in the initial mixture is shown in Fig. 3a. It can be seen from the graph that when the content of $SiCl_4$ in the mixture is from 0 to 3%, the degree of conversion of $CCl_4$ is 38%. With an increase in the concentration of $SiCl_4$ from 3 to 8%, the degree of conversion begins to increase and is 57% when the concentration of $SiCl_4$ in the mixture is 8%.

Gas chromatographic analysis showed that in the absence of $SiCl_4$ in the initial mixture, the reaction products contain $CCl_4$, HCl and a small amount (less than 5% mole.) of chloroform. During the experiment, there is also an intensive release of solid deposits on the surface of the electrodes in the form of dendrites (see Fig. 3b,c). It was shown that the main solid-phase product are C and SiC.

## 4. Thermodynamic analysis

In order to analyze the equilibrium composition of the main transformation products at a pressure of 1 atm., thermodynamic modeling was carried out for $H_2/SiCl_4/CCl_4$ mixture. The software package Chemical Thermodynamics Calculator, which implements the Gibbs method with fixed $T$ and $P$, was used. Formulas (7) obtained in the conditional minimization of Gibbs energy (4) express the component concentrations through Lagrange non-identified multipliers. Solving the balance equations (5) and (6) in terms of phases $f_0$ and elements $j_0$ relative to $\{\lambda_j\}$ after substituting (7) into (5) and (6) determines the equilibrium composition of the heterophase system.

$$G(T,P,\{n_j\})/RT = \sum_i n_i(g_i + \ln n_i) - \sum_f \overline{n_f} \ln \overline{n_f} \tag{4}$$

$$\sum_i n_i a_{if} - \overline{n_f} = 0, f = 1C \dots f_0 \tag{5}$$

$$\sum_i n_i a_{ij} - b_j = 0, j = 1C \dots f_0 \tag{6}$$

$$n_i = \overline{n_f} \exp(\sum_j a_{ij}\lambda_j) \tag{7}$$

$$g_i = \Delta_f H^o(298)/RT - \Phi^o_{298}/R + \begin{cases} \ln \tilde{P}, & \text{vapor} \\ 0, & \text{condensate} \end{cases} \tag{8}$$

The values $g_i$ contain the initial thermodynamic information for each $i$-th component. The matrix $A_{ij} = \{a_{ij}, a_{if}\}$, in addition to the stoichiometric coefficients of the participation of the $j$-th element in the chemical formula of the $i$-th substance, includes the coefficients of belonging of the $i$-th component to phase $f$ $a_{if} = \{1, 0\}$. According to the initial chemical composition of the reagents, the elemental composition $b_j$ of the system in the balance equations (6) is calculated.

## 5. Discussion of results

Fig. 4a shows the temperature dependence of the equilibrium composition of the main conversion products at a pressure of 1 atm. for $SiCl_4$+$CCl_4$+$10H_2$ system. It can be seen that in the region of $T = 1600$ K there are condensed phases C and SiC. Thus, the results of experimental studies show the possibility of formation of C and SiC in the realized experimental conditions. Thermodynamic analysis confirms that the formation of these products is observed at $T = 1600$ K, i.e. at a temperature determined using emission thermometry which is characteristic of the ends of the electrodes and the near-electrode region.

Deposition of carbide occurs at the electrodes in the zone of high electric field strength due to which the molecules acquire an induced dipole moment. In the process of burning of the discharge, the electrode acquires a negative charge relative to the plasma volume. At the same time, due to electrostatic interaction, positively charged carbon or silicon atoms, which are part of the dipole, are attracted to the electrode surface. During adsorbtion on the surface of the electrode, the atoms of carbon and silicon form phases of C and SiC. The proposed mechanism of deposition is shown in Fig. 4b.

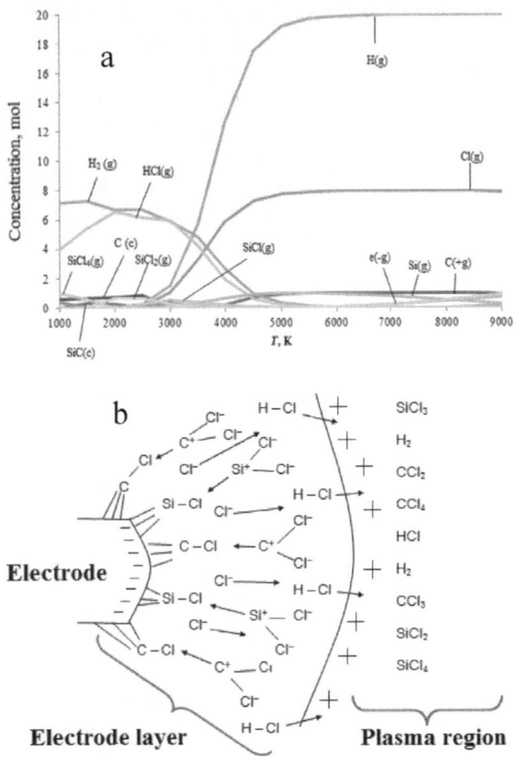


*Figure 4. a - Temperature dependence of composition of the main conversion products at 1 atm. pressure for $SiCl_4+CCl_4+10H_2$ system, b - Mechanism of deposition of SiC.*

## 6. Conclusions

As a result of the research, the conditions for breakdown in RF (40.68 MHz) contracted arc discharge of atmospheric pressure on hydrogen stabilized between two rod electrodes were determined. The method of emission thermometry was used to determine the temperature of the heated ends of the electrodes in contact with the plasma and the electrode areas, that was 1600 K. Heating the electrodes to the given temperature contributes to the deposition of the phases C and SiC on them which is confirmed by thermodynamic analysis. It can be assumed that in the process of plasma-chemical hydrogen reduction of $SiCl_4$ the impurities of chlorohydrocarbons behave in a similar way which leads to the effect of additional purification.

### Acknowledgements

The work was conducted on the topic of government task No 0095-2016-0006.

### References

[1] A. V. Gusev, R. A. Kornev, and A. Y. Sukhanov. Preparation of trichlorosilane by plasma hydrogenation of silicon tetrachloride. *Inorganic Materials*, 42(9):1023–1026, 2006. doi:10.1134/S0020168506090172.

[2] R. A. Kornev, V. A. Shaposhnikov, and A. M. Kuz'min. Gas-dynamic and thermal processes in the synthesis of trichlorosilane by hydrogen reduction of silicon tetrachloride in a high-frequency discharge. *Russian Journal of Applied Chemistry*, 87(9):1246–1250, 2005. doi:10.1134/S1070427214090079.

[3] A. G. Bubnov, V. I. Grinevich, and V. V. Kostrov. *High energy chemistry*, 25(4):365–371 (in Russian), 1991.

[4] I. Pazinkevich. Abstract of papers. In *11th Int. Gas Congress, Moscow*, page 19 (in Russian), 1970.

[5] L. Mochalov, R. Kornev, A. Logunov, and et al. Behavior of carbon-containing impurities in the prrocess of plasma-chemical distillation of sulfur. *Plasma Chemistry and Plasma Processing*, 38:387–598, 2018. doi:10.1007/s11090-018-9879-1.

[6] A. V. Gusev, R. A. Kornev, and A. Y. Sukhanov. Behavior of organochlorine impurities in the process of plasma-enhanced synthesis of trichlorosilane. *High Energy Chemistry*, 42(4):324–325, 2008.

[7] A. V. Gusev, R. A. Kornev, and A. Y. Sukhanov. Behavior of carbon-containing impurities during plasma synthesis of trichlorosilane. *High Energy Chemistry*, 42(1):56–58, 2008. doi:10.1134/S0018143908010116.

[8] Y. P. Raizer. *Gas Discharge Physics*. Springer, Berlin, 1991.

[9] R. A. Kornev, P. G. Sennikov, and V. V. Nazarov. Application of RF (13.56 MHz) arc discharge for plasma chemical conversion of volatile fluorides. *Plasma Physics and Technology*, 4(2):169–172, 2017. doi:10.14311/ppt.2017.2.169.

[10] A. N. Magunov. Temperature measurement of objects with unknown emissivity by spectral pyrometry. *Scientific instrument engineering*, 20(3):22–26 (in Russian), 2010.

doi:10.14311/ppt.2019.2.115
Plasma Physics and Technology 6(2):115–118, 2019 

# A COMPUTATIONAL MODEL FOR HIGH CURRENT DENSITY ARC PLASMAS

M. Alija*, M. Kurrat

*Faculty of Electrical Engineering, Information Technology, Physics, Technische Universität Braunschweig, 38106 Braunschweig, Germany*

* m.alija@tu-braunschweig.de

**Abstract.** A computational model for high current density arc plasmas is developed. Under the assumption of thermodynamic equilibrium the arc plasma is described as a compressible laminar fluid based on the magnetohydrodynamic (MHD) equations and the transport and thermodynamic properties of air. The arc plasma is studied in time and space on macroscopic values such as the temperature and the pressure. The simulation results are discussed and future research work is identified addressing the scientific domain of high current density arc plasmas.

**Keywords:** computational plasma physics, high current density arc plasma, magnetohydrodynamics.

## 1. Introduction

Computational models are essential for the understanding of the physics of arc plasmas. To encapsulate the behaviour of arc plasmas in time and space several physical domains are addressed in the computational models. Promising models studying high current densities arc plasmas are embraced in [1–3]. The arc plasma is described as a single fluid using the MHD equations and studied in time and space. A key parameter for the development of the models is the current density. The coupling and the influence of the magnetic field is addressed in [2, 3] for current densites of $j \approx 10^6\,\mathrm{A{\cdot}m^{-2}}$. In [1] the focus lies on the influence of the heat transfer by radiation for a current density of $j \approx 10^9\,\mathrm{A{\cdot}m^{-2}}$. Investigations of [4] shows even higher current densities reaching values of approximately $j \approx 10^{11}\,\mathrm{A{\cdot}m^{-2}}$ in the arc plasma fall regions. Higher current densities leads to higher energy densities in the plasma and subsequently to a wider range of operating conditions. For this the phenomena in an arc plasmas are still not fully understood. To advance the understanding and motivated by the studies [1–3], in this paper a computational model for high current density arc plasmas for $j = 1 \cdot 10^8\,\mathrm{A{\cdot}m^{-2}}$ is presented. Starting from the Boltzmann equation the computational model is presented. The arc plasma is studied and future research work is identified.

## 2. Computational model

Starting from Liouville theorem the development of the distribution of the species in a plasma is described by the generalized Boltzmann equation [2, 3, 5, 6] with

$$\frac{\partial f}{\partial t} + \mathbf{v} \cdot \frac{\partial f}{\partial \mathbf{x}} + \frac{q}{m}(\mathbf{E} + \mathbf{v} \times \mathbf{B}) \cdot \frac{\partial f}{\partial \mathbf{v}} = \sum C(f) \quad (1)$$

Covering the interaction of the species as elastic or inelastic collisions by $C(f)$ and the influence of electromagnetic fields by $\mathbf{E}$ and $\mathbf{B}$, $f$ presents the distribution of the species, where $\mathbf{x}$ is the spatial and $\mathbf{v}$ is the velocity phase space coordinate. This leads to a six dimensional description of the species. Using the moments of equation 1 the MHD equations are derived, which describes the arc plasma as a fluid with electrons and ions. Under the assumption of thermal equilibrium and quasi neutrality the one-fluid model is derived. The plasma is considered as a compressible and laminar flow, where the electrons and ions are not separated. The equations of the computational model are expressed in a conservative form, where the behaviour of the mass and energy is of importance. The conservation of mass is expressed by [1, 3, 5, 7]

$$\frac{\partial \rho}{\partial t} + \nabla \cdot (\rho \mathbf{v}) = 0 \quad (2)$$

where $\rho$ is the density and $\mathbf{v}$ the velocity. As the density of the plasma can change with pressure, the time derivative of the density is considered in the equation. The modeling of the mass is supplemented by formulating the equation of the impuls for the particles with [1, 3, 5, 7]

$$\frac{\partial \rho \mathbf{v}}{\partial t} + \rho(\mathbf{v} \cdot \nabla)\mathbf{v} = \nabla \cdot (\boldsymbol{\tau} - p\mathbf{I}) + \mathbf{j} \times \mathbf{B} \quad (3)$$

where $\boldsymbol{\tau}$ the viscous tensor, $p$ is the pressure, $\mathbf{I}$ the identity tensor, $\mathbf{j}$ the electric current density and $\mathbf{B}$ the magentic field. A weak coupling is performed by implementing the crossproduct of $\mathbf{j}$ and $\mathbf{B}$. This sets of equations are extended by formulating the conservation for the energy with [3, 5]

$$\rho\, c_p \left(\frac{\partial T}{\partial t} + \mathbf{v} \cdot \nabla T\right) = \nabla \cdot (\lambda \nabla T) + (\mathbf{j} \cdot \mathbf{E}) - W_{\mathrm{R}} \quad (4)$$

where the weak coupling is implemented by the Joule heating per unit volume by the product of the current density $\mathbf{j}$ and electric field strength $\mathbf{E}$. $c_p$ is the specific heat capacitance, $T$ the temperature, $\lambda$ the thermal conductivity and $W_R$ the radiation losses per unit volume. Pressure and visocus work are considered as negligible small. The electromagnetic fields are described by using Maxwell's equations. Assuming quasistatic behaviour, the electric potential is calculated by the scalar definiton of $\phi$ as a Laplace equation with

$$\nabla \cdot (\kappa \nabla \phi) = 0 \tag{5}$$

By using the electric potential $\phi$ the electric field is solved using $\mathbf{E} = -\nabla\phi$ and the electric flow field is calculated by $\mathbf{j} = -\kappa\nabla\phi$. The magnetic field is modeled with a further differential equation using the magnetic vectorpotential. The vectorpotential is formulated as a Poisson equation with

$$\Delta \mathbf{A} = -\mu_0 \mathbf{j} \tag{6}$$

where each component of $\mathbf{A}$ is calculated by the current density $\mathbf{j}$. With the vector potential the magnetic field is solved by

$$\mathbf{B} = \nabla \times \mathbf{A} \tag{7}$$

The heat distribution in the electrodes is modeled by Fourier's law. For this the differential equation with

$$\rho_{\text{Elec}}\, c_{\text{Elec}} \frac{\partial T}{\partial t} = \nabla \cdot (\lambda_{\text{Elec}} \nabla T) + (\kappa_{\text{Elec}}\, \mathbf{E}^2) \tag{8}$$

is solved for the electrodes. The properties of the electrodes in equation 8 are selected for copper [8], with the $\rho_{\text{Elec}} = 8900\,\text{kg}\cdot\text{m}^{-3}$ as the density, $c_{\text{Elec}}$ is the specific heat capacitance with $385\,\text{J}\cdot(\text{kg}\cdot\text{K})^{-1}$ and $\lambda_{\text{Elec}}$ the thermal conductivity with $400\,\text{W}\cdot(\text{m}\cdot\text{K})^{-1}$. $\kappa_{\text{Elec}}$ is the electrical conductivity with $5.99\,\text{MS}\cdot\text{m}^{-1}$. Energy transport by radiation in the plasma is described by the net emission coefficient $W_R$ as it offers an acceptable computational performance and describes the physical process of absoprtion and emission in the arc plasma. $W_R$ is derived by building the divergence of the heat flux [9] and modeling the arc plasma as a cylindrical isothermal channel with a radius $r$. By using this assumption $W_R$ is implemented by

$$W_R = \int_0^\infty B_f\, \kappa_f\, e^{(-\kappa_f r)}\, \text{d}f \tag{9}$$

where $B_f$ is the intensity of a black body radiator according to Planck and $\kappa_f$ the spectral absorption coefficient. The calculated values for $W_R$ in dependence of $T$ and $p$ are used based on [10, 11]. The computational model is solved by using the finite element algorithm of Comsol Multiphysics.

## 3. Computational geometry, boundary and initial conditions

In this part the geometry and the boundary conditions of the computational model are presented. The selected geometry is shown in Figure 1. The calculation is performed in 2D in order to improve the computational time and to encapsulate the physical compelxities. The plasma chamber has width of 25 mm and a height of 40 mm. The electrode A and electrode B are 10 mm × 15 mm. The high density arc plasma is ignited between the electrodes and evolves in the chamber.

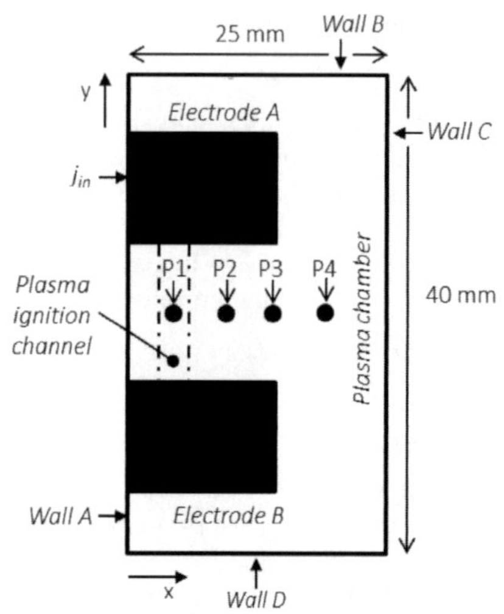


*Figure 1. 2D geometry for which the computational model is solved*

A high current density with $j_{\text{in}} = 1\cdot10^8\,\text{A}\cdot\text{m}^{-2}$ is set to electrode A by implementing $j_{\text{in}} = \mathbf{n}\cdot\mathbf{j}$. For a 1 mm thick electrode this corresponds to a 1 kA current. The boundaries of the plasma chamber are implemented with the condition $\mathbf{n}\cdot\mathbf{j} = 0$. For the boundaries at the interface between the electrodes and the plasma, continuity is implemented for the normal component of $\mathbf{j}$. The tangential component of $\mathbf{j}$ at the interface is formulated by the ratio of $\kappa$ of the electrodes and the air plasma. For the electrode B the potential is fixed to $\varphi = 0$. The magnetic field is initially set to $\mathbf{B} = \mathbf{0}$. It is assumed that the magnetic field decreases at the boundaries of the plasma chamber nearly to zero. This is implemented by building the cross-product of $\mathbf{n}\times\mathbf{A} = \mathbf{0}$ for the boundaries of the plasma chamber. The temperature of the plasma chamber and the electrodes is set initially to $T = 293.15\,\text{K}$ [12]. The pressure in the plasma chamber is initially defined with $p = 101325\,\text{Pa}$ [12]. The initial temperature for the plasma ignition is set to $T \approx 15\,\text{kK}$, as the electrical conductivity $\kappa(T, p)$ is nearly constant and can be simplified in this way. The evolvment of the plasma in the chamber is implemented by setting the heat flux at the boundaries of the chamber to $\mathbf{n}\cdot\mathbf{q} = 0$. A movement of the plasma along the electrodes and the walls is assumed to be zero and implemented by setting the boundaries of the electrodes and the walls to $\mathbf{v} = \mathbf{0}\,\text{m}\cdot\text{s}^{-1}$. The plasma velocity field is implemented initially with $v_x = 0\,\text{m}\cdot\text{s}^{-1}$ and $v_y = 0\,\text{m}\cdot\text{s}^{-1}$.

## 4. Air plasma properties

The computational model is completed by the physical properties of an air plasma. They can be divided into the thermodynamic properties and transport properties. The density $\rho(T,p)$, the specific heat capacitance $c_p(T,p)$ and the adiabatic coefficient $ac(T,p)$ belong to the thermodynamic properties. The electrical conductivity $\kappa(T,p)$, the thermal conductivity $\lambda(T,p)$ and the dynamic viscosity $\mu(T,p)$ belong to the transport properties. The properties are calculated for a wide temperature range of up to 100 kK and a wide pressure range of up to 123 bar based on [13]. In Figure 2 the electrical conductivity over the temperature for three different pressures is depicted.

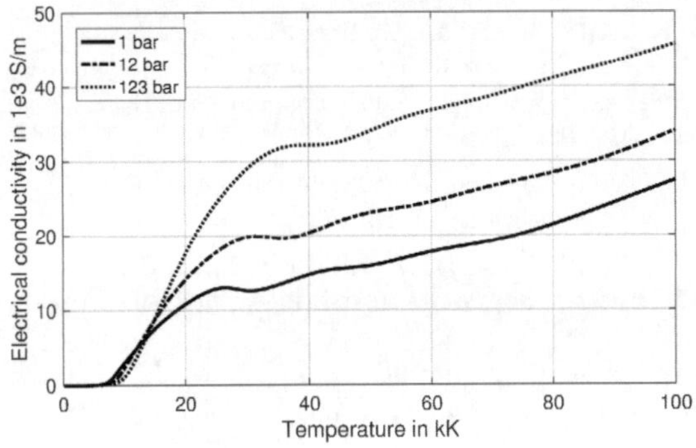


*Figure 2. Electrical conductiviy $\kappa(T,p)$ of an air plasma in dependence of the temperature and the pressure*

## 5. Computational experiments

In this part the results of the computational experiments are presented. For the calculation times with $t = 5$ µs, 10 µs, 20 µs and 30 µs the distribution of the temperature is depicted in Figure 3.

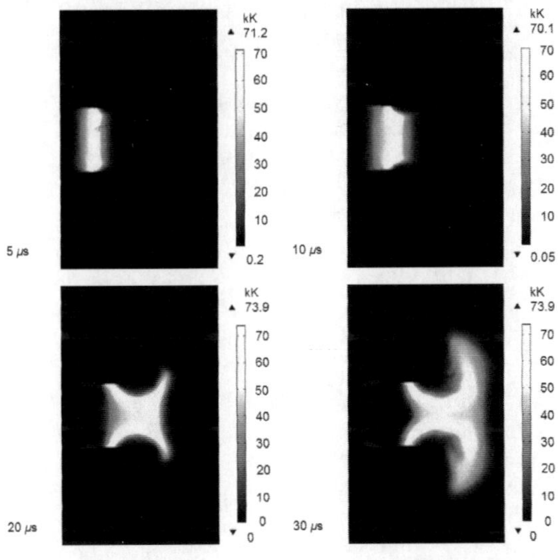


*Figure 3. Distribution of the plasma temperature for selected calculation times.*

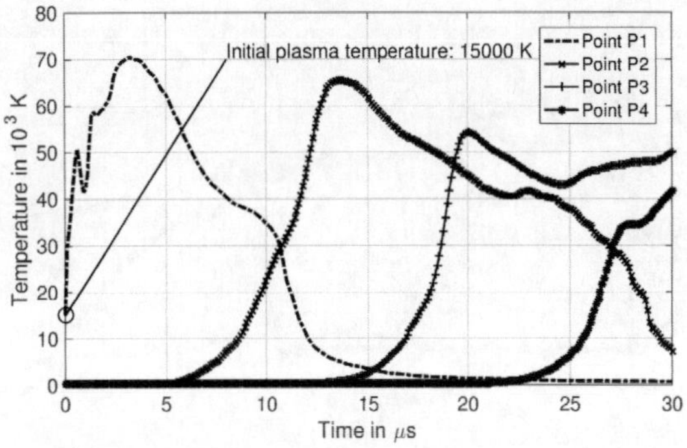


*Figure 4. Temperature behaviour of the plasma in dependence of the time.*

The energy, that is fed into the plasma chamber by Joule heating per unit volume, can be associated with a change of the physical state values of the air plasma. Considering the temperature distribution in Figure 3, the plasma is heated up. Starting from the ambient temperature, a temperature increase can be observed in Figure 3 ranging up to 73923 K. For $t = 5$ µs the temperature increase is concentrated in the plasma channel. For $t = 10$ µs, 20 µs and 30 µs the temperature increase encompasses a wider area in the plasma chamber. In order to study the distribution of plasma in detail, four evaluation points along the positive x axis are selected. The selected points are marked in Figure 1 with P1, P2, P3 and P4. The calculated temperatures over the time for the points are depcited in Figure 4. For each of the selected points the temperature shows a similiar behaviour.

As a consequence of the Joule heating the temperature raises continously till to a maximum value. For P1 the maximum is $T_{\text{max, P1}} = 70393$ K. After $T_{\text{max}}$ the temperature decreases continously to the intial temperature $T = 293.15$ K. Compared to P1, the temperature increase for P2, P3 and P4 in Figure 4 starts with delay. This can be associated with a movement of the plasma in the plasma chamber along the positive $x$ axis. The movement of the plasma can also be observed in the distribution of the temperature in Figure 3. In order to study the plausibility of the computational experiments, the pressure and density behviour over the time are considered. For the selected points the pressure is depicted in Figure 5.

Comparing the pressure and temperature curve for P1 a proprotional behaviour can be observed as expected. For P1 a maximum pressure with $p_{\text{max, P1}} = 904.5$ kPa is calculated. After $p_{\text{max}}$ the pressure decreases below the initial pressure. This can be explained by the increase of $T$ and $p$ along the selected points. The pressure curves for P2, P3 and P4 show a steeper increase compared to the temperature curves in Figure 5. This is a result of the reflected pressure at the chamber wall A. In Figure 6 the calculated densities for the selected points over the time are shown. For a given pressure, the densities

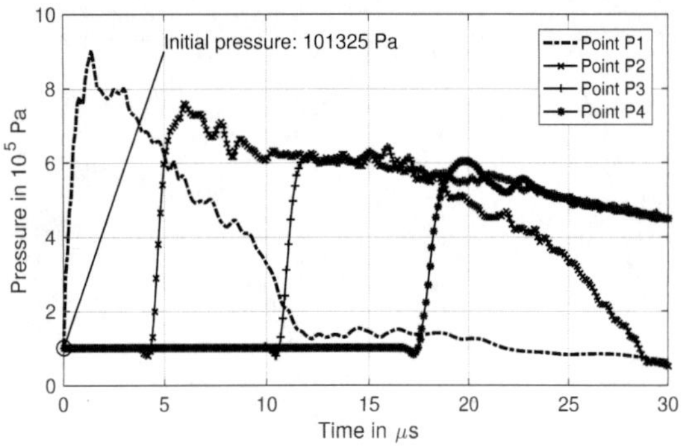


*Figure 5. Pressure behaviour of the plasma in dependence of the time*

in Figure 6 show an antiproprotional behaviour to the temperature curves in Figure 4 as expected.

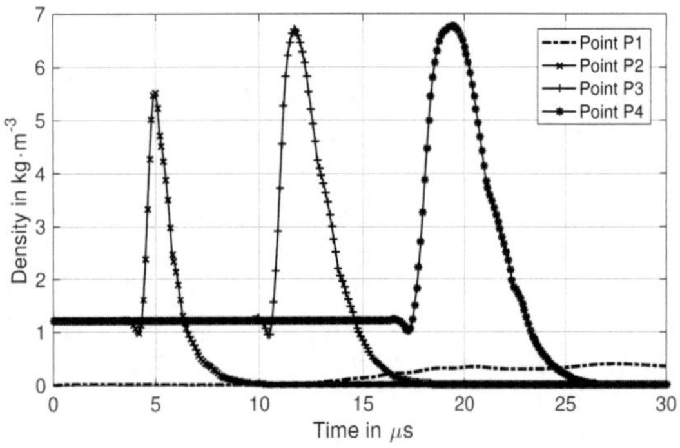


*Figure 6. Density of the plasma in dependence of the time for the selected points*

Considering the temperature and pressure curve for example for P2 for $t \leq 5\,\mu$s, ambient conditions with $T = 293.15$ K and $p = 101325$ Pa are given at this point in the plasma chamber. The density for this point at the same time is $\rho = 1.217\,\mathrm{kg{\cdot}m^{-3}}$, which is approximately the density of air with $\rho = 1.293\,\mathrm{kg{\cdot}m^{-3}}$ [14] at ambient conditions. The computational results seem to be generally plausible, when comparing the results qualitatively with the empirical model of an ideal gas [14], where $T \sim p$ and $T \sim 1/\rho$.

## 6. Conclusions and future work

A computational model for high current density arc plasmas is developed and the computational experiments are presented. With the derived equations a high current density arc plasma is simulated for a 2D geometry. Plausibility of the computational results is fostered by the qualitative comparison with an empirical model. Further investigations are motivated, addressing the modeling of the plasma and electrode interaction and the description of the plasma ignition. For this wide temperature range attention should be also paid to the radiation, which is evident when considering the Stefan-Boltzmann law. Furthermore a quantitative validation of the computational results should be addressed.

## References

[1] C. Sander, J.-E. Schmutz, and M. Kurrat. Analysis for radiation discretization for modelling spark gap for surge currents. *Plasma Physics and Technology*, 4(1):56–59, 2017. doi:10.14311/ppt.2017.1.56.

[2] P. Huguenot. *Axisymmetric high current arc simulations in generator circuit breakers based on realgas magnetohydrodynamics models.* ETH Zurich, Zurich, 2008. doi:10.3929/ethz-a-005579457.

[3] M. Kumar. *Three dimensional high current arc simulations for circuit breakers using real gas resistive magnetohydrodynamics.* ETH Zurich, Zurich, 2009. doi:10.3929/ethz-a-005927654.

[4] S. Coulombe and J. Meunier. Arc-cold cathode interactions: parametric dependence on local pressure. *Plasma Sources Science and Technology*, 6(4):11, 1997. doi:10.1088/0963-0252/6/4/008.

[5] J. H. Ferziger and M. Perić. *Numerische Strömungsmechanik.* 3. Springer, 2008.

[6] H. Goedbloed and S. Poedts. *Principles of Magnetohydrodynamics.* Cambridge University Press, 2004.

[7] F. F. Chen. *Introduction to Plasma Physics and Controlled Fusion.* Springer International Publishing, 2016.

[8] E. Ivers-Tiffée and W. von Münch. *Werkstoffe der Elektrotechnik.* B.G. Teubner Verlag, 2007.

[9] M. F. Modest. *Radiative Heat Transfer.* Academic Press, Elsevier, 2013.

[10] Y. Naghizadeh-Kashani, Y. Cressault, and A. Gleizes. Net emission coefficient of air thermal plasmas. *Journal of Physics*, 35(2925):10, 2002. doi:10.1088/0022-3727/35/22/306.

[11] T. Billoux, Y. Cressault, P. Teulet, and A. Gleizes. Calculation of the net emission coefficient of an air thermal plasma at very high pressure. *Journal of Physics*, 406(012010):11, 2012. doi:10.1088/1742-6596/406/1/012010.

[12] P. J. Mohr, D. B. Newell, and B. N. Taylor. Codata recommended values of the fundamental physical constants: 2014. *Journal of Physical and Chemical Reference Data*, 45(4):74, 2016. doi:10.1063/1.4954402.

[13] D. Angola, G. Colonna, C. Gorse, and M. Capitelli. Thermodynamic and transport properties in equilibrium air plasmas in a wide pressure and temperature range. *European Physical Journal*, 46:129–150, 2008. doi:10.1140/epjd/e2007-00305-4.

[14] M. Bartelmann, B. Feuerbacher, T. Krüger, D. Lüst, A. Rebhan, and A. Wipf. *Theoretische Physik 4 Thermodynmaik und Statistische Physik.* Springer Spektrum, 2018.

doi:10.14311/ppt.2019.2.119
*Plasma Physics and Technology* 6(2):119–122, 2019 

# INFLUENCE OF RESONANCE RADIATION TRANSPORT ON CHEMICAL EQUILIBRIUM IN AN ARGON ARC

S. GORTSCHAKOW[a,*], D. KALANOV[b], YU. GOLUBOVSKII[c]

[a] *Leibniz Institute for Plasma Science and Technology (INP), Felix-Hausdorff-Strasse 2, 17489 Greifswald, Germany*

[b] *Leibniz Institute of Surface Engineering (IOM), Permoserstrasse 15, 04318 Leipzig, Germany*

[c] *Saint-Petersburg State University, Ulyanovskaya 3, 198504 St. Petersburg, Russia*

[*] sergey.gortschakow@inp-greifswald.de

**Abstract.** Deviations from chemical equilibrium in argon arc plasma are analysed by means of collisional-radiative model. Corresponding comprehensive kinetic scheme has been developed and applied form study of free-burning arc at the conditions typical for welding applications. While the natural lifetime have been used for radiation emitted from highly excited argon states, the resonance radiation was described taking into account the radiation transport effects. Resulting spatial distributions of excited argon atoms are compared for the cases of LTE and two-temperature plasma using different approaches for the description of the resonance radiation transport.

**Keywords:** radiation transport, collisional-radiative model, argon arc.

## 1. Introduction

Modern arc simulations take into account the deviations from equilibrium plasma state in order to reach a better agreement with experiments. In addition to thermal non-equilibrium the departures from chemical equilibrium are of specific interest since they are essential for plasma properties in the colder regions. In general, the description of plasma chemistry requires application of collisional-radiative models. In recent papers devoted to studies of the free-burning Ar arc [1–3] a self-consistent two-temperature fluid model with accounting for non-equilibrium effects was presented. The modelling approach is based on the solution of the Navier-Stokes equations for mass, energy, and momentum conservation along with the equations of electric continuity, Ohm's and Ampere's law, which self-consistently determine electric and self-induced magnetic plasma fields. Furthermore, the model comprises of the balance equations for different species, taking into account transport processes due to diffusion and convection, and various collisional-radiative transitions. The inclusion of radiation transport equations in the balance equations for individual particles in the frames of such description is a very challenging task. Therefore, the models [1–3] use effective lifetime approximation in order to take into account the corresponding radiation trapping. In the work [4], which used the input data from model [2], a matrix method for description of resonance radiation transport was proposed. Transport of resonance radiation is a notable process, which influences the formation of various spatially-temporal structures in the gas-discharge plasma. This type of the plasma transport competes with other transport mechanisms, such as diffusion and convection. Therefore, a correct consideration of the radiation trapping is an important point in plasma modelling.

In the present contribution the results obtained by collisional-radiative model with extended kinetic scheme are presented. As an extension of approach [4], the radiation transport accounts for all resonance states of argon atom, also several additional highly-excited levels have been taken into account. Furthermore, the model includes both the axial and radial dependency for the conditions typical for tungsten inert gas welding (TIG). Input parameters, such as gas and electron temperature profiles, electron density are predicted by a full non-equilibrium MHD model of TIG arc. Spatial distributions of excited atoms densities are compared with Boltzmann distribution, which characterizes the equilibrium plasma state. The contributions of collisional processes, radiative processes as well as of radiation transport for these regions are analysed with focus on deviations from equilibrium state.

## 2. Description of the model

Analysis of the deviations from equilibrium plasma state was performed through the comparison between three different models. Model A assumes that the chemical equilibrium state holds in the arc plasma and in the arc fringes, while models B and C use a comprehensive kinetic scheme for the determination of spatial profiles of species densities. The difference between the model B and model C arises from the different consideration of transport effects. Model B uses a widely applied approach of effective lifetime, i.e. the effect of radiation transport is summarized in prolonged lifetime of corresponding excited level, which is constant for all spatial positions. In contrast to this, model C included the precise consideration of radiation transport process, which is space-dependent.

The collisional-radiative model use several input parameters determined by MHD model from [2, 3]. In present contribution the electron and gas temperature, as well as of the electron density obtained from the simulations of a TIG arc in argon are used.

## 2.1. LTE plasma

The majority of the arc models assumes that the plasma is in LTE state. First improvement of plasma description is provided by so-called two-temperature (2T) models. The two-temperature approach considers a separate energy balances for the electrons and heavy particles. However, the 2T approach still assumes that the distribution over excited levels follows the equilibrium rules. The chemical equilibrium in 2T plasma is determined by the electron temperature $T_e$ and is described by Boltzmann distribution

$$\frac{N^*}{N_0} = \frac{g^*}{g_0} e^{-\frac{E^*}{k_B T_e}} \tag{1}$$

where $N^*$ and $N_0$ are the densities of excited and ground state, $g^*$ and $g_0$ are the corresponding statistical weights and $E^*$ is the excitation threshold of corresponding excited level.

The ion density $n_i$ (and, therefore, due to the assumption of quasi-neutrality and presence of single-charged ions only, the electron density) can be determined from Saha equilibrium relation

$$n_i^2 = \frac{2}{\Lambda^3} \frac{Z_i}{Z_0} N_0 e^{-\frac{E_{io}}{k_B T_e}} \tag{2}$$

with Debye length $\Lambda$, partition functions $Z_0, Z_i$ and ionization energy $E_{io}$.

## 2.2. Collisional-radiative model

Collisional-radiative model takes into account the plasma-chemical processes in a free-burning argon arc using a 24 level model for argon. The included levels are: the ground state $1p_0$, metastable levels $1s_5$ and $1s_3$, resonance levels $1s_4$ and $1s_2$, individual radiating levels $2p_{10} - 2p_1$, lumped levels $3d$, $5s$, $5p$, $4d+6s$, $4f$, $6p$, $5d+7s$, $5f+5g$, as well as the ion in the ground state. Figure 1 shows the major plasma-chemical processes considered in the model. The kinetic scheme includes 611 plasma-chemical processes, among others excitation and de-excitation electron-atom collisions, collisional and photo-ionization, recombination, as well as various radiative processes. Corresponding data for reaction rates can be found in [2, 3]. While the natural lifetime has been used for radiation emitted from highly excited argon states the resonance radiation was described taking into account the radiation transport effects. For other radiation processes the plasma has been assumed to be optically thin and spontaneous transition probabilities were applied to calculate the radiation losses. The diffusive and convective transport was neglected since the frequencies of the collisional-radiative processes are dominant at considered conditions. The balance equation for resonance states Ar($1s_4$,$1s_2$), (Holstein-Biberman equation) in the case of stationary plasma reads

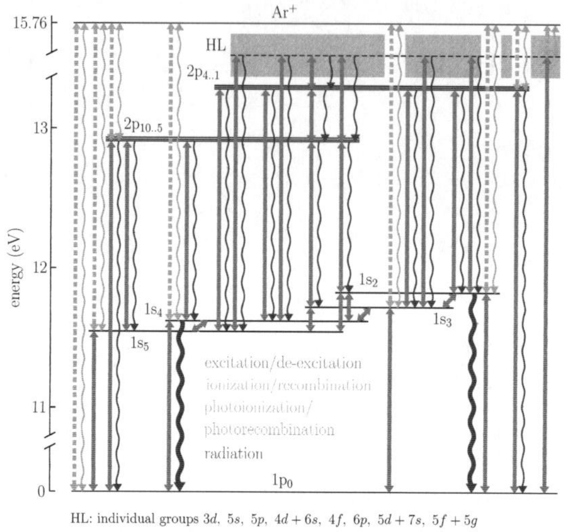


*Figure 1. Argon kinetic scheme used in the model.*

$$A N_r(\vec{r}) - A \int_V N_r(\vec{r}) K(\vec{r}, \vec{r'}) d^3 r' = Z_r^+(\vec{r}) - Z_r^-(\vec{r}) + W_r(\vec{r}). \tag{3}$$

Here $A$ is a spontaneous emission probability, $N_r$ is the density of the resonance atoms, $Z_r^-$ is total de-excitation per unit of time at the unit volume, $Z_r^+$ is the total excitation due to transitions from other states of considered system, $W_r$ is a summary excitation due to transitions from ground state and recombination. The kernel of radiation transport operator $K(\vec{r}, \vec{r'})$ is the probability for a photon emitted in the point $\vec{r}$ to be absorbed in the point $\vec{r'}$. Considering spatial inhomogeneity of the absorption coefficient the kernel can be represented in the form

$$K(\vec{r}, \vec{r}') = \frac{1}{4\pi} \int_0^\infty d\nu \frac{\epsilon_\nu(\vec{r}') \kappa_\nu(\vec{r})}{|\vec{r} - \vec{r}'|^2} e^{-\int_{\vec{r}'}^{\vec{r}} \kappa_\nu(\xi) d\xi}, \tag{4}$$

where $\epsilon_\nu(\vec{r}')$ and $\kappa_\nu(\vec{r})$ are the line profiles of emission and absorption, and $\nu$ is the photon frequency. The expression under the integral describes the probability of the photon travel between the points $\vec{r}$ and $\vec{r}'$ without being absorbed.

The matrix solution approach [4] consists of discretization of the integral equation (3) by dividing the entire plasma volume into a number of small cells (n) in accordance with the chosen geometry. The discretization procedure is similar to that described in detail in the previous work [4]. In addition the axial dependency was taken into account by consideration of finite number of axial positions. After discretization the left side of equation (3) at the position $r_k$ takes the representation

$$AN_r(\vec{r}) - A\int_V N_r(\vec{r})K(\vec{r},\vec{r}')\mathrm{d}^3r' = \sum_{i=0}^{n-1} A_{\mathrm{eff}}N_r(r_i)b_{i,k} \quad (5)$$

Here $b_{i,k}$ denotes the transport matrix and $A_{\mathrm{eff}}$ is the effective probability of radiative transition by Biberman [5, 6]. Finally, the equation system at the position $r_k$ takes the following form

$$\sum_{q\neq m}\sum_{i=0}^{n} N_m(r_i)\left[A_{\mathrm{eff}}b_{k,i}\delta_{m=\mathrm{res},q} + Z^-_{m,q}(r_i)\right] = W_m(r_k) + \sum_{q\neq m} Z^+_{q,m}(r_k) \quad (6)$$

Here $Z^+, Z^-$ denote the rates of excitations and de-excitations due to collisional-radiative transitions from states with index $q$, to state with index $m$. Kronecker delta $\delta(m = \mathrm{res}, q)$ allows for to formalize the equation for the case of radiation transport inclusion, e.g. in considered case the resonance radiation trapping process, relating to the Ar($1s_4$) or Ar($1s_2$) state. Notice, that in case of non-resonant atoms the radiation transport matrix reduces to an unit matrix and $A_{\mathrm{eff}} \equiv A$. Furthermore, in the case of non-radiating levels, like e.g. metastable, only the second term in the parenthesis in the left part of equation (6) remains.

Two different approaches have been used, namely effective lifetime approximation (model B, matrix $b_{i,k}$ is replaced by unity matrix) and solution of radiation transport equation by the matrix method (model C).

## 3. Results and discussion

Schematic view of the arc temperature distribution is given in Fig. 2. The input parameters correspond to following experimental conditions: arc length 8 mm, arc current 200 A DC, argon at atmospheric pressure between a conical tungsten cathode and a water cooled flat copper anode. Gas is fed through a nozzle near the cathode with a flow rate 12 slpm [1].

The results will be presented on example of excited levels $1s_4$, $2p_{10}$ and $2p_1$ (Fig. 3). Analysis has been done for the case with homogeneous absorption coefficient ($k$ =1), since the spatial dependency of the absorption coefficient has much lower impact on the results in comparison with transport effects [4]. Determination of the density was performed for regions separated from electrodes by a distance of 1 mm in order to exclude the processes in the cathode and the anode layers where the application of the model is restricted. Figure 3 clarifies that the density profiles becomes more narrow for the states with higher excitation energy for all models. Introduction of radiation transport causes significant changes in the density

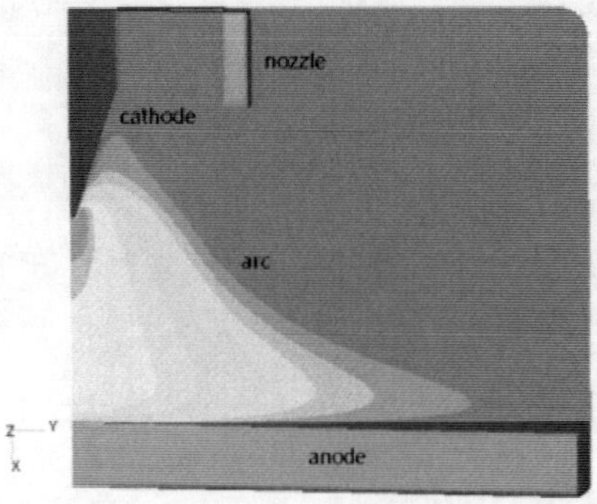


*Figure 2. Schematic view of temperature distribution in a TIG arc.*

distributions (cf. results from model A vs models B and C). (i) The radial profiles becomes much broader. (ii) Close to the cathode, the region with higher population reduces due the transport of excitation toward the outer regions (cf. for example, radial profiles at spatial position $z$ = 1 mm). (iii) There are clear changes in axial direction, which arise both from application of collisional-radiative model and radiation transport effects. The differences are less pronounced for higher excitation energies.

Formally, the changes should appear for the species which are directly involved into radiation transport, i.e. Ar($1s_4$) and Ar($1s_2$). However, there are two major kinetic pathways which connect various excited species (levels between each other). The first pathway is the excitation in electron collisions, for example

$$\mathrm{Ar}(1s_4) + e \rightarrow \mathrm{Ar}(2p_{10}) + e \quad (7)$$

In such a way the changes in density distributions of $1s$ levels becomes „transported“ to $2p$ levels. However, this channel is weaker comparing with the second pathway. Ar($1s$) levels are closely connected with each other via excitation processes in the electron collision and subsequent radiative deexcitation. An example of corresponding reaction pathway is as follows:

$$\mathrm{Ar}(1s_4) \xrightarrow{e} \mathrm{Ar}(2p_{10}) \xrightarrow{h\nu} \mathrm{Ar}(1s_5) \quad (8)$$

Therefore, the changes in population of resonance states lead to corresponding deviation in populations of all excited states. In the case of lowest excited states ($1s$ states) the differences arise up to seven orders of magnitude in the arc fringes. The broadening of the $2p$ states profiles is less pronounced in the central part of the arc, but achieve about two orders of magnitude in the arc fringes.

Inclusion of the full radiation transport in the equation system (model C) causes the spatial redistribution of Ar($1s$) excited states densities. For higher excited states, like e.g. $2p$ and higher, the approach of effective lifetime can be applied.

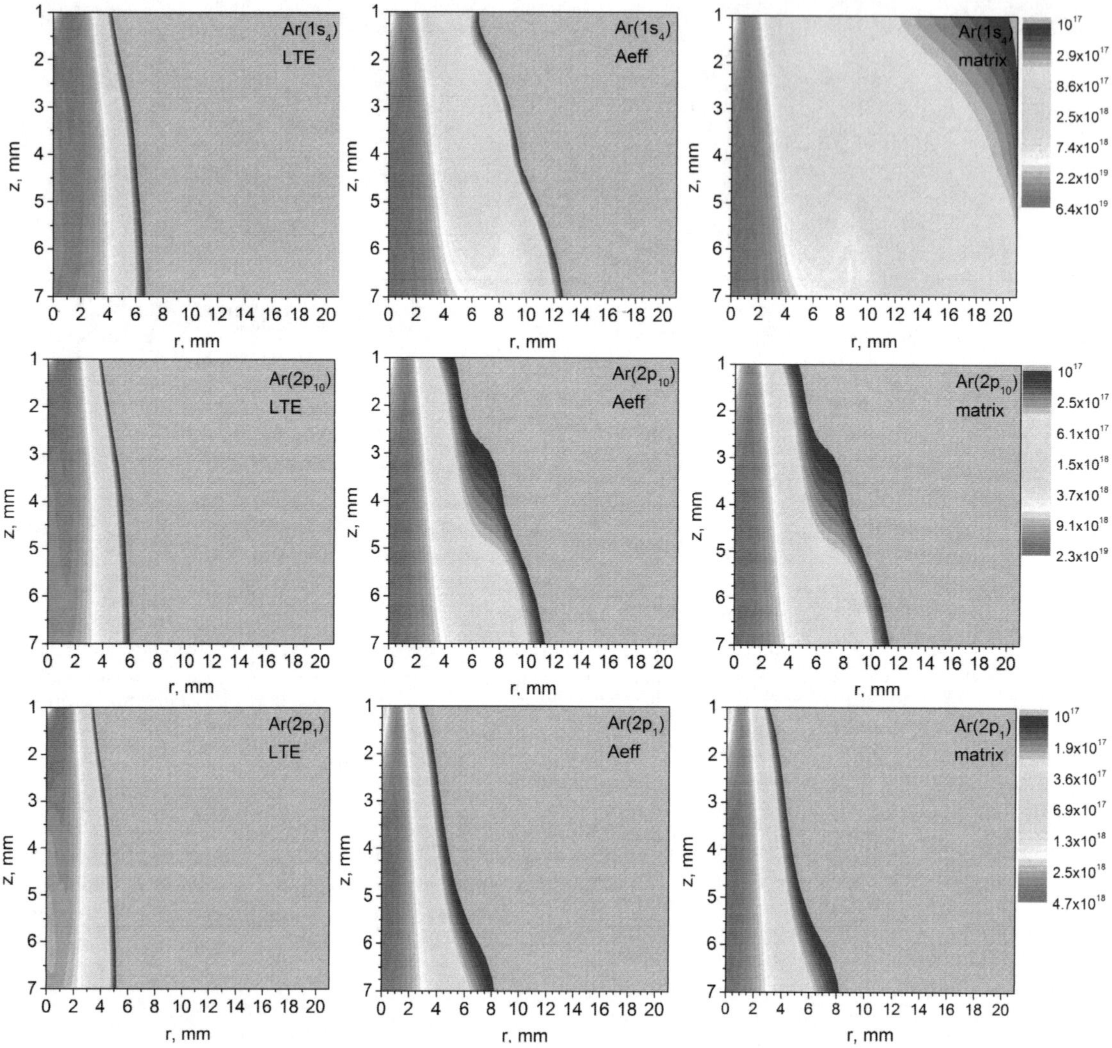


*Figure 3. Spatial distributions of excited atom density (logarithmic scale, in $m^{-3}$) for different states predicted by model A (left row), model B (middle row) and model C (right row).*

## 4. Conclusions

Resonance radiation trapping plays an important role in formation of spatial density profiles of excited atoms in a free-burning Ar arc. A noticeable growth of the $1s$ and lower lying $2p$ states population in the arc fringes takes place due to resonance radiation transport. These states are highly mixed due to collisional-radiative processes. For the higher excited levels which have a short lifetime and less connections to other states via excitation-de-excitation reactions this effect becomes less pronounced with increasing excitation energy of the level. Finally, the broadening of radial profiles due to the resonance radiation transport becomes negligible for high-lying excited states. Notice that the effect of profile broadening has a certain dependency on spatial distribution of absorption coefficient $\kappa$, which was not accounted for. Consideration of correct $\kappa$ values can decrease the absolute values of excited species densities in the arc fringes [4].

## References

[1] M. Baeva et al. Two-temperature chemically non-equilibrium modelling of transferred arcs. *Plasma Sources Sci. Technol.*, 21(5), 2012. doi:10.1088/0963-0252/21/5/055027.

[2] M. Baeva and D. Uhrlandt. Plasma chemistry in the free-burning ar arc. *J. Phys. D: Appl. Phys.*, 46(32), 2013. doi:10.1088/0022-3727/46/32/325202.

[3] M. Baeva. Thermal and chemical nonequilibrium effects in free-burning arcs. *Plasma Chem. Plasma Proc.*, 36(1):151–157, 2016. doi:10.1007/s11090-015-9650-9.

[4] Y. Golubovskii et al. Effects of traping of resonance radiation in free-burning ar arc. *J. Phys. D: Appl. Phys.*, 48(22), 2015. doi:10.1088/0022-3727/48/22/225203.

[5] L. M. Biberman. On the theory of the diffusion of resonance radiation. *Zh. Eksp. Teor. Fiz*, 17:417, 1947.

[6] T. Holstein. Imprisonment of resonance radiation in gases. *Phys. Rev.*, 72(12):1212–1233, 1947. doi:10.1103/PhysRev.72.1212.

doi:10.14311/ppt.2019.2.123
*Plasma Physics and Technology* **6**(2):123–126, 2019 

# ANALYSIS OF EROSION RESISTANCE OF CuC ARCING CONTACTS MANUFACTURED BY PLASMA SPRAYING TECHNOLOGY

P. Derevyankin[a], V. Frolov[a,*], D. Gonzalez[b], S. Gortschakow[b], R. Methling[b], D. Uhrlandt[b]

[a] *Peter the Great St. Petersburg Polytechnic University, Polytechnicheskaya 29, 195251 St. Petersburg, Russia*

[b] *Leibniz Institute for Plasma Science and Technology (INP), Felix-Hausdorff-Strasse 2, 17489 Greifswald, Germany*

* frolov.eed@gmail.com

**Abstract.** Results of the erosion resistance study of the CuC electrodes manufactured by plasma spraying technology are presented. The diagnostics has have been performed by means of high-speed camera technique and optical emission spectroscopy which has been synchronised with electrical diagnostics. After the load material analyses, namely determination of mass loss and analysis of micro sections, have been done. It has been found that the erosion rate depends on conditions applied for the formation of coating layer, which was deposited either in air or in argon plasma. Furthermore, the erosion rate is significantly lower in the case of air plasma spraying.

**Keywords:** electrode erosion, low-voltage circuit breaker, coating.

## 1. Introduction

An increase of the lifetime of the low-voltage high-current switching devices (LVSD) requires the guaranty that the arc transfers from the main contacts to arcing contacts. One of the main factors affecting the erosion of arcing contacts during arc burning are plasma jets [1], which starts from in the arc roots [2]. The arc movement under the action of electromagnetic forces can be analysed both by experiment and modelling. Modern research relies on mathematical simulation due to the complexity of experimental analysis of the processes occurring inside the switching devices. When analysing the arc movement in a switching device, several aspects have to be addressed: position of the moving contact, influence of the arc chamber material [3], arc forces [4], contact erosion, arc power and the position of the arc in the arc chamber [5]. Furthermore, the behaviour of the electric arc depends on the electrode surface material, electromagnetic blowing conditions and electrode material composition.

The circuit breakers use contact systems with additional arcing contacts with better erosion resistance than the main contacts. At the same time, the arc moves along the surface of the arcing contacts toward the external elements of the arc chamber. The arc duration in DC and AC circuit breakers does not exceed 10–12 ms, so arcing contacts covered with a layer (or overlays) of erosion-resistant materials can be used. Erosion-resistant coating can be formed by plasma spraying using the copper-based composite materials in argon or air.

Promising materials for use in the arc contacts of low voltage circuit breakers are copper based composite materials. Various alloys have been proposed with refractory metals [6] or graphite as LVSD contact materials. In particular, in studies of the erosion of tool electrodes, an analysis of CuCr, CuMo, CuW and CuC alloys was carried out [7]. The analysis demonstrated a high erosion resistance in the cited materials. According to [7], composites using graphite showed the best erosion properties. In this case material porosity also contributed to that effect, in addition to refractory phase content. It should also be noted that the electron output energy of graphite is higher than that of copper [8]. This can have a significant impact on the behaviour of the arc at the arc roots, in particular, to improve the transition of the arc from the main contact to the arcing contact.

When an arc is burning between the breaker contacts, specific areas of material ejection from the surface and plasma jets can be distinguished [2]. The main causes of arc contact wear are: evaporation triggered by arc burning, splashing of metal as a result of explosive processes, removal of molten droplets by external forces. Plasma jets significantly affect arc contact wear and have a significant role in all of these processes. Therefore, the analysis of plasma jet behaviour is relevant for finding ways to reduce the erosion of LVSD contacts. The formation and development of plasma jets and some other processes in an electric arc depends on spatial distribution of temperature, current density, mass flow and energy.

The increase in erosion resistance of circuit breaker arcing contacts requires solving a set of problems. In this article only the influence the conditions of plasma coating formation on electrode erosion is considered. The analysis of the behaviour of plasma jets emerging from arc roots is also provided.

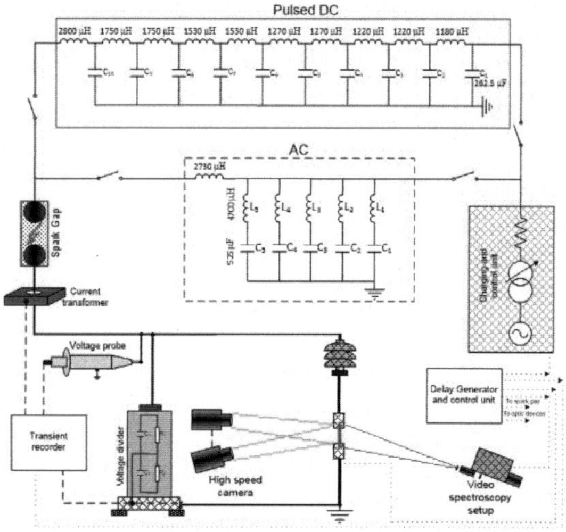


*Figure 1. Principal scheme of experimental setup [9].*

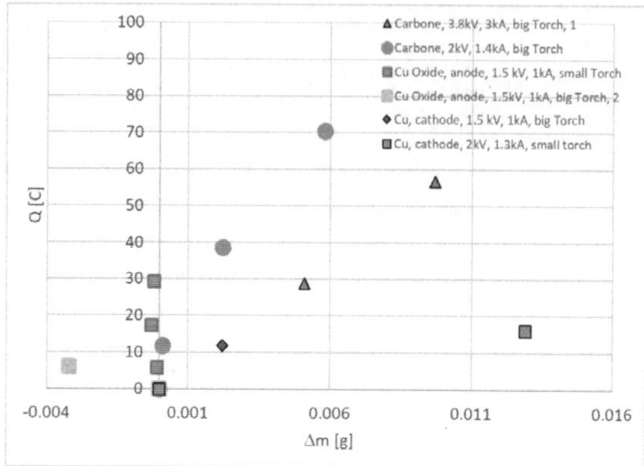


*Figure 2. Mass loss $\Delta m$ of various electrodes with different coatings in dependence on transferred charge Q. Parameters of electrode load (current and voltage) are given in the legend.*

## 2. Experimental setup

The principal scheme of experimental setup is shown in Fig. 1. Detailed description of the current source and diagnostics is given elsewhere [9]. This setup provides both AC pulsed DC waveforms at the current level below 5 kA and DC duration up to 20 ms. In the present experiments the current was restricted to maximum 3 kA and the arc duration was about 10 ms. The setup was extended by high-current diodes in order to restrict the polarity of load current.

Several electrode prototypes were studied: uncoated copper electrode, electrode with CuC coating in argon and electrode samples obtained using different fractions of CuC (95/5 by volume) powder material deposited in air plasma. A feature of the last electrodes is a high content of copper oxides in the coating.

The electrodes were loaded by current in ambient air. Besides the electrical measurements, spectroscopy setup was used to register the arc radiation spectrum and a high speed camera has supplied the arc images.

## 3. Results and discussion

### 3.1. Analysis of electrode erosion

The value of electrical erosion has been characterized by the mass loss under given experimental conditions.

The electrodes were arranged vertically (see Fig. 1) for the purpose of electrode erosion analysis for different coatings and recording of plasma jets dynamics. Each test electrode with different coatings was always used as the top electrode. The bottom electrode was a CuW (80/20) electrode in the form of a hemisphere of significantly larger diameter. This material was used in order to reduce the material transfer from the bottom electrode.

Figure 2 shows the analysis of the mass loss depending on the transferred charge for electrodes with copper-graphite coatings deposited in argon (marked as «Carbon» in the figure) and in air (marked as «Cu Oxide» in the figure).

The largest mass loss was registered for uncoated copper electrodes. It amounts to about 12 mg after a charge transmission of about 20 C. The electrodes coated with CuC have shown high erosion resistance at currents below 1.4 kA. The mass loss was less than 4 mg at transmitted charge of more than 70 C. However, at higher currents (3 kA) a significant decrease in erosion resistance is observed. Thus, value of $\Delta m$ was about 9 mg at a transferred charge of about 55 C. The best erosion resistance has been achieved by the electrodes coated in air plasma environment. These electrodes have high levels of oxides in their coating. An electrode with a similar coating when used as a cathode at a current of 1 kA showed an insignificant (up to 4 mg) increase in mass, probably due to the condensation of gaseous copper released from the anode surface. When the current value was raised a slight increase in erosion was observed.

The analysis of photographs of electrodes and micro sections has shown significant surface damage for the electrodes coated in argon. The electrodes coated in an air plasma environment have greater amount of gas inclusions inside the coating. However, these inclusions do not show such a significant expansion as for electrodes made in argon. For any type of spraying, detachment of this coating from the electrode is observed, which is an adverse factor that significantly reduces the reliability of electrodes with similar coatings. In future, it is necessary to consider the possibility of using a substrate for spraying to improve the bond between electrode materials and coating.

### 3.2. Analysis of plasma jets

The occurrence and development of plasma jets in an electric arc depends on the spatial distribution of temperature, current density, as well as on mass and energy flow.

In high-current arcs with plasma jets the temperature reaches its highest value at the points close to the arc root. Acquired arc spectra have been analysed using the basic data from [10]. The arc temperature was estimated by Boltzmann plot for the electrodes with different coatings. The analysis shows that the average arc temperature close to the electrode surface with a CuC is about 3000K higher than that for pure copper electrode. This difference may arise from the higher energy of electrons leaving the surface. Furthermore, in case of a copper electrode without powder coating the amount of metal vapour is much larger. This can also reduce the arc temperature.

The simplified equation that describes the energy balance in a DC arc with plasma jets reads [2]:

$$\nabla \cdot (\mathbf{j}\phi - \lambda \nabla T + h\rho \mathbf{v}) = 0 \qquad (1)$$

Here $\mathbf{j}$ is the current density, $\phi$ - potential, $\lambda$ - thermal conductivity, $T$ - temperature, $h$ - enthalpy, $\rho$ - density, $\mathbf{v}$ - plasma velocity.

Three components of the energy flow are Joule heating, heat conduction and energy brought by mass inflow. In the case of stationary arc the velocity of the particles depends on the location of the particles in the arc. It has a maximum in the near-electrode region and decreases for the positions away from the arc axis. Furthermore, the particle velocity is influenced by the output energy of the particles, which they gain leaving the electrode surface. Particle motion are attributed to electromagnetic forces and thermal processes in the arc. The directed plasma jets create an injection effect and draw electrode material into the arc.

Thus, the energy and temperature of the arc depend on the current density and the applied potential and on the characteristics of the electrode material. As it follows from Eq. 1, an increase in the flow velocity in a plasma jet will lead to a growth of temperature assuming a constant current density and thermal conductivity.

The parameters of plasma jets between opposing electrodes depend on the location of the electrodes, the distance between them, their materials and current characteristics.

The major reasons for arcing contact wear of high-current switches are metal evaporation under the action of energy released at the contacts or in the arc; metal spraying as a result of accelerated melting, explosive processes and gas evolution; removal of melted droplets by various external forces.

When the arc is burning between two spherical surfaces, the anode and cathode plasma jets are directed not strictly aligned, but in a certain angle to each other due to the action of electromagnetic forces acting on the jets. In this embodiment, a portion of the arc is supported by plasma jet instead of electrodes.

The mass loss of the electrodes caused by plasma jets strongly depends on metal melting temperature and evaporation. In some cases described below a mass increase was found instead of mass loss. A possible reason is that the coating components partially have a high melting point and do not evaporate under the action of a plasma jets, but rather condense the copper vapour from incoming flux.

Various jet shapes should be considered in the study of evaporation of various materials. When analysing high-speed images of arc burning on a copper electrode a large amount of ionized gases surrounding the jet can be seen. A similar phenomenon is not observed on electrodes with an arc-resistant coating.

The lengths and diameters of the jets also depend on the characteristics of the electrode coatings. Materials with higher electrical conductivity, thermal conductivity and low boiling point create plasma jets of larger length. Copper is also characterized by a longer anode jet. When testing the copper-graphite electrodes over the copper hemisphere (Figure 3), the upper jet of CuC dominates independent on electrode polarity. The reason of such behaviour is, first, due the smaller radius of coated electrode and secondly, graphite has a higher electron output energy comparing to copper.

Figure 3 shows the behaviour of an alternating current arc when testing an electrode with a copper-graphite coating. Initially, the development of the arc (frames 1–4) can be distinguished, then the plasma jets starts to move along the surface of the lower electrode, the cathode, and the anodic jet deviates in the opposite direction (frames 3–5). The arc length increases and its resistance grows, and ionized gas with high conductivity take the previous arc position. A new jet occurs on the surface of the cathode near the anode jet. Anode and cathode jets attract to each other (frames 5–7). After this instant the arc stretches again stretching (frames 9–11) up to the instant of current zero crossing, when the plasma disappears (frame 12). After the arc re-ignition, which is caused by the presence of residual conductivity in the electrode gap filled with ionized gas (frames 13–16) the jets are formed again.

Figure 3 also shows that after the current passes through zero and the upper electrode becomes an anode, smaller anode jets occurs, which indicates a change in the intensity of the output of the electrodes. However, since the diameters of the electrodes are not the same, the dominant jet always develops from smaller electrode, regardless of polarity and electrode material.

The arc ignites with nearly coaxial anodic and cathodic jets. With increasing current the jet length also increases and the jet tips shift toward the lateral side away from each other. The anode jet becomes dominant when the current enhances. This process is accompanied by a sharp increase in erosion caused by the thermal and dynamic effects of the anode jet. When the anode jet reach the cathode its tip spreads and suppress the cathode jet. The anode jet promotes the cathode melting and splashing of molten metal. In such a situation the erosion of small electrode is

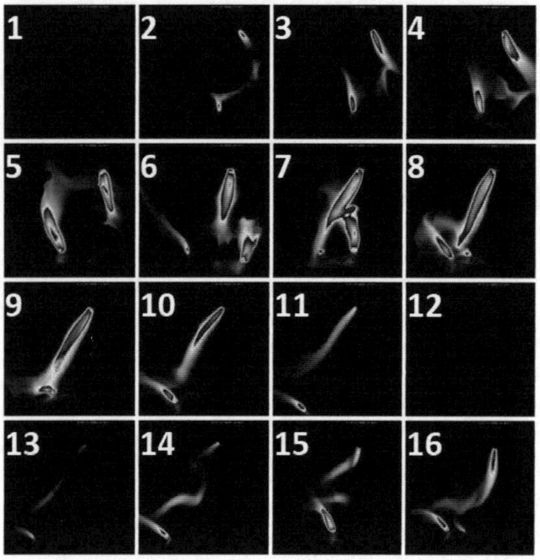


*Figure 3. Arc images in case of copper-graphite coating. AC operation with $I_{\max} = 1.4\,kA$, voltage of $2\,kV$.*

mainly determined by the metal evaporation, while the ablation processes prevail in erosion of big electrode.

During the research of various coatings, characteristics of plasma jets for each coating were identified. In the case of copper electrode well visible metal vapour of evaporating metal, which indicate both the erosion of the test electrode and the subsequent contamination of the surface of the opposite electrode by deposited copper was registered. Copper-graphite electrodes coated in the argon environment shows the highest jet intensity, which is associated with an increased electron energy during the graphite evaporation. The electrodes, which provide the lowest erosion were coated in an air environment, have less intense plasma jets.

It should be noted that the movement of the arc along the surface of the electrodes of various materials was almost the same. However, in addition to the coating material, the intensity of the jets was significantly affected by the characteristics and polarity of the applied current.

Further research is necessary to analyse the dependence of the jet intensity on electrode polarity.

## 4. Conclusions

The erosion behaviour of various coatings which can be applied for arcing contacts of high-current circuit breakers was analysed. It has been found that the contacts with a coating deposited in an air plasma environment, i.e. having a large amount of copper oxides, show higher erosion resistance. The analysis of the behaviour of plasma jets emitted from the surfaces of the electrodes was carried out. The most intense jets were observed from the surface of copper-graphite coatings deposited in argon. The dependence of the intensity of the jets on the strength and polarity of the applied current is analysed. The analysis of the movement of plasma flows along the electrode surface was carried out. For further analysis of the behaviour of plasma flows, it is necessary to carry out studies using electrodes of various configurations, since the radius of the «lower» electrode had a significant impact on the characteristics of plasma jets. The plasma jets on the electrodes with a copper-graphite coating have a higher temperature in the near-electrode area.

Further research is planned for analysis of the influence of oxides on the behaviour of the jets and erosion dependence on the coating structure.

## Acknowledgements

The authors from INP thank miss Uta Haeder-Schhult for valuable technical support in preparation and analysis of micro sections.

## References

[1] D. Shin, I. Golosnoy, and J. McBride. Development of switching performance evaluator and arc modelling tool for low-voltage switching devices. *COMPEL*, 37(6):1943–1957, 2018. doi:10.1108/COMPEL-03-2016-0112.

[2] O. B. Bron and L. K. Sushkov. *Plasma flows in LVSD electric arc*. Energiya: Leningrad, 1975.

[3] I. Murashov et al. Analysis of electromagnetic processes inside the arc interrupting system of a high-current circuit breaker. *Plasma Physics and Technology*, 4(2):161–164, 2017. doi:10.14311/ppt.2017.2.161.

[4] V. Y. Frolov, A. O. Kvashnin, and I. V. Murashov. Nonstationary mathematical model of a magnetic arc blast system. In *Proceedings of EIConRus 2018*, pages 622–625, 2018. doi:10.1109/EIConRus.2018.8317173.

[5] Y. Li et al. Numerical simulation on pressure field in chamber of low-voltage circuit breaker in different conditions. In *2015 3rd International Conference on Electric Power Equipment - Switching Technology (ICEPE-ST)*, pages 12–17, 2015. doi:10.1109/ICEPE-ST.2015.7368326.

[6] L. Chen et al. Erosion and surface morphology of the graphite electrodes in high-current, high-coulomb transfer gas switch. *IEEE Trans. Plasma Sci.*, 46(10):3320–3324, 2018. doi:10.1109/TPS.2018.2851307.

[7] S. A. Oglezneva, N. D. Ogleznev, and L. D. Sirotenko. Investigation of the relationship between the structure and properties of electrode tools for EDM cutting systems "copper - metal" and "copper - graphite". *South Ural State University Register. Series: mechanical engineering*, 16(1):63–71, 2016.

[8] A. I. Pushkarev, Y. N. Novoselov, and R. V. Sazonov. Efficiency of operation of a planar diode with an explosive emission cathode during plasma formation delay.

[9] A. Khakpour et al. Impact of temperature changing on voltage and power of an electric arc. *Electric Power Systems Research*, 143:73–83, 2017. doi:10.1016/j.epsr.2016.10.009.

[10] National Institute of Standards and Technology. *Atomic Spectra Database*. https://www.nist.gov/pml/atomic-spectra-database.

doi:10.14311/ppt.2019.2.127
Plasma Physics and Technology 6(2):127–130, 2019


# APPLICATION OF RF (40.68 MHZ) ARC DISCHARGE FOR PLASMA CHEMICAL CONVERSION OF VOLATILE CHLORIDES OF SILICON AND GERMANIUM

R. Kornev[a,*], P. Sennikov[a], V. Nazarov[a,b], A. Sukhanov[a], L. Shabarova[a]

[a] *G.G. Devyatykh Institute of Chemistry of High-Purity Substances of RAS, 49 Tropinin Str., Nizhny Novgorod, 603951, Russia*

[b] *Institute of Applied Physics of RAS, 46 Ulyanova Str., Nizhny Novgorod, 603950, Russia*

* kornev@ihps.nnov.ru

**Abstract.** The processes of hydrogen reduction of silicon and germanium chlorides are studied under RF conditions (40.68 MHz) of contracted atmospheric pressure arc discharge stabilized between two rod electrodes. The main gas-phase and solid products of plasma-chemical transformations are determined. Using the hydrogen reduction of $SiCl_4$ as an example, the numerical simulation of gas-dynamic and heat exchange processes for this type of discharge was performed. The studies were carried out for the optimal conditions for obtaining trichlorosilane using modern computational technologies of fluid dynamics which made it possible to detail the velocity and temperature fields, the gas flow rates in the plasma region, and also determine the main temperature zones affecting the hydrogen reduction process.

**Keywords:** RF arc discharge, reduction of volatile chlorides, simulation of gas-dynamic.

## 1. Introduction

Silicon and germanium remain the most popular material for microelectronics, photovoltaics and optics. The technologies of their industrial production, based on the synthesis and purification of their chlorides, require constant improvement in connection with the increased requirements for the environmental safety of production. Thus, for silicon tetrachloride, there is a need to process it into chlorosilanes - the initial products of the synthesis of silicon. The production of germanium from its tetrachloride is characterized by multistage which leads to an additional flow of impurities into the target product. In this regard, the study of the applicability of plasma-chemical technologies for the direct synthesis of silicon and germanium from $SiCl_4$ and $GeCl_4$ is an actual task. In the development of plasma methods, much attention is paid to the diagnosis of reactive plasma, which allows determining the gas temperature, classifying intermediate particles and establishing the mechanism of chemical reactions [1]. In some cases, when the plasma zone is small or unavailable for physical diagnostic methods, information about the temperature and the basic mechanisms of chemical reactions can be obtained by examining the gas-dynamic and heat exchange features of the process [2, 3]. One of the ways to study gas-dynamic and heat exchange processes is their numerical simulation [4–6]. The purpose of this work was to study the influence of the main technological parameters of the process of hydrogen reduction of $SiCl_4$ and $GeCl_4$, such as pressure and the ratio of reagents on the degree of conversion, as well as an estimate of the gas temperature in the gas discharge zone using computational fluid dynamics.

## 2. Experimental

Experiments on the investigation of the reduction processes of silicon and germanium tetrachlorides in hydrogen plasma were carried out on installation which schematic diagram was described in detail in [7, 8]. The power of high-frequency oscillator was 340 W, the frequency was 40.68 MHz. The flow of plasma-forming gas $H_2$+$SiCl_4$($GeCl_4$) was varied in the range of 100–700 $cm^3$/min. The pressure during the experiment was kept constant at 760 Torr. The molar ratio $H_2/SiCl_4(GeCl_4)$ was changed in the range of 2–15. The plasma-chemical reactor was a tube of silica glass, along the axis of which the electrodes were placed. A high-frequency voltage was applied to the electrodes from the generator through a matching device and the discharge was ignited between the electrodes. A vapor-gas mixture of silicon tetrachloride and hydrogen was supplied to the discharge volume. The power supplied to the plasma discharge zone was determined by the calorimetric method according to the method [9]. The dependence of the yield of chlorosilanes and chlorogermans as well as silicon and germanium on the conditions of plasma-chemical reduction was experimentally investigated. The yield of silicon and germanium was determined by the gravimetric method with an accuracy of $1 \cdot 10^{-4}$ g. The content of chlorosilanes and chlorogermanes was determined by gas chromatography with the detection limit of 0.1% and IR spectroscopy with a resolution of 0.1 $cm^{-1}$.

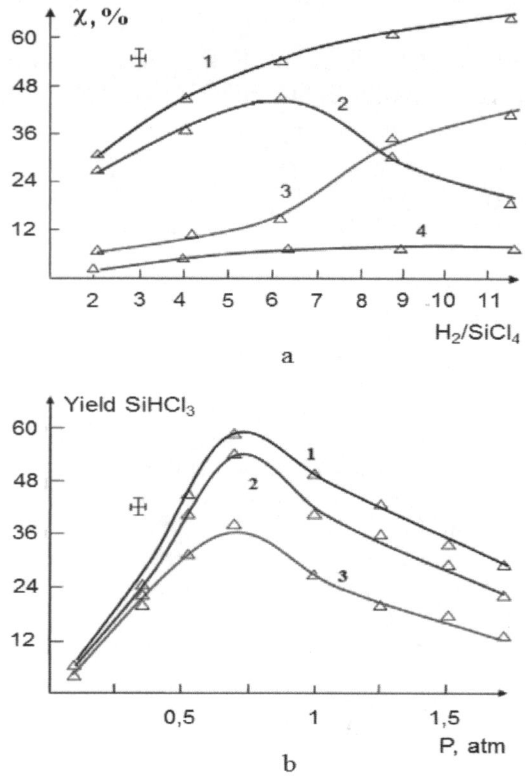


*Figure 1. a is the dependence of the total degree of conversion of $SiCl_4$(1), the percentage of yield of $SiHCl_3$ (2), Si (3) and $SiH_2Cl_2$ (4) on $H_2/SiCl_4$ ratio; b is the dependence of the yield of $SiHCl_3$ on the pressure in reactor, $H_2/SiCl_4$ = 6 - (1), 4 - (2), 9 - (3).*

## 3. Mathematical Model

The mathematical model was built in homogeneous formulation of the problem and included five different media: $H_2$, $SiCl_4$, air - liquids and solids - silicon and silica. The thermophysical and transport characteristics of these media are taken from [10, 11]. The reactor was a silica tube with a diameter 60 mm where nozzles-electrodes were coaxially arranged to meet each other. The diameter and length of the nozzle 10 mm and 50 mm, the length of the end portion of the nozzle is 15 mm. The diameter and length of the electrode 6 mm and 70 mm, the distance to which the electrode leaves the nozzle is 10 mm. The distance between the electrodes is 7 mm, the diameter of the outlet of the reactor is 10 mm. The problem of heat transfer in the hydrodynamic flow of a mixture of ideal gases is considered in the mathematical formulation. The mode of flow of gases is turbulent. The mixture interacting in the flow is considered taking into account heat transfer by means of convection and heat conduction and is described by the following system of partial differential equations:

$$\frac{\partial \rho_i}{\partial t} + \mathrm{div}(\rho_i \vec{V_i}) = 0, \tag{1}$$

$$\frac{\partial(\rho_i \vec{V_i})}{\partial t} + \nabla(\rho_i \vec{V_i} \otimes \vec{V_i}) = -\nabla p + \nabla \tau_i, \tag{2}$$

$$\begin{aligned} &\frac{\partial(\rho_i c_{pi} T)}{\partial t} + \mathrm{div}(\rho_i c_{pi} T \vec{V_i}) = \\ &= \mathrm{div}(\lambda_{ie} \nabla T) + \mathrm{div}(\vec{V_i} \tau_i) + \frac{\partial p}{\partial t} + S_E, \end{aligned} \tag{3}$$

where index $i = 1$ refers to $H_2$, $i = 2$ - to $SiCl_4$, $i = 3$ - to air, $i = 4$ - to silica, $p$ is the absolute pressure, $T$ is the absolute temperature, $\rho_i$ is the density of corresponding environment, $\rho_1$,$\rho_2$ are calculated from the equation of ideal gas, $\rho_3$, $\rho_4$ are constants,$\vec{V_i}$ is the velocity vector of corresponding environment, $S_E$ is the power induced by discharge, $\tau_i$ is the stress tensor connected with the rates of deformation generalized by Newton/s hypothesis $\tau_i = \mu_{ie}(\mathrm{div}(\vec{V_i}) + \mathrm{div}(\vec{V_i}))^{\mathrm{T}} - \frac{2}{3}\delta \mathrm{div}(\vec{V_i}))$, $\mu_{ie}$, $\lambda_{ie}$ are effective viscosity and thermal conductivity of environment, correspondingly, $\mu_{ie} = \mu_i + \mu_{it}$, $\lambda_{ie} = \lambda_i + \lambda_{it}$, $\mu_i$, $\lambda_i$ are the laminar viscosity and thermal conductivity, $\mu_{it}$, $\lambda_{it}$ are turbulent (vortex) viscosity and thermal conductivity which are interconnected through the turbulent Prandtl number and determined by the chosen model of turbulence [12]. The plasma region was a ball with a radius of $R/2 \approx 7$ mm in contact with the electrodes. The temperature of the electrode in the places of their contact with the plasma is set at $\approx 1760$ K. To simulate the plasma, an energy source is created in the reactor. The region of energy release is a cylinder at the base of which a circle with a radius of $P = 1.375$ mm and a height $H = 2.5$ mm. The density of energy release is given by the formula

$$q = 10^9(1 - 8 \cdot 10^5 (h + H)^2)\pi P^2), \tag{4}$$

where $h$ is the cylinder height.

The density of energy release is set thus so that the total value in the field of energy release corresponds to 15 W. For silica walls of the reactor, only the energy equation is solved which reduces to the heat conduction equation for solid walls. Note that the gases in the problem are considered to be the compressible liquids. The system of equations (1)-(3) is closed by the transport SST model of Menter turbulence [13]. This model is a combination of the $k-\omega$ model (a more accurate description of flows near the walls) and the $k-\varepsilon$ model (simulation of currents far from solid boundaries).

## 4. Discussion of the results

The dependence of the degree of conversion of $SiCl_4$ to chlorosilanes and silicon on the ratio of reagents in the initial mixture is shown in Figure 1a. The maximum yield of $SiHCl_3$ (44%) is observed when the ratio $[H_2]$:$[SiCl_4]$=6.9. With further increase in the concentration of hydrogen, the yield of trichlorosilane decreases. The yield of Si with increasing hydrogen concentration monotonously increases from 8 to 47%, and $SiH_2Cl_2$ from 3 to 8%. It should be noted that the results on the yield of trichlorosilane (Fig. 1a)

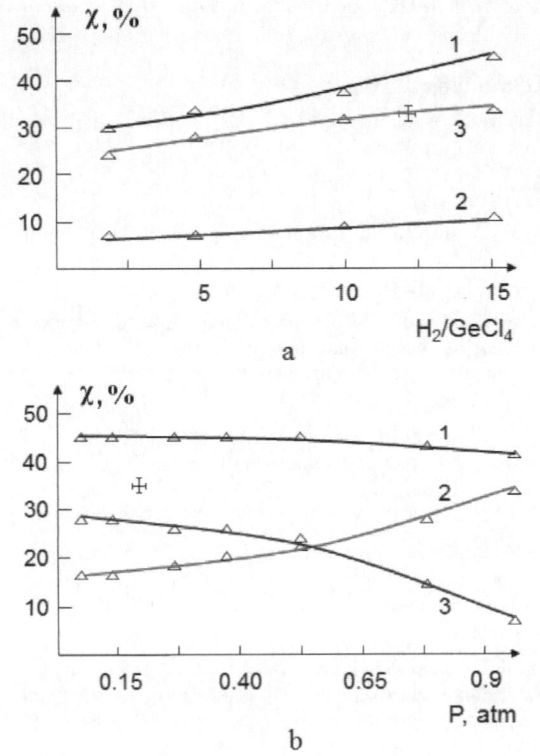


*Figure 2. a is the dependence of the total degree of conversion of $GeCl_4$ (1) and percentage yield of Ge (2) and $GeHCl_3$ (3) on $H_2/GeCl_4$ ratio; b is the dependence of the total degree of conversion of $GeCl_4$ (1) and the percentage yield of $GeHCl_3$ (2) and Ge (3) on the pressure in reactor.*

are in good agreement with the data of [14]. The main, in the realized experimental conditions, are the reactions:

$$SiCl_4 + H_2 \rightarrow SiHCl_3 + HCl, \quad (5)$$

$$2SiHCl_3 \rightarrow SiCl_4 + SiH_2Cl_2, \quad (6)$$

$$SiHCl_3 + H_2 \rightarrow Si + 3HCl, \quad (7)$$

With an increase in pressure above 1 atm., a greater discharge contraction is observed and, at constant power, the discharge ignition stops at a pressure of 1.75 atm. Depending on the pressure (Fig. 1b), an extreme view of the dependence of the yield of trichlorosilane is observed with the maximum at pressure of 0.75 atm. With an increase in pressure above 1 atm., a greater discharge contraction is observed and, at constant power, the discharge stops at a pressure of 1.75 atm. For the process of hydrogen reduction $GeCl_4$, depending on the increase in the ratio of reagents (Fig. 2a) in the initial mixture, the degree of conversion also increases. The main products $GeHCl_3$ and Ge are formed by the reactions:

$$GeCl_4 + H_2 \rightarrow GeHCl_3 + HCl, \quad (8)$$

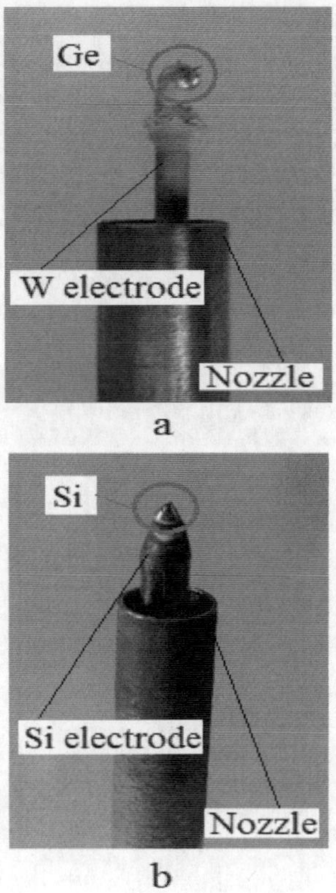


*Figure 3. View of electrodes after hydrogenation processes reduction of a - $GeCl_4$, b - $SiCl_4$*

$$GeCl_4 + 2H_2 \rightarrow Ge + 4HCl, \quad (9)$$

The dependence of $GeCl_4$ conversion on pressure (Fig. 2b) is not expressed by the maxima. A decrease in $GeCl_4$ conversion and Ge yield is observed while the yield of $GeHCl_3$ increases. The deposition of Si and Ge on the electrodes is shown in Fig. 3. It is seen that in both cases the formation of ingot-like, compacted are formed at the ends of the electrodes. Fig. 4 a shows a view of the gas discharge at atmospheric pressure. In the physical experiment, the plasma formation zone is a ball with a radius of $\sim 7$ mm, slightly offset to the upper surface of the reaction chamber due to the fact that the output of the reacted gas mixture is located on top. This zone, according to the simulation (Fig. 4b), is in the temperature range 900–1700 K. The temperature value of 1700 K corresponds to the area of contact of the electrode with the plasma. It is in this area, as well as in the area surrounding it, during the experiment the deposition of silicon and germanium is observed. Simultaneously with the deposition, the material melts, resulting in a compacted, ingot-like sample (Fig. 3). 900–1700 K high-temperature zone is also a source of atomic hydrogen which contributes to the reduction of $SiCl_4$ and $GeCl_4$. Fig. 4c, d show an isothermal surface with $T = 900$ K corresponding to the plasma region and the streamline of the gas mixture through this surface, respectively. According to the conducted numerical experiments, 95% of the

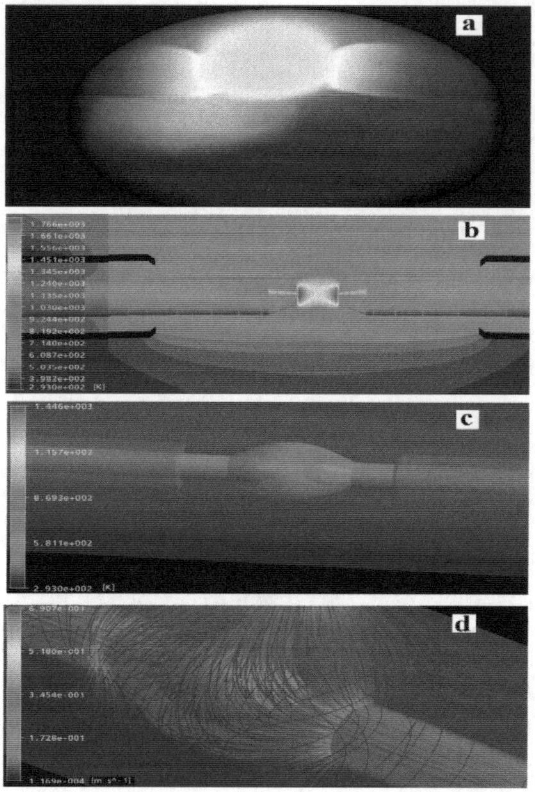


*Figure 4. a is the view of gas discharge; b is the distribution of temperature field in plasma zone as well as along the electrode; c is the nozzle-electrode system with isothermal surface 900 K; d are the lines of gas mixture flows in nozzle-electrode-isotherm system 900 K.*

gas mixture passes through this surface. Therefore, it can be argued that in the reaction chamber of this configuration there is a full interaction of the gas mixture with the plasma, in which the reaction of reduction of silicon tetrachloride to trichlorosilane takes place. The obtained data on the coefficient of the gas mixture entering the plasma region allow us to state that the yield of trichlorosilane equal to 45% is close to thermodynamically equilibrium at $T = 900\,\mathrm{K}$.

## 5. Conclusions

Under the conditions of RF (40.68 MHz) of a contracted arc discharge of atmospheric pressure, hydrogen reduction processes of silicon and germanium chlorides are investigated. The main products of the hydrogen reduction process are Si, Ge, $SiHCl_3$, $SiH_2Cl_2$ and $GeHCl_3$. Based on the results obtained using numerical simulation, it can be concluded that the temperature of the reaction area of a chemically active plasma is 900–1700 K, and the amount of the gas mixture passing through this region is close to 100%. In this case, the formation of condensed phase is observed at the ends of the electrodes at a temperature close to 1700 K, and the formation of chlorosilanes and chlorogermanes at a temperature of about 900 K.

## Acknowledgements

This work was supported by the grant of the Russian Science Foundation No. 17-13-01027.

## References

[1] Y. P. Raizer. *Gas Discharge Physics*. Springer, Berlin, 1991.

[2] B. R. Greene et al. Characterization of a 50 kW inductively coupled plasma torch for testing of ablative termal protection materials. In *55-th AIAA Aerospace Sciences Meeting*, number P.0394, 2017. doi:10.2514/6.2017-0394.

[3] B. Giesen et al. Formation of Si-nanoparticles in a microwave reactor: Comparison between experiments and modelling. *Journal of Nanoparticle Research*, 7(1):29–41, 2005. doi:10.1007/s11051-005-0316-z.

[4] V. Frolov et al. Experimental investigations of the hybrid plasma torch with reverse vortex stabilization. *Romanian Journal of Physics*, 56:36–40, 2011.

[5] L. V. Shabarova, R. A. Kornev, and P. G. Sennikov. Simulation of gas-dynamic and thermal processes in vortex-stabilized, inductively coupled argon-hydrogen plasma. *High Energy Chemistry*, 52(5):423–428, 2011. doi:10.1134/S0018143918050120.

[6] D. V. Ivanov and S. G. Zverev. Mathematical simulation of processes in ICP/RF plasma torch for plasma chemical reactions. *IEEE Transactions on plasma science*, 45(15):3125–3129, 2017. doi:10.1109/TPS.2017.2773140.

[7] R. A. Kornev, P. G. Sennikov, and A. I. Shishkin. Preparation of boron carbide from $BF_3$ and $BCl_3$ in hydrogen plasma of arc RF discharge. *Plasma Chemistry and Plasma Processing*, 37(4):997–1008, 2017. doi:10.1007/s11090-017-9821-y.

[8] R. A. Kornev, V. M. Vorotyntsev, A. N. Petukhov, and et al. Catalytic effects of electrode material on the silicon tetrachloride hydrogenation in RF-arc-discharge. *RSC Advances*, 6:99816–99824, 2016. doi:10.1039/C6RA21023F.

[9] A. V. Gusev, R. A. Kornev, and A. Y. Sukhanov. Preparation of trichlorosilane by plasma hydrogenation of silicon tetrachloride. *Inorganic Materials*, 42(9):1023–1026, 2006. doi:10.1134/S0020168506090172.

[10] A. B. Murphy. Transport coefficients of hydrogen and argon-hydrogen plasmas. *Plasma Chemistry and Plasma Processing*, 20(3):279–297, 2000. doi:10.1023/A:1007099926249.

[11] A. A. Furman. Inorganic chlorides. *M.: Chemistry*, page 416 (in Russian), 1980.

[12] R. I. Nigmatulin. Dynamics of multiphase media. part 1. *M.: Science*, page 416 (in Russian), 1980.

[13] F. R. Menter. Two-equation eddy-viscosity turbulence models for engineering applications. *The Amer. Inst. of Aeron. and Astron. Journal*, 32(8):1598–1610, 1994. doi:10.2514/3.12149.

[14] K. R. Sarma and C. S. Chanley. Hydrogenation method silicon tetrachloride. Patent 4542004 USA, 1985.

doi:10.14311/ppt.2019.2.131
Plasma Physics and Technology **6**(2):131–134, 2019 

# TRANSPORT PROPERTIES OF THERMAL PLASMA CONTAINING FLUORO-NITRILE ($C_4F_7N$)-BASED GAS MIXTURES

V.R.T. Narayanan[a,*], Ch. Rümpler[b], M. Gnybida[a], P. Slavíček[c]

[a] *Eaton European Innovation Center, Bořivojova 2380, 252 63 Roztoky, Czech Republic*

[b] *Eaton Industries GmbH, Hein-Moeller-Straße 7-11, 53115 Bonn, Germany*

[c] *Department of Physical Chemistry, University of Chemistry and Technology, Technická 5, 166 28 Prague, Czech Republic*

[*] VenkatRNarayanan@Eaton.com

**Abstract.** Gas mixtures containing fluoro-nitriles $C_4F_7N$ or fluoro-ketones $C_5F_{10}O$ as minority components (<20%) have been identified as promising alternatives to $SF_6$ in medium voltage gas-insulated switchgear (GIS) applications, because of their low Global Warming Potential together with their dielectric strength values being comparable to $SF_6$. The buffer gases in such fluoro-nitrile or fluoro-ketone based gas mixtures are usually $N_2$, $O_2$, $CO_2$, or air. In this contribution, we provide calculation results of transport properties, assuming local thermodynamic equilibrium (LTE), of thermal plasma containing following gas mixtures: $C_4F_7N - CO_2$ and $C_4F_7N - CO_2 - O_2$. The modifications in the thermodynamic and transport properties upon the addition of oxygen to the $C_4F_7N - CO_2$ mixtures in the temperature range 300 K–30 kK at 1 bar are provided and discussed. These properties have been utilized to calculate the plasma temperature profile for a free-burning arc in a companion paper.

**Keywords:** Fluoro-nitrile, thermal plasma, transport properties, $SF_6$ alternative.

## 1. Introduction

In the decades following the Kyoto Protocol (1997), there has been an active industry-wide interest in analyzing possible replacements to $SF_6$ in high voltage gas-insulated switchgear applications with voltage ratings greater than 52 kV [1]. In the last 5 years, focus has been oriented towards replacing $SF_6$ in secondary distribution networks possessing medium voltage gas-insulated switchgear with substantially lower voltage ratings $\leq$ 24 kV. Unlike the high pressures greater than 5 bar observed in high voltage switchgear designs, typical pressurization level in medium voltage GIS equipment is close to atmospheric. Nonetheless, owing to the high boiling points (at 1 bar) of the promising main alternatives – fluoro-ketones ($C_5F_{10}O$, 27°C) and fluoro-nitriles ($C_4F_7N$, −4.7°C), suitable buffer gases like air or $CO_2$ for example, are added to prevent liquefaction at temperatures lower than −30°C. The main alternative to $SF_6$ gas has a mole fraction of less than 20%, so that the overall mixture possesses a significantly lower Global Warming Potential (GWP) compared to $SF_6$. Furthermore, depending on the specific application, $SF_6$ or the proposed alternative gas mixtures could either be utilized as: (a) an insulation medium and/or, (b) a nominal current switching medium. During switching high temperature thermal plasma with temperatures in the order of 10 kK is generated and thermodynamic and transport properties of the arc plasma are needed to accurately model it's behavior within the switchgear geometry. Fluoro-nitrile based mixtures are in the focus of this contribution. Here $CO_2$ has been identified as a suitable buffer gas in the literature and $O_2$ is added to significantly reduce the generation of carbon monoxide (CO).

| Elements | N, O, C, F |
|---|---|
| Polyatomic species | $C_2N_2$, $C_2N$, $CN_2$, $NF_2$, $NF_3$, $N_2F_2$, $N_2F_4$, $CF_2$, $CF_3$, $CF_4$, $C_2F$, $C_2F_2$, $C_2F_3$, $C_2F_4$, $C_2F_5$, $C_2F_6$, $C_3F_6$, $C_3F_5N$, $C_3F_4N$, $C_2F_3N$, CFN, $C_2FN$, $C_4F_7N$, $N_2O$, $NO_2$, NOF, $NFO_2$, $NFO_3$, $NOF_3$, $F_2O$, $F_2O_2$, CFO, $COF_2$, $C_2F_4O$, $C_3O_2$, $C_2O$, $CO_2$, $O_3$ |
| Diatomic species | $F_2$, $N_2$, $N_2^+$, CN, $CN^+$, NF, CF, NO, $NO^+$, CO, $O_2$, $O_2^+$ |
| Monatomic species and ions | F, $F^+$, $F^{2+}$, $F^{3+}$, N, $N^+$, $N^{2+}$, $N^{3+}$, C, $C^+$, $C^{2+}$, $C^{3+}$, O, $O^+$, $O^{2+}$, $O^{3+}$, $e^-$ |

*Table 1. List of species used in species composition calculations.*

In this work, we calculate the relevant properties for the fluoro-nitrile-based $C_4F_7N - CO_2 - O_2$ mixtures for the temperature range 300 K–30 kK and pressure of 1 bar, with maximum mole fractions of $C_4F_7N$ and $O_2$ being 20%. We have independently validated our results for 10% and 20% $C_4F_7N$ mixed with $CO_2$ mixtures at 5 atm for the different properties of interest and have observed very good agreement with recently published data [2], except for dynamic vis-

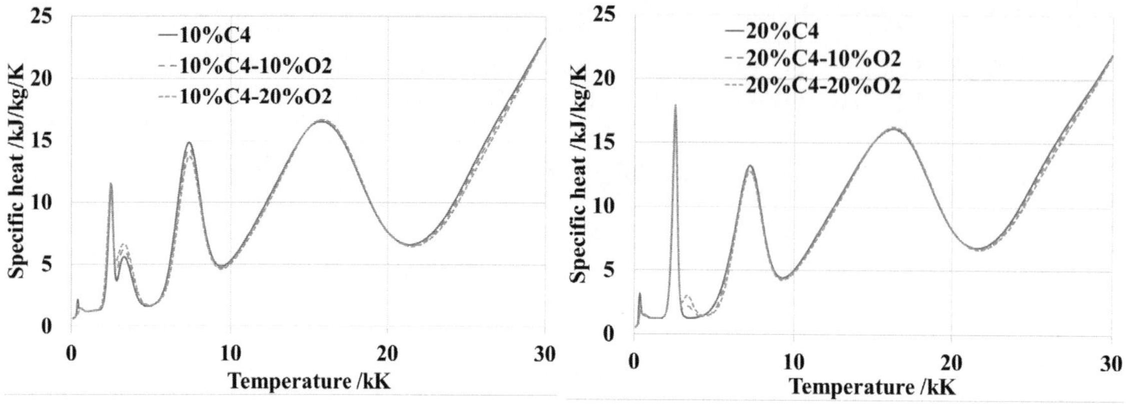


(a) 10% $C_4F_7N$ (b) 20% $C_4F_7N$

*Figure 1. Specific heat values for different* $C_4F_7N$ *(C4) mixtures at 1 bar.*

cosity. These validation results are not provided here since the properties results at high pressures (>1 atm) are beyond the scope of current work.

## 2. Species composition

The list of species considered for the species composition calculations have been provided in Table 1. The following species were omitted in this work, to facilitate a direct comparison with recent results from the literature [2]: $C_4F_6$, $C_3F_8$, $C_3F_7$, $C_3F_4$, $C_2$, $C_3$, $C_4$, $COF_4$, $CO^+$. Nonetheless, in a future work, results including these species will be considered, since the species composition and hence the properties at temperatures lower than 5 kK are expected to be different in the presence of the omitted species. Also, the condensed species (graphite) has not been included in the species list and consequently, the species composition is not expected to be accurate at temperatures lower than 2 kK. This discrepancy will also be addressed in a future work.

The molecular spectroscopic data – moments of inertia, vibrational frequencies, and the corresponding degeneracies – are necessary to calculate the internal partition functions and finally, the equilibrium composition of gas mixtures through the minimization of Gibbs free energy. However, such data is unavailable in the literature for several complex polyatomic species included in this current work. For these species we have utilized composite methods: CBS–QB3 by Petersson [3] and G4 method by Curtiss (B3LYP/aug–cc–pVTZ basis) [4] to evaluate the relevant molecular data. For the sake of brevity, the species composition results for the $C_4F_7N - CO_2 - O_2$ mixtures will not be provided. With all the afore mentioned input data, the computationally efficient algorithm of Godin and Trepanier [5] was employed to determine the equilibrium composition.

## 3. Thermodynamic properties

Once the species composition has been obtained, the total enthalpy and the specific heat at constant pressure ($C_p$) of the gas mixtures can be calculated through the calculation of mass density and partial derivatives of the total partition functions with respect to temperature. The variation of specific heat with temperature for 10% and 20% $C_4F_7N$-based mixtures are depicted in figure 1. The peaks below 1 kK are observed to have contributions from bigger polyatomic species - $C_2F_6$ and $CF_4$. The peaks at 2.5 kK, 3.5 kK, and 7 kK are owing to the contributions from dissociation of $COF_2$, $CO_2$ and CO respectively, while the peak at 16 kK pertains to the contributions from first ionization of different monatomic species. In the absence of $O_2$ in the mixture, the magnitude of the peak at 2.5 kK is significantly higher for 20% compared to 10% $C_4F_7N$. Increasing the mole fraction of $C_4F_7N$ in $C_4F_7N - CO_2$ mixtures from 10% to 20% results in an increase in the formation of CO and $COF_2$ at temperatures below 4 kK. However, with the addition of up to 20% $O_2$ to these mixtures, the mole fractions of CO are significantly reduced, while those of $CO_2$ and $COF_2$ are increased. With increasing %$O_2$ in the original mixture, the peaks at 3.5 kK for both 10% and 20% $C_4F_7N$ mixed with $CO_2$ gain small contributions from the dissociation of $O_2$, which can be observed through the tiny peaks on top of the peak in the absence of $O_2$. The specific heat results for 10% and 20% $C_4F_7N$ mixed with $CO_2$ and without $O_2$, at 5 atm have been observed to be in excellent agreement with published data [2].

## 4. Transport properties

The transport properties are calculated using the classical Chapman-Enskog method [6] involving the calculation of collision integrals, which requires primarily as input relevant inter-molecular/atomic potentials for interactions between different species. The neutral-neutral and non-parent neutral-charged species in-

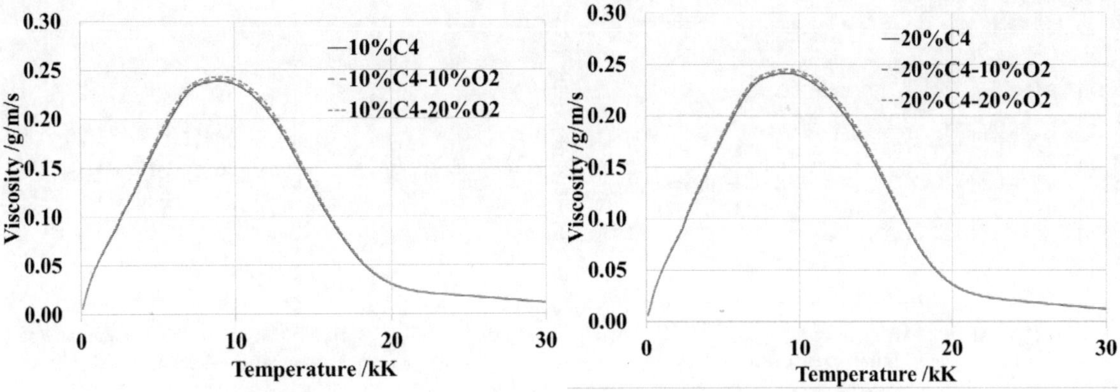


(a) 10% $C_4F_7N$ (b) 20% $C_4F_7N$

*Figure 2. Viscosity values for different* $C_4F_7N$ *(C4) mixtures at 1 bar.*

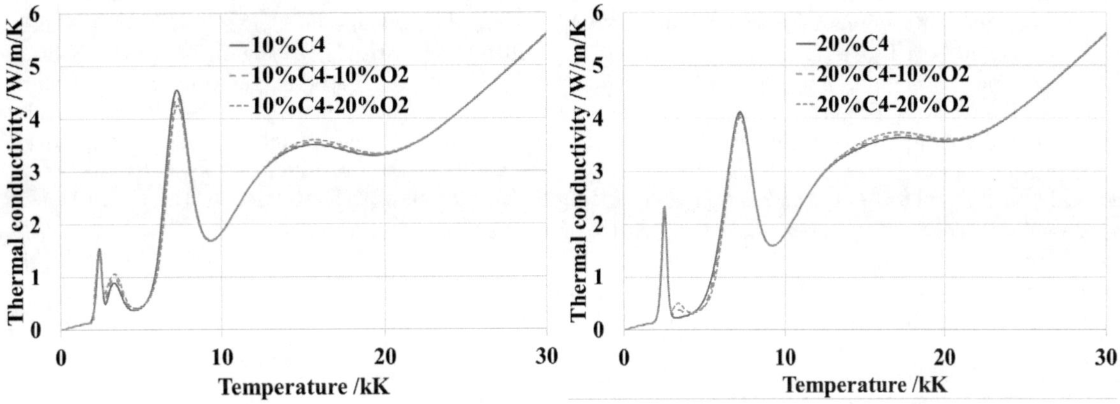


(a) 10% $C_4F_7N$ (b) 20% $C_4F_7N$

*Figure 3. Thermal conductivity values for different* $C_4F_7N$ *(C4) mixtures at 1 bar.*

teractions are described using the phenomenological Lennard-Jones Potential, which require as input the static polarizabilities and effective electron numbers. For polyatomic species with data unavailable from the literature, the dipole polarizabilities are typically solved via coupled-perturbed Hartree-Fock calculations (used for DFT methods) or using a linear response approximation. According to a benchmark study, both dipole moments and polarizabilities exhibit a slow convergence with the size of the basis set; for aug-cc-pVTZ basis set, the results appear to be converged. We have calculated average polarizability ($\alpha$) and polarizability anisotropy ($\Delta\alpha$) of $C_4F_7N$ in the gas phase for the global minimum (B3LYP/aug-cc-pVTZ) and along an NVT molecular dynamics (B3LYP/6-31+g*) trajectory. The parent neutral-charged species interactions are handled using the charge-transfer cross section approximation of the form: $Q_{ex} = (A - B\ \ln E)^2$, where $E$ is the collisional energy. The charged-charged species interactions are described using the Debye-length based screened Coulomb potential. The Debye-length has been calculated without including the contributions from ionic species. Finally, the momentum-transfer cross section data available from the literature is utilized for electron-neutral species interactions. The cross-section data for several complex polyatomic species, $C_4F_7N$ for example, are currently unavailable and have been neglected in our calculations. The assumption is valid since $C_4F_7N$ dissociates into minor molecular fragments before the appearance of electrons at temperatures greater than 4 kK.

The results for the dynamic viscosity variation with temperature are depicted in figure 2. Neither the temperature at the viscosity peak nor the magnitude of the viscosity peak are significantly altered by increasing either the %$C_4F_7N$ or %$O_2$. However, we have observed significant differences in the location and the magnitude of the peak, compared to the published data for 10% and 20% $C_4F_7N - CO_2 - O_2$ mixtures at 5 atm. The temperature at the viscosity peak and the peak viscosity magnitude from our work are approximately: 9.6 kK and 0.24 mPa-s respectively, while these values from published data are approximately: 15 kK and 0.175 mPa-s respectively. The reasons for these discrepancies are currently under investigation.

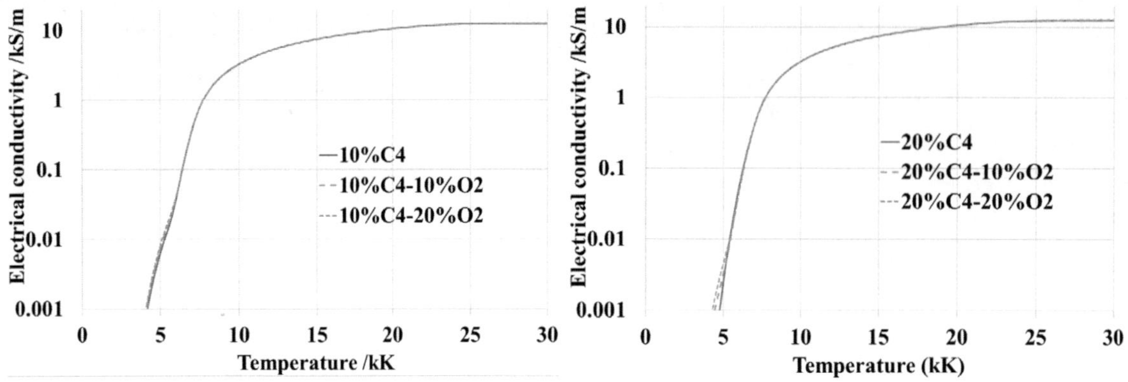


(a) 10% $C_4F_7N$ (b) 20% $C_4F_7N$

*Figure 4. Electrical conductivity values for different* $C_4F_7N$ *(C4) mixtures at 1 bar.*

Nonetheless, our dynamic viscosity results for pure $CO_2$ have been observed to be in excellent agreement with those reported in an earlier work [7] compared to those in the recent work [2].

The thermal conductivity results for the different mixtures are plotted in figure 3. It can be clearly observed that the profiles, especially the temperatures corresponding to the different peaks, mirror closely those from the specific heat. The only distinct difference between 10% and 20% $C_4F_7N - CO_2 - O_2$ mixtures is the significantly higher peak at around 2.6 kK for the latter case compared to the former. This peak is expected to be arising from the dissociation of $COF_2$ and $COF_2$ formation is enhanced by increasing the mole fraction of $C_4F_7N$ from 10% to 20%. Similar to those for the specific heat, the peaks at 3.5 kK are expected to contain contributions from dissociation of $CO_2$ and $O_2$, while the peaks at 7 kK correspond to CO dissociation. Nonetheless, the maximum relative error between our results with data from the literature [2] for 10% and 20% $C_4F_7N - CO_2 - O_2$ mixtures at 5 atm is approximately 25%, corresponding to the magnitude of the peak at 7 kK: 4.2 W/m/K calculated vs. 3.2 W/m/K in the published data. This discrepancy is expected to be a consequence of utilization of different interaction potentials corresponding to CO-related interactions.

The results of electrical conductivity variation with temperature are shown in figure 4. Similar to the dynamic viscosity, there are no significant changes to the electrical conductivity values upon the addition of $O_2$. Moreover, the results for 10% and 20% $C_4F_7N$ mixed with $CO_2$ at 5 atm are in excellent agreement with published data [2].

## 5. Conclusions

The calculation of thermodynamic and transport properties are necessary for analyzing the performance of promising $SF_6$ alternatives as an arc switching medium within medium voltage switchgear geometries. In this contribution, these properties have been calculated for two specific $C_4F_7N - CO_2$ gas mixtures, with 10% and 20% $C_4F_7N$ together with the addition of up to 20% $O_2$. Based on the equilibrium chemical composition results, increase in the $\%C_4F_7N$ results in an increase in the formation of CO. Even though the addition of $O_2$ results in significant reduction in the formation of CO, it enhances the production of $COF_2$.

### Acknowledgements

This work is part of the project Nr. TH03020063 of the Technological Agency of the Czech Republic.

### References

[1] H. Nechmi et al. Fluoronitriles/$CO_2$ gas mixture as promising substitute to $SF_6$ for insulation in high voltage applications. *IEEE Trans. Dielectr. Electr. Insul*, 23(5):2587–2593, 2016. doi:10.1109/TDEI.2016.7736816.

[2] Y. Wu et al. Properties of $C_4F_7N$-$CO_2$ thermal plasmas: thermodynamic properties, transport coefficients and emission coefficients. *Journal of Physics D: Applied Physics*, 51(15):1552061–12, 2018. doi:10.1088/1361-6463/aab421.

[3] G. Petersson et al. A complete basis set model chemistry. I. The total energies of closed-shell atoms and hydrides of the first-row atoms. *J. Chem. Phys.*, 89(4):2193–218, 1988. doi:10.1063/1.455064.

[4] L. Curtiss et al. Gaussian-4 theory. *J. Chem. Phys.*, 126(8):0841081–12, 2007. doi:10.1063/1.2436888.

[5] G. Godin and Y. Trepanier. A robust and efficient method for the computation of equilibrium composition in gaseous mixtures. *Plasma Chem Plasma Process*, 24(3):447–473, 2004. doi:10.1007/s11090-004-2279-8.

[6] S. Chapman and T. Cowling. *The mathematical theory of non-uniform gases.* Third edition. John Wiley Inc., 1970.

[7] A. Yang et al. Thermodynamic properties and transport coefficients of $CO_2$-Cu thermal plasmas. *Plasma Chem Plasma Process*, 36(4):1141–1160, 2016. doi:10.1007/s11090-016-9709-2.

doi:10.14311/ppt.2019.2.135
*Plasma Physics and Technology* **6**(2):135–139, 2019 

# DEVELOPMENT OF TWO-TEMPERATURE MATHEMATICAL MODEL OF PROCESSES IN DISCHARGE CHAMBER OF MULTI-CHAMBER ARRESTER OPERATING IN CONDITIONS OF MOUNTAIN AREAS

V. Frolov[a,*], D. Ivanov[a], A. Sivaev[b], A. Chusov[b]

[a] *Peter the Great St. Petersburg Polytechnic University, Polytechnicheskaya, 29, 195251, St. Petersburg, Russia*
[b] *Streamer Electric AG, Nevsky Prospect, 147, 191024, St. Petersburg, Russia*
[*] frolov.eed@gmail.com

**Abstract.** This article is devoted to a development of a mathematical model of plasma processes in a discharge chamber of a multi-chamber arrester operating under reduced pressure. The results obtained using this model will make it possible to develop recommendations on modifying the design of multi-chamber arresters for operation in conditions of mountain areas. A composition, thermodynamic and transport properties of plasma for the two-temperature model were calculated on the basis of the data on the materials of the discharge chamber and on the basis of the experiments performed earlier. The paper presents the results of calculations i.e. distributions of discharge voltage and electrical conductivity on time at various pressures. A comparison of obtained results was carried out.

**Keywords:** 2T model, simulation, pulse arc discharge, multi-chamber arrester.

## 1. Introduction

Protection of important energy facilities from direct lightning impact requires the creation of special devices for lightning protection.

Currently, a new promising way to protect overhead power lines from lightning impact is the use of multi-chamber arresters [1–3]. The multi-chamber arrester consists of a large number of series-connected chambers in which an arc discharge occurs during breakdown (Figure 1). Such a discharge is accompanied by erosion of the electrodes material and ablation of the material of the discharge chamber. During that process an increased pressure occurs in the chamber leading to the formation of a plasma jet from the discharge chamber and to the extinction of the electric arc (Figure 2).

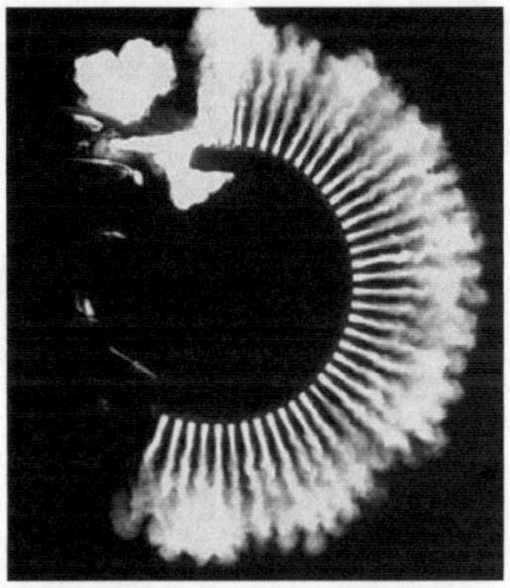

*Figure 1. Discharge in a multi-chamber arrester.*

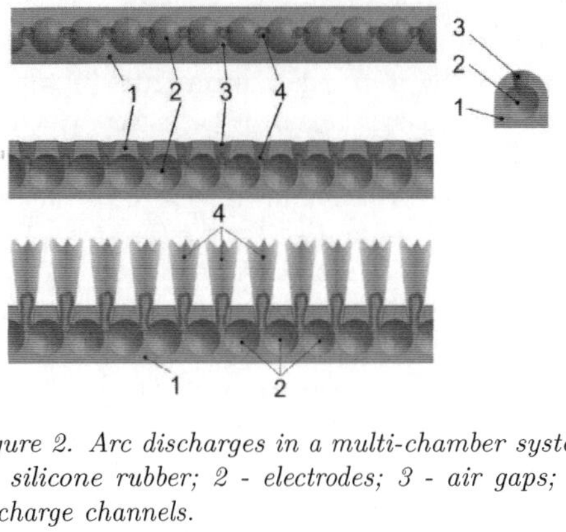


*Figure 2. Arc discharges in a multi-chamber system: 1 - silicone rubber; 2 - electrodes; 3 - air gaps; 4 - discharge channels.*

Investigations of the physical processes that take place inside such devices have been carried out over the years to improve the efficiency of multi-chamber arrester operation. Along with experimental methods [4], theoretical studies using mathematical modeling become widespread [5–10]. The data obtained during experiments are used in the mathematical model as initial data as well as for the purpose of correcting the model.

Currently, there is the problem of application of arresters for power lines located in mountain areas where in some cases the pressure may be 0.3–0.8 of the atmospheric pressure. The extinction of an electric arc in the arresters under such conditions becomes difficult, which is caused by the stable maintenance of the arc under reduced pressure. Therefore, the creation of lightning arresters requires significant adjustments.

At reduced pressure plasma in the presence of an electric field is characterized by a disturbance of thermal equilibrium. Therefore, plasma in such conditions is more adequately described by the two-temperature model [11].

The purpose of this article is to develop a mathematical model of non-equilibrium plasma processes of a pulsed arc discharge of complex chemical composition using the example of processes in a discharge chamber of a multi-chamber arrester operating in mountainous areas.

## 2. Methods

### 2.1. Calculation of plasma properties

To develop a non-equilibrium mathematical model of a pulsed arc discharge it is necessary to know the thermodynamic and transport properties of a plasma as a function of electron temperature, heavy particle temperature, and pressure.

A method of calculation of the composition, thermodynamic and transport properties of both equilibrium and two-temperature plasma is described in sufficient detail in the references [11–15].

In the case of the two-temperature approximation the plasma composition depends on the electron temperature $T_e$, the temperature $T_h$ of the heavy species and the pressure $p$. In practice, the dependence of plasma composition on temperature $T_e$ and the parameter $\theta = T_e/T_h$ is usually calculated at constant pressure.

The method of formation a system of equations describing composition in a chemically reacting medium is usually founded on two different approaches, one of which is based on the equations of mass action law using equilibrium constants [13, 14], and the other one - on finding the minimum of the thermodynamic potential of the system [11, 12]. The first method was chosen for calculation.

Silicone rubber has the following chemical formula:

$$(C_2H_6SiO)_n$$

Using this information and considering possible cases of the electrode material (Cu, W, Fe) the 85 components were taken into calculation.

The equilibrium constants for the equations of the mass action law were taken from [16, 17], and the partition functions of atoms and ions were calculated based on NIST data [18]. Enthalpies of individual components were taken from [16, 17] to calculate the thermodynamic properties.

To calculate the transport properties, it is necessary to find data on collision cross sections for each pair of species. The book [19] was used as the main source for specifying collisions between two neutral species and between neutral and charged components. If the required cross section was not found in the book [19] then we acted as follows.

When considering collisions between two neutral components it was assumed that the cross section can be described by the Lennard-Jones interaction potential [20], the main interaction parameters were taken from [21, 22] using rules given in [22].

When considering collisions between neutral and charged components it was assumed that the cross section can be described by the polarization potential, the parameters of which were taken from [21, 22].

The collision of two charged components was described by a screened Coulomb potential [23].

A series of calculations were carried out for a pressure range of 0.3–1.0 atm and a temperature range of 1000–30000 K with the following proportions of basic chemical elements included into the system:

$$\text{Si:O:C:H}=1:1:2:6;\ \text{Fe:O}=1:10.$$

These ratios were determined using a condition that the walls of the discharge chamber are made of silicone rubber and the electrodes are made of steel.

The ratio Fe:O=1:10 was made on the basis of experimental studies namely a spectral analysis of the plasma jet at the outlet of the chamber of the multi-chamber arrester. Comparison of the results of subsequent calculations with experimental data of electrical parameters of the discharge confirms the obtained ratio Fe:O.

The obtained dependences of plasma properties on the electron temperature and the heavy particles temperature [24] were used in the further mathematical model.

### 2.2. Mathematical model

A mathematical model of non-equilibrium plasma processes of a pulsed arc discharge in a discharge chamber of a multi-chamber arrester was developed using the software *ANSYS Fluent*.

In this paper, a two-temperature plasma model was used. This model was described in detail in [25], a chemical equilibrium was assumed. In addition to the equations of gas dynamics and the energy equation of heavy particles, which are standardly solved in *ANSYS Fluent*, the following equations were included into the mathematical model using the *UDF* (user-defined functions) interface: the energy equation for electrons, the equation for scalar electric potential and the equation for vector magnetic potential [25].

A calculation domain including both the space inside the discharge chamber and the plasma jet region is shown in Figure 3. The task was solved in a two-dimensional formulation.

The computational mesh consisted of 55000 triangular control volumes.

The boundary conditions were defined as follows. A value of scalar potential difference was set at the electrodes (Figure 3, boundaries 1) so that to provide the necessary discharge current (see below). The ablation of the walls of the discharge chamber was taken into account (Figure 3, boundaries 2), i.e. a source of mass was considered on the wall. Its value was calculated as follows.

The power $P_{pl}$ transmitted from the plasma into the wall region $\Delta S$ due to thermal conductivity is determined by the formula:

$$P_{pl} = -\lambda \frac{dT}{dn} \Delta S \qquad (1)$$

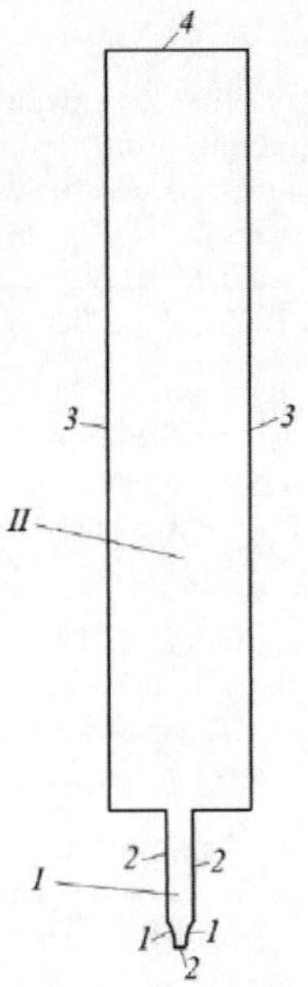


*Figure 3. Calculation domain: I - the space inside the discharge chamber; II - the plasma jet region; 1 - electrodes; 2 - walls of the discharge chamber; 3 - symmetry boundary; 4 - outlet (ambient pressure).*

where $\lambda$ is thermal conductivity, $\mathrm{d}T/\mathrm{d}n$ is temperature gradient in the direction that is normal to the wall.

The energy consumption $P_{\text{vap}}$ for the evaporation of the wall material is determined by the formula:

$$P_{\text{vap}} = \Delta H_{\text{vap}} \frac{\mathrm{d}m}{\mathrm{d}t} \tag{2}$$

where $\Delta H_{\text{vap}}$ is a specific heat of evaporation, $\mathrm{d}m/\mathrm{d}t$ is mass rate of evaporation that is source of mass for the calculation domain.

From the condition that all power transferred to the wall was spent on the evaporation of the wall material

$$P_{\text{pl}} = P_{\text{vap}} \tag{3}$$

one can calculate a source of mass $\mathrm{d}m/\mathrm{d}t$.

Boundary 3 (see Figure 3) were considered as symmetry boundaries since the arrester has a multi-chamber design (it means that the same chambers are located to the left and right of the considered discharge chamber). The ambient pressure was set at the boundary 4 (Figure 3).

## 3. Results

Simulation was carried out for the time range from 0 to 250 µs for two cases: ambient pressure was 1 atm and 0.3 atm.

In both cases the experimental dependence of the discharge current on time was used. To reduce the computation time, only the first 250 µs and 150 µs before current zero were used. The graph of the used dependence of the discharge current on time is shown in Figure 4.

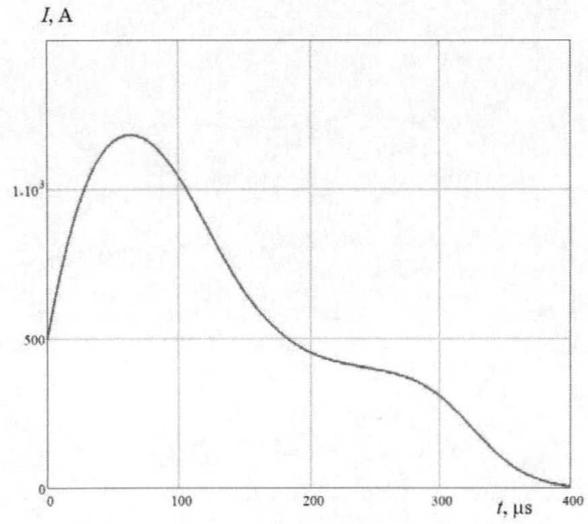


*Figure 4. A dependence of the discharge current on time.*

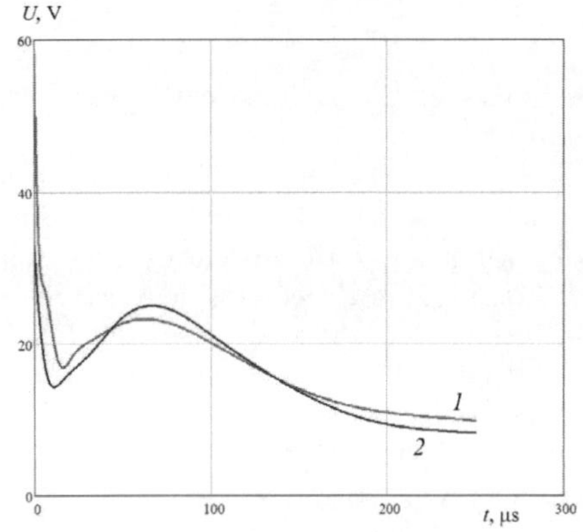


*Figure 5. The dependence of the discharge voltage on time for operation at different pressures: 1 - 1.0 atm, 2 - 0.3 atm.*

The time dependences of the voltage providing the given current for pressure of 1 atm and 0.3 atm are shown in Figure 5. It is seen that at the end of the time dependences the voltage on the discharge chamber is greater in the case of a pressure of 1 atm than in the case of a pressure of 0.3 atm. This suggests that with decreasing pressure the breaking capacity of multi-chamber arresters decreases.

A conductivity of the discharge gap (see Figure 6) was calculated on the base of the obtained data:

$$G = \frac{I}{U} \tag{4}$$

It can be seen (Figure 6) that at the end of the time dependences the conductivity of the discharge gap at a pressure of 0.3 atm is 20% higher than the conductivity at a pressure of 1 atm.

Thus, the results of performed calculations showed that the conductivity of the discharge gap increased by 20% with a decrease in pressure from 1 to 0.3 atm (the range of pressure variation is due to the operating conditions of the arresters). This means that when

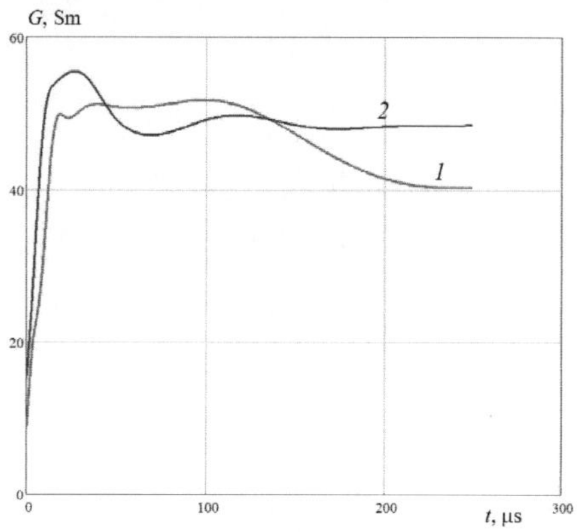


*Figure 6. The dependence of the discharge conductivity on time for operation at different pressures: 1 - 1.0 atm, 2 - 0.3 atm.*

the pressure decreases, conditions of the arc quenching degenerate.

Therefore, in order to increase the breaking capacity of multi-chamber arresters it is proposed to introduce insulating barriers into the regions of the discharge channel 4 (see Figure 2) between the electrodes. As a result, the length of the arc discharge will increase which contributes to an increase in evaporation of the chamber walls; the arc voltage increases which ultimately will increase the breaking capacity of a multi-chamber arrester.

It should also be noted that the investigating plasma processes have short time of existence. Under these conditions (high specific dissipated power, high arc velocity, reduced pressure) chemical reactions may not have time to finish, and chemical equilibrium may not be reached. Therefore, to clarify the characteristics of the breaking capacity of multi-chamber arresters it is necessary to improve the mathematical model, namely, to take into account the disturbance of chemical equilibrium.

## 4. Conclusions

Thus, the thermodynamic and transport properties of the plasma (air with vapors of materials of the discharge chamber wall and of electrodes material) were calculated and a two-dimensional two-temperature model of plasma processes in the discharge chamber of the multi-chamber arrester was developed. The calculation results showed a decreasing in the breaking capacity of the arrester by 20% with a decrease in pressure from 1 to 0.3 atm. To increase the arrester's breaking capacity, it is proposed to introduce an insulating barriers into the regions of the discharge channel 4 (see Figure 2) between the electrodes.

To clarify the characteristics of the breaking capacity of multi-chamber arresters it is necessary to improve the mathematical model, namely, to take into account the disturbance of chemical equilibrium.

## References


[1] G. Podporkin, E. Enkin, E. Kalakutsky, V. Pilshikov, and A. Sivaev. Lightning protection of overhead lines by multi-chamber arresters and insulator-arresters. In *Proceedings of the 30th Int. Symp. on Lightning Protection*, page 7845759, 2010. doi:10.1109/ICLP.2010.7845759.

[2] R. Kozakov, A. Khakpour, S. Gorchakov, D. Uhrlandt, D. Ivanov, I. Murashov, G. Podporkin, and V. Frolov. Investigation of a multi-chamber system for lightning protection at overhead power lines. *Plasma Physics and Technology*, 2(2):150–154, 2015.

[3] G. Podporkin. Development of long flashover and multi-chamber arresters and insulator-arresters for lightning protection of overhead distribution and transmission lines. *Plasma Physics and Technology*, 2(3):241–250, 2015.

[4] M. Pinchuk, A. Budin, I. Kumkova, A. Bogomaz, A. Sivaev, A. Chusov, and R. Zaynalov. Energy deposition in discharge chamber of lightning protection multichamber system. *Journal of Physics: Conference Series*, 774(1):012187, 2016. doi:10.1088/1742-6596/774/1/012187.

[5] V. Frolov, D. Ivanov, I. Murashov, and A. Sivaev. Calculation of the composition of plasma of an arc pulsed discharge in a multi-chamber arrester. *Technical Physics Letters*, 41(4):310–313, 2015. doi:10.1134/S1063785015040069.

[6] V. Frolov, D. Ivanov, I. Murashov, and A. Sivaev. Mathematical simulation of processes in discharge chamber of multi-chamber system for lightning protection at overhead power lines. In *Proceedings of the 2016 IEEE North West Russia section young researchers in electrical and electronic engineering conference, ELCONRUSNW 2016*, pages 562–565, 2016. doi:10.1109/EIConRusNW.2016.7448245.

[7] A. Chusov, G. Podporkin, M. Pinchuk, D. Ivanov, I. Murashov, and V. Frolov. Development of a physical 2-D model for arc quenching chamber of lightning protection multichamber systems. In *Proceedings of the 2016 33rd International Conference On Lightning Protection, ICLP 2016*, page 7791509, 2016. doi:10.1109/ICLP.2016.7791509.

[8] Z. Guo, X. Long, Z. Qian, and N. Qiu. Three dimensional simulation of the arc inside an insulator-arrester with a multichamber system. *AIP Advances*, 6(4):045117, 2016. doi:10.1063/1.4948453.

[9] V. Frolov, D. Ivanov, and R. Belsky. Increasing of operation security and of breaking capacity of surge arresters. In *Proceedings of the 2017 IEEE Russia Section Young Researchers In Electrical And Electronic Engineering Conference, ELCONRUS 2017*, pages 1520–1523, 2017. doi:10.1109/EIConRus.2017.7910861.

[10] V. Frolov, D. Ivanov, G. Podporkin, and A. Sivaev. Development of mathematical model of processes in multi-chamber arrester for identification of criteria of arc extinction. In *Proceedings of the 2017 International Symposium On Lightning Protection, XIV SIPDA 2017*,

pages 240–243, 2017.
doi:10.1109/SIPDA.2017.8116930.

[11] M. I. Boulos, P. Fauchais, and E. Pfender. *Thermal Plasmas: Fundamentals and Applications*. New York: Plenum, 1994.

[12] M. Capitelli, G. Colonna, and A. D'Angola. *Fundamental Aspects of Plasma Chemical Physics: Thermodynamics*. Springer Series on Atomic, Optical, and Plasma Physics. Springer-Verlag New York, 2012.

[13] V. P. Glushko et al. *Thermodynamic and Thermophysical Properties of Combustion Products, Ed. by V. P. Glushko, Vol. 1*. Nauka Moskva, 1971.

[14] A. Suris. *Thermodynamics of High-Temperature Processes*. Metallurgiya Moskva, 1985.

[15] V. Frolov and D. Ivanov. Calculation of a plasma composition and its thermophysical properties in cases of maintaining or quenching of electric arcs. *Journal of Physics: Conference Series*, page 012040, 2018. doi:10.1088/1742-6596/1058/1/012040.

[16] NIST atomic spectra database. Official cite. http://physics.nist.gov/PhysRefData/ASD/index.html.

[17] V. P. Glushko et al. *Thermodynamic Properties of Individual Substances: Reference Book, 3rd ed., Vols. 1-4*. Nauka Moskva, 1978-1982.

[18] M. W. Chase Jr. *NIST-JANAF Thermochemical Tables, Fourth Edition, Part I and II*. Journal of Physical and Chemical Reference Data. 1998.

[19] M. Capitelli, D. Bruno, and A. Laricchiuta. *Fundamental Aspects of Plasma Chemical Physics: Transport*. Springer Series on Atomic, Optical, and Plasma Physics. Springer-Verlag New York, 2013.

[20] J. Hirschfelder, C. Curtiss, and R. Bird. *Molecular theory of gases and Liquids*. Wiley New York, 1954.

[21] Y. Cressault, R. Hannachi, P. Teulet, A. Gleizes, J.-P. Gonnet, and J.-Y. Battandier. Influence of metallic vapours on the properties of air thermal plasmas. *Plasma Sources Science and Technology*, 17(3):035016, 2008. doi:10.1088/0963-0252/17/3/035016.

[22] P. Andre, J. Aubreton, S. Clain, M. Dudeck, E. Duffour, M. Elchinger, B. Izrar, D. Rochette, R. Touzani, and D. Vacher. Transport coefficients in thermal plasma. applications to mars and titan atmospheres. *The European Physical Journal D*, 57(2):227–234, 2010. doi:10.1140/epjd/e2010-00036-5.

[23] E. Mason, R. Munn, and F. Smith. Transport coefficients of ionized gases. *The Physics of Fluids*, 10(8):1827–1832, 1967. doi:10.1063/1.1762365.

[24] D. Ivanov, V. Skornyakov, I. Savelieva, M. Korotkikh, V. Shestakov, D. Uhrlandt, and G. Podporkin. Mathematical simulation of operation of multi-chamber arrester for lightning protection of power lines: calculation of thermophysical properties of nonequilibrium plasma. In *MATEC Web of Conferences*, volume 245, page 07003, 2018. doi:10.1051/matecconf/201824507003.

[25] P. Freton, J. Gonzalez, Z. Ranarijaona, and J. Mougenot. Energy equation formulations for two-temperature modelling of 'thermal' plasmas. *Journal of Physics D: Applied Physics*, 45(46):465206, 2012. doi:10.1088/0022-3727/45/46/465206.

doi:10.14311/ppt.2019.2.140
*Plasma Physics and Technology* **6**(2):140–143, 2019 

# ARC BEHAVIOURS IN THE HV $SF_6$ GAS-BLAST INTERRUPTER

S. Averyanova*, E. Tonkonogov

*Institute of Energy and Transport System, Peter the Great St. Petersburg Polytechnic University, 29, Polytechnicheskaya st., St. Petersburg, 195251, Russia*

* averly@yandex.ru

**Abstract.** A comprehension of the dielectric strength recovery processes during the interruption of short-circuit currents in the high-voltage $SF_6$ gas-blast circuit breakers is necessary for their modernisation in order to increase the rated voltage and short circuit breaking current per one break. This paper presents numerical results of the turbulence effects on the interruption ability in the $SF_6$ extinguishing arc chamber.

**Keywords:** dielectric strength recovery, turbulence, nozzle geometry, $SF_6$ circuit breaker.

## 1. Introduction

The breakdown probability during the interruption mode increases with increasing the rated voltage. It is necessary to reach a good comprehension of physical processes occurring in high-voltage (HV) gas-blast circuit breakers (CBs) during the interruption mode for solving the above mentioned problem. It is possible to distinguish four phases of the dielectric strength recovery after the current zero [1].

The first phase is related to the thermal mode. The hot-gas channel remains conductive. The thermal balance between heating and cooling processes, primarily due to the turbulent dissipation in the nozzle throat, plays the important role during this mode. This mode lasts several µs [1].

The second phase is related to the dielectric mode. For this mode, the volume of the hot-gas channel is negligible in the downstream zone of the insulation nozzle, but in the upstream zone of the nozzle there is a considerable amount of hot-gas portions, especially in the stagnation zone. As well known, the thermal clogging is used for the additional pressure rise during the interruption process in the puffer-type HV gas-blast CB. On the one hand, the pressure rise helps to increase this interruption ability. On the other hand, the temperature in the compression volume increases too, and this leads to reduction in the upstream dielectric strength. Indeed, the experimental data [1, 2] show that the presence of conducting gas in the upstream zone of the nozzle may lead to the breakdown in the vicinity of the $u_1$ point (the four-parameter reference line for the transient recovery voltage (TRV)) for BTF (the breaker terminal fault 100%) [3]. At the same time, the experimental data [4] for the puffer-type CB with rated parameters of 50 kA, 300 kV ($u_1$=239 kV, $t_1$=119 µs [3]), and insulation nozzle throat diameters $d$ = 35, 40 (= $d_0$, Fig. 1), 45 mm show a clear correlation between dielectric strength recovery and insulation nozzle throat diameter. This correlation is nonlinear, and the smaller nozzle diameter leads to the worst interruption ability.

The third phase is related with the additional gas heating due to recombination effects. The turbulent kinetic energy and the dielectric strength in the upstream zone are increasing rapidly during this mode, but there is a probability of the breakdown through the stagnation zones.

The final phase begins after ending the recombination. The dielectric strength is related to that of cold gas.

This paper analyses the arc and flow interface parameters that may influence the puffer-type HV $SF_6$ CB interruption ability in the vicinity of the $u_1$ point (between the 2nd and the 3rd modes) for different diameters of the insulation nozzle throat.

## 2. Mathematical Simulation

Fig. 1 shows the schematic diagram of the puffer-type HV $SF_6$ CB that was used for the numerical simulation and the computational domain is plotted with dotted lines.

The insulation throat diameter varies within $d = (0.85 - 1.15)d_0$. The diameter of the metal nozzle throat, $d_m$, is 2 mm smaller than that of the insulation nozzle throat. The rated pressure is equal to $p_0$ = 7 atm. The current is changing with time according to the synthetic test with the stationary value $I_{max}$ = 2.5 kA, and $dI/dt$ = -17.7 A µs$^{-1}$.

For numerical analysis of the insulation throat diameter impact on the dielectric strength recovery, the time-averaged system of equations was used, including the Navier-Stokes equations, the continuity and energy equations [5]. The Ohm law, models of the turbulent viscosity and radiation transfer are added to the system of differential equations.

The turbulence model for the steady state analysis takes into account the turbulence effects in the shear layer between arc and gas flow produced by two types of instability: both the shear instability and Rayleigh-Taylor instability [5]. For the unsteady state analysis, the "freeze"-theory for the kinematic turbulent viscosity, $\nu_t$, was used [5]. It means that at all moments

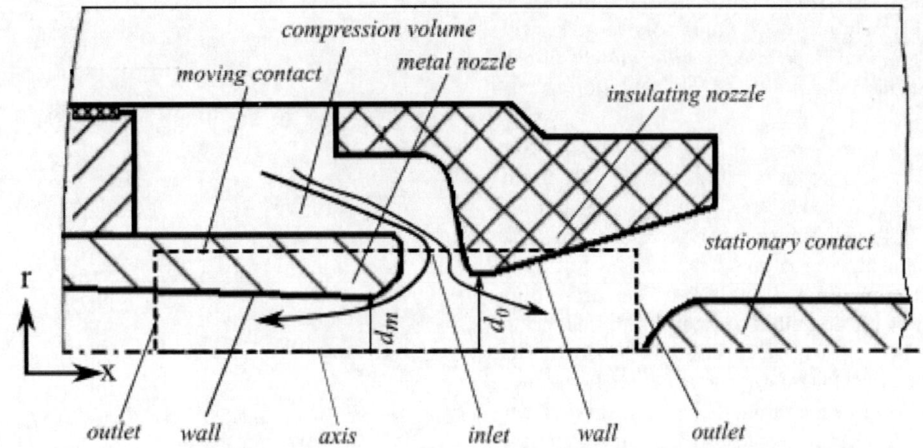


*Figure 1. The simplified geometry and the computational domain of the puffer-type $SF_6$ CB.*

of time when the current is rapidly falling down, the kinematic turbulent viscosity is assumed constant and its numerical value is equal to its stationary value. The dynamic coefficient of the turbulent viscosity, $\mu_t$, comprises the geometric mean density calculated in terms of the point density and that of the cold gas in this section. When the current is falling down, the arc density and the turbulent viscosity are growing after the current zero.

For the radiation transfer, the "radiation conductivity"approximation was used [6]. The radiation transfer plays an important role in the heat transfer at the maximum current, but it does not matter after the current zero.

As known from experiments [7], after the current zero there is the post-arc part where the temperature is falling down more quickly than in other parts. Therefore, we may assume that the gas dielectric strength will start to recover if the temperature on the arc axis reaches values of 3000 K at least at two points [8].

For the numerical simulation, the home-made 2D CFD code was used based on the algorithm given in [9] together with the convective fluxes approximation by Roe's method [10, 11]. The gas material functions are computed based on the data given in [8]. The boundary conditions are presented in Fig. 1. The unsteady state simulation time is equal to 300 µs, including the current falldown 141 µs period and dielectric strength recovery 159 µs period. Based on the experimental pressure curves [12, 13], the inlet static pressure assumed constant during the simulation. The walls ablation and its influence on the pressure in the compression volume is not considered in this simulation.

## 3. Results and Discussion

Results of the numerical simulation are presented in Figs. 2–5. So, considered is the influence of the arc and flow interface parameters on the dielectric strength recovery in the vicinity of the $u_1$ point. Tested are three moments of time related to the stationary arc parameters (-141 µs), the thermal breakdown phase (3 µs), and the dielectric breakdown phase in the vicinity of the $u_1$ point (119 µs).

Fig. 2 shows the normalised turbulent viscosity (the ratio of turbulent viscosity, $\mu_t$, to the maximum value of $\mu_t$ at $d_0$ and -141 µs) as the function of the normalised radial coordinate, $r/d_0$, for the insulation nozzle throat at the steady state of the arc and flow. Obviously, the geometry with insulation nozzle throat being equal to $0.85\,d_0$ ensures the largest of the turbulence viscosity in the shear layer between the arc and flow. The difference between maximum values on the curve of this figure and those related to the insulation nozzle throat diameter being equal to $d_0$ is about 40%. The maximum value of the turbulent viscosity, $\mu_t$, decreases with increasing the nozzle throat diameter. It is due to the decrease in the pressure ratio leading to the decrease in the shear instability component of the turbulent viscosity, $\mu_t$.

The difference between nozzle throat diameters $0.85\,d_0$ and $d_0$ remains equal to about 40% at all moments of time. After the current zero, the value of the turbulent viscosity for $d > d_0$ increases more slowly than for $d \leqslant d_0$, and is practically independent of the nozzle throat diameter. This is due to the small value of the turbulent viscosity in the shear layer between the arc and flow. It means that after the current zero the turbulent heat transfer is weak and the temperature falls down (while the density grows up) more slowly than for geometries with $d \leqslant d_0$.

Fig. 3 shows the normalised turbulent viscosity (the ratio of turbulent viscosity, $\mu_t$, to the maximum value of $\mu_t$ at $d_\mathrm{m}$ and -141 µs ) as the function of the normalised radial coordinate, $r/d_0$, for the metal nozzle throat at the steady state of the arc and flow. Obviously, the turbulent viscosity in the metal nozzle throat reveals the same behaviour as in the insulation one. The difference between $0.85\,d_0$ and $d_0$ nozzle throat diameters is equal to about 33% at all mo-

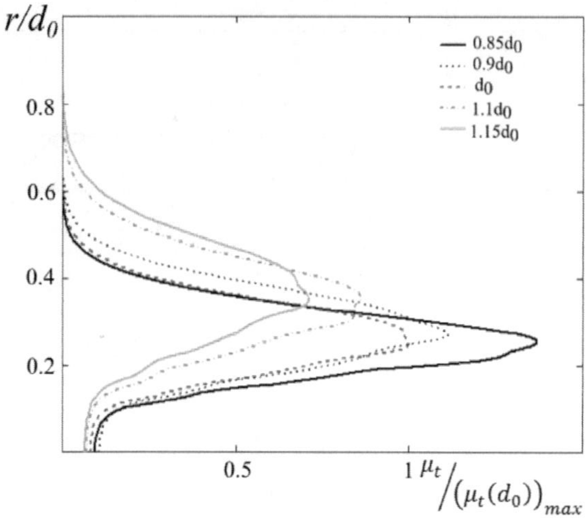


*Figure 2. The normalised turbulent viscosity as the function of the normalised radial coordinate for the insulation nozzle throat at the steady state flow.*

ments of time. After the current zero, the value of the turbulent viscosity for $d > d_0$ increases more slowly than for $d \leqslant d_0$ and is practically independent of the nozzle throat diameter similar of the insulation nozzle throat.

Fig. 4 shows the average temperature on the axis, $T_{\text{axis, mid}}$, as the function of time, $t$. The maximum temperature relates to the maximum nozzle diameter, $1.15\,d_0$. In [7] it is shown, that the arc radius is in the inverse relationship to the pressure. For the geometry considered, the pressure is smaller and the arc radius is maximum in all presented geometries. Considering that the synthetic test is used, the Joule heating is the same for all cases. Therefore, the lowest temperature is reached when the arc radius is largest. However, it should be noted that the steady-state temperature for all geometries are about 18000 K, that is consistent with the experimental data [7]. In the vicinity of the current zero the average axial temperature is about 10000 K, except for the $1.15\,d_0$ insulation nozzle throat diameter. The minimum turbulent viscosity in this nozzle is related to the weak decay of the conductivity in the post-arc channel. Moreover, in this case, the pressure ratio is less than the critical value for this geometry, and the convective heat transfer is weaker than in other computations. In the vicinity of the $u_1$ point all curves show temperatures smaller than 3000 K, except for the $1.15\,d_0$ insulation nozzle throat diameter. It means that the most part of the gas on the axis is non-conductive, but gas conductivity zones exist near the outlet boundaries in the computational domain.

Fig. 5 shows the normalised radius of the conductivity zone in the insulation nozzle throats (the ratio of the conductivity zone radius, $r_{\text{cond}}$, to that for the insulation nozzle throat diameter, $r_{\text{cond}}(d_0)$, at every moment of testing) as the function of the normalised insulation nozzle throat diameter, $d/d_0$. This figure presents only two curves because in several microseconds the temperature in the insulation nozzle reaches 3000 K and the gas becomes non-conductive. The maximum radius is related to the $1.15\,d_0$. Moreover, the falldown rate for $d > d_0$ is smaller than that for $d \leqslant d_0$.

Considering the above mentioned, we assume that the minimum diameter ($0.85\,d_0$) ensures the best arc and flow interface parameters for arc interruption and dielectric strength recovery. Nevertheless, this analysis does not take into account the wall ablation. The arc radiation is very intensive at the current peak in the HV CB [4]. The reduction in the nozzle throat diameter leads to the increase in the thermal clogging time. The long-time clogging leads to the nozzle wall erosion and the increase in the nozzle throat diameter due to the long-time arc filling of the nozzle. Therefore, the gas high temperature in the compression volume may lead to the slower falldown of the post-arc temperature and the increase in the breakdown probability [4]. The approximate estimate [7] gives the arc radius about 35 mm for simulations. In this case, the thermal clogging can be strong enough, that is confirmed by the data [4]. Therefore, it may be assumed that the insulation nozzle throat diameter should exceed the arc radius by 10–15%. In this range of throat diameters, the thermal clogging ensures the pressure ratio leading to the flow conditions under which the arc interruption is successfully fulfilled. On the other hand, the temperature rise in the compression volume due to the nozzle clogging has no significant effect on the dielectric strength recovery. If the nozzle diameter exceeds this range, the pressure rise in the compression volume is not great enough to ensure the successful arc interruption.

## 4. Conclusions

The numerical analysis shows that the arc interruption and dielectric strength recovery in the puffer-type HV CB (BTF 100%) are closely related to the arc and flow interface parameters, which ensures the dominant cooling mechanisms, such as turbulent and convective heat transfer. The above mentioned range of the insulation nozzle throat diameter range may be used when we provide arc and flow interface conditions for successful arc interruption. In this paper, the wall ablation was not taken into account, but this is one of the causes of the breakdown through the wall boundary layer with the rated voltage per one break increasing in the puffer-type HV CB. When considering the impact of the insulation nozzle ablation on the dielectric strength recovery it will help us to predict BTF 100% test results more qualitatively.

## References

[1] E. Schade. Similarity of the dielectric recovery characteristic of axially blown arcs in $SF_6$. In *Proc. 8-th Conf. on Gas Discharges, Oxford*, pages 50–53, 1985.

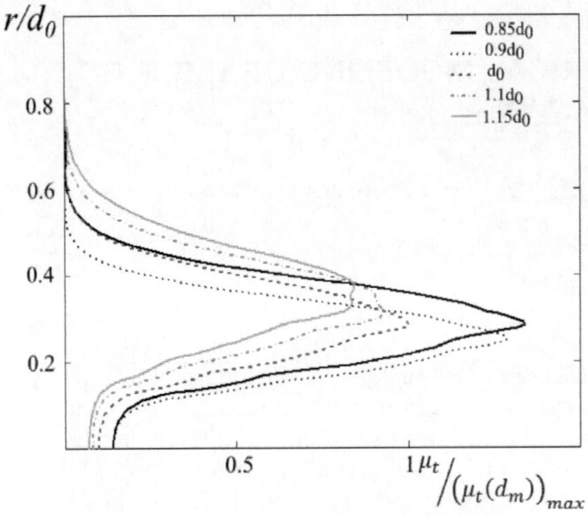


*Figure 3. The normalised turbulent viscosity as the function of the normalized radial coordinate for the metal nozzle throat at the steady state flow.*

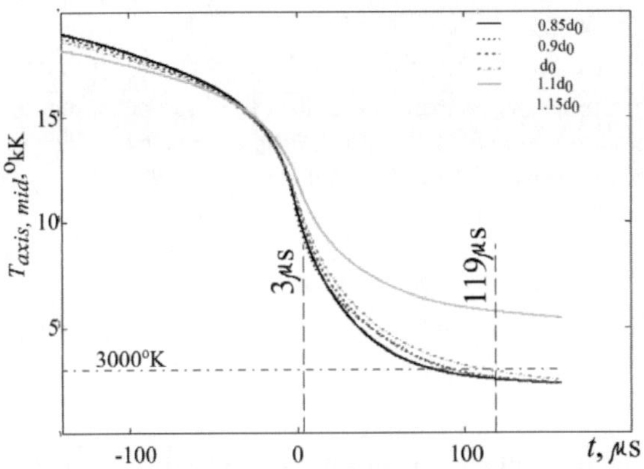


*Figure 4. Average axial temperature as the function of time.*

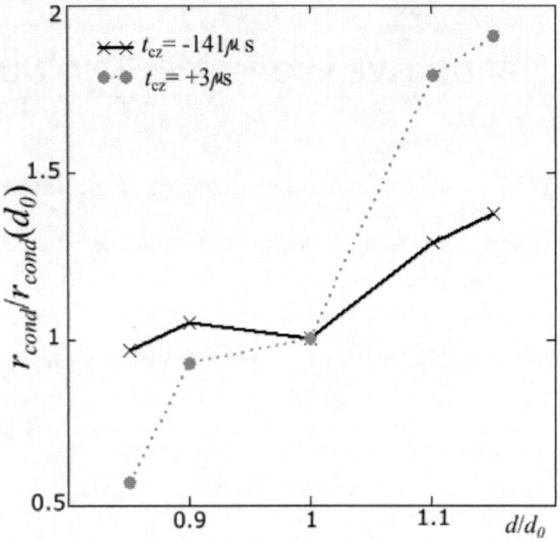


*Figure 5. The normalised radius of the conductivity zone in the insulation nozzle throats as the function of the normalised insulation nozzle throat diameter.*

[2] S. Yanabu, H. Mizoguchi, A. Kobayashi, Y. Ozaki, and Y. Murakami. Factor influencing the interruption ability of $SF_6$ puffer breaker and development of 300 kV-50 kA one-break circuit breaker. *IEEE Trans. on Power App. and Syst.*, PAS-101(6):1511–1518, 1982. doi:10.1109/TPAS.1982.317199.

[3] High-voltage switchgear and controlgear - part 100: Alternating current circuit-breakers. IEC 62271-100:2008 ed2.0.

[4] H. Ikeda, T. Ueda, A. Kobayashi, M. Yamamoto, and Y. S. Development of large-capacity, $SF_6$ gas interruption chamber and its application to GIS. *IEEE Trans. on Power App. and Syst.*, PAS-103(10):3038–3043, 1984. doi:10.1109/TPAS.1984.318308.

[5] S. Averianova, N. Akatnov, and E. Tonkonogov. Numerical modeling of the decaying arc. In *VII Int. Conf. "Plasma Physics and Plasma Technology", Minsk (Belarus)*, pages 655–658, 17-21 Sept., 2012.

[6] S. Averianova, N. Akatnov, and E. Tonkonogov. Numerical modelling of the interrupting arc in HV $SF_6$-blast interrupters. In *XV Symp. on Physics of Switching Arc, Brno (Czech Republic)*, volume 1, pages 3–6, 22-26 Sept., 2003.

[7] K. Ragaller (editor). *Current interruption in high voltage networks.* Plenum Press, NewYork, 1978.

[8] L. S. Frost and R. W. Liebermann. Composition and transport properties of $SF_6$ and their use in a simplified enthalpy flow arc model. *Proceedings of the IEEE*, 59(4):474–485, 1971. doi:10.1109/PROC.1971.8206.

[9] R. M. Beam and R. F. Warming. An implicit factored scheme for the compressible Navier-Stokes equations. *AIAA Journal*, 16:393–401, 1978. doi:10.2514/3.60901.

[10] P. L. Roe. Approximate Riemann solvers, parameter vectors and difference scheme. *J. Comput. Phys.*, 43:357–372, 1981. doi:10.1016/0021-9991(81)90128-5.

[11] C. Hirsch. *Numerical Computation of Internal and External flows, v.2.* Wiley, 1990.

[12] A. Kornhaas, F. Reichert, C. Leu, and F. Berger. Simulation of interaction between switching arc and switching mechanics in $SF_6$ self-blast circuit breakers. In *XVIII Symp. on Physics of Switching Arc, Brno (Czech Republic)*, pages 276–279, 7-11 Sept., 2009.

[13] J. Zhong, Z. Li, and Y. Guo. Pressure and arc voltage measurements in a 252 kV $SF_6$ puffer circuit breaker. In *XXth Symp. on Physics of Switching Arc, Brno (Czech Republic)*, volume Contributed Papers, pages 332–335, 2-6 Sept., 2013.

doi:10.14311/ppt.2019.2.144
*Plasma Physics and Technology* **6**(2):144–147, 2019 

# RADIATIVE PROPERTIES AND NUMERICAL MODELING OF $C_4F_7N$-$CO_2$ THERMAL PLASMA

M. Gnybida[a,*], Ch. Rümpler[b], V.R.T. Narayanan[a]

[a] *Eaton European Innovation Center, Bořivojova 2380, 252 63 Roztoky, Czech Republic*
[b] *Eaton Industries GmbH, Hein-Moeller-Straße 7-11, 53115 Bonn, Germany*
[*] MykhailoGnybida@Eaton.com

**Abstract.** $C_4F_7N$ and $C_4F_7N - CO_2$ mixtures are considered as alternatives to $SF_6$ for use in medium voltage gas insulated switchgear applications (GIS), due to the low global warming potential and good dielectric properties of $C_4F_7N$. Current work is focused on the calculation of radiative properties (absorption coefficients) of $C_4F_7N - CO_2$ thermal plasma and computational fluid dynamics (CFD) simulations of free burning $C_4F_7N - CO_2$ arcs that are stabilized by natural convection. Absorption coefficients of $C_4F_7N - CO_2$ plasma used in the CFD model are derived from spectral absorption coefficients by Planck averaging. An optimization procedure has been applied to find the optimal number of spectral bands as well as spectral band interval boundaries. Radiation and flow model results for $C_4F_7N - CO_2$ in comparison to $SF_6$ and air are provided and discussed.

**Keywords:** radiation, thermal plasma, gas insulated switchgear, $SF_6$ alternative.

## 1. Introduction

Sulphur hexafluoride ($SF_6$) is commonly used in gas insulated switchgear applications (GIS) for medium voltage and high voltage applications. In case of medium voltage (1000 V to 52 kV), $SF_6$ is usually used as an electric insulation medium that provides significantly enhanced dielectric strength compared to air as insulation medium, which can be utilized to shrink the GIS switchgear design compared to air insulated designs. Switching in $SF_6$ is performed in some medium voltage applications too, e.g. load break switches. In case of high voltage switchgear applications (above 52 kV), $SF_6$ provides excellent electric insulation as well as interruption capabilities, which are used in GIS but also in circuit breaker applications.

One important drawback of $SF_6$ is the 23 500 times larger global warming potential (GWP) compared to $CO_2$. Because of this very high GWP, many alternative gases have been considered in pursue of replacing $SF_6$ in medium and high voltage switchgear and circuit breaker applications. One promising alternative gas is $C_4$ perfluoro-nitrile $C_4F_7N$, which has a GWP of 2 200 and its dielectric strengths is 2 times higher than $SF_6$. Because of the high boiling point of $C_4F_7N$, which is -4.7 °C at atmospheric pressure, this $SF_6$ alternative needs to be diluted with some other gas to avoid liquefaction. Otherwise the lower operating temperature limit might not be sufficient.

In this contribution, the radiative properties of $C_4F_7N$ are calculated and applied in CFD models of free burning arcs that are stabilized by natural convection. Besides radiative properties, transport, and thermodynamic properties are needed as well to setup a complete model. The calculation approach and results are introduced in a companion paper of this conference [1]. Results of the radiation properties calculation and the CFD model for $C_4F_7N - CO_2$ will be compared with those of air and $SF_6$.

## 2. Radiative properties

Radiation transport is a very important part of power balance of a thermal plasma. The simplest approach to model the effect of radiation is the net emission coefficient (NEC) [2], which describes the net radiation from an isothermal sphere or cylinder. Although the knowledge of the NEC allows to estimate the radiation losses in the hottest regions of the arc, it cannot predict the reabsorption of radiation in the colder arc fringes and the radiative flux impinging on surrounding surfaces.

More capable radiation transfer models are the P1 and the discrete-ordinates method (DOM) [3]. With these models the radiation spectrum can be divided into several gray bands to take into account the spectral variation of radiative transport. A constant Planck or Rosseland averaged absorption coefficient is derived from the spectral absorption coefficient for the particular band. Instead of resolving the whole spectrum, the radiation transfer is solved for a small number of bands only, which does reduce the computational effort dramatically while providing reasonable accuracy. In this paper, we calculate Planck mean absorption coefficient $k_{\mathrm{Planck},i}$ for each band $i$ according to:

$$k_{\mathrm{Planck},i} = \frac{\int_{\nu_i}^{\nu_{i+1}} k_\nu B_\nu \mathrm{d}\nu}{\int_{\nu_i}^{\nu_{i+1}} B_\nu \mathrm{d}\nu}, \quad (1)$$

where the lower and upper interval boundaries (band limits) are denoted by $\nu_i$ and $\nu_{i+1}$ respectively and $B_\nu$ represents the black body radiative intensity. We renormalize the line contribution of the spectral absorption coefficients according to [4] with a characteristic plasma absorption length $H = 3 \cdot R_p$, as suggested in [5]. The plasma radius $R_p$ is defined as radius value where the temperature decreased to 50% of the core temperature.

To determine the necessary number of bands and location of the band limits in the spectral range, optimization and verification procedures as introduced in [5–7] are applied to a specific mixture of $C_4F_7N$ and $CO_2$. In the optimization procedure two infinitely long cylindrical calculation domains are considered, one with 3 cm and the other one having a 10 cm radius. Both profiles are discretized using 30 points. The plasma inside these domains shall be represented by a fixed temperature profile along the radius, as shown in Fig. 1.

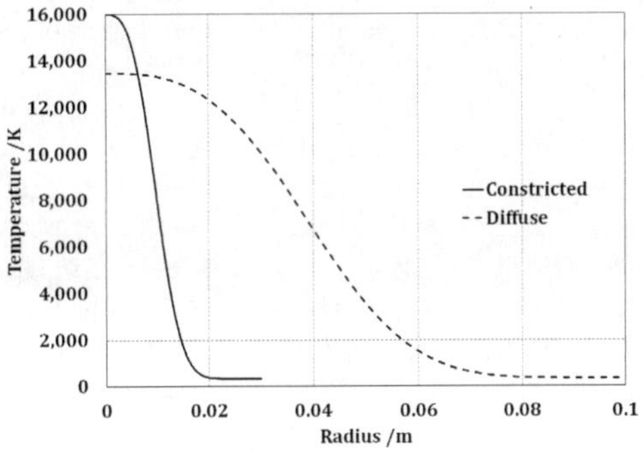


*Figure 1. Assumed plasma temperature profiles.*

The more constricted profile has a temperature maximum at 16 000 K while the diffuse profile is defined with a 13 460 K temperature maximum in the arc core. Each domain is evenly filled with a 90% $CO_2$ – 10% $C_4F_7N$ gas mixture at atmospheric pressure (1 atm). For both profiles optimization runs were performed. The two temperature profiles are used to calculate the species compositions first, which are used as input for the calculation of the spectral absorption coefficients $k_\nu$. The frequency range in the calculation covers $10^{12}$ Hz to $10^{16}$ Hz with a frequency step size of $2 \cdot 10^{10}$ Hz. An example of the calculated spectral absorption coefficient with line and continuum contributions for a 90% $CO_2$ – 10% $C_4F_7N$ gas mixture at 10 000 K and 1 atm is shown in Fig. 2.

A relative error norm $\Delta F'_\nu$ is used to quantify the difference between the spectrally resolved and the band averaged solution of the radiation transfer equation [6]:

$$\Delta F'_\nu = \sqrt{\frac{\sum_i \left(\nabla \cdot F_{\mathrm{exact},i} - \nabla \cdot F_{\mathrm{mean},i}\right)^2}{\sum_i \left(\nabla \cdot F_{\mathrm{exact},i}\right)^2}}, \qquad (2)$$

where $\nabla \cdot F_{\mathrm{exact},i}$ represents the divergence of the radiative flux calculated in the spectral solution for each point $i$ and $\nabla \cdot F_{\mathrm{mean},i}$ represents the divergence of radiative flux calculated using the line limited Planck mean absorption coefficients.

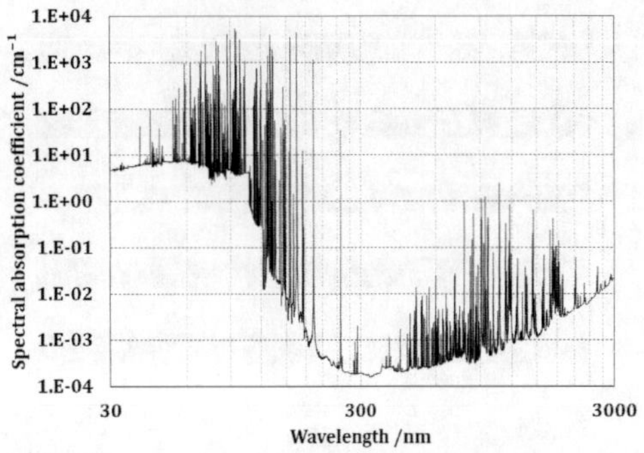


*Figure 2. Spectral absorption coefficient for a 90% $CO_2$ and 10% $C_4F_7N$ gas mixture at 10 000 K and 1 atm.*

In this paper contributions from atomic and molecular radiation as well as continuum radiation are taken into account. Table 1 provides the list of species that are considered in the radiation calculations. The optimization procedure was performed for a different total number of bands (2 to 6) to analyze the dependency of accuracy of the band averaged model on the number of bands. An example of the optimization results for a 90% $CO_2$ – 10% $C_4F_7N$ gas mixture and the diffuse temperature profile at 1 atm are shown in Fig. 3. As a major result the derived band boundaries for the different total number of bands are all located in UV region of the spectrum.

| | |
|---|---|
| Atoms | N, O, C, F |
| Molecules | $N_2$, $O_2$, $O_3$, NO, $C_2$, $C_3$, $C_4$, CO, CN, $CO_2$, $CN_2$, $C_2N$, $C_2N_2$, $C_2O$, $C_3O_2$, $F_2$, CF, $C_2F_2$, $C_2F_4$, $C_2F_6$, $CF_2$, $CF_3$,$CF_4$, NF, NOF, $NO_2F$, $FNO_3$, $NF_2$, $N_2F_2$, $F_2O$, $F_2O_2$, $NF_3$, $NOF_3$, $N_2F_4$, $NO_2$, $N_2O$, $COF_2$ |
| Electron and ions | e, $N^+$, $N^{2+}$, $N^{3+}$, $N_2^+$, $O^+$, $O^{2+}$, $O^{3+}$, $O_2^+$, $NO^+$, $C^+$, $C^{2+}$, $C^{3+}$, $CO^+$, $CN^+$, $F^+$, $F^{2+}$, $F^{3+}$ |

*Table 1. List of species used in radiation calculations.*

Corresponding error norm values according to Eq. 2 are provided in Fig. 4. The 2–band approximation results in an error of 12.98 % that is decreasing with increasing number of bands down to 2.55 % for 6 bands. A larger number of bands does improve the accuracy of the radiation model as expected, but the gain of accuracy using 6 bands rather than and 5 bands is only about 0.13 %. For this specific gas mixture and temperature profile, solving a 6–band model would not provide a much benefit over a 5–band model.

Although the 6–band approximation provides higher accuracy over the 3–band approximation, the 3–band approximation is used in subsequent arc modeling since a larger number of bands does increase the computational effort when solving radiation transfer. For this case following band limits in

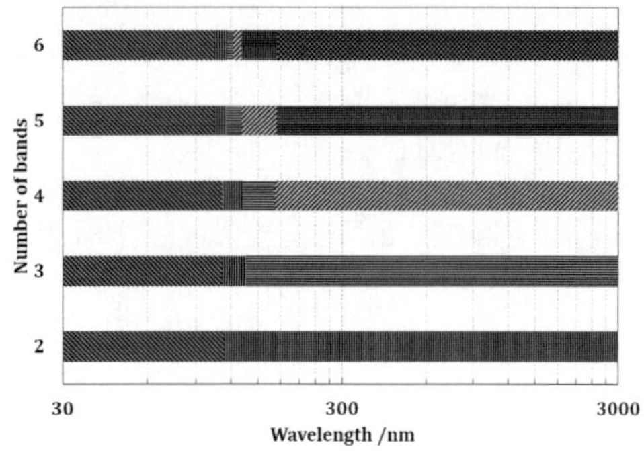


*Figure 3. Optimized band boundaries of mean absorption coefficients for a 90% $CO_2$ – 10% $C_4F_7N$ gas mixture and diffuse temperature profile at 1 atm.*

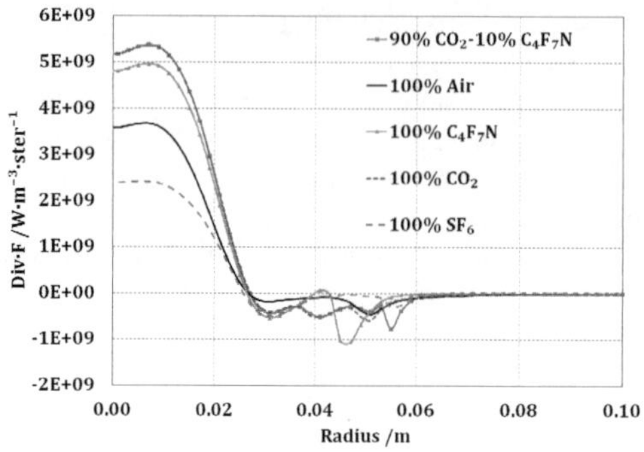


*Figure 5. Divergence of radiative flux for various gas mixtures, spectral resolved model.*

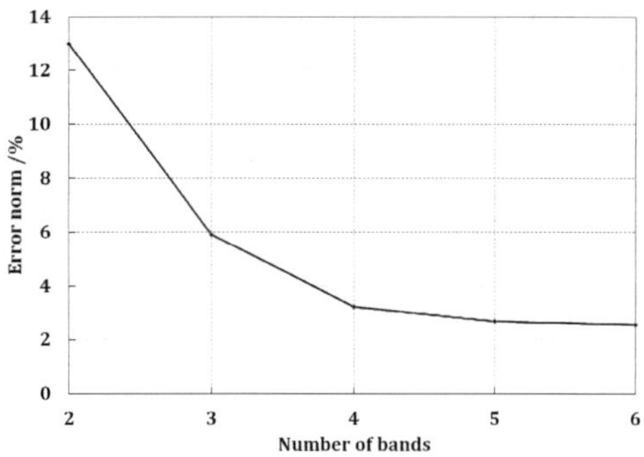


*Figure 4. Error norm for a 90% $CO_2$ – 10% $C_4F_7N$ gas mixture and diffuse temperature profile at 1 atm.*

terms of wavelength are considered: the first band covers $\lambda = 30\,\text{nm}$ to $\lambda = 111.6849\,\text{nm}$, the second band covers $\lambda = 111.6849\,\text{nm}$ to $\lambda = 135.7897\,\text{nm}$, and the third band covers $\lambda = 135.7897\,\text{nm}$ to $\lambda = 29864.21\,\text{nm}$.

The divergence of the radiative flux along the domain radius as result of the spectrally resolved model is shown in Fig. 5 for the 90% $CO_2$ – 10% $C_4F_7N$ gas mixture as well as for pure $C_4F_7N$, $CO_2$, $SF_6$, and air. The divergence of the radiative flux in the arc core for pure $CO_2$ is 6.9% larger compared to pure $C_4F_7N$ . The 90% $CO_2$ – 10% $C_4F_7N$ mixture shows the same profile of the divergence of the radiative flux in the arc core compared to pure $CO_2$ due to small mass fraction of $C_4F_7N$ in the gas mixture. Small differences between profiles are observed only in the reabsorption region. Compared to pure $SF_6$, the cooling term is increased by almost 100 % for pure $C_4F_7N$.

## 3. CFD simulations

The commercial finite-volume solver *ANSYS® Fluent* [8] is used to model free burning $SF_6$, air, and 90% $CO_2$ – 10% $C_4F_7N$ arcs that are stabilized by natural convection. The governing equations consist of the compressible hydrodynamic equations including the $k-\varepsilon$ turbulence model and the simplified set of Maxwell equations [9]. The discrete ordinate method (DOM) in *ANSYS® Fluent* is used to solve the radiation transport in the plasma using Planck mean absorption coefficients as presented in above section. The plasma is treated as a single-fluid at local thermodynamic equilibrium (LTE) and the necessary thermodynamic and transport properties of the pure gases and gas mixtures are calculated according to [9].

A 2D axi-symmetric geometry was defined to model free burning arcs as investigated in [10]. The simulation domain has a radius of $R = 1.5\,\text{cm}$ ($y$-direction, see Fig. 6), the distance between the electrodes is $d = 1.5\,\text{cm}$ ($x$-direction, see Fig. 6), and the distance between gas inlet (max. $x$-coordinate, see Fig. 6) and outlet (min. $x$-coordinate, see Fig. 6) is $l = 2\,\text{cm}$. A constant current of 10 A is applied on vertical surfaces of electrodes ($y$-direction, see Fig. 6). The effect of gravitation is considered by means of a -1.34 Pa gauge pressure boundary condition at the outlet (gravity is acting in $x$-direction, see Fig. 6).

The resulting stationary arc temperature distributions for $SF_6$, air, and $C_4F_7N - CO_2$ arcs are shown in Fig. 6. Due to the pressure gradient between inlet and outlet, natural convection leads to a broadening of the plasma temperature profile close to outlet boundary condition. The calculated radial temperature profiles for $SF_6$, air, and $C_4F_7N - CO_2$ arcs at the midpoint between electrodes are given in Fig. 7. Additionally, radial temperature profiles from Lowke's paper are shown. The model results are in good agreement with Lowke's model results for $SF_6$ and air arcs in the central regions of the discharge. Slight deviation can be observed in the outer region of the plasma mainly because of different transport, thermodynamic, and radiation properties used in the models. It is well known that $SF_6$ is superior to air as an arc quenching medium, where the constricted radial temperature profile result in higher temperature gradients with larger heat conduction losses. This is one reason for a better performance of $SF_6$, besides a faster recovery compared to air due to its electronegativity.

The $C_4F_7N - CO_2$ arc has more constricted radial profile than the arc in air, but it is more diffuse in

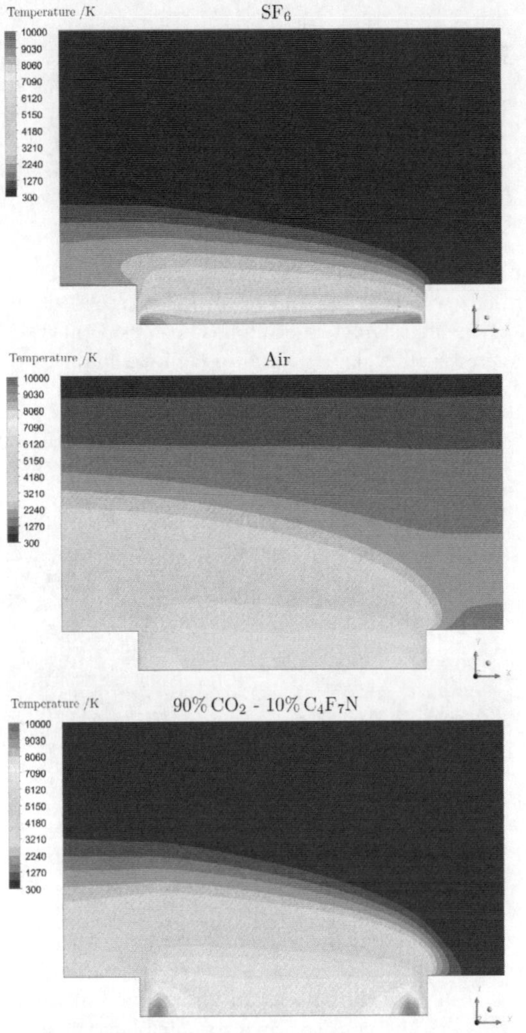


*Figure 6. Temperature distribution of 10 A free burning arcs in various gases.*

comparison to $SF_6$ arc. This is an indication for a weaker arc quenching performance of a $C_4F_7N - CO_2$ gas mixture compared to $SF_6$.

## 4. Conclusions

$C_4F_7N - CO_2$ gas mixtures are promising candidates to replace $SF_6$ in medium and high voltage switchgear applications. In this contribution, radiative properties of a specific $C_4F_7N - CO_2$ gas mixture have been calculated that can be utilized as input data for arc models of medium and high voltage switchgear. An optimization procedure was performed for a 90% $CO_2$ – 10% $C_4F_7N$ gas mixture and defined temperature profiles to find the optimal number of spectral bands and spectral band interval limits. CFD simulations of free burning 90% $CO_2$ – 10% $C_4F_7N$ arcs that are stabilized by natural convection, are carried out and compared with arcs burning in $SF_6$ and air. The calculated temperature profile of the arc in 90% $CO_2$

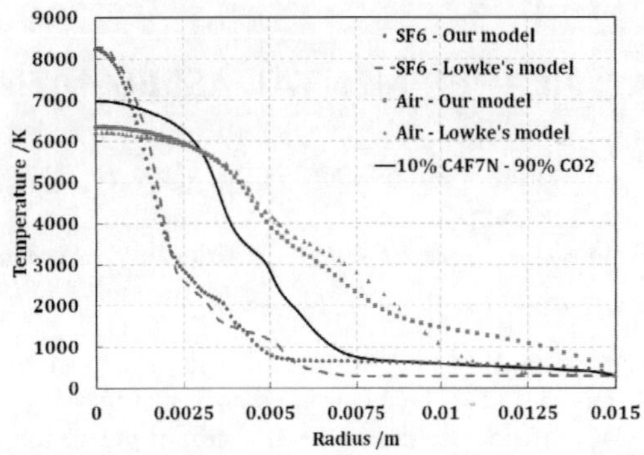


*Figure 7. Calculated temperature profiles for the 10 A arcs at midpoint between electrodes.*

– 10% $C_4F_7N$ gas mixture is more diffuse compared to the $SF_6$, which indicates a weaker arc extinguishing performance.

## Acknowledgements

The authors thank P. Kloc and M. Bartlova for providing the radiation database. This work is part of the project Nr. TH03020063 of the Technological Agency of the Czech Republic.

## References

[1] V. Narayanan et al. Transport properties of thermal plasma containing fluoro-nitrile ($C_4F_7N$)-based gas mixtures. *Plasma Physics and Technology*, 6(2):131, 2019. doi:10.14311/ppt.2019.2.131.

[2] A. Gleizes et al. Thermal plasma modelling. *J Phys D: Appl Phys*, 38(9):R153, 2005. doi:10.1088/0022-3727/38/9/R01.

[3] M. F. Modest. *Radiative Heat Transfer*. Third edition. Academic Press, 2013.

[4] H. Nordborg and A. Iordanidis. Self-consistent radiation based modelling of electric arcs: I. efficient radiation approximations. *J Phys D: Appl Phys*, 41(13):135205, 2008. doi:10.1088/0022-3727/41/13/135205.

[5] P. Kloc, V. Aubrecht, and M. Bartlova. Numerically optimized band boundaries of Planck mean absorption coefficients in air plasma. *J Phys D: Appl Phys*, 50(30):305201, 2017. doi:10.1088/1361-6463/aa7627.

[6] P. Kloc et al. On the selection of integration intervals for the calculation of mean absorption coefficients. *Plasma Chem Plasma Process*, 35:1097–1110, 2015. doi:10.1007/s11090-015-9648-3.

[7] M. Gnybida, C. Rümpler, and V. Narayanan. Optimal band selection for the calculation of Planck mean absorption coefficients. *Plasma Physics and Technology*, 4(1):70, 2017. doi:10.14311/ppt.2017.1.70.

[8] ANSYS®. Release 17.1, ANSYS, Inc., 2016.

[9] C. Rümpler and V. Narayanan. Arc modeling challenges. *Plasma Physics and Technology*, 2(3):261–270, 2015.

[10] J. J. Lowke. Calculated properties of vertical arcs stabilized by natural convection. *J. Appl. Phys.*, 50(1):147, 1978. doi:10.1063/1.325698.

doi:10.14311/ppt.2019.2.148
*Plasma Physics and Technology* **6**(2):148–151, 2019 

# EXPERIMENTAL ASSESSMENT OF PTFE POST-ARC ABLATION

Y. Babou*, P. Corfdir, R.-P. Suetterlin

*Corporate Research, ABB Switzerland Ltd., Segelhofstrasse 1K, CH-5405 Baden-Dättwil, Switzerland*

* yacine.babou@ch.abb.com

**Abstract.** The study addresses the post arc ablation (PAA) of PTFE (Polytetrafluoroethylene) material after being stressed by high current arcs. Arcs were generated in ambient air applying AC or DC current profiles to reach energy input in the range 7–22 kJ. The characterization has been performed essentially based on standard optical measurement techniques. The shadowgraph technique enabled us to show that the PAA flow is composed by a significant amount of carbonaceous soots lasting for several dozens of milliseconds after current interruption. The pyrometry technique allowed to estimate the soots temperature in the range 1400–2200 K.

**Keywords:** post arc ablation, pyrometry, shadowgraphy.

## 1. Introduction

The understanding of the ablation process of PTFE (Polytetrafluoroethylene) material undergoing intense incident heat fluxes is of critical importance for the design and optimization of various systems such as space propulsion systems [1, 2], thermal protection systems for atmospheric entries [3, 4] and ablation-assisted current interruption systems [5, 6]. In this later case, the ablation of the PTFE material interacting with high current arcs needs to be determined with enough accuracy in order to enable reliable prediction of the interrupting performances. Especially in the case of high voltage circuit breakers based on self-blast approach [7], the knowledge of the PTFE ablative behavior after current interruption remains of considerable importance since it critically affects the arc breakdown voltage during the hot dielectric recovery phase. So far, the post-arc ablation (PAA) of the PTFE has been poorly investigated while it drives significantly the flow properties (its composition and pressure distributions) and consequently the hot dielectric recovery performances. The experimental investigations presented hereafter are aimed at characterizing the PTFE PAA after being stressed by high current arcs.

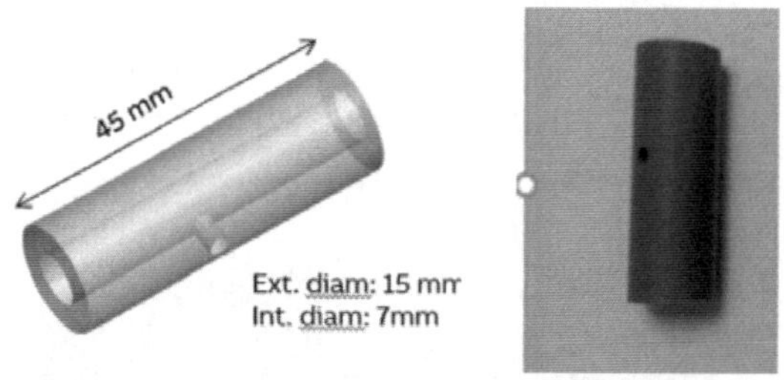


*Figure 1. Tested PTFE tubes. (Left) Tube design. (Right) Photograph of actual tube.*

## 2. Experimental setup

The experiments have consisted of generating high current arcs in a tube made of grey PTFE to assess the post-arc ablation phenomenon in ambient air. We have used PTFE cylindrical tubes of 45 mm length with an internal diameter of 7 mm (see Figure 1). The PTFE tubes were tested by applying either AC or DC current profile in order to assess the effect of the current steepness (transition from peak value to zero) on the PAA. The arc was ignited by means of a copper ignition wire. The generic experimental setup adopted for assessing PTFE PAA is sketched in Figure 2.

The setup includes a capacitor bank able to store energy up to 240 kJ. To achieve DC current profiles, we have used an interleaved Buck converter with three parallel lines as power supply that allows one to generate an arbitrary current profiles with an intensity of up to 2 kA per line, resulting in a total maximum current of 6 kA. In order to provide the desired current intensity, the voltage measured between the electrodes as well as the actual current intensity are used to compute the switching times of insulated gate bipolar transistors (ABB HiPak IGBTs) located at each of the three lines of the converter. Using three parallel paths with specific offsets in the switching phase makes it possible to achieve a considerable reduction in the current ripple. A thyristor is mounted in parallel to the test object for fast current switching ($< 20\,\mu s$). To obtain AC pulses, the capacitor bank is discharged through an inductance and the arc. The values of the capacitance and inductance are selected to result in a roughly 10 ms sine half-wave, corresponding to a 50 Hz grid frequency. Note that for all AC experiments, the thyristor remains connected to the test object but is not triggered. The arc voltage has been measured with a regular differential voltage probe and the current flowing through the test object has been measured by means of a Rogowski probe.

The diagnostic setup comprises a shadowgraph system, a two-colors pyrometer and a high-speed camera for direct visualization during the arc phase. For this

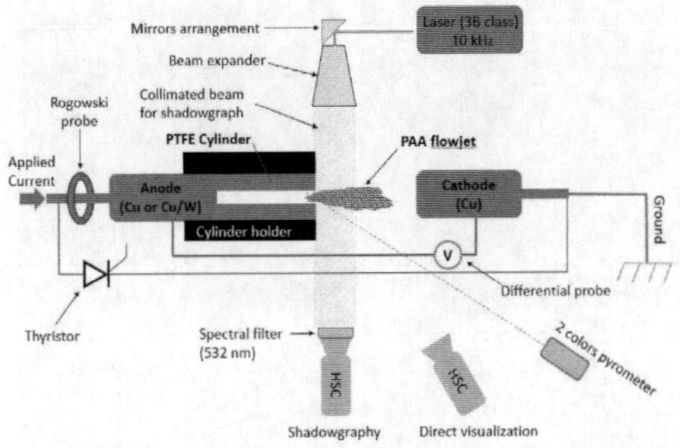


*Figure 2. Sketch of the experimental setup.*

camera, a 1.6 neutral density filter was mounted on the objective lens and the frame rate was set to 41 kHz with an exposure time of 0.9 µs. The shadowgraphy system combines a frequency doubled Nd:YAG laser, emitting at 532 nm and operated at 10 kHz, with a high-speed camera (HSC) to record shadowgrams during and after arcing. A neutral density filter providing an optical density of 2.6 and a bandpass filter centered at 532 nm are mounted on the objective lens of the HSC, and the exposure time was set to 1 µs. Before the test campaign, we tuned finely the shadowgraph setup, using a supersonic jet of cold air as object of reference to adjust the camera settings to obtain images of highest quality. The two-colors pyrometer is equipped with Si and Ge detectors, measuring infrared light intensity in the spectral ranges 0.9–1.04 µm and 1.1–1.8 µm, respectively. The pyrometer is coupled with an optical fiber which was arranged to collect the emission from the inner surface of the PTFE tube. Note that this goal can be reached only for an exhaust flow which is optically transparent in the spectral range detected by the pyrometer while the actual exhaust flow after the current interruption is composed of dense stream of solid particles as we will see. As a consequence the pyrometer measures the temperature of that stream of particle.

## 3. Results

Typical shadowgraph records taken in the post-arc phase after the application of 2 kA DC current for 10 ms are shown in Figure 3. The displayed situation is fairly representative of the overall considered test conditions. After current zero, a directional jet of small particles is streaming from the PTFE tube lasting for several tens of milliseconds. The fact that this jet appears black stems to the fact that it is opaque at the laser wavelength (532 nm). With time after current zero, the aggregates of bigger size are released and gradually the stream of particles slows down and its dynamics is dominated by diffusion transport, characterized by a strong mixing with ambient air. A visual inspection of the PTFE tube and electrodes after a single test reveals the presence of large amount of black powder (see Figure 4). The composition of the powder has been analyzed by means of X rays spectroscopy and a typical image is reported in Figure 5. It turned out that this powder consists in its larger extent of carbonaceous soots (of about 100 nm size) with traces of oxygen and fluor and in its lesser extent of large copper clusters (of few micrometers size) resulting from the anode erosion.

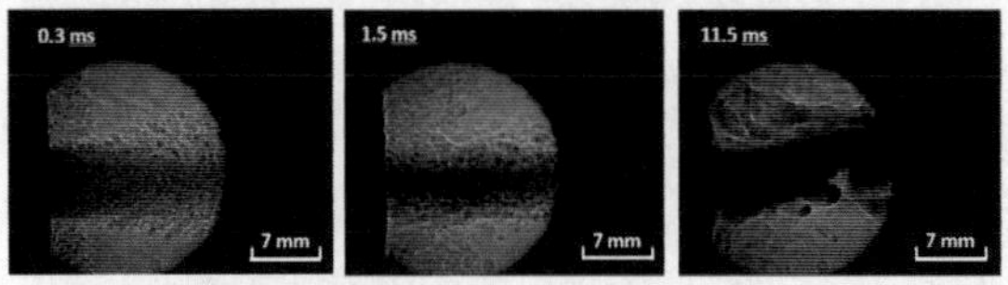


*Figure 3. Typical shadowgraphs records at different time after current zero.*

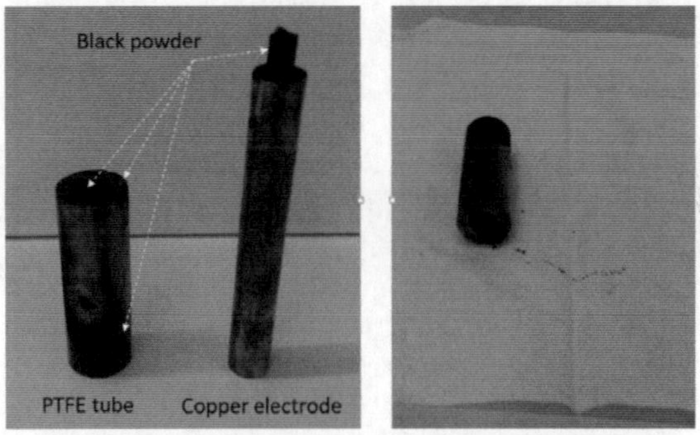


*Figure 4. Post test visual inspection.*

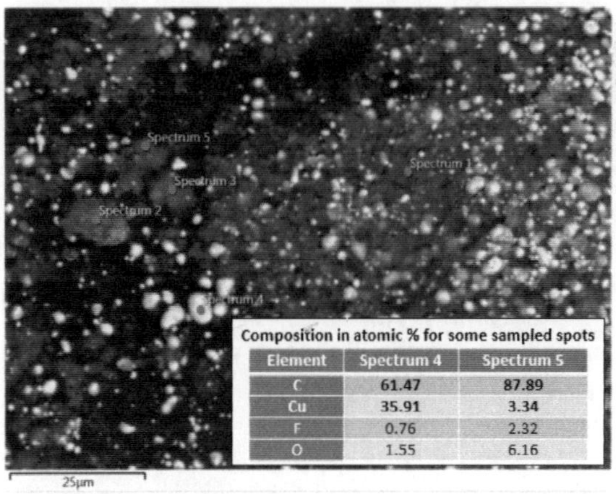


Composition in atomic % for some sampled spots

| Element | Spectrum 4 | Spectrum 5 |
|---|---|---|
| C | 61.47 | 87.89 |
| Cu | 35.91 | 3.34 |
| F | 0.76 | 2.32 |
| O | 1.55 | 6.16 |

*Figure 5. Energy dispersive X-ray spectroscopy mapping of a sample of black powder. The white spots correspond to the copper agreggates. The grey background corresponds to the carbonaceous soot. (In insert) Typical composition of the black powder.*

### 3.1. DC current tests

In a first step, we investigate the PAA following on the application for 10 ms of a DC current pulse of amplitude ranging from 1.5 to 3 kA, see Table 1. The

| **Current (kA)** | **1.5** | **2.0** | **2.5** | **3.0** |
|---|---|---|---|---|
| **Energy (kJ)** | 7.5 | 11.4 | 15.1 | 20.2 |

*Table 1. DC test series settings.*

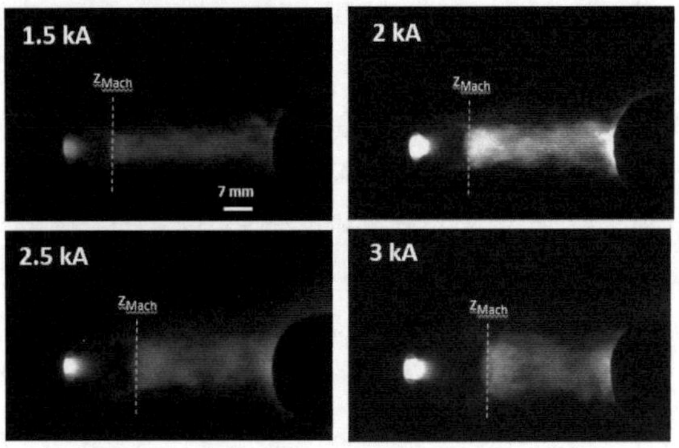


*Figure 6. Mach disk position for the DC test series.*

corresponding energy inputs were determined on the basis of the arc voltage measurements.

The pressure built-up in the tube during the arc phase has been estimated on the basis of the HSC images taken during the arc phase. The exhaust flow exhibits the typical structure of a supersonic jet characterized by stationary shockwave patterns with the presence of so-called Mach disk as seen in Figure 6. With increasing current intensity, the position of the first Mach disk ($z_M$) with respect to the tube exhaust increases, indicating that the pressure within the PTFE tube increases as well. From the Mach disk position, the total pressure in the tube is estimated using the empirical relation reported in [8] for ideal gas expansion. Although such estimate in our conditions is debatable, it gives a fair indication of the pressure evolution with applied current, see Table 2.

| **Current (kA)** | **1.5** | **2.0** | **2.5** | **3.0** |
|---|---|---|---|---|
| $z_M$ (mm) | 9 | 12 | 15 | 21 |
| Pressure (bar) | 3.7 | 6.5 | 10.2 | 20 |

*Table 2. Estimates of the tube pressure during DC arcing phase.*

The effect of the applied DC current on the PAA can be appreciated from the shadowgraph records reported in Figure 7 for 1.2 ms after current interruption. With increasing applied current (i.e. with increasing energy input), the diameter of the PAA jet tends to broaden, the jet of particles appears to be denser and lasts for longer times.

The temperature time profiles rebuilt from the pyrometer measurements in the post-arc phase are displayed in Figure 8. The measurements start from about 1 ms after current interruption because the pyrometer was operated during the arc phase and

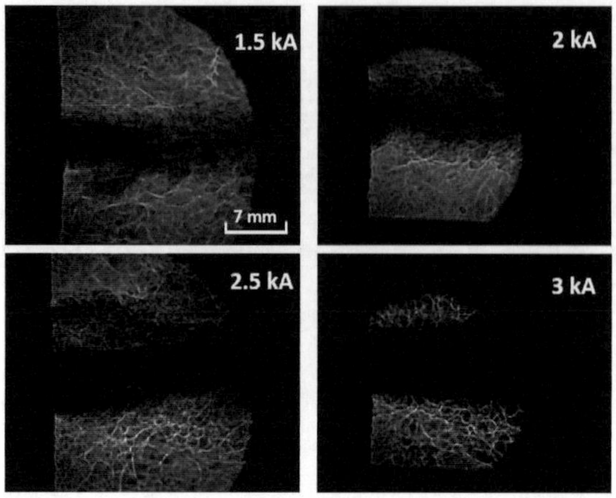


*Figure 7. Shadowgraph records for the DC test series taken about 1 ms after current interruption.*

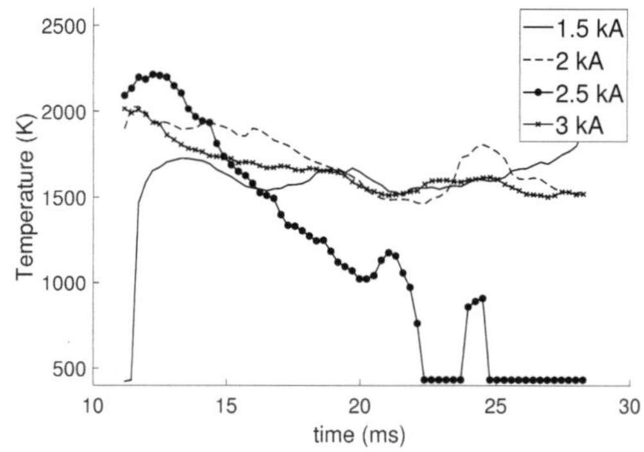


*Figure 8. Temperature profiles for the DC test series.*

saturated by the arc jet light. It took about 1 ms to go out from saturation. The evolution of the temperatures does not follow a net trend but exhibits instead a strongly irregular plot because the stream of particle itself exhibits a fluctuating dynamics characterized by a kind of intermittent or "bursty"behavior. This make difficult to interpret the trends followed by the temperature with increasing current and to draw a definitive statement. But still one can note that the temperatures lie in the range 1500–2200 K for several milliseconds. Above 20 ms after the current injection, the signal is close to the noise level therefore the rebuilt temperature profiles are no more reliable.

### 3.2. AC current tests

In a second step, we investigate the post arc ablation following on the application for 10 ms of a 50Hz AC current profile. For this test series, the peak values of the applied current pulses are set to achieve pulse energies comparable to that of the DC experiments, see Table 3. The effect of the applied AC current on the PAA flow is shown in Figure 9 for 1 ms after current interruption. The overall situation is similar to the DC test series: increasing the applied current leads to a more prominent stream of particle due

| Current peak (kA) | 2.3 | 3.1 | 3.8 | 4.5 |
|---|---|---|---|---|
| Energy (kJ) | 8.6 | 14 | 18.2 | 22.4 |

*Table 3. AC test series setting.*

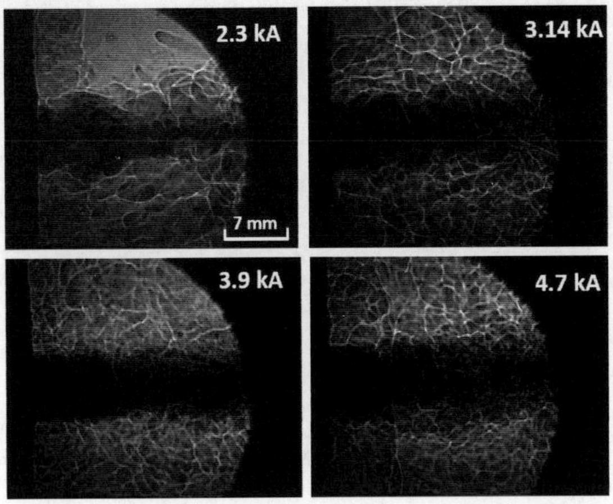


*Figure 9. Shadowgraph records for the AC test series taken about 1 ms after current interruption.*

to a more intense PAA process. The temperature profiles rebuilt from the pyrometer records also exhibit similar features to the DC tests and cover the same temperature range. However, for the AC test series, the profiles seem to pass by a plateau or a weak local minimum just after current interruption, before steadily decreasing. While for the DC test series, one could say that, in the main, the temperature decays without passing by a plateau. Further tests are required to confirm the observations and assess possible reasons for temperature differences depending on the applied current profiles.

## 4. Conclusions

The experiments reported here enabled us to characterize the PAA phenomenon taking place after stressing a PTFE material with high current arcs in ambient air. As a matter of fact, the PAA is characterized by the synthesis of a significant amount of carbonaceous soots also containing a non-negligible amount of oxygen and fluor. The shadowgraphy in combination with pyrometry technique were found to be well-suited to characterize the PAA flow main features. In our test conditions, the PAA flow consists of a stream of particles lasting for several tens of milliseconds and of temperature in the range 1400–2200 K. The time evolution of the temperature seems to be affected by the applied current profile (either AC or DC) however supplementary investigations are required to substantiate this point. Beyond such initial assessment of the PAA, by combining these measurements in the frame of appropriate assumptions, it is also possible to infer the ablation mass rate of PTFE against the input energy. Ultimately it should be mentioned that in actual circuit breaker arrangement, the PTFE made nozzle throat is typically cleared by colder gas blown from the heating or compression volumes, that would modify the PAA characteristics with respect to the present results.

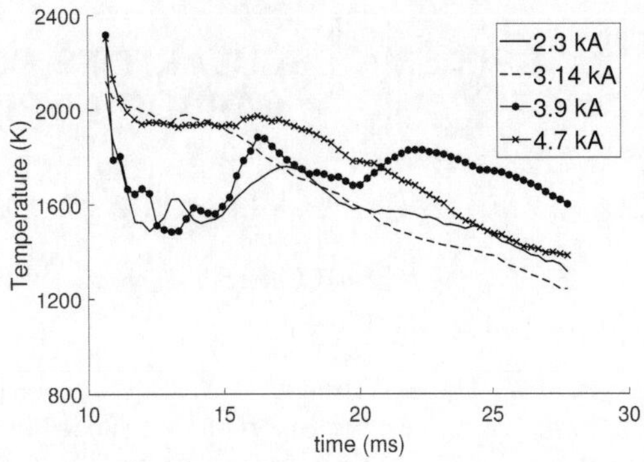


*Figure 10. Temperature profiles for the AC test series.*

### Acknowledgements

The authors acknowledge Martin Seeger for his fruitful inputs and discussion and Maud Martinon for performing the analysis of the soot sample.

### References

[1] W. J. Guman and P. E. Peko. Solid-propellant pulsed plasma microthruster studies. *Journal of Spacecraft and Rockets*, 5(6):732–733, 1968. doi:10.2514/3.29340.

[2] Y. Chan, C. Montag, G. Herdrich, and T. Schönherr. Review of thermal pulsed plasma thruster-Design, characterization, and application. In *Proc. 34th Int. Electr. Propuls. Conf.*, pages 1–10, 2015.

[3] N. Diaconis, J. B. Fanucci, and G. W. Sutton. The heat protection potential of several ablation materials for satellite and ballistic re-entry into the earth's atmosphere. *Planetary and Space Science*, 4:463–478, 1961. doi:10.1016/0032-0633(61)90152-0.

[4] J. G. Marvin and R. B. Pope. Laminar convective heating and ablation in the Mars atmosphere. *AIAA Journal*, 5(2):240–248, 1967. doi:10.2514/3.3948.

[5] H. Taxt. Ablation-assisted current interruption in medium voltage switchgear–Development and prospect. In *Proceedings of the Nordic Insulation Symposium*, volume 24, 2015. doi:10.5324/nordis.v0i24.2281.

[6] M. Seeger. Perspectives on research on high voltage gas circuit breakers. *Plasma Chemistry and Plasma Processing*, 35(3):527–541, 2015. doi:10.1007/s11090-014-9595-4.

[7] M. Seeger, J. Tepper, T. Christen, and J. Abrahamson. Experimental study on PTFE ablation in high voltage circuit-breakers. *Journal of Physics D: Applied Physics*, 39(23):5016, 2006. doi:10.1088/0022-3727/39/23/018.

[8] H. Ashkenas and F. S. Sherman. Structure and utilization of supersonic free jets in low density wind tunnels, 1965. Technical Report NASA-CR-60423.

doi:10.14311/ppt.2019.2.152
*Plasma Physics and Technology* **6**(2):152–155, 2019 

# PLASMA PECULARITIES OF ARC DISCHARGE BETWEEN CARBON-COPPER ELECTRODES

A. Veklich*, S. Fesenko, V. Boretskij

*Taras Shevchenko National University of Kyiv, 64/13, Volodymyrska str., Kyiv, Ukraine*

* an.n.veklich@gmail.com

**Abstract.** The experimental investigations of plasma parameters of electric arc discharge between composite carbon-copper electrodes are carried out by optical emission spectroscopy. It was shown, that arc channel width of electric arc between C-Cu composite electrodes tends to be wider in comparison with those between one-component copper electrodes. The experimentally obtained radial profiles of electron density in plasma channel of investigated arc discharges confirm such conclusion.

**Keywords:** Plasma, electron density, carbon-copper composite.

## 1. Introduction

The transfer of electrical signals and power between two components of electric circuits that are in relative motion to each other, is usually realized through sliding electrical contacts [1].

Nowadays, composite materials on carbon base are usually used as pantograph contact inserts, in particular, in high-speed railway [1–3]. On the one hand, these composites have advanced electric and exploitation characteristics. On the other hand, erosion properties of such materials are still under careful examination in developing processes of this composite. The matter is that the arc discharges appear during pantograph lowering process in pantograph-catenary system, so injection of contact material into discharge gap has place. Therefore, investigations of electric arc plasma between such composite electrodes can be useful for further optimization of materials.

It is well-known, that graphite, coke, copper-graphite or copper alloys, as materials in producing of inserts, have some advantages and disadvantages [4].

In particular, in spite of good lubricating properties of graphite and coke inserts, they have a comparatively high electrical resistivity. So, deterioration of tribological parameters of copper trolley wire or catenary takes place due to annealing in a result of significant Joule heating [5]. Moreover, such inserts have a sufficiently low hardness, so their service life is reduced.

The metal contact inserts on a base of copper alloys are characterized by a significantly less resistivity and a longer life. However, as a result of their affinity with the copper wire, the last one is significantly abraded by such inserts. Additionally, it must be noted that cost of copper wire is significantly higher in comparison with the cost of inserts.

Therefore, design engineers of such inserts usually recommend a copper-graphite composite as an optimal material for their produce [4]. Such composite has good lubricating properties due to its graphite component. The additions of copper in these materials provide the much higher electrical conductivity in comparison with graphite inserts. To avoid an affinity of these composite inserts with a conducting wire, the copper content in such material is insignificant.

The aim of this work is a study of plasma peculiarities of model electric arc ignited in air atmosphere between C-Cu composite electrodes. Namely, this paper is focused on the experimental investigations of parameters of thermal plasma mixture and calculations of electrical conductivity as well as electron density in discharge column.

## 2. Experimental investigations

### 2.1. Arc discharge arrangement

The free burning electric arc was operated in air between the end surfaces of C-Cu composite non-cooled vertically arranged electrodes.

The diameter of the rod electrodes was 6 mm, the discharge gap was 8 mm, arc current was 3.5 or 30 A. To avoid the metal droplet appearing, a pulsing high current mode was used: namely, the rectangular current pulse of 30 A was put on the "duty" low-current (3.5 A) discharge. The high-current pulse duration was of 30 ms.

### 2.2. Experimental techniques

#### 2.2.1. Temperature measurement

Boltzmann plot techniques was used for plasma temperature determination. The one-pass tomographic recording of the spatial distribution of spectral line intensities was proceeded [6–8]. Copper spectral lines Cu I 510.5, 515.3, 521.8, 570.0 and 578.2 nm were used for plasma diagnostics in this study. Monochromator MDR-12 with 3000-pixels CCD linear image sensor (B/W) Sony ILX526A accomplished fast scanning of spatial distribution of radial intensity. The registration of arc plasma emission was performed at 7 ms after current pulse rise when a steady-state mode of electric arc discharge was realized. Due to the instability of the discharge, statistical averaging

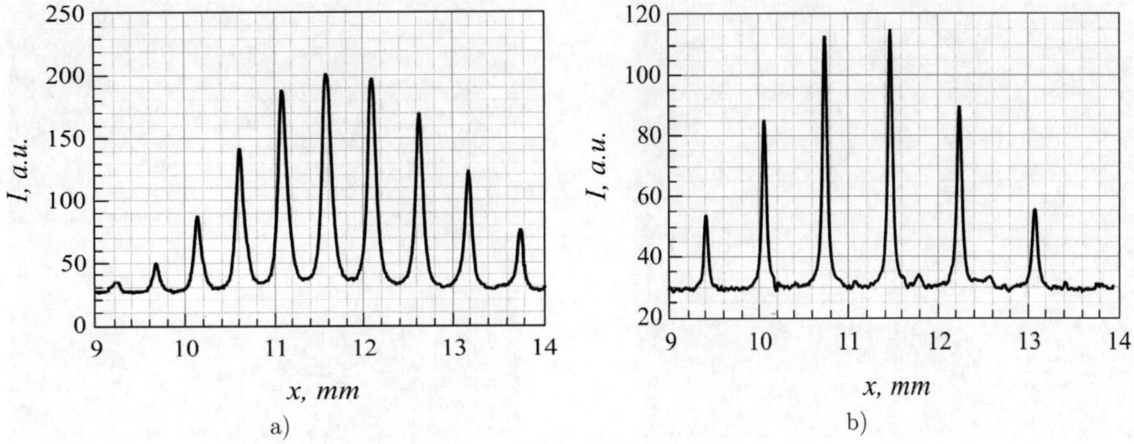


*Figure 1. Interferograms of the Cu I 515.3 nm spectral line radiated by plasma of arc discharges of 30 A current between copper (a) and C-Cu (b) electrodes.*

of the recorded spatial distributions of the radiation characteristics was carried out.

#### 2.2.2. Electron density measurement

Radial profiles of electron density were obtained from half-width of spectral line Cu I 515.3 nm in assumption of dominating quadratic Stark effect. The spectral device combined with Fabry-Perot interferometer in etalon mode was used for registration of spectral line profiles at 30 A [9].

As it was turned out, such Fabry-Perot interferometer can't be used for half-width measurements at arc of 3.5 A. So, electron density in this case was calculated by algorithm based on previously obtained plasma parameters, namely: temperature distributions for 3.5 and 30 A, electron density for 30 A, intensities ratio of Cu I spectral lines. Comprehensive description of this algorithm is proposed in paper [8].

In addition, the electron density was obtained by solving of the energy balance equation in an arc discharge positive column. The preliminary determination of the temperature radial distribution and electric field strength was carried out [9]. A more detail description of experimental setup and measurement procedure one can find in [10, 11].

## 3. Results and Discussions

The electron density in the arc discharge plasma by 30 A current between pure copper electrodes was determined from the half-width of the spectral line Cu I 515.3 nm at each local point, which corresponds to the maximum on the interferogram (Fig. 1a). Unfortunately, the electron density in the arc plasma by 30 A current between the copper-graphite electrodes can not be determined in this way, since the width of the spectral line Cu I 515.3 nm is comparable with the instrumental function of the Fabry-Perot interferometer (Fig. 1b).

Therefore, the electron density of arc discharge plasma between C-Cu electrodes of current of 30 and 3.5 A was determined from the electric field strength of the positive column. For this, the energy balance equation was solved using a predetermined plasma temperature distribution, in assumption that the thermal conductivity of the plasma mixture corresponds to the thermal conductivity of the pure air in the given temperature range. The electron density of arc discharge plasma of current of 3.5 A between pure copper electrodes was determined from the ratio of Cu I lines for currents of 3.5 A and 30 A, as it was mentioned above. The results of the electron density determination are shown in Fig. 2.

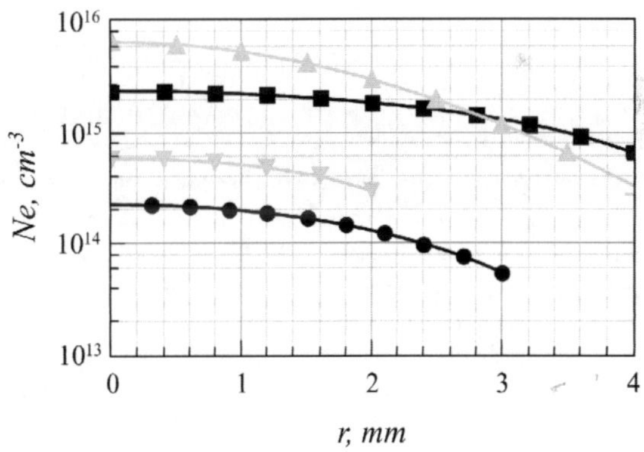


*Figure 2. Radial distribution of electron density in arc discharge plasma between copper (3.5 A: "▼"; 30 A: "▲") and C-Cu (3.5 A: "●"; 30 A: "■") electrodes.*

It follows from of the electron density distributions, that its absolute value between C-Cu electrodes is less than between pure copper electrodes, but it has less gradient, which suggests a greater width of the discharge column. To test of this assumption, the visualization of arc discharge is performed by CCD camera (1/8000 s) from the same solid angle for each electrodes composition.

From the comparison of figures 3 one can conclude that the increase of the current from 3.5 A to 30 A

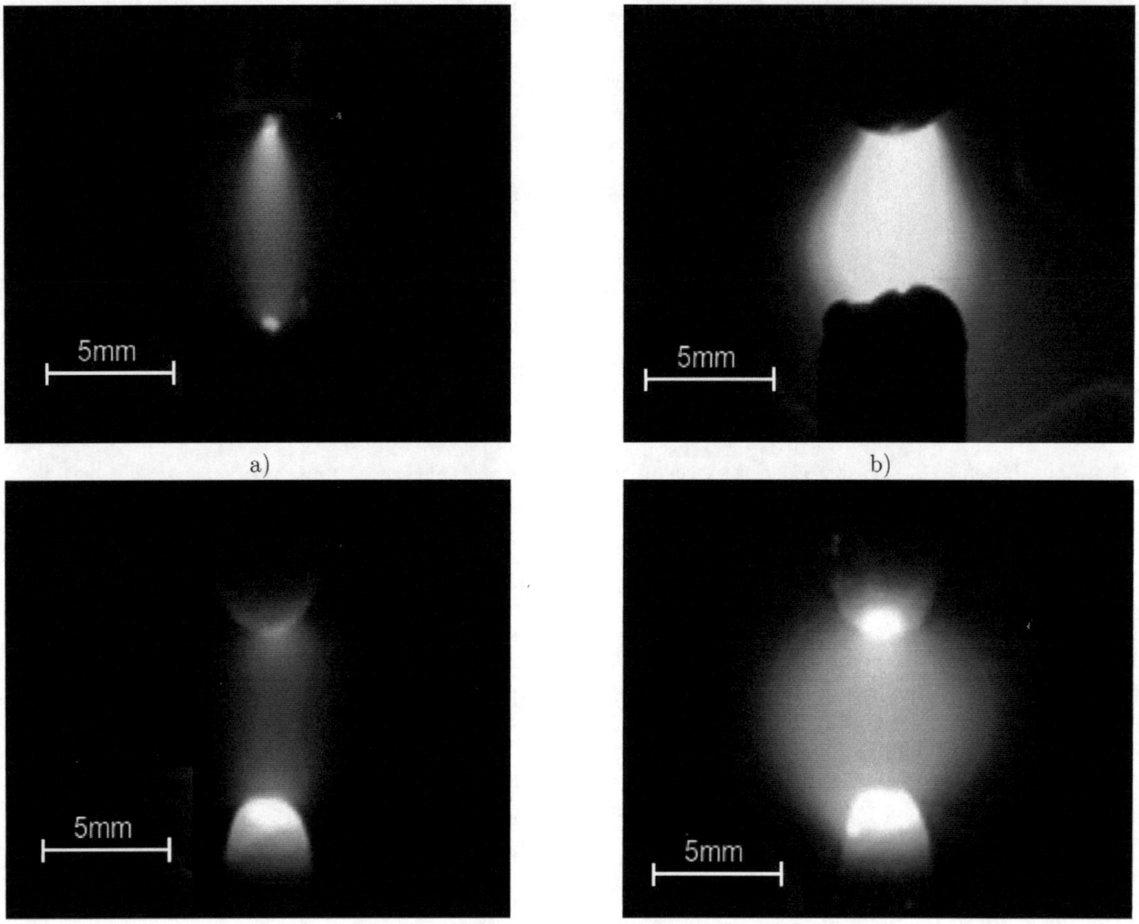


a) b) c) d)

*Figure 3. The images of arc discharge channels between copper (3.5 A - (a), 30 A - (b)) and C-Cu (3.5 A - (c), 30 A - (d)) electrodes.*

leads to the expansion of the arc channel in the case of copper (Fig. 3a and 3b) and composite C-Cu (Fig. 3c and 3d) electrodes as well. At the same time, the discharge channel between the composite C-Cu electrodes is wider than between of copper electrodes. It must be stressed that this difference is more obviously observed for the current of 30 A (see, Fig. 3b – Cu and 3d – C-Cu). Such behaviour of arc channel is not observed so clearly for the current of 3.5 A.

To explain the reason of this phenomenon, namely, the growth of the channel width of discharge between C-Cu electrodes, the working surface of electrodes was carefully analyzed after extinction of the arc discharge (Fig. 4). In the arc impact on composite C-Cu electrodes they much more heated than copper electrodes, because of their relatively low thermal conductivity. Since the sublimation temperature of graphite (4073 K) is much higher than the melting and boiling temperatures of copper (1357 K and 2848 K, respectively), then the superheated copper is percolated through the pores of the graphite matrix. So, the hardened copper droplets are clearly observed on the electrode surface after switch-off of the arc discharge. Therefore, the emission of copper vapor occurs at the discharge operation not only within the narrow molten area the arc spots, as in the case of copper electrodes, but within the much larger surface area of composite electrodes.

Consequently, composite C-Cu electrodes have a much higher effective copper vapor emission area than Cu electrodes, which is the reason of the expansion of the discharge channel of the arc between this type contacts due to the effective thermal ionization of these vapors.

One can conclude, finally, the total arc current in case of composite C-Cu electrodes is provided by the according electrical conductivity of discharge gap because of natural realization of the wider radial profile of electron density of relative low value in comparison with copper electrodes.

## 4. Conclusions

Plasma peculiarities of model source of electric arc ignited in air atmosphere between C-Cu composite

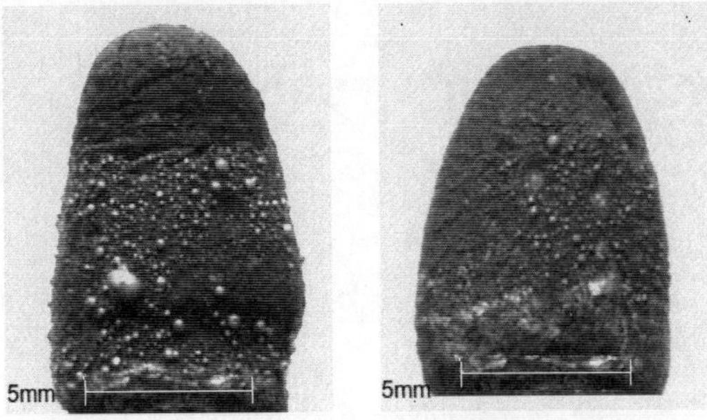


a) b)

*Figure 4. The images of composite C-Cu electrodes treated by arc discharge plasma: (a) - anode, (b) - cathode.*

electrodes were studied. The experimental investigations of parameters of thermal plasma mixture and calculations of electrical conductivity as well as electron density in discharge column are carried out.

It was found, that arc channel width of electric arc between C-Cu composite electrodes tends to be wider in comparison with those between one-component copper electrodes. In fact, the experimentally obtained radial profiles of electron density in plasma channel of investigated arc discharges confirm such conclusion.

The additional visualization of discharge gap and electrodes during arcs operation was used to clarify this phenomenon. It turned out, the composite C-Cu electrodes have a much larger effective copper vapor emission area than Cu electrodes. It is the reason of the expansion of the discharge channel of the arc between this type contacts due to the effective thermal ionization of these vapors. So, the total arc current in case of composite C-Cu electrodes is provided by the according electrical conductivity of discharge gap because of natural realization of the wider radial profile of electron density of relative low value in comparison with one-component copper electrodes.

## Acknowledgements

Authors wish to thanks Mr. A.Murmantsev for useful help in the preparation of this paper.

## References

[1] P. G. Slade. *Electrical contacts, principles and applications.* CRC Press Taylor & Francis Group, 2014.

[2] S. Fesenko, A. Veklich, V. Boretskij, Y. Cressault, A. Gleizes, and P. Teulet. Properties of thermal air plasma with admixing of copper and carbon. *J. Phys.: Conf. Ser.*, 550:1–8, 2014. doi:10.1088/1742-6596/550/1/012008.

[3] G. Wu, Y. Zhou, G. Gao, J. Wu, and W. Wei. Arc erosion characteristics of Cu-impregnated carbon materials used for current collection in high-speed railways. *IEEE Transactions on Components, Packaging and Manufacturing Technology*, 6(8):1014–1023, 2018. doi:10.1109/TCPMT.2018.2811801.

[4] V. Berent and S. Gnezdilov. Improvement of performance of current collectors on the carbon base. *Friction & Lubrication in machines and mechanisms*, 2:18–23, 2009.

[5] B. Miedzinski, W. Dzierzanowski, J. Wandzio, N. Grechanyuk, and V. Shoffa. Dynamics of sliding contacts in mine slow-speed railway transportation. *Electrical contacts and electrodes. - Kyiv: "Frantsevich Institute for Problems of Materials Science"*, 63–69, 2012.

[6] I. Babich, V. Boretskij, A. Veklich, and R. Semenyshyn. Spectroscopic data and Stark broadening of Cu I and Ag I spectral lines: Selection and analysis. *Advances in Space Research*, 54:1254–1263, 2014. doi:10.1016/j.asr.2013.10.034:

[7] A. Veklich and V. Zhovtyanskii. Fast tomographic plasma spectrometer. *Journal of Applied Spectroscopy*, 50(4):565–570, 1989. doi:10.1007/BF00659475.

[8] I. Babich, V. Boretskij, A. Veklich, A. Ivanisik, R. Semenyshyn, L. Kryachko, and R. Minakova. Spectroscopy of electric arc plasma between composite electrodes Ag-CuO. *Electrical contacts and electrodes. - Kyiv: "Frantsevich Institute for Problems of Materials Science"*, 82–115, 2010.

[9] A. Veklich, S. Fesenko, V. Boretskij, Y. Cressault, A. Gleizes, P. Teulet, Y. Bondarenko, and L. Kryachko. Thermal plasma of electric arc discharge in air between composite Cu-C electrodes. *Problems of Atomic Science and Technology. Series: Plasma Physics*, 6:226–229, 2014.

[10] R. Semenyshyn, A. Veklich, I. Babich, and V. Boretskij. Spectroscopy peculiarities of thermal plasma of electric arc discharge between electrodes with Zn admixtures. *Advances in Space Research*, 54:1235–1241, 2014. doi:10.1016/j.asr.2013.11.042.

[11] A. Veklich, A. Lebid, P. Soroka, V. Boretskij, and I. Babich. Investigations of thermal plasma with metal impurities. part ii: Peculiarities of spectroscopy by W I, Mo I, Cu I spectral lines. *Problems of Atomic Science and Technology. Series: Plasma Physics*, 19:213–215, 2013.

doi:10.14311/ppt.2019.2.156
*Plasma Physics and Technology* **6**(2):156–160, 2019 

# ROTATING GLIDING ARC: INNOVATIVE SOURCE FOR VOC REMEDIATION

J. Čech[a,*], L. Prokeš[a], M. Zemánek[a], L. Dostál[b], D. Šimek[b], J. Valenta[b], R. Žebrák[c], L. Zápotocký[c], P. Sťahel[a]

[a] *CEPLANT, Faculty of Science, Masaryk University, Kotlarska 2, 611 37 Brno, Czech Republic*

[b] *Faculty of Electrical Engineering and Communication, Brno University of Technology, Technicka 12, 616 00 Brno, Czech Republic*

[c] *DEKONTA a.s., Dretovice 109, Stehelceves, Czech Republic*

[*] cech@physics.muni.cz

**Abstract.** The large-scale plasma treatment of waste gas in industrial or municipal conditions requires high efficiency of plasma conversion process at high processing speed, i.e., large volumetric flow. The integration of the plasma unit into existing systems puts demands on the pipe-system compatibility and minimal pressure drop due to adoption of plasma processing step. These conditions are met at the innovative rotating electrode gliding arc plasma unit described in this article. The system consists of propeller-shaped high voltage electrode inside grounded metallic tube. The design of HV electrode eliminates the pressure drop inside the air system, contrary the plasma unit itself is capable of driving the waste gas at volumetric flow up to 300 $m^3$/hr for 20 cm pipe diameter. In the article the first results on pilot study of waste air treatment will be given for selected volatile organic compounds together with basic characteristic of the plasma unit used.

**Keywords:** gliding arc, rotating electrode, waste gas, VOC decomposition, plasma diagnostics.

## 1. Introduction

The electric arc discharge [1] represents the very basic electric discharge. However, its electrode lifetime is limited as the highly energetic arc plasma causes significant heating/erosion of the electrodes. To reduce the these adverse effects various methods of forced arc channel motion were invented (e.g., [2], to name one of them). One of the modification of arc discharge forcing the arc to move is so-called 'gliding arc' [3, 4], which is studied due to its favourable properties given by transient nature of this plasma source consisting of both thermal equilibrium and thermal non-equilibrium plasma [5–7].

The original invention of gliding arc used horn electrodes design, where the inter-electrode distance was gradually increased and the gas flow along the electrodes' axis forces the arc channel to 'glide'along the electrodes, while extending the channel length leading finally to discharge extinction [3]. This classic way of arc gliding enforcement, which utilizes dragging of arc channel using gas flow, could be used for plasma treatment of the gas stream. However the nozzle used to generate the gas stream could represent a gas flow bottleneck, which limits the amount of gas to be treated and requires a high pressure of the gas on the nozzle entry. The research on the utilization of gliding arc for gas treatment lead to invention of several discharge designs, where the discharge distribution could be further shaped by e.g. the rotation of conical electrode [8] or creation of gas vortex in discharge chamber [9, 10].

However, there exists the other approach of dragging the arc using the motion of electrodes themselves, resulting in changing the effective electrode distance (arc length), in combination with active transport and vectoring the gas stream to plasma region using the propeller-shaped discharge electrode. This originates the invention of 'rotating gliding arc' plasma source [11, 12], which is suitable for large-scale plasma treatment of gasses, or powders.

In our contribution the results of waste gas decontamination studies performed on the volatile organic compounds model species will be presented together with the results of discharge diagnostics.

## 2. Experimental

For the experiments the rotating electrode gliding arc plasma unit (RGA) was used, see Figs. 1 and 2. The treatment chamber (see Fig. 3) composed of closed-loop pipe system was constructed for controlled waste air plasma treatment experiments. The chamber was equipped with the diagnostics ports and the pumping system for the cleaning of the chamber and control of the gas composition. The RGA unit compose of the specially crafted high-voltage electrode setup of the propeller shape and high-voltage power supply. For pilot experiments the 5-blade axial blower (made of metal) was adapted to serve as the HV electrode and the grounded electrode was made of metallic pipe of 20 cm in diameter. The HV electrode was driven at the 700–1400 rpm, resulting in the peripheral velocity of the blade edge of approx. 7–14 $m.s^{-1}$. With this

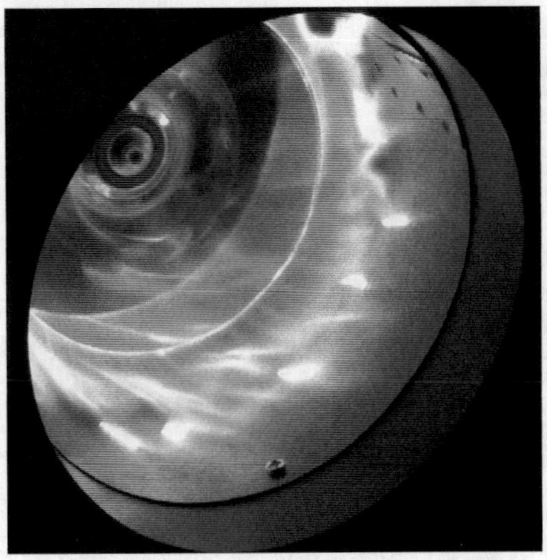

Figure 1. The RGA in operation.

Figure 2. The propeller-shaped electrode system.

Figure 3. The closed-loop decontamination chamber.

setup the 'self-propelled' volumetric flow of waste gas through the RGA unit tops at approx. 300 $m^3/hr$. The HV electrode was powered with HV power supply, that is capable to deliver up to 1.5 kW peak power at 1–10 kV voltage amplitude, with frequency ranging from 50 Hz to 50 kHz. See experimental setup on Fig. 4.

For the diagnostics the two groups of techniques were used: i) the plasma diagnostics, consisting of fast-framing camera, spectrometer and digital storage oscilloscope, and ii) the gas analysis system consisting of gas chromatography system and gas analyser.

For the analysis of the gliding arc channels the fast-framing camera Photron FASTCAM SA-X2 type 1000K-M2 equipped with Nikkor 180/2.8 lens stopped to f/8 was used. The optical resolution of the camera was approx. 0.2 mm, base on the optical setup and the camera resolution (1024x1024 px). The plasma treatment process was monitored using optical emission spectroscopy, the portable spectrometer Avantes AvaSpec-ULS3648TEC with fixed grating (200–1100 nm), slit width of 25 µm and mean spectral resolution of 1.4 nm was used. The electrical parameters were recorded using digital storage oscilloscope LeCroy WaveRunner 6100A with HV-probe Tektronix P6015A and current monitor Pearson 2877 probe.

For the waste air simulation, the controlled contamination of volatile organic compounds (VOCs), such as toluene ($C_6H_5CH_3$) or acetone ($CH_3COCH_3$) was introduced to the closed loop of the chamber and RGA plasma treated in several short plasma exposures. The residual contamination analysis was performed using photoacoustic gas monitor Innova 1412i (LumaSense Technologies) and gas chromatograph Agilent 7890B connected with flame ionization detector and quadrupole mass spectrometer Agilent 5977A (Agilent Technologies).

## 3. Results and Discussion

The aim of the rotating electrode gliding arc development was the inclusion of the plasma decomposition unit into existing waste gas systems with minimal additional cost and high remediation efficiency. The design of axial ventilator as a self-propelling electrode was proven successful, as the inner tube diameter of 20 cm was sufficient for 300 $m^3/hr$ gas throughput (and prototypes with more than 30 cm diameter were tested with up to 1000 $m^3/hr$ gas throughput). For the gas remediation efficiency test the closed-loop system was used. To follow the plasma decomposition dynamics the plasma was operated in burst mode with effective plasma exposure time of 3 s during each burst and the treated gas was regularly sampled after each burst. The typical result of the plasma treatment is given in Fig. 5. The graph shows the reduction of pollutant present in treated gas in relative units (percent). It can be seen that the plasma decomposition to less than 30% of initial pollutant contamination was achieved in 3 s of plasma operation. After another 3 s of plasma operation the reduction of pollutant

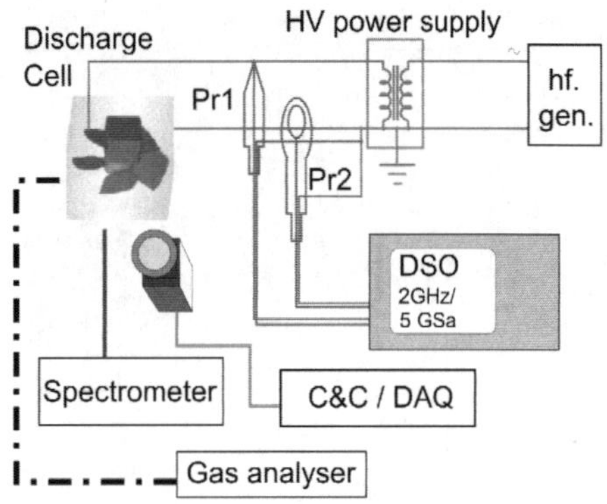


*Figure 4. The experimental setup diagram.*

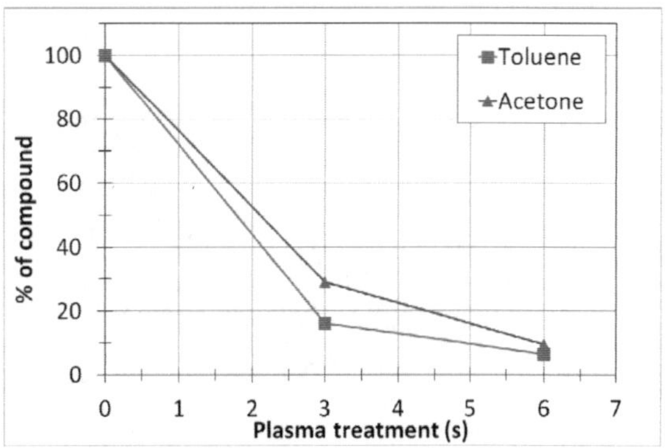


*Figure 5. The results on plasma remediation of polluted gas for high initial concentration VOC removal ($5200\,\mu m/m^3$ for toluene and $990\,\mu m/m^3$ for acetone) at 1.2 kW input power at 50 kHz frequency.*

reached more than 90%. For these results we have to note, that the reduction was performed on rather high initial concentration of $5200\,\mu m/m^3$ for toluene and $990\,\mu m/m^3$ for acetone.

We have seen, that the plasma remediation efficiency was dependent on the input power to the RGA unit and the efficiency increases with increasing input power. In order to follow the changes in discharge parameters the plasma diagnostics was performed. The discharge structure was observed using fast camera at approx. 2× the input voltage frequency. With this setup we were able to follow the evolution of the discharge channel in subsequent half-periods, but we were not able not follow the single channel evolution within the single half-period of the discharge due to limited frame rate of the camera used. The RGA plasma consists of a bright plasma channel between the HV and grounded electrode. This narrow plasma channel is connecting the edge of propeller HV electrode blade and some point at the surface of outer grounded electrode, see Fig. 6. The plasma channel is periodically reignited between the points of connection, similar to classic horn electrode design [13]. However, in RGA design there is no need to introduce the additional gas flow through the narrow nozzle of gas input to force the arc to glide along the electrodes. This results in low pressure drop in the RGA unit, beneficial to proposed industrial application.

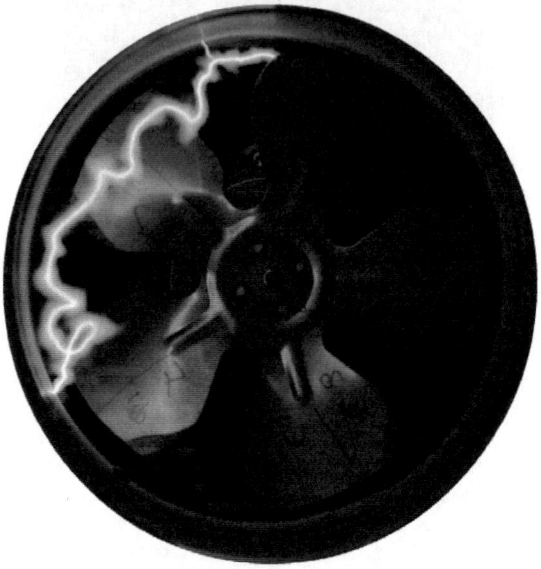

*Figure 6. The fast camera image of RGA plasma channel input power of 1500 W and 5 kHz operational frequency.*

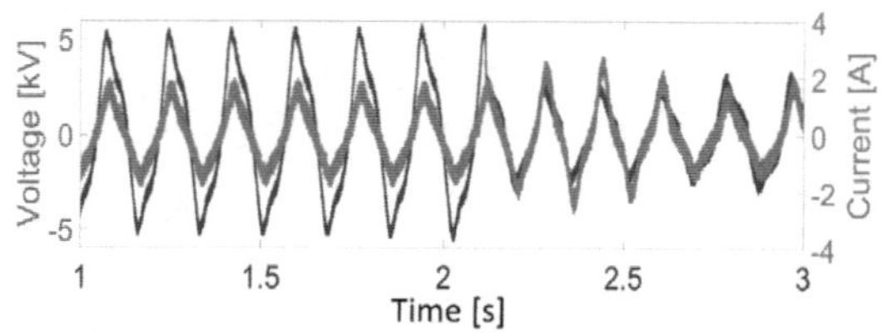


*Figure 7. The current-voltage (VA) waveforms of the RGA unit operated at input power of 1500 W and 5 kHz operational frequency.*

The discharge channels once ignited evolves in time being stretched by the rotation of HV electrode and the gas flow in the RGA unit. We have seen only a single plasma channel being ignited between the edge of HV blade electrode and outer grounded electrode during a single half-period of input voltage. As the plasma channel is dragged by the rotating HV electrode the length of plasma channel increases and the shape of the plasma channel becomes distorted as the result of turbulent gas flow drag in the direction perpendicular to electrode motion. As the channel evolves the contact point of the plasma channel at outer electrode also evolves and slowly moves in the same direction as the HV electrode. The plasma channel collapses when the input power is no longer sufficient to maintain the discharge (energy losses) or when the channel touches the grounded outer electrode. When the input power is increased the channel length increases and the plasma channel could undergo a series of hops, where the new connection point is established at the surface of outer electrode prolonging the lifetime of plasma channel. As the channel length is correlated with the channel lifetime, the longer the channel is and the longer it can hop in the discharge zone, the longer the effective plasma treatment of the gas will be sustained.

The typical current-voltage (VA) characteristics is given in Fig. 7 for the 1.5 kW input power at 50 kHz frequency. The sudden drop in the amplitude of the voltage could be attributed to the ignition of the discharge. The typical emission spectrum of the discharge consists of highly intense emission from the cathode fall region, where the discharge contacts the electrodes. There the emission of the atomic lines of iron, zinc and aluminium lines were identified, indicating hot spots at the electrodes. The plasma diagnostics was able to give us insight into the changes of plasma channel parameters with respect to outer conditions and to tailor the discharge towards optimal operating conditions and also to modify the design of electrodes in order to optimize the gas flow inside the plasma region.

The last key parameter of RGA unit was the durability of the system. Especially considering the harsh conditions of plasma operation in potentially corrosive environment of treated waste gas and the reactivity of created plasma species, or potentially even the remediation products. For this purpose the long-term stress-tests of the RGA plasma unit were performed. We have checked the condition of the plasma unit after assembly and after the RGA unit was exposed to corrosive environment. Even after 1000 hours in corrosive environment the RGA units does evidence significant degradation, which indicates its potential for real industrial application.

## 4. Conclusions

The novel design of gliding arc plasma source was constructed and tested for the environmental application of waste air pollution reduction. The advantage of propeller-shaped rotating electrode design proved to be efficient in waste air pollution removal. Devices based on rotating electrode gliding arc technology (RGA) were developed in the laboratories of CEPLANT center and they were applied for the tests of the efficiency of decomposition of selected model VOCs (acetone and toluene) in large volumes of polluted air (hundreds or thousands of $m^3$/hr). As the final decomposition products, only $H_2O$ and $CO_2$ were detected in quantifiable amounts.

The RGA plasma cleaning technology of VOCs contaminated air is advantageous for its low operating costs, and intrinsic compatibility with commonly used ventilation systems and also with some other decontamination techniques. The peak reduction of high concentration contaminants was more than 90% in less than 6 second, which means concentration reduction from 5200 $\mu m/m^3$ of toluene, resp. 990 $\mu m/m^3$ of acetone, to less than 350 $\mu m/m^3$ of toluene, resp. 95 $\mu m/m^3$ of acetone. The RGA plasma unit proved to be compatible with existing waste gas pipe systems, regarding the minimal pressure drop due to active gas transport in RGA unit and the scalability of the RGA system (tube diameter, possibility of tandem work, etc.). We have performed also the plasma diagnostics which shows the correlation of contamination removal efficiency using the gliding arc with the discharge occupied area/the channel lifetime. The RGA unit and its applications were patent protected (CZ patent No. 2015 - 310 A3).

## Acknowledgements

Authors JC, LP, MZ and PS gratefully acknowledge financial support from the project LO1411 (NPU I) funded by Ministry of Education, Youth and Sports of the Czech Republic; project CZ.01.1.02/0.0/0.0/15_019/0004703 Operational Programme Enterprise and Innovations for Competitiveness; project TH02030332, Technology Agency of the Czech Republic.

Authors LD, DS and JV gratefully acknowledge financial support from the Ministry of Education, Youth and Sports of the Czech Republic under NPU I program (project No. LO1210) and OP VVV Programme (project No. CZ.02.1.01/0.0/0.0/16_013/0001638 CVVOZE Power Laboratories - Modernization of Research Infrastructure).

## References

[1] Y. P. Raizer. *Gas Discharge Physics*. Berlin: Springer Verlag, 1991.

[2] S. D. Raezer. Arc apparatus employing three dimensional arc motion and dynamic balancing, 1966. Patent No.: US 3,416,021.

[3] H. Lesueur, A. Czernichowski, and J. Chapelle. Dispositif de génération de plasmas basse température par formation de décharges électriques glissantes, 1988. Patent No.: FR 2639172.

[4] H. Lesueur, A. Czernichowski, and J. Chapelle. Électro-brûleurs à arcs glissants. *Journal de Physique Colloques*, 51(C5):C5–57–C5–64, 1990. doi:10.1051/jphyscol:1990508.

[5] A. Fridman, S. Nester, L. Kennedy, A. Saveliev, and O. Mutaf-Yardimci. Gliding arc gas discharge. *Progress in Energy and Combustion Science*, 25(2):211–231, 1999. doi:10.1016/S0360-1285(98)00021-5.

[6] A. Czernichowski. Gliding arc: Applications to engineering and environment control. *Pure and Applied Chemistry*, 66(6):1301–1310, 1994. doi:10.1351/pac199466061301.

[7] F. Richard, J. M. Cormier, S. Pellerin, and J. Chapelle. Physical study of a gliding arc discharge. *Journal of Applied Physics*, 79(5):2245–2250, 1996. doi:10.1063/1.361188.

[8] T. Ombrello, X. Qin, Y. Ju, A. Gutsol, A. Fridman, and C. Carter. Combustion enhancement via stabilized piecewise nonequilibrium gliding arc plasma discharge. *AIAA Journal*, 44(1):142–150, 2006. doi:10.2514/1.17018.

[9] M. McNall and S. Coulombe. Characterization of a rotating gliding arc in argon at atmospheric pressure. *Journal of Physics D: Applied Physics*, 51(44):445203, 2018. doi:10.1088/1361-6463/aade44.

[10] M. Ramakers, J. A. Medrano, G. Trenchev, F. Gallucci, and A. Bogaerts. Revealing the arc dynamics in a gliding arc plasmatron: a better insight to improve $CO_2$ conversion. *Plasma Sources Science and Technology*, 26(12):125002, 2017. doi:10.1088/1361-6595/aa9531.

[11] P. Stahel, Z. Navratil, M. Zemanek, and M. Cernak. A method of generating low-temperature plasma, a method of plasma modification of liquid, powder, and solid materials with this low-temperature plasma, and a device for carrying out these methods, 2016. Patent application No.: WO2016177353A1.

[12] P. Stahel, Z. Navratil, M. Cernak, and M. Zemanek. Method of generating low-temperature plasma, plasma treatment process of fluids, powder materials and solid substances by making use of such low-temperature plasma and apparatus for making the same, 2015. Patent No.: CZ2015310A3.

[13] L. Potočňáková, J. Šperka, P. Zikán, J. J. W. A. Van Loon, J. Beckers, and V. Kudrle. Experimental study of gliding arc plasma channel motion: buoyancy and gas flow phenomena under normal and hypergravity conditions. *Plasma Sources Science and Technology*, 26(4):045014, 2017. doi:10.1088/1361-6595/aa5ee8.

doi:10.14311/ppt.2019.2.161
*Plasma Physics and Technology* **6**(2):161–164, 2019


# INFLUENCE OF COPPER VAPOURS IN $SF_6$ PLASMA

P. FRETON[a,*], J-J. GONZALEZ[a], A. HARRY SOLO [a], F. REICHERT[b], A. PETCHANKA[b]

[a] *Laplace, UMR 5213 CNRS-UPS-INP, Université Paul Sabatier 118 rte de Narbonne, bat3R2, 31062 Toulouse Cedex France*

[b] *Siemens AG, E T HP CB PLM RD B PB, Nonnendammallee 104, 13629 Berlin, Germany*

[*] freton@laplace.univ-tlse.fr

**Abstract.** In this study a theoretical approach allows estimating the ablation mass flux of copper from a corrected Hertz-Knudsen flux. The influence of the copper vapours coming from the anode electrode to an $SF_6$ plasma is studied in a simplified 2D configuration. Depending on the plasma pressure an ablation or a diffusion state is considered. The amount of copper versus time is presented. An RMS current $I = 10\,\mathrm{kA}$ is applied leading at $t = 10\,\mathrm{ms}$ to an amount of copper equal to 0.55 mg. The vapours change the plasma properties mainly the electrical conductivity and radiation and so the plasma behaviour. At time $t = 5\,\mathrm{ms}$ the electrode erosion leads to a copper plasma. This simple case shows the necessity to well consider the copper erosion in plasma modelling as in High Voltage Circuit Breaker (HVCB) where higher current are considered.

**Keywords:** Copper ablation, High Voltage Circuit Breaker, Electric arc.

## 1. Introduction

In HVCB modelling, ablation of electrodes is rarely taken into account whereas in reality, at high current it could be an important process. Modelling such ablation is quite complex due to numerous mechanisms involved. In this paper, an approach is proposed to represent this phenomenon. In a first part, the basis of the ablation model are discussed. Then a magneto hydrodynamic model and the specific source terms due to the ablation are presented. A "simple" geometry is used to test this new model. Results show a strong presence of copper within the plasma even with a small part of the electrode ablated. The vapours should change strongly the properties of the plasma and then the importance of mechanisms involved in the HVCB.

## 2. Theory

### 2.1. Ablation model

When an electrode is submitted to the heat and radiation fluxes coming from the arc, its temperature can increase and then ablation occurs. The description of the ablation process is quite difficult to describe and depends on the pressure vapour $P_{\mathrm{vap}}$ of the ablated metal. In our case, metal is copper. $P_{\mathrm{vap}}$ is given in (1). It depends on the surface temperature $T_{\mathrm{s}}$, the temperature of vaporisation of the metal $T_{\mathrm{vap}}$, the mass of the metal particle $m_{\mathrm{Cu}}$ and its heat of vaporisation $L_{\mathrm{vap}}$, $k_{\mathrm{b}}$ is the Boltzmann constant.

$$\frac{P_{\mathrm{vap}}}{P_0} = \exp\left(\frac{L_{\mathrm{vap}}\, m_{\mathrm{Cu}}}{k_{\mathrm{b}}}\left(\frac{1}{T_{\mathrm{vap}}} - \frac{1}{T_{\mathrm{s}}}\right)\right) \qquad (1)$$

Depending on the pressure of the plasma $P_0$ two different cases have to be considered:

☐ Case 1: The "diffusion case" where $\frac{P_{\mathrm{vap}}}{P_0} \leq 1$

☐ Case 2: The "ablation case" where $\frac{P_{\mathrm{vap}}}{P_0} > 1$

In the first case, the temperature of the surface $T_{\mathrm{s}}$ is under the temperature of vaporisation of the metal $T_{\mathrm{vap}}$. Then, the surface is in equilibrium with the plasma and metal vapours diffuse from the surface to the plasma. The mass fraction of copper on the surface can be calculated from equation (2). $M_{\mathrm{Cu}}$ and $M_{\mathrm{Pl}}$ are repectively molar mass of copper and plasma.

$$\omega_{\mathrm{Cu}} = \frac{P_{\mathrm{vap}}\, M_{\mathrm{Cu}}}{P_{\mathrm{vap}}\, M_{\mathrm{Cu}} + (P_0 - P_{\mathrm{vap}})\, M_{\mathrm{Pl}}} \qquad (2)$$

In the second case, an ablation process occurs. Semenov [1] explains that in this situation, a strong flow of copper vapours goes away from the surface and that, close to the surface, the vapour is no more in equilibrium in the plasma. Indeed, a Knudsen layer appears where the distribution function of the vapours is no more Maxwellian. This layer is followed by another one in overpressure (compared with the one of the plasma) but in equilibrium. Semenov [1] has proposed a 1D approach to describe the phenomena of ablation in this context in order to obtain the ablated mass flux $\Phi_{\mathrm{Cu}}$ (kg/(m$^2$·s)) of copper and also the mean velocity of the copper flow in the plasma $v_{\mathrm{Cu}}$. Even if this approach is very interesting, its implementation is difficult. Benilov [2] proposes another approach which consists to estimate the ablation mass flux of copper as proportional to the Hertz-Knudsen flux $\Phi_{\mathrm{HK}}$ (3)

$$\Phi_{\mathrm{HK}} = P_{\mathrm{vap}} \sqrt{\frac{m_{\mathrm{Cu}}}{2\pi k_{\mathrm{b}} T_{\mathrm{s}}}} \qquad (3)$$

Then copper mass flux is given by equation (4) where $\alpha_v$ can be calculated from material and plasma data (see [2] for more details).

$$\Phi_{\mathrm{Cu}} = \alpha_v \, \Phi_{\mathrm{HK}} \qquad (4)$$

We compared the two approaches which lead for copper to the same results [3] for the ablation mass flux. In his article, Benilov estimates the velocity of the particles coming inside the plasma from the ratio of the mass flux to the density at the edge of the Knudsen Layer $n_\infty$ as reported in equation (5).

$$v_{\mathrm{Cu}} = \frac{\Phi_{\mathrm{Cu}}}{m_{\mathrm{Cu}} \, n_\infty} \qquad (5)$$

## 2.2. The magneto hydrodynamic model

### 2.2.1. geometry

To analyse the results, and to study the influence of the copper vapours, we have used the geometry given in Figure 1. The geometry is axisymetric with axis AI. In the upper part, AB is an electrode considered as an anode. BC and HI are pressure inlet/outlet, CDEFGH are walls.

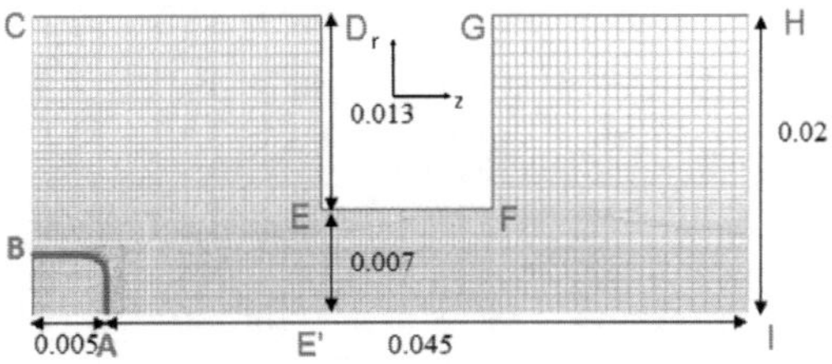


*Figure 1. Geometry used for the simulation (dimensions in m).*

### 2.2.2. Hypotheses

We assume that the configuration can be describe in two dimensions (2D) with an axis of symmetry. The top electrode is pure copper. As the problem is 2D, a porous electrode located on segment (EE') is used. The fluid is assumed to be Newtonian and laminar, the gas is $SF_6$ with copper (coming from the electrode) and the plasma is assumed to be in Local Thermal Equilibrium (LTE). The problem is transient.

### 2.2.3. Equations

In order to describe the arc fluid model the Navier Stokes equations are solved coupled with the mass and energy conservation equations. All these equations can be written in the form of the generalized conservation equation (6).

$$a \frac{\partial}{\partial t}(\rho\phi) + b \, \nabla \cdot (\rho \vec{v} \phi) = \nabla \cdot \big(\Gamma \nabla(\phi)\big) + S_\phi \qquad (6)$$

For the fluid, the equations solved are mass conservation, momentum equations for axial and radial velocity $v_z$ and $v_r$, energy conservation for enthalpy $h$ and then temperature $T$. The equation for the mass fraction of $SF_6$, $\omega_{SF_6}$, is solved and mass fraction of copper $\omega_{\mathrm{Cu}}$ deduced. Thermodynamic and transport properties of pure $SF_6$ and copper are used and Wilke laws [4] are applied to obtain the mixture properties. Electrical properties of the plasma (current density $\vec{j}$, magnetic field $\vec{B}$) are obtained repectively from the equation for electric potential $V$ and potential vector components $A_r$ and $A_z$. The radiation of the plasma is treated by Discrete Ordinate method on 7 bands where mean absorption coefficient of $SF_6$ and copper have been pre-calculated. This enables to obtain the radiative flux and its divergence. Additional details on the equations set and boundary conditions can be found in previous papers [5–7]. With copper vapours, specific source terms for ablation model have to be taken into account. Furthermore, the copper vapours coming from the electrode are transported in the plasma. In the real case, they can re-solidify when the gas temperature becomes lower than the one of vaporisation. If this condition occurs during calculation, we assume that copper is re-solidified and that it disappears from the fluid part. The following source terms are then written for ablation model and re-solidification of copper:

- ☐ Source term in the equation of mass conservation. For the mass conservation source term $S_m$ given in equation (7) is used in order to take into account the exiting vapours at low temperature from a surface $S$ of normal $\vec{e}_s$ close to a cell of volume $V_{\mathrm{cell}}$. Losses of copper in the volume are described by the second part of the equation.

$$S_m = \beta \left[\frac{\Phi_{\mathrm{Cu}} \, S}{V_{\mathrm{cell}}}\right] - \alpha \left[\frac{\omega_{\mathrm{Cu}} \, \rho_{\mathrm{Cu}}(T)}{\Delta t}\right] \qquad (7)$$

With $\alpha = 0$ if $T > T_{\mathrm{Vap}}$ and 1 otherwise. $\Delta t$ is the time step of the simulation. $\beta$ is equal to 1 in the layer close the electrode $\sim 0.1\,\mathrm{mm}$, 0 elsewhere.

- ☐ Source terms in momentum equations. It is necessary to take into account the momentum of vapours flowing from the anode surface to the plasma. The source terms of equations (8) and (9) are used. Re-solidification of copper leads also to a decrease of momentum in the fluid domain (second part of the equations)

$$S_{v_z} = \beta \left[(\rho_{\mathrm{Cu}}(T_{\mathrm{vap}}) \, v_{\mathrm{Cu}} \, \vec{e}_s \cdot \vec{e}_z\right] - \alpha \left[\omega_{\mathrm{Cu}} \, \rho_{\mathrm{Cu}}(T) \, v_z\right] \qquad (8)$$

$$S_{v_r} = \beta \left[\rho_{\mathrm{Cu}}(T_{\mathrm{vap}}) \, v_{\mathrm{Cu}} \, \vec{e}_s \cdot \vec{e}_r\right] - \alpha \left[\omega_{\mathrm{Cu}} \, \rho_{\mathrm{Cu}}(T) \, v_r\right] \qquad (9)$$

- ☐ Source term for energy equation. Concerning the vapours arriving in the plasma, we assumed that if their enthalpy is lower than the plasma one, they are pre-heated in the anode sheath by collisions with the electrons and by radiation. If this is not the case, they are injected with their enthalpy at the temperature of vaporization (It should be underlined that in the reference energy of the system, enthalpy of copper gas should contain $L_{\text{vap}}$ for temperatures over $T_{\text{vap}}$). This assumption has a consequence on the energy balance at the anode. This point will be presented in the next paragraph. Then, the energy source term can be estimated from equation (10). $h_{\text{mean}}$ is the mean enthalpy of the plasma in the layer close to the electrode. If only one cell is present in the layer (as in our geometry) this is the enthalpy of the plasma in this cell. It depends on temperature $T$, pressure $P$ and mass fraction $\omega_{SF_6}$ of $SF_6$. The second part concerns the "re-solidification" of the copper if temperature decrease under the temperature of vaporisation

$$\begin{aligned} S_{nrj} &= \beta \left[ \frac{\Phi_{\text{Cu}} S}{V_{\text{cell}}} \operatorname{Max}\left(h_{\text{mean}}, h(T_{\text{vap}})\right) \right] \\ &-\alpha \left[ \omega_{\text{Cu}} \rho_{\text{Cu}}(T) \left[ h(T_{\text{plasma}}) - h(300\,\text{K}) \right] \right] \end{aligned} \tag{10}$$

#### 2.2.4. Modelling of the anode

The anode is made of copper. Properties function of temperature for thermal conductivity, mass density and specific heat are used [8]. We don't take into account the liquid copper movements but liquefaction of copper is taken into account by adapting the specific heat close to the liquefaction temperature $T_{\text{liq}}$. Radiation emissivity of copper given by [9] is used.

In the electrode, electric potential and vector potential equations are solved.The source term of Joule effect is taken into account. In order to model the anode, we use the assumption proposed by Lowke and Tanaka [10]: a small layer of 0.1 mm with an electric conductivity of $10^4$ S/m is used in the sheath. The energy flux at the anode surface is given in equation (11) where $q_{\text{cond,plasma}}$ is the conduction flux from the plasma, $q_{\text{rad}}$ the radiation flux estimated from DOM method, $W_{\text{Cu}}$ the electron work function of the copper. No voltage in the sheath is considered and depending on the case it can be as positive as negative. In the energy flux on the anode, the heating of vapours coming from anode to plasma is taken into account. On the surface, this flux is balanced by the conduction flux towards the anode.

$$\begin{aligned} q_{\text{anode}} &= q_{\text{cond,plasma}} - q_{\text{rad}} \\ &+ \frac{\vec{J} \cdot \vec{e}_s}{e} \left[ \frac{5}{2} k_{\text{b}} \left( T_{\text{s}} - T_{\text{plasma}} \right) + W_{\text{s,Cu}} \right] \\ &- \Phi_{\text{Abl}} \left[ h(T_{\text{vap}}) - h_{\text{mean}} + L_{\text{vap}} \right] \end{aligned} \tag{11}$$

#### 2.2.5. Parameters for the calculation

The problem is solved by Ansys Fluent software. Specific User Defined Subroutines have been developed for taking into account plasma characteristics [7] and

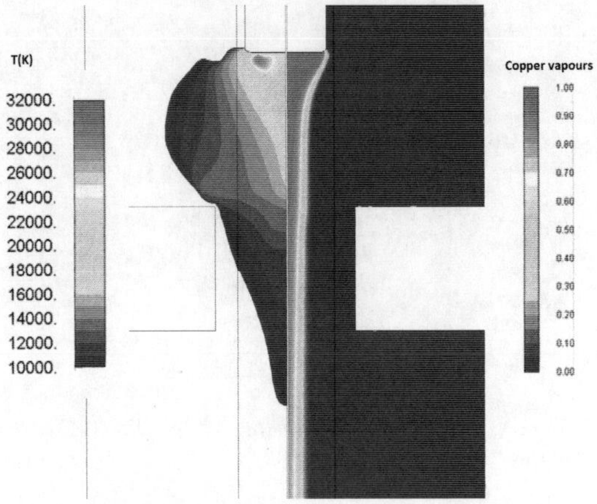


*Figure 2. Temperature and copper mass fraction for time t=5.2 ms.*

the interaction with the anode. The imposed current density is a half wave sinus curve, with a period of 10 ms, and a maximum intensity equal to 10 kA. The time step for the calculations is $\Delta t = 10\,\mu s$.

## 3. Results

### 3.1. Temperature and copper mass fraction

Temperature and mass fraction fields obtained at time $t = 5.2$ ms are presented in figure 2. Concerning mass fraction, we can observe that, at this time, the whole core of the plasma is composed of pure copper. The temperature is quite high with a maximum around 30 kK. One can observe that this maximum is located out of the axis, close to the corner of the electrode. This effect is due to the current path which occurs preferentially on the electrode corner. To verify this point, the current density vector is plotted in figure 3. Where the current density is maximum, the Joule effect is high leading to the out of axis maximum of temperature. This out of axis maximum could sound strange but similar situation can be reported in MIG welding configuration where there is a strong presence of metal vapours in the center of the plasma [11]. This shows the importance of well describing the current path from the electrode to the plasma.

### 3.2. Ablation process

Another interesting field to plot is the mass source term due to copper ablation. It is plotted in figure (4) with the temperature in the anode. At this time, the temperature is high close to the electrode surface in contact with the plasma and propagation within the electrode can be observed. Nevertheless, only the corner of the electrode is over the vaporisation temperature which leads to ablation (case $2\,\frac{P_{\text{vap}}}{P_0} > 1$). The amount of copper coming from the ablation is then located very close to this corner. Nevertheless it is interesting to observe that the copper has propagated in the whole core of the plasma, changing its properties compared to pure $SF_6$.

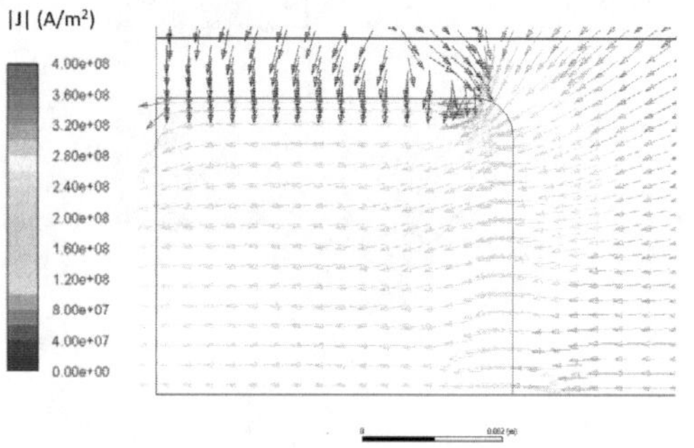

*Figure 3. Current density vector close to the electrode for time t=5.2 ms.*

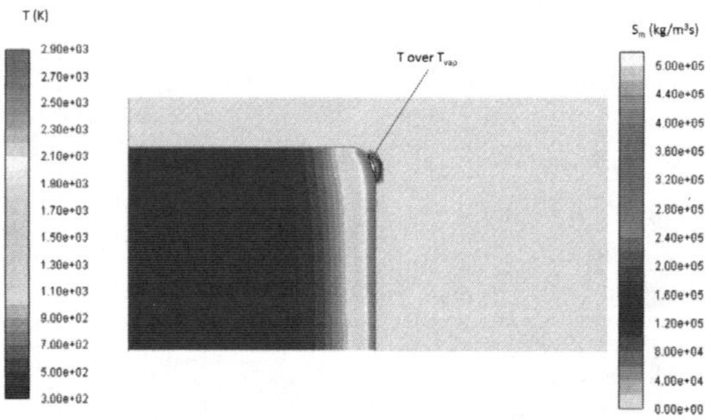

*Figure 4. Temperature in the electrode (left) and mass source term due to copper ablation (equation (7)) at t=5.2 ms.*

The calculated case leads to an energy injected of 5.3 kJ. For this energy the amount of ablated copper with time is plotted in figure 5. A total amount of 0.6 mg was obtained. Of course this quantity should increase with the current intensity value.

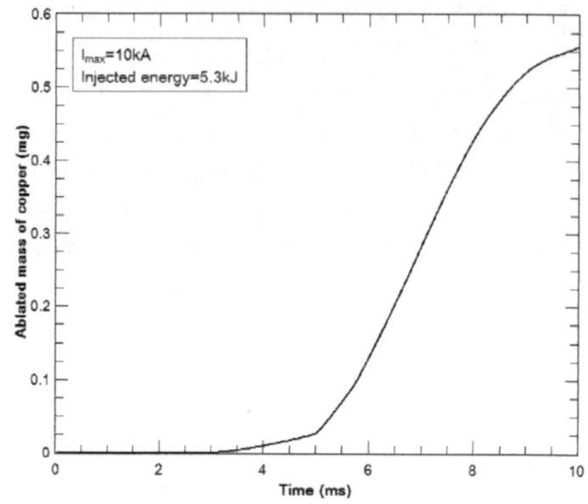

*Figure 5. Ablation of copper material with time t.*

## 4. Conclusions

In this paper we presented a model in $SF_6$ plasma taking into account the ablation of a copper electrode. For the ablation process, two main cases must be distinguished: One corresponding to a diffusion process and another corresponding to a "convective" ablation. We detailed how to take into account the amount of copper linked with hydrodynamic equations. A transient case with a sinus half current wave at 10 kA was presented. This case shows the capacity of the model to describe the ablation process. Even if a small part of the electrode is ablated, the plasma totally burns in a copper medium without any $SF_6$ in its core. This means that in a real HVCB configuration, the fact to consider the copper ablation should change the properties of the plasma and then the models predictions. This effect will be more important increasing the current value.

### References


[1] I. Semenov et al. Modelling of binary alloy (Al-Mg) anode evaporation. *Modelling Simul. Mater. Sci. Eng.*, 20(5):055009, 2012. doi:10.1088/0965-0393/20/5/055009.

[2] M. Benilov et al. Vaporization of a solid surface in an ambient gas. *J. Phys. D: Appl. Phys.*, 34(13):1993–1999, 2001. doi:10.1088/0022-3727/34/13/310.

[3] P. Freton et al. Proposition for a Cu/W ablation model for HVCB electrodes. In *21st International Conference on Gas Discharges and Their Appplications (GD2016)*, Sept 11-16 2016. Nagoya (JAPAN).

[4] A. Gleizes et al. Mixing rules for thermal plasma properties in mixtures of argon, air and metallic vapours. *Plasma Sources Science and Technology*, 19(5):055013, 2010. doi:10.1088/0963-0252/19/5/055013.

[5] J.-J. Gonzalez et al. Turbulence and magnetic field calculations in high-voltage circuit breakers. *IEEE Trans. Plasma. Sci.*, 40(3):936–945, 2012. doi:10.1109/TPS.2011.2180404.

[6] F. Reichert et al. Modelling and simulation of radiative energy transfer in high-voltage circuit breakers. *J. Phys. D, Appl. Phys.*, 45(37):375201–375212, 2012. doi:10.1088/0022-3727/45/37/375201.

[7] P. Freton et al. Magnetic field approaches in dc thermal plasma modelling. *J. Phys. D, Appl. Phys.*, 44(34):202–345, 2011. doi:10.1088/0022-3727/44/34/345202.

[8] Y. S. Touloukian. *ThermophysicalProperties of Matter: The TPRC Data Series; a Comprehensive Compilation of Data.* New York: IFI/Plenum, 1970.

[9] C. Cagran and G. Pottlacher. Normal spectral emissivities of liquid copper, liquid goldand liquid silver at 684.5 nm. *Journal of Non-Crystalline Solids*, 353(32):3582–3586, 2007. doi:10.1016/j.jnoncrysol.2007.05.117.

[10] J. J. Lowke and M. Tanaka. LTE-diffusion approximation for arc calculations. *Journal of Physics D: Applied Physics*, 39(16):3634, 2006. doi:10.1088/0022-3727/39/16/017.

[11] J. J. Lowke et al. Metal vapour in mig arcs can cause (1) minima in central arc temperatures and (2) increased arc voltages. *Welding in the World*, 54:R292–R297, 2013. doi:10.1007/BF03266742.

doi:10.14311/ppt.2019.2.165
*Plasma Physics and Technology* **6**(2):165–169, 2019 

# IMPACT OF THE METAL EVAPORATION RATE IN VACUUM INTERRUPTERS ON VAPOUR EXPANSION AND DEPOSITION

A. E. GEISLER*, N. WENZEL

*Siemens AG, Corporate Technology, Research in Energy and Electronics, 91050, Erlangen, Germany*

* andreas.ag.geisler@siemens.com

**Abstract.** The emission of contact material into vacuum after switching operation of a vacuum interrupter is crucial for the metallisation of the ceramic surfaces. This work focuses on the simulation of various regimes of metal vapour pressure using an extended version of an existing DSMC code that now allows the visualisation of the interaction types and locations during the vapour expansion. The model was applied to a typical vacuum interrupter geometry at different current levels between 3 A and 100 kA. The simulations show that in the low current case the likelihood for a particle hitting a ceramic surface can be more than a factor of 5 higher than in the high current case. An explanation of this observation will be given by analysing the interaction history of the respective particles.

**Keywords:** vacuum interrupter, DSMC method, vapour deposition, insulator coating, vacuum arc.

## 1. Introduction

Arcs in vacuum emit metal vapour; sometimes this is a desired effect (e.g. in vacuum arc deposition applications), in other cases it can result in an undesired performance degradation of the arcing device (e.g. in vacuum interrupters (VIs)) [1–8]). In terms of computational simulations of this process not only is the high-pressure gradient a numerical challenge, but also the absolute pressure regimes that exist in parallel such as in vacuum interrupters during the switching operation. The Direct Simulated Monte Carlo (DSMC) method [9], however, can be applied for all pressure regimes. In this respect, one important parameter in fluid dynamics is the Knudsen number $K$, defined as the ratio of the mean free path of the gas particles and the typical dimension of the geometry. Even though the DSMC method is less efficient for $K \ll 1$, where the Navier-Stokes equations can be applied, this method can be used for all $K$ as long as the particle interaction time is short compared to the time of the particle flight between two interactions. In our simulations the flight between two collisions is a free flight of neutral copper atoms, but basically the method allows to account for an external force acting on each particle.

This work focuses on the simulation of various regimes of metal vapour pressure using an extended version of an existing DSMC code [10] that now allows the visualization of the interaction types and locations during the vapour expansion.

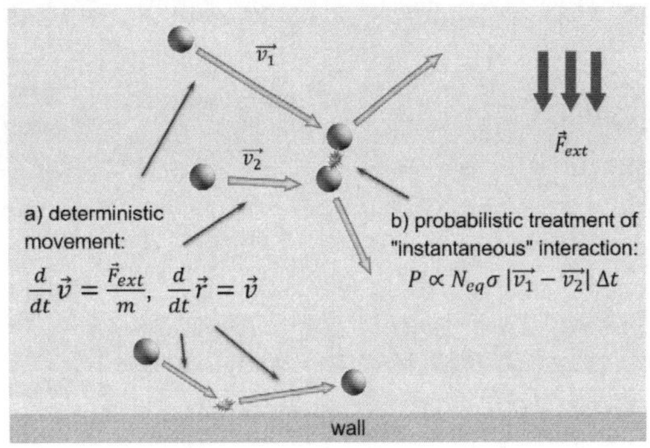


*Figure 1. The basic concept of the DSMC method. Decoupling of free flight and collisions during a time step: a) 'conventional' integration of the equation of motion between the interactions taking a possible external force into account, and b) 'dice rolling' for the interactions treated as an instantaneous stochastic process. The likelihood $P$ of an interaction and - in case of an successfull hit - the energy transfer to internal degrees of freedom as well as the post-collisional particle velocities are calculated by random numbers obeying a probability distribution function that corresponds to a selectable interaction model.*

## 2. The DSMC method

The DSMC method is well described and validated in [11–13]. It is a standard numerical method for rarefied gas dynamics. Rarefied gases are those for which the total volume of the atoms or molecules is small compared to the system volume, i.e. $n\,V_{\text{atom}} \ll 1$, with $n$ being the atom/molecule density. This is fulfilled even for air at normal conditions. In this case the interaction can be approximated by an instant change of momentum. The basic idea of DSMC is now to introduce a sort of super-particles (DSMC particles), each representing an equivalent number $N_{\text{eq}}$ of real atoms or molecules with an adapted collision cross-section $\sigma_{\text{DSMC}} = N_{\text{eq}}\sigma$, and to divide the processes in a deterministic movement between two collisions and a probabilistic (Monte Carlo driven) interaction beween the particles (fig. 1). The collision of two particles is simulated based on a selectable interaction

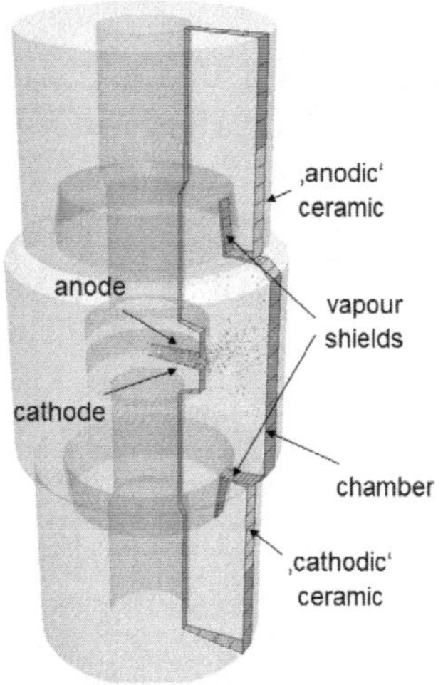


*Figure 2. 10°-sector of a simplified VI being simulated in this study with the typical components: contacts formed by the cathode (emits the particles) and the anode, the metallic switching chamber, two vapour shields, and the insulating ceramics.*

model. Currently the Variable Hard Sphere model (VHS) [11] and Variable Soft Sphere model (VSS) [14]) are available as both had been taken over from the basic underlying OpenFOAM® DSMC code [15].

## 3. Phenomena for different erosion rates/pressure regimes

Vapour expansion simulations in VIs are helpful for example, when the amount or distribution of the metal deposition on the insulating ceramics for different operation modes and geometries shall be investigated. In the course of such studies we observed that in the low current case the likelihood of particles ejected from the contacts is often higher by factors than that in case of high currents. Also, this likelihood is more sensitive to the wall properties in the low current case. Fig. 2 shows a simplified VI design being used for all simulation results presented in this work and fig. 3 reflects the above mentioned phenomena showing the likelihood for a copper atom to hit the insulator of the VI as function of a constant unipolar current and of the sticking coefficient of the metallic wall chamber.

A possible explanation is that the volume interaction of the metal vapour plays an increasing role with higher erosion rates and surface properties such as the sticking coefficient become more and more unimportant. These assumptions might be quite obvious, but with our previous simulation code there were no means to visualise and quantify the effects. Also when looking to the adsorption of metal vapour on the ceramics it is unclear whether those particles originate directly from wall interactions, e.g. bounce off the stem or the shield surface, or from volume interactions.

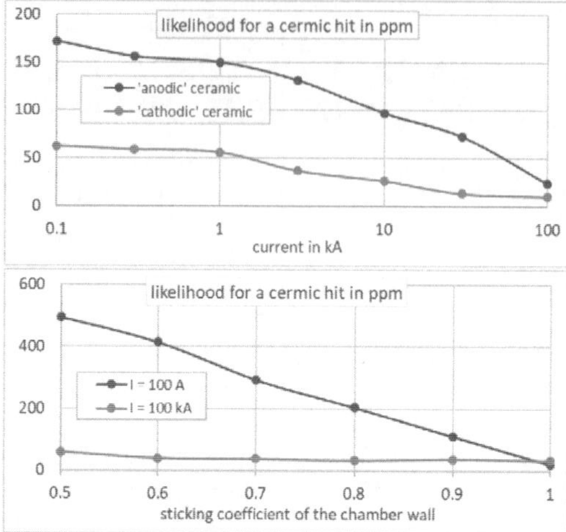


*Figure 3. Dependencies of the likelihood for a particle leaving the cathode surface to hit the ceramic insulator on the 'anodic' or 'cathodic' side for the VI geometry as depicted in fig. 2*

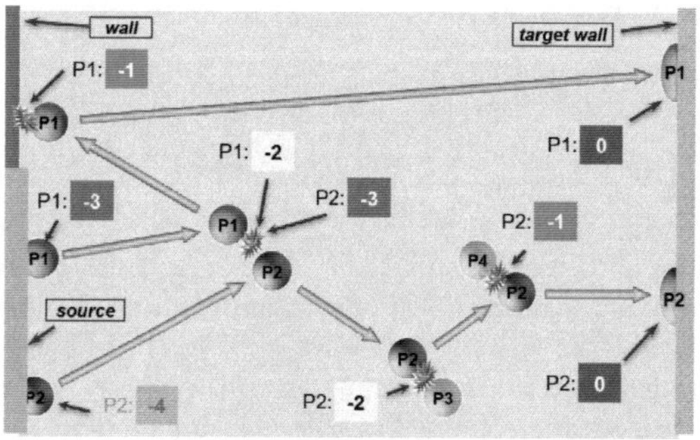


*Figure 4. Examples of interaction sequences: Going forward in time, the history for particle P1 is particle creation (c) in the source region (the cathode in all simulations presented in this paper), particle collision (p), wall reflection (w), and finally wall adsorption (a); For particle P2 the total sequence reads c,p,p,p,a. In the following, only the four last interactions (at most) are considered. The indices 0, -1, -2, -3 are described in the text.*

## 4. Code Extension

In order to get a better understanding of the mechanisms causing the observed numerical results and to check that these can be explained by physical effects, we extended the existing code by a sort of interaction history functionality. Now various information of each adsorbed particle on a definable target wall are stored and can be visualised. It is possible to specify how many interactions shall be stored for each successful target wall adsorption event. An interaction can be one of the following: (i) particle creation, i.e. the particle is ejected from the particle source, the cathode, (ii) particle-particle interaction, meaning a collision with another particle in the volume, (iii) particle-surface reflection, happening whenever a particle hits a surface, but does not stick to it, (iv) wall adsorption. Together

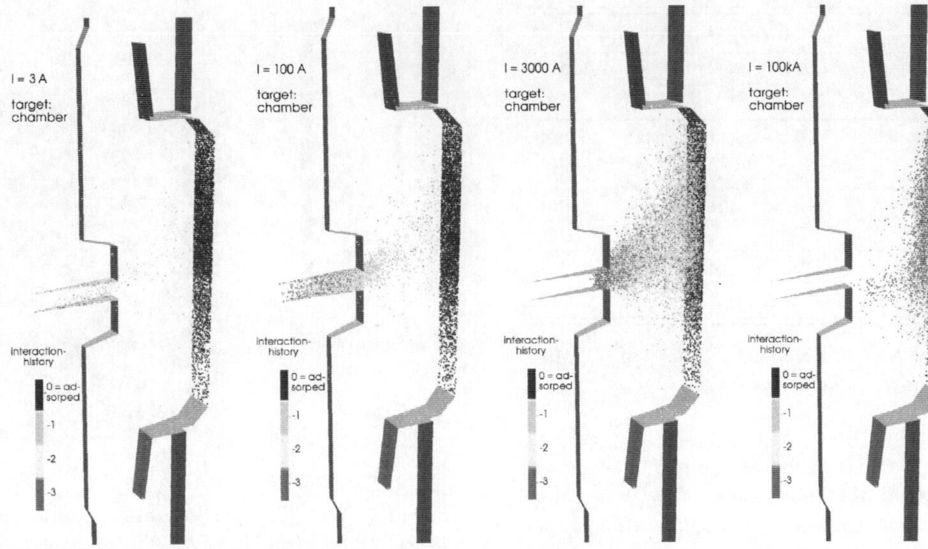


*Figure 5. Visualisation of the particle interaction history for all particles ending on the metallic chamber wall after 3 ms for constant currents of 3 A (hypothetical), 100 A, 3 kA and 100 kA.*

with the interaction type, the position and the time of the event is recorded. Also the interaction partner (the patch being hit or the particle number in case of a collision) is stored as an additional information. In fig. 4, two exemplary paths with different interaction histories are depicted. In all following figures, interaction sequences are coloured in the same manner; in chronological order: "red" denotes the last but two interaction before sticking (index -3), "yellow" the last but one (-2), and "green" the last interaction (-1) before sticking. The very last interaction (index 0) is always the adsorption itself which is colored "blue". From the location one can deduce whether a certain interaction was a particle creation, a wall reflection, or a volume (particle-particle) interaction.

| I | DSMCs | adsorption on | | |
|---|---|---|---|---|
| (kA) | created | anode | chamber | 'anodic' ceramic |
| 0.1 | 7736327 | 5968957 | 1359772 | 1347 |
| 100 | 7786058 | 6129140 | 1344588 | 217 |

*Table 1. Number of DSMC-particles created and adsorbed on various surfaces for $N_{eq} = 1 \cdot 10^7 \cdot I/A$, a setting that generates enough particles hits on the 'anodic' ceramic.*

## 5. Results and Discussion

The following simulation results were obtained for the VI design as depicted in fig. 2. The contact radius was 50 mm, the gap between cathode and anode was 10 mm. The chamber inner wall had a radius of 100 mm, the ceramic inner wall radius was 95 mm. The arc is assumed to be a well-behaved diffuse arc burning in an axial magnetic field (AMF) contact. Thus, in all simulations the cathode was considered to be the sole particle source and it's erosion rate $\dot{m}$ was calculated by $\dot{m} = 100\,\mu\text{g/C} \cdot I$, [16, 17] where $I$ is the electric current assumed to be constant for the simulated time of 3 ms and evenly distributed over the cathode. The created particles are neutral copper atoms only, and the VHS interaction model was applied.

The sticking coefficients were set to 0 for the cathode and 0.8 for the anode. These settings lead to about 78% of the created particles to get re-adsorbed on the anode (cf. the columns under 'anode' and 'created' in tab. 1 that will be discussed below), and thus about 22% leaving the gap and being adsorbed on the chamber wall, the 'anodic', and the 'cathodic' ceramic surfaces. The obtained ratio of created and escaping particles appears to be reasonable if one compares it with results reported e.g. in [16]. For all other surfaces the sticking coefficient was set to 0.9. Only a 10°-sector has been simulated with cyclic boundary conditions on the intersections. These boundary conditions make sure a particle leaving the simulation domain through one of the intersections re-enters the domain on the opposite side at the same radial and axial position with unchanged radial and azimthal velocity components.

As an example of the evaluation of the aforementioned history information, fig. 5 shows the interaction history of those particles which were adsorbed on the chamber wall. From left to the right the particle flux created by the cathode was each time increased by a factor of about 30 starting with a hypothetical value of 3 A (not reflecting a real copper cathode spot distribution) and ending at 100 kA. The corresponding pressures inside the gap were calculated to be 50 mPa for 3 A to 1.6 kPa for 100 kA at temperatures of about 2500 K. In accordance with [18], less than 25% of the

| history index | I (kA) | interaction type: creation | wall: chamber | wall: other | collision |
|---|---|---|---|---|---|
| -1 | 0.1 | 0 | 811 | 291 | 245 |
| | 100 | 0 | 0 | 16 | 201 |
| -2 | 0.1 | 452 | 297 | 139 | 459 |
| | 100 | 0 | 0 | 26 | 191 |
| -3 | 0.1 | 356 | 123 | 121 | 295 |
| | 100 | 0 | 0 | 18 | 199 |

*Table 2. Details of the interaction history for DSMC particles being adsorbed on the 'anodic' ceramic.*

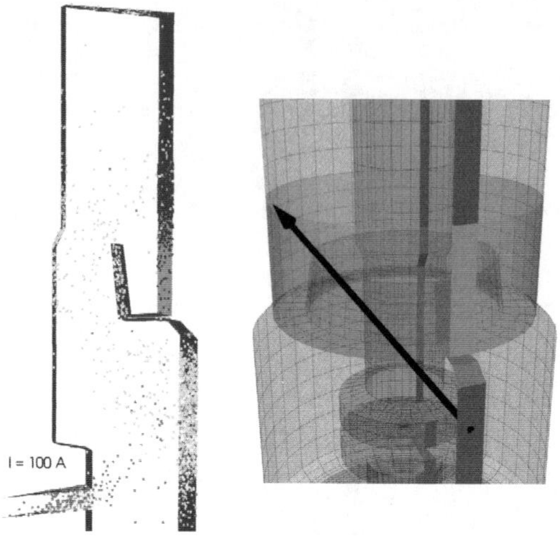


*Figure 6. Left: The particle interaction history for all particles sticking to the 'anodic' ceramic wall after 3 ms for a constant current of 100 A. Right: The arrow shows the direct line of sight between contact stem and shield from the chamber wall to the ceramic.*

emitted particles are leaving the gap region as most of the particles are adsorbed on the anode. The asymmetry of the particle emission and the deposition on the walls is a result of the fact that only the cathode (the lower contact face) is emitting particles. Note that $N_{eq}$ was set to $2 \cdot 10^9 \cdot I/\mathrm{A}$ and is thereby varied linearly with the current (i.e. $N_{eq} = 6 \cdot 10^9$ in case of 3 A and $2 \cdot 10^{14}$ for 100 kA) leading roughly to $4 \cdot 10^4$ created DSMC-particles independent of the actual current in order to obtain similar statistics and numerical noise levels for all runs.

Looking on the green dots in fig. 5, marking the position of the last interaction before the DSMC-particle hits the chamber, we observe the expected. For an extremely low and hypothetical current, the particles originating from the cathode would directly hit the chamber more or less without any further interaction. This corresponds to a surface to surface radiation without any volume scattering or absorption. At $I = 100$ A, the virtual source of the particles in this geometry is the transition area from the gap between the contacts and the chamber volume. Starting from there, hardly any further volume interaction takes place. Increasing the current further, the number of volume interactions increases, too. And finally at $I = 100$ kA, there is not a single particle that originates directly from the cathode, the primary particle source.

Performing a similar analysis, but this time with respect to the 'anodic' ceramic, we get the results tabulated in tab. 1 and tab. 2 (low and high current cases) and shown in fig. 6 (low current case). $N_{eq}$ was set to $10^7 \cdot I/\mathrm{A}$, so again the number of tracked DSMC-particles is the same for both currents, but this time a factor 200 higher. This is needed in order to get a reasonable number of particles on the ceramics. Tab. 1 lists more general numbers with respect to the different hit rates: At 0.1 kA the likelihood to hit the anodic ceramic for a particle leaving the cathode is a factor of 6 higher than in case of 100 kA. However, since $N_{eq}(100\,\mathrm{kA})$ is a factor of $10^3$ higher than $N_{eq}(0.1\,\mathrm{kA})$, in absolute terms the deposition produced per unit time is a factor of 160 higher in the high current case.

From tab. 2 one can see that for both current levels no particle from the contact surface hits the ceramic directly as we have zero particles created for the history index '−1'. This of course is a quite general goal in VI design, namely, to avoid direct sputtering of the insulator by the primary particle source. In case of a high evaporation rate, however, it takes more than three interactions for a particle originating from the chamber wall to reach the 'anodic' ceramic (0 particles for index -3 under 'chamber'), which accommodates for the fact of a large pressure of the order of Pa in that region, corresponding to a mean free path of the order of millimetres. The main source of particles hitting the examined ceramic in this arrangement are volume (collision) interactions. In contrast to this, tracing back two interactions in the low current case, one finds 452 particles that are created, meaning one third of all particles being adsorbed on the 'anodic' ceramic (1347) do have a single interaction after being emitted from the cathode: most likely this is the reflection at the chamber wall, explaining the influence of the impact of sticking coefficient here. This is supported by the fact that 811 particles of 1347 originate directly from the chamber.

As previously assumed for the low current case (100 A), the main source of the particles hitting the ceramic is the switching chamber which appears greenish in the left of fig. 6, indicating that this was the last interaction before adsorption. The vast majority of particles has gone through the vacuum without any further volume interaction. This is possible since there exists a direct line of sight from the chamber wall to the ceramic as shown on the right of fig. 6. This fact might not be evident when looking on the cross section of the VI only. However, one has to bear in mind that whenever a particle leaves the calculation domain via one of the intersection faces, the cartesian velocity components change as a result of the 10° angle between the intersection surface normals. As a consequence, when tracking a non interacting particle moving on a straight line, it appears to be bend radially outward each time it crosses the intersection. Hence, this particle apparently moves on a curved line reaching areas that seem to be occluded in the 2D cross section.

## 6. Conclusions

The results are obtained for a particular VI geometry with AMF contacts assuming the cathode surface as vapour source region. They document that at high currents volume interaction (particle-particle collisions) plays the most important role and wall interactions (reflection and desorption by inner surfaces) are less important. At low currents, on the contrary, the most likely sputter mechanism is the reflection of the particles at the metallic VI chamber wall. Volume interaction and reflection by other surfaces are less important for the VI geometry in this case.

The new feature of our DSMC code improves the understanding of metal vapour expansion and deposition in a VI stressed by operational and short circuit currents, and it helps to identify critical issues occurring in the course of VI development. However, up to now, the simulation is not accurate by its own, but rather needs to be compared and calibrated with experiments. Care should be taken to interpret the simulation of real switching operations, because the erosion rate might be different and the source region can be non-uniform, particularly if the VI is close to its breaking limit and the anode takes over an active particle emitter role.

### References

[1] P. Slade. *The Vacuum Interrupter: Theory, Design, and Application.* 1st Edition. CRC Press, Taylor and Francis Group, London, 2008.

[2] H. Fink, D. Gentsch, and M. Heimbach. Condensed metal vapor on alumina ceramic in vacuum interrupters. *IEEE Trans. Dielectr. Electr. Insul.*, 9(2), 2002. doi:10.1109/94.993736.

[3] B. Kuehn, M. Kurrat, M. Hilbert, I. Gramberg, and D. Gentsch. Characterization of metal vapor deposition on vacuum interrupter ceramics and its impact on electric field distribution. *IEEE Trans. Dielectr. Electr. Insul.*, 24(6), 2017. doi:10.1109/TDEI.2017.006486.

[4] H. Schellekens and J. Battandier. Metal vapor deposition on ceramic insulators and vacuum interrupter lifetime prediction. In *Proc. 19th Inter'nl Symp. on Disch. and Electr. Insulation in Vac.*, volume 1, 2000. doi:10.1109/DEIV.2000.877270.

[5] I. S. Gramberg, M. Kurrat, and D. Gentsch. Investigations of copper chrome coatings on vacuum circuit breaker ceramics by electron probe microanalysis and electric field simulation. *IEEE Transactions on Plasma Science*, 41, 2013. doi:10.1109/TPS.2013.2273260.

[6] B. Kuehn, M. Kurrat, M. Hilbert, and D. Gentsch. Multiple shield arrangements breakdown model in vacuum interrupters. In *2016 27th International Symposium on Discharges and Electrical Insulation in Vacuum (ISDEIV)*, volume 2, pages 1–4, 2016. doi:10.1109/DEIV.2016.7763948.

[7] M. Kurrat. Vacuum arc ion flux from vacuum interrupter contact gap. *Plasma Physics and Technology Journal*, 4, 2017. doi:10.14311/ppt.2017.2.186.

[8] E. D. Taylor. Application of research in field emitter arrays to the breakdown of contacts in vacuum. In *2012 25th International Symposium on Discharges and Electrical Insulation in Vacuum (ISDEIV)*, pages 41–44, 2012. doi:10.1109/DEIV.2012.6412445.

[9] G. Bird. Approach to translational equilibrium in a rigid sphere gas. *Physics of Fluids*, 6, 1963. doi:10.1063/1.1710976.

[10] A. Geisler and N. Wenzel. Metal vapor deposition in vacuum interrupters: Utilization of dsmc methods and considerations on vapor sources. *28th International Symposium on Discharges and Electrical Insulation in Vacuum*, 2, 2018. doi:10.1109/DEIV.2018.8537030.

[11] G. Bird. *Molecular Gas Dynamics and the Direct Simulation of Gas Flows.* Number Bd. 1 in Molecular Gas Dynamics and the Direct Simulation of Gas Flows. Clarendon Press, 1994.

[12] T. Scanlon, E. Roohi, C. White, M. Darbandi, and J. Reese. An open source, parallel dsmc code for rarefied gas flows in arbitrary geometries. *Computers and Fluids*, 39, 2010. doi:10.1016/j.compfluid.2010.07.014.

[13] K. Hencken. Investigation of metallic vapor condensation in a vacuum interrupter using dsmcfoam. *9th OpenFOAM® Workshop*, Zagrep, Croatia, 2014.

[14] K. Koura and H. Matsumoto. Variable soft sphere molecular model for air species. *Physics of Fluids A: Fluid Dynamics*, 4(5), 1992. doi:10.1063/1.858262.

[15] H. G. Weller, G. Tabor, H. Jasak, and C. Fureby. A tensorial approach to computational continuum mechanics using object-oriented techniques. *Computers in Physics*, 12(6), 1998. doi:10.1063/1.168744.

[16] G. Farrall, F. Hudda, and J. Toney. The time-resolved characterization of erosion products from high-current, copper vacuum arcs. *Plasma Science, IEEE Transactions on*, 11, 1983. doi:10.1109/TPS.1983.4316240.

[17] J. Kutzner, R. Batura, and Z. Stachowiak. Experimental analysis of the electrode products emitted by high-current vacuum arcs. *IEEE Trans. Plasma Sci.*, 27(4), 1999. doi:10.1109/27.782255.

[18] M. B. Schulman, P. G. Slade, and J. A. Bindas. Effective erosion rates for selected contact materials in low-voltage contactors. *IEEE Transactions on Components, Packaging, and Manufacturing Technology: Part A*, 18(2), 1995. doi:10.1109/95.390312.

doi:10.14311/ppt.2019.2.170
*Plasma Physics and Technology* **6**(2):170–174, 2019 

# POST-ARC CHARACTERISTICS OF HIGH-CURRENT ARCS IN VACUUM CIRCUIT-BREAKERS

N. Wenzel[a,*], W. Haas[b]

[a] *Siemens AG, Corporate Technology, Research in Energy and Electronics, 91050 Erlangen, Germany*

[b] *Retired employee of Siemens AG, Zentralabteilung Forschung und Entwicklung, 91050 Erlangen, Germany*

[*] norbert.wenzel@siemens.com

**Abstract.** The post-arc (PA) characteristics of vacuum arcs in transverse magnetic field contacts are studied for short-circuit currents of up to 123 kA peak and transient recovery voltages below 875 V. The measured PA currents are interpreted in terms of an Electric Resistance Model and the models of Andrews-Varey, Langmuir-Child, and Slepian-Schmelzle. Whereas in the late PA period, the calculations do not agree well with the measurements, the PA behavior is well described in the early period after current-zero. It is concluded that the PA discharge is amplified by ionization of metal vapor particles in the boundary sheath due to electron impact.

**Keywords:** vacuum arc, vacuum circuit-breaker, transient recovery voltage, post-arc current, boundary sheath, transverse magnetic field contact.

## 1. Introduction

Vacuum arcs are successfully applied in switching technology to interrupt high short-circuit (SC) currents. The transient recovery voltage (TRV) of the electrical grid after current-zero removes the residual electric charge carriers of the switching arc plasma from the contact gap region. This results in a "post-arc current" (PAC) that is still flowing after vacuum arc extinction (e.g. early papers [1, 2] or recent papers [3–5]). The PAC behavior of the contact gap is of central importance for the breaking performance of a vacuum interrupter.

Due to the fast removal of the electrons within microseconds after current-zero, the (heavy) metal ions (copper and chromium) establish a positive space charge sheath near the post-arc (PA) cathode, the "post-arc boundary sheath", which is expanding towards the PA anode. The movement of the ions and the ionization of neutral metal vapor particles within the sheath sensitively influence the PAC characteristics.

Former published studies of the PAC behavior were restricted to medium SC currents ranging from 36 kA rms to 60 kA rms [1, 2, 6–8]. They were mostly concerned with high TRV's up to 80 kV [6, 8] and 126 kV [9]. Studies concerned with low TRV's ($\leq$ 960 V) reported in [10] were restricted to medium SC currents ($\leq$ 50 kA rms) and focused to axial magnetic field contacts.

In this contribution, the PAC behavior of a transverse magnetic field (TMF) contact in a high-current low-voltage vacuum interrupter is investigated. Experiments in a synthetic test circuit and physical models of the PA boundary sheath are described, and analytical computations are used to interpret the measurements. The studies resume and critically revalidate investigations, that had been reported in [11], but not yet published in written form.

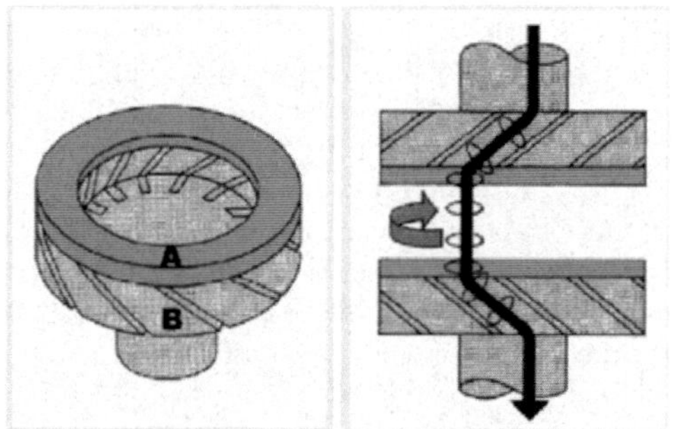


*Figure 1. TMF contact [18]. Left: Cu contact piece (B) with CuCr contact ring (A); contact diameter = 90 mm; width of contact ring = 18 mm. Right: Current path and moving constricted vacuum arc.*

## 2. Experiments

The experiments are performed with a conventional high-current low-voltage vacuum interrupter. A moving constricted metal vapor arc is drawn by separating the electrodes of a cup-shaped (contrate) TMF contact to a gap of 3 mm at a velocity of typically 1 m/s. The contact system is sketched in Figure 1.

Single sinusoidal SC currents (AC half period 12 ms) with amplitudes of up to 123 kA peak corresponding to 87 kA rms are produced by a capacitor bank, supplemented by a DC pre-current of 100 A. Recovery voltages with the same relative step response waveform and varying steady-state values ranging from $U_W = 385$ V to 875 V (absolute values) are generated in a synthetic test circuit. Wide-band transformers (Pearson Electronics) are used to monitor the tremendously differing current amplitudes in the SC path

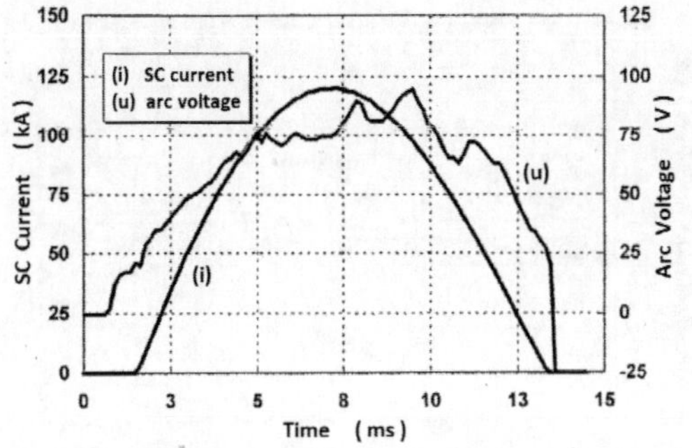


*Figure 2. Measured time evolutions of SC current and vacuum arc voltage. Time t = 0: start of DC pre-current; t = 1 ms: contact separation; t = 2 ms: start of current half-period.*

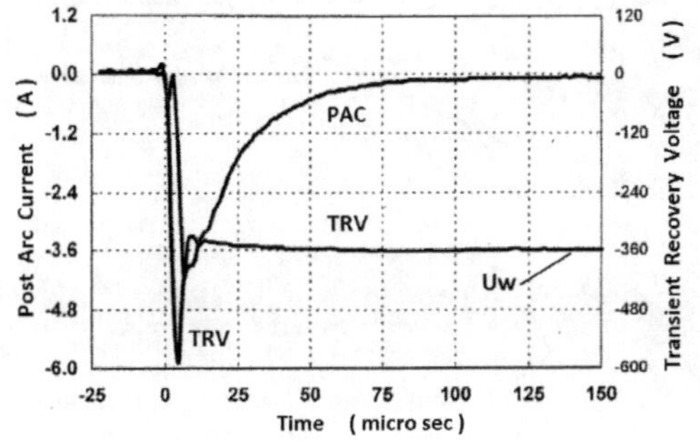


*Figure 3. Measured time evolutions of applied TRV ($U_S$) and resulting PAC current-zero. After the initial voltage peak (-595 V), the TRV converges to the steady-state value (-360 V) denoted as $U_W$.*

and the PAC path of the synthetic circuit. Figure 2 displays a typical example of the measured time evolutions of SC current and arc voltage. Figure 3 shows a typical example of the measured time evolutions of the TRV (denoted as $U_S$ in the next chapters) and the resulting PAC. The relative TRV waveform remains the same, when the steady-state voltage $U_W$ changes.

## 3. Modeling

Different one-dimensional analytical models are used to describe the expansion of the PA boundary sheath within the contact gap towards the (new) PA anode under the impact of the TRV. The sheath expansion is sketched in Figure 4. Though primarly appropriate for diffuse vacuum arcs in axial magnetic field contacts, one-dimensional computation is also applicable to TMF contacts, since constricted vacuum arcs become diffuse near current-zero (below 15 kA).

The models formulate relations between the instantaneous values of the PAC density, $j$, the expansion velocity of the sheath boundary, $\partial s/\partial t$, and the TRV, $U_S$. The expansion velocity is linked to the PAC density via Equation (1). The quantities $q$ and $n_0$, respectively, denote the electric charge and the density of the ions in the transition zone between boundary

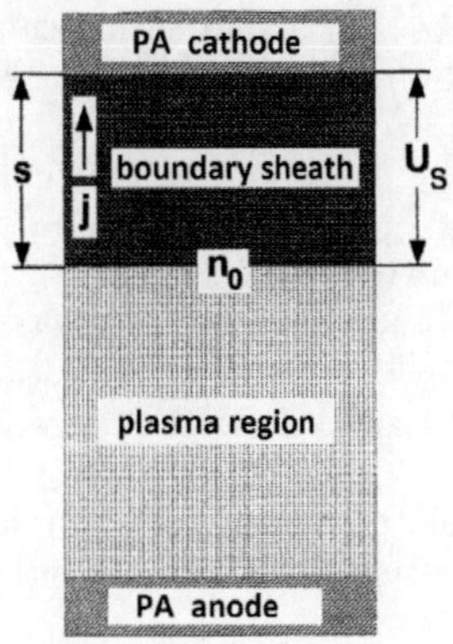


*Figure 4. Sketch of PA boundary sheath (width s) expanding from PA cathode towards PA anode under the impact of the TRV ($U_S$). j = PAC density; $n_0$ = ion density in transition zone between boundary sheath and PA plasma region. Details see text.*

sheath and PA plasma region. The function $g(U_S)$ reflects the influence of $U_S$ and is specified by the respective model of the boundary sheath.

$$\frac{\partial s}{\partial t} = \frac{j}{q\,n_0}\,g(U_S) \tag{1}$$

The measured PAC evolutions are interpreted in terms of an "Electric Resistance Model" of the sheath developed in [11]. And they are compared with simulations based on the sheath models of Andrews-Varey, Langmuir-Child, Slepian, and Schmelzle.

The Transition Model of Andrews-Varey [12, 13], and the model according to the Langmuir-Child Law [14] describe the "free" flight of the ions in the boundary sheath. The model of Andrews-Varey takes the displacement and the conduction current into account, and it assumes only small changes of the ion parameters over a transition of the boundary sheath. The model yields Equation (2), an implicit relation between the TRV, $U_S$, and the PAC density, $j$, both linked by the width $s$ of the boundary sheath ($M$: ion mass, $\varepsilon_0$: vacuum permittivity constant).

$$s^2 = \frac{4}{9}\left(\frac{\varepsilon_0 M}{2q^4}\frac{j^2}{n_0^3}\right)\left[\left(1+\frac{2q^3}{M}U_S\frac{n_0^2}{j^2}\right)^{\frac{3}{2}} + 3\frac{2q^3}{M}U_S\frac{n_0^2}{j^2} - 1\right] \tag{2}$$

The model of Langmuir-Child results in Equation (3), an explicit relation between $U_S$ and $j$. It is derived from Equation (2) by neglecting the kinetic energy of the ions, $0.5\,M\,(\partial s/\partial t)^2$, against the energy they gain due to the TRV, $q\,U_S$.

$$U_S = \left(\frac{9}{4\varepsilon_0}\right)^{\frac{2}{3}}\left(\frac{M}{2q}\right)\,s^{\frac{4}{3}}\,j^{\frac{2}{3}} \tag{3}$$

The models of Slepian and Schmelzle [15, 16], and the "Electric Resistance Model" [11] describe the motion of the ions under the assumption that the motion is dominated by collisions in the boundary sheath. The models neglect the displacement current and yield explicit relations between $U_S$ and $j$.

The models of Slepian and Schmelzle, respectively, result in the (precise) first and the (approximate) second term of Equation (4) in terms of the mobility of the ions, $\mu$, whereupon the (second) term according to Slepian neglects the summands proportional to $j/(q n_0)^2$ by assuming high ion densities $n_0$ in the transition zone.

$$U_S = \frac{2}{3}\left(\frac{2j}{\varepsilon_0 \mu}\left[\left(\frac{j\varepsilon_0}{2q^2n_0^2}+s\right)^{\frac{3}{2}} - \left(\frac{j\varepsilon_0}{2q^2n_0^2}\right)^{\frac{3}{2}}\right]\right)^{\frac{1}{2}} \approx \frac{2}{3}\left(\frac{2}{\varepsilon_0}\frac{j}{\mu}s^{\frac{3}{2}}\right)^{\frac{1}{2}} \tag{4}$$

The "Electric Resistance Model" reflects the local Ohm Law. It connects the PAC density, $j$, with the average electric field, $U_S/s$, in the boundary sheath via Equation (5) in terms of the ion mobility, $\mu = \sigma/(q\langle n\rangle)$, where $\langle n\rangle$ denotes the average ion density in the boundary sheath.

$$U_S = \frac{j\,s}{q\,\langle n\rangle\,\mu} \tag{5}$$

The described models are applicable provided that evaporation on the PA cathode is negligible. In the case of TMF arcs, particularly at higher currents, the surface of the (old) anode during SC current flow, i.e. the (new) PA cathode, could be largely molten supplying ions even after SC current-zero into the switching gap. The model descriptions are thus justified for switching operations well below the SC breaking limit of a TMF contact system.

## 4. Results and discussion

The experiments document that high SC currents, large SC arcing times, and high TRV's cause high values of PAC amplitude and residual electric PA charge, the time integral of the PAC. Figure (5) shows the peak PAC amplitude and the residual PA charge as function of the steady-state value of the TRV waveform, $U_W$, for a current of 87 kA rms and an arcing time of 10 ms. The PA quantities measured for successful interruptions can be well fitted to an exponential function. The PA quantities referring to SC failures exhibit significantly higher amplitudes than those for successful switching operations.

Figure (6) shows a comparison of the measured time evolutions of TRV (points) and PAC (points) with simulations (lines) based on the "Electric Resistance Model" and the models of Langmuir-Child and

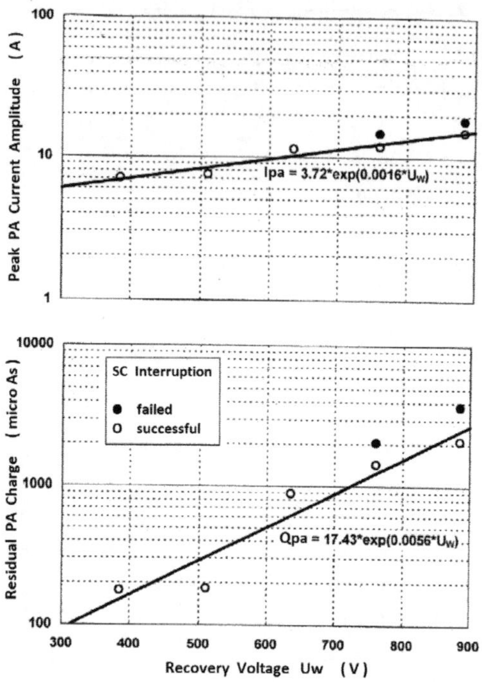


*Figure 5. Measured average values of peak PAC amplitude $I_{PA}$ (top) and residual PA charge $Q_{PA}$ (bottom) as function of TRV $U_W$. Open circles: successful interruptions, full circles: failed interruptions, solid line: fit to successful interruptions. SC current = 87 kA$_{rms}$, arcing time = 10 ms.*

Slepian-Schmelzle. The model parameter is the average plasma density, $n$, in the contact gap after SC arc extinction. The simulations according to Andrews-Varey (not shown here) and Langmuir-Child yield nearly identical results.

In the early PA period (between 1 µs and 4 µs after current-zero), the simulations display a good description of the measured evolution of the PAC, if an average plasma density of several $10^{19}\,\mathrm{m}^{-3}$ is assumed after the extinction of the SC arc. In the late PA period (> 5 µs) however, the calculations do not agree well with the experiment. This finding indicates that the PA discharge in the contact gap after high-current interruption is amplified by ionization of neutral metal vapor particles in the boundary sheath due to electron impact. The PA amplification starts at sufficiently broad boundary sheath, even under the influence of the applied low recovery voltages (< 1000 V). The ionizing electrons in the boundary sheath can be generated at the PA cathode by thermofield emission ("hot spots") and by secondary emission (ion impact). These physical processes are not described by the used simulation models. In the very early PA period (< 1 µs), the experimental PAC signals indicate that high-energy particles are penetrating the boundary sheath at the very beginning of sheath development. This phenomenon reported by [17], e.g., is also not described by the used simulation models

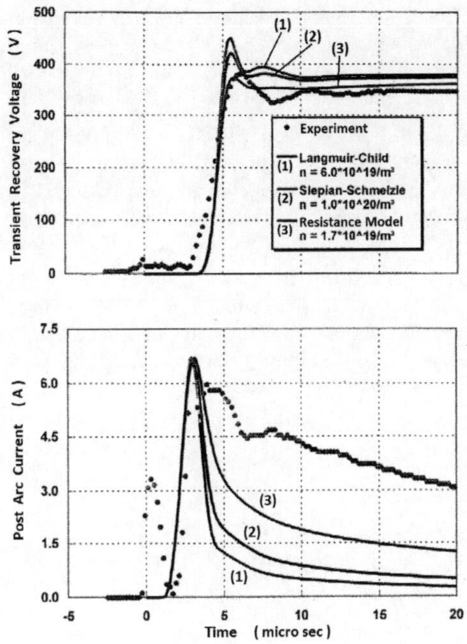


*Figure 6. Comparison of measured time evolutions of TRV (points) and PAC (points) with simulations (lines). Boundary sheath models: Langmuir-Child (1), Slepian-Schmelzle (2), and Electric Resistance Model (3). $n$ = plasma density in contact gap after SC arc extinction. Time $t = 0$: SC current-zero. SC current = $87\,kA_{rms}$, arcing time = 10 ms.*

## 5. Conclusion

In this contribution, the PA characteristics of constricted vacuum arcs moving in a small-gap cup-shaped TMF contact system (CuCr) have been investigated for SC currents (42 Hz) of up to 123 kA peak and for TRV's below 875 V (steady-state values). The measured PAC evolutions are strongly related to SC current amplitude, arcing time, and applied TRV. The experiments are compared with simulations of the PA boundary sheath based on the models of Andrews-Varey, Langmuir-Child, and Slepian-Schmelzle, and on an "Electric Resistance Model". In the early PA period (between 1 µs and 4 µs after SC current-zero), the simulations display a good description of the PAC evolution. In the very early (below 1 µs) and in the late PA period (later than 5 µs after SC current-zero), the calculations do not agree well with the experiment.

The following conclusions can be drawn: After high-current interruption, the PA discharge in the contact gap is amplified by ionization of neutral metal vapor particles in the boundary sheath due to electron impact. The ionizing electrons in the boundary sheath are generated at the PA cathode by thermo-field emission and by secondary emission. High-energy particles are penetrating the boundary sheath at the very beginning of sheath development.

Diagnosing and one-dimensional analytical modeling of the PA behavior allow the evaluation of withstand voltage characteristics and re-ignition behavior of a small-gap TMF vacuum interrupter contact, at least in a crucial time interval after SC interruption. This offers the possibility of a purposeful optimization of high-current vacuum circuit-breakers without applying complex multidimensional numerical simulation methods.

## References

[1] K. Matsumoto. Post arc current after high current interruption in vacuum. In *Proc. 11th ISDEIV, Berlin, vol.2, sections 3-6*, 1984.

[2] S. Yanabu et al. Post arc current of vacuum interrupters. *IEEE Trans. Power Apparatus and Systems*, 104(1):166–172, 1985. doi:10.1109/TPAS.1985.318890.

[3] G. Ge, X. Cheng, M. Liao, X. Duan, and J. Zou. Vacuum arcs and postarc characteristic of vacuum interrupters with external AMF at current zero. *IEEE Trans. Plasma Sci.*, 46:1003–1009, 2018. doi:10.1109/TPS.2018.2816028.

[4] Y. Mo, Z. Shi, J. Li, S. Jia, and L. Wang. Experimental research of post-arc currents in vacuum circuit breakers. In *Proc. 28th ISDEIV, Greifswald*, volume 1, pages 193–196, 2018.

[5] A. Shemshadi. Modeling of plasma dispersion process in vacuum interrupters during postarc interval based on FEM. *IEEE Trans. Plasma Sci.*, 47(1):647–653, 2019. doi:10.1109/TPS.2018.2879519.

[6] K.-D. Weltmann, E. Huber, and K. Froehlich. Arc memory effect on postarc current in vacuum circuit breaker - simulation and experiment. In *Proc. 18th ISDEIV, Eindhoven*, pages 545–548, 1998. doi:10.1109/DEIV.1998.738710.

[7] S. Hakamata, Y. Kurosawa, K. Natsui, K. Hirasawa, and K. Kurita. Post arc current of vacuum interrupter after large current interruption. *IEEE Trans. Power Delivery*, 3(4):1692–1697, 1988. doi:10.1109/61.193973.

[8] K. Steinke, M. Lindmayer, and K.-D. Weltmann. Post-arc currents of vacuum interrupters with axial magnetic field contacts under high current and voltage stress. In *Proc. 19th ISDEIV, Xi'an*, pages 475–480, 2000. doi:10.1109/DEIV.2000.879030.

[9] E. Dullni, E. Schade, and B. Gellert. Dielectric recovery of vacuum arcs after strong anode spot activity. *IEEE Trans. Plasma Sci.*, 15(5):538–544, 1987. doi:10.1109/TPS.1987.4316750.

[10] D. Pavelescu, S. Nitu, G. Dumitrescu, and P. Anghelita. Post-arc current in low-voltage vacuum circuit breaker: Measurements and physical peculiarities. *IEEE Trans. Plasma Sci.*, 31(5):869–876, 2003. doi:10.1109/TPS.2003.818421.

[11] W. Haas and N. Wenzel. Das Nachstromverhalten stromstarker Metalldampf-Lichtboegen in Vakuum (The post-arc current behavior of high-current metal vapor arcs in vacuum). In *Fachtagung der Deutschen Gesellschaft für Plasmatechnologie, Bochum, Poster*, 1996.

[12] J. Andrews and R. Varey. Sheath growth in a low pressure plasma. *The Physics of Fluids*, 14:339–343, 1971. doi:10.1063/1.1693433.

[13] R. Holmes and S. Yanabu. Post-arc current mechanism in vacuum interrupters. *J. Phys. D: Appl. Phys.*, 6(10):1217–1231, 1973. doi:10.1088/0022-3727/6/10/306.

[14] S. Brown. *Introduction to Electrical Discharges in Gases. Chapter 12, pp. 200-225.* Wiley series in plasma physics. John Wiley and Sons, New York, 1966.

[15] J. Slepian. Extinction of an A-C arc. *Transactions on American Institute of Electric Engineering (AIEE)*, 47(4):1398–1408, 1928. doi:10.1109/T-AIEE.1928.5055155.

[16] M. Schmelzle. *Grenzen der Selbstloeschung kurzer Lichtbogenstrecken bei Wechselstrombelastung (Limits of self-extinction of short arc gaps under AC stress).* PhD Thesis, Technical University of Braunschweig, 1968.

[17] X. Li, J. Zou, W. Dong, D. Liang, and T. Qin. Experiment research on post-arc current in DC vacuum circuit breakers. In *Proc. 4th Int. Conf. Electric Power Equipment - Switching Technology, Xi'an*, pages 375–378, 2017. doi:10.1109/ICEPE-ST.2017.8188867.

[18] D. W. Branston, W. Haas, W. Hartmann, R. Renz, and N. Wenzel. Thermodynamic model of moving vacuum arcs on contrate contacts. In *Proc. 22nd ISDEIV, Matsue*, volume 1, pages 277–280, 2006. doi:10.1109/DEIV.2006.357285.

doi:10.14311/ppt.2019.2.175
*Plasma Physics and Technology* **6**(2):175–179, 2019 

# DETERMINING THE TIME CONSTANT OF ARCS AT ARBITRARY CURRENT LEVELS

L. S. J. Bort*, T. Schultz, C. M. Franck

*High Voltage Laboratory, ETH Zurich, Physikstrasse 3, 8092, Zurich, Switzerland*

* lorenz.bort@alumni.ethz.ch

**Abstract.**

For the development and optimization of gas circuit breakers and switchgear, a detailed understanding of the arc related processes is of great importance. Ideally, analytical or numerical models with predicitive capability can be found and used during the design process preceding costly and time-consuming experiments.

In the present contribution, we report on a novel measurement and evaluation technique to determine the thermal arc time constant ("thermal inertia") that is commonly used in simple black-box models to describe the arc's dynamical properties. The method is introduced and applied to example arcs under varying blow gas conditions in air.

**Keywords:** switching arc, gas circuit breakers, black-box model, arc time constant.

## 1. Introduction

The understanding of electric arcs is crucial for the design and improvement of mechanical switches in the power system. This applies for a wide range of switches, from low voltage AC to HVDC. One important internal parameter of arcs is their time constant. It describes a characteristic time for the exchange of energy between arcs and their surrounding. This has a considerable impact on the electrical behavior of the arc.

It is well established that interruption becomes more difficult for higher current gradients before zero crossing. This is due to the fact, that the conductance of the arc at zero crossing is not at its steady state for zero current, i.e. zero, but at a finite value. For current zero crossing, this means that the arc still contains considerable thermal energy and thus still conducts current. The conductance at current zero (CZ) depends on the arc time constant, as well as on the current gradient before CZ. For steep current gradients, or large time constants, the conductance at current zero is high, which can lead to reignitions. As the time constant varies with gas flow conditions, nozzles and gas types, it is one important performance parameter of mechanical interrupters.

In HVDC, current interruption can be realized using a mechanical circuit breaker and a counter current injection circuit. In these current injection topologies, the current wave shape (especially at CZ) is determined by the circuit breaker itself. Consequently, it is possible to implement circuits that drastically decrease the current gradient shortly before CZ to improve interruption [1, 2]. The length of the phase with decreased current gradient needed to improve interruption performance is defined by the arc time constant in this case. The arc time constant is thus an optimization parameter.

Another field of application where it is important to know the arc time constant is the optimization of passive oscillation switches, which are commonly used as transfer switches (most notably the metal return transfer switch) in HVDC applications. Here, the voltage-current characteristic of interrupters leads to the growth of an oscillating current. For fast interruption, high current oscillation frequency is helpful, as long as the the oscillation period is longer than $20\tau$. This means in this application, the arc time constant of the should be minimized.

To determine the characteristic time constant $\tau$ of a given setup, several possibilities exist [3–7], which were evaluated in previous work [8, 9]. All of these have certain drawbacks, and do not allow the determination of thermal inertia at arbitrary current levels.

According to [8, 9] a staircase current waveform would be ideal, but short current impulses superimposed to quasi-DC current are proposed if a step with sufficiently steep slope is impossible to create. In this paper, a method to determine the time constant of arcs, based on triangular impulse currents, is developed and implemented. As validation case, tests showing the transition between free-burning and blown arc are presented.

## 2. Theory

The response of an arc to dynamic changes in current is generally described by black box models, namely the ones of Mayr and Cassie [10, 11]. These models assume that the loss terms are a function of the arc conductance $g$ and cannot change instantaneously. Some time is needed to change to the new equilibrium state, which can be described by a time constant. In case of a current increase from $i_0$ to $i_1$, it must be a change that leads to a greater conductance $g_1$. This change requires a certain change of the arc enthalpy, to either increase the arc radius, or the temperature. The required energy has to be supplied by the difference of the instantaneous power the arc loses to its surroundings $P_{cool}$ (due to radiation, convection and thermal conduction) and instantaneous heating power $u \cdot i$. The Mayr model is usually used to describe arcs during a few microseconds before and after current zero in AC circuit breakers. It assumes the instantaneous cooling power $P_{cool}$ is constant and there is one time constant $\tau$ which describes the relaxation process. Cassie, Habedank, Shavemaker and many others [11–13] generalized this, by making $P_{cool}$ and possibly $\tau$ a function of $g$, to describe the high current phase as well. The most general form is therefore

$$\frac{1}{g}\frac{dg}{dt} = \frac{1}{\tau(g)}\left(\frac{ui}{P_{cool}(g)} - 1\right). \qquad (1)$$

The physical interpretation of this model is that the arc consists of one energy reservoir (the enthalpy of the heated plasma column) and that the conductance is the fundamental variable to describe its state. In steady state, ohmic heating equals the sum of all cooling terms, i.e. $P_{cool} = u \cdot i$, and therefore $g$ (and with it the arc enthalpy) doesn't change. In transient conditions, for example at falling current, there is a difference between heating and cooling power, leading to heating or cooling of the plasma, i.e. the arc adapts to changing conditions by shrinking for example. For simplicity, the most general form of the transient arc model equation (1) will be called *Mayr Equation* throughout the manuscript, although *Cassie Equation* would be equally right or wrong, because the two only differ in how $P_{cool}$ depends on $g$ (see [14–16]).

The literature mentioned above provides different modified models, which assume certain analytic functions for $P_{cool}(g)$ and $\tau(g)$. Cassie for example assumes a constant voltage during high currents. Schwarz [3] introduced some additional variables to parameterize $P_{cool}(g)$ and $\tau(g)$, and more recently also other modifications were published. A summary of several models can be found in [16]. In the present work, those more specific models were not studied, but the generalized form was used. The only constraint this general form has is that both $P_{cool}(g)$ and $\tau(g)$ only depend on $g$ and not on $u$ or $i$, without specifying a priori what that dependence should be. Other external factors, like contact separation distance and blowing gas flow obviously also impact the arc. The Mayr equation (1) assumes those external conditions remain constant.

This arc model leads to the following behaviour: All changes of external conditions — namely how much current flows through the arc — that take much longer than $5\,\tau$ can be considered quasi-static. The difference between $P_{cool}$ and $u \cdot i$ will always be virtually zero, and the system will be in equilibrium any time.

On the other hand, current variations that happen on a time scale much shorter than $\tau$ do not allow the arc to react. The conductance will not change, and

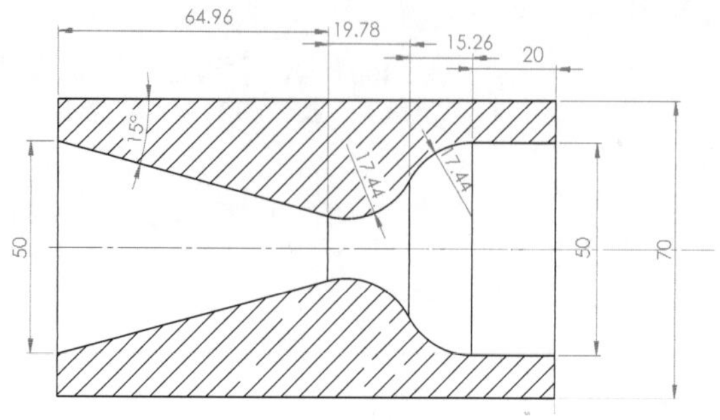


*Figure 1. Nozzle used for the experiments, made from PMMA to allow optical observation. Gas flow is from right to left. The downstream contact tip is flush with the nozzle exit plane, the upstream contact protrudes a few* mm *into the nozzle segment that is still straight.*

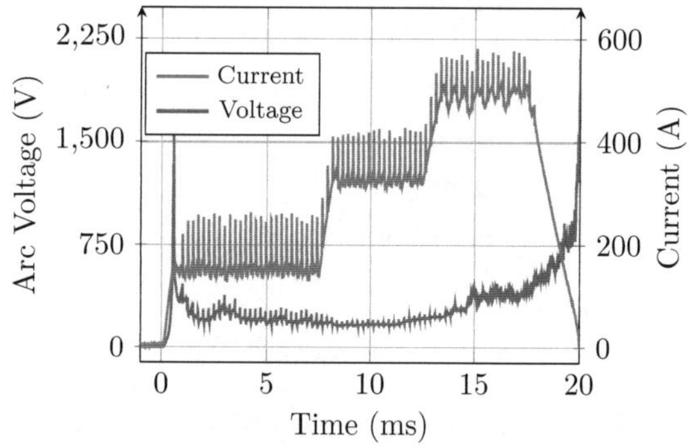


*Figure 2. Current and arc voltage during measurement to determine arc constant $\tau$ at three current levels.*

the arc will behave like a resistor [10]. Any method that aims to determine the arc time constant should create current changes on a time scale that is between these two extreme regions.

## 3. Experimental Methods

The DUT is a model interrupter using air, described in more detail in [2]. A converging-diverging nozzle with single flow direction is used, as shown in figure 1. The contacts are stationary and have a diameter of 22 mm. The arc is initiated using an ignition wire. The blowing system uses pressure stored in gas bottles and fast acting valves.

A flexible pulsed direct current source (FPDCS) [9, 17] is used to generate the current waveforms for testing. The current shape can be formed by up to three independently acting buck-converter modules.

In the performed tests, a base current level was kept by a module with large inductance (4.5 mH). On this, fast rising spikes are superimposed by a second module with low inductance (0.2 mH). The voltage and current traces for a measurement with three different base current levels are illustrated in Figure 2.

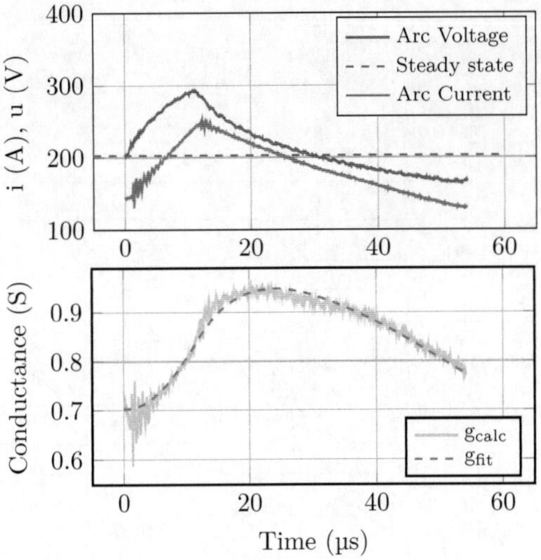


*Figure 3. Close-up of a single spike, used to determine arc time constant $\tau$.. Top plot shows measured values u and i, bottom plot shows $g_{calc} = i/u$ and the fit for $\tau = 36\,\mu s$, $P_{cool,1}(0.71\,S) = 29\,kW$ and $P_{cool,2}(0.95\,S) = 45\,kW$.*

Figure 3 shows measured voltage and current for a representative single spike. The conductance calculated from voltage and current, as well as the fit by the described algorithm are illustrated as well.

## 4. Fitting Algorithm

Determination of $\tau$ means finding a solution for which equation (1) accurately reproduces the measured results, without using any analytical formula that expresses $\tau$ as a function of other parameters. This requires experimental data of dynamic arc behavior, i.e. current and voltage waveforms of an arc that is perturbed from a steady state. In the present study, a quasi-static DC current (gradient lower than $1\,\text{A}\,\mu\text{s}^{-1}$) is superimposed with transient current impulses of triangular wave-shape of high gradient (greater then $10\,\text{A}\,\mu\text{s}^{-1}$), as shown in Figure 2. If the rise time and fall time of the impulse are in the order of magnitude of $\tau$, it is possible to determine $\tau$, by a method proposed in [9].

First, a value of $\tau$ is assumed. Then, the conductance and voltage response of an arc with the assumed value of $\tau$ is calculated with equation (1). In a third step, the calculated conductance waveform is compared to the measured values $i/u$, by calculating the root mean square difference. This process is repeated for different values of $\tau$, until a minimum difference is found.

In the present work, an algorithm to solve the Mayr equation was implemented, tested and used to perform the fit. A key requirement to the algorithm is to avoid the need to calculate derivatives of measured signals. Differentiating measured current and voltage signals accurately and with sufficient bandwidth is impossible, due to the inherent decrease of signal to noise ratio of derivatives. The optimized fitting algorithm instead uses the forward Euler method to solve equation (1), which makes the following assumptions:

- □ The arc is in steady state before the current transient, therefore all variables at $t_0$ are known $g_0 = i_0/u_0$, $P_{cool,0}(g_0) = u_0 i_0$
- □ During the current transient changes in $g$ are small enough so $\tau(g)$ can be considered constant.
- □ During the current transient, $P_{\text{cool}}(g)$ can be linearized, by specifying a second point $P_{\text{cool},1}(g_1)$ additional to the starting point $P_{cool,0}(g_0)$.
- □ There are no random fluctuations during the transient period due to other reasons that would also alter $P_{\text{cool}}(g)$, like arc root movement etc.

With these assumptions, initial conditions and assumed values for $\tau$ and $P_{\text{cool},1}(g_1)$, equation (1) can be used to solve for the conductance waveform $g(t)$, if the current waveform $i(t)$ is given. This is done by iteratively calculating $\dot{g}(t)$, $g(t)$ and $u(t)$ for one point in time, using the values from the previous time step. To minimize the error, time steps much smaller than $\tau$ have to be used. Since the sample rate of current and voltage is high (>50 MS) to reject common mode noise of the source (see [17]), this is not a problem.

The result for idealized current waveforms is shown in Figure 4. The response of the arc depends on the slope and duration of the current spikes, relative to $\tau$. For the first spike $\tau$ was set to 1/4 of the rise time, which means the conductance lags behind the instantaneous value, but has a similar shape. For the second spike $\tau$ is equal to the rise time. Here, the lag of the conductance is so significant that the shape of the waveform starts to differ from the hypothetical steady state curve. This also leads to bigger changes of the arc voltage. The last spike has a $\tau$ four times longer than the rise time, meaning the current rises and falls so fast that the conductance can hardly follow. Peak conductance is reached almost at the end of the spike, and afterwards exponential decay follows.

To aid interpretation, dashed lines show the voltage and conductance values one would obtain for an arc that is always in steady state, i.e. for $\tau = 0$. Dotted lines represent the resistive case, in which the arc conductance does not change at all, i.e. for $\tau = \infty$. In this picture it becomes clear why a rise time around $\tau$ is favoured. For this spike, the conductance changes significantly, but at the same time it is also significantly delayed compared to the steady curve. The voltage curve is in between the two extremes during the rising edge, and shows significant undershoot on the falling edge. The left spike with rise time $4\tau$ is too slow for the arc to be transient. Its conductance follows the steady state curve closely, which leads only to a small deviation of the voltage from the static $u(i)$ value. This makes it hard to determine a numeric value for $\tau$, especially if the deviation of the voltage

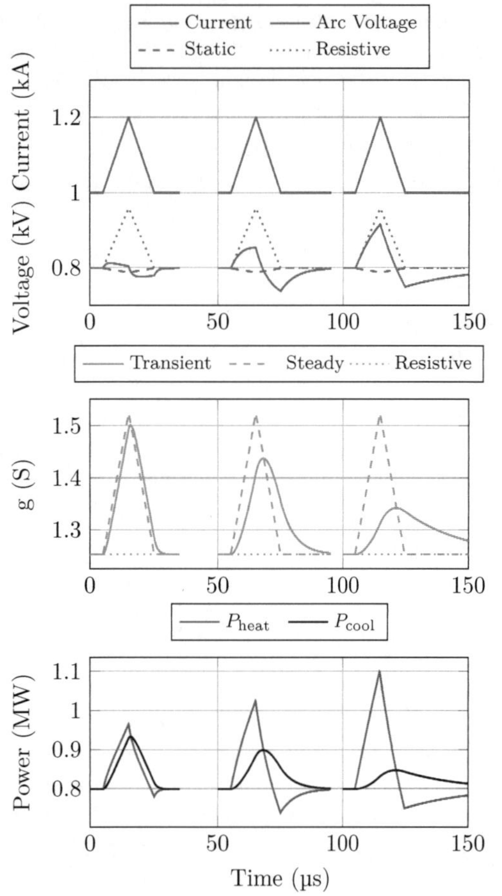


*Figure 4. Simulated response of an arc to triangular current impulses with 10 µs rise and fall time. Arc time constant τ is changed for each spike (2.5 / 10 / 40 µs).*

curve becomes small enough to be in the order of the inductive voltage drop between the two voltage probes. The rightmost spike is too fast for the arc to properly react at all. Its conductance does not change as much, and the voltage follows the resistive curve.

## 5. Results

To test the sensitivity of the presented method, several measurements have been conducted and analyzed.

To introduce a considerable change in the arc's thermal inertia, in the first experiment the transition between free burning and blown conditions at fixed base current level are observed. Figure 5 illustrates pressure build-up in the nozzle throat, arc voltage and current as well as $\tau$ for this measurement. Compared to the blown phase, cooling is considerably lower in the free burning phase. As expected, this manifests in a lower arc voltage. Optical investigations with a high speed camera show a large luminescent region. When gas flow is initiated at around 15 ms, this region shrinks which coincides with a decreasing time constant, even before there is significant pressure change at the throat. As pressure builds up, the arc voltage

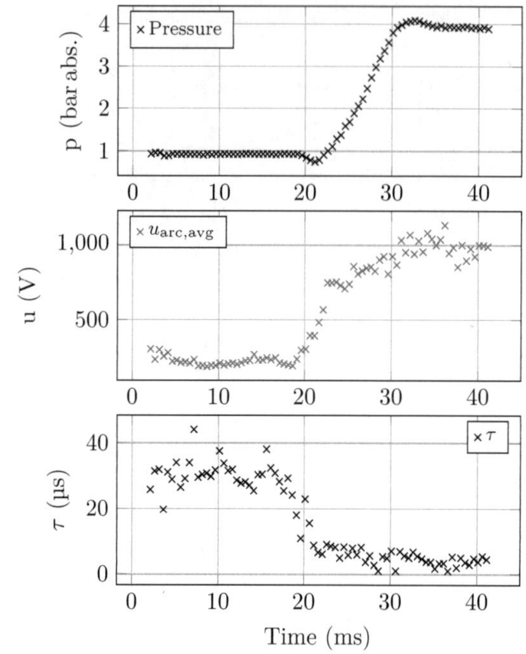


*Figure 5. τ for variable pressure for a base current of 130 A and a spike amplitude of 100 A. Gasflow starts accellerating around 16 ms as seen by highspeed footage.*

increases until the pressure reached its new steady state value.

The development of $\tau$ is reasonable. A smaller, more confined arc is able to exchange energy with its surrounding faster, resulting in a reduced thermal inertia. Arc radius as well as measured $\tau$ start to decrease considerably as soon as gas flow is observed on the highspeed video, which is already a few ms before the flow chockes and pressure build-up is measured, as shown in figure 6. Consequently, the measurements can be seen as a proof of concept for the presented method.

In a second step, blowing pressure is kept constant and the arc's time constant is evaluated for a fixed current level. The results for 115 spikes are illustrated in in Figure 7. It can be seen that the arc time constant under these conditions is mainly around $\tau = 5\,\mu s$ with values ranging from $\tau = 0\,\mu s \ldots 10\,\mu s$. Without further investigations it cannot be concluded if this spread represents real variations in arc time constants or the uncertainty of the method itself.

## 6. Conclusions

An important characteristic of (switching) arcs is the time constant with which the arc can exchange thermal energy with its surrounding and therefore change its conductance.

For this article, a method to determine thermal inertia (i.e. time constant) of arcs by using triangular impulse currents has been implemented. In example

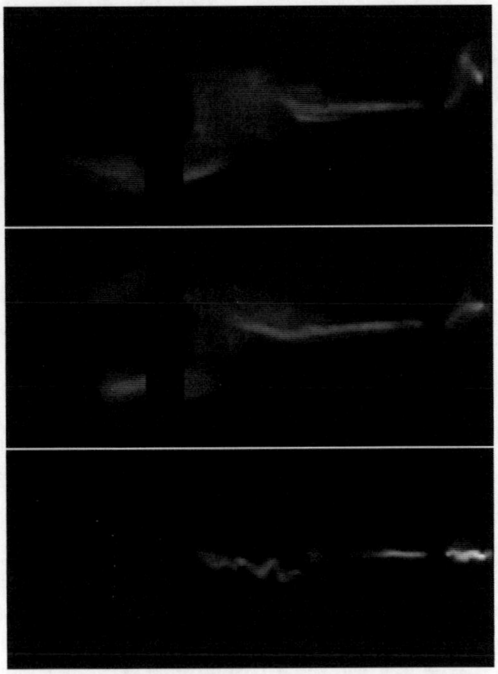

*Figure 6. Three frames form highspeed video, showing appoximately the same part as figure 1. Top: Freeburning stage, captured at 13.120 ms, Center: transition, flow already visible upstream arc starts contracting, captured at 19.125 ms. Bottom: Chocked flow conditions reached, arc radius fully contracted, caputured at 24.620 ms. Exposure time was 0.16 µs in all cases.*

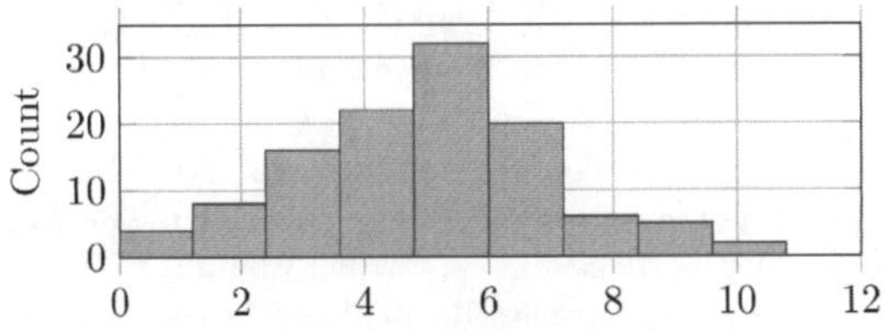


*Figure 7. Histogram of the fitted arc time constant $\tau$ for a base current of 140 A ($N_{total}$ = 115 spikes).*

measurements, thermal inertia of an axially blown switching arc has been investigated for the transition from free-burning to blown. The results align well with the expected development, which makes this method a promising tool for future investigations. The logical next improvement of the method is a more detailed analysis of the scattering of $\tau$ shown in figure 7, especially to distinguish between real arc properties and experimental uncertainty.

## Acknowledgements

The authors would like to thank ABB Switzerland for financial support, and the master student Benjamin Hammerich for his help during experiments. This research is part of the activities of the Swiss Centre for Competence in Energy Research on the Future Swiss Electrical Infrastructure (SCCER-FURIES).

## References

[1] A. N. Greenwood and T. H. Lee. Theory and application of the commutation principle for hvdc circuit breakers. *IEEE PAS*, PAS-91(4):1570–1574, July 1972. doi:10.1109/TPAS.1972.293310.

[2] T. Schultz et al. Improving interruption performance of mechanical circuit breakers by controlling pre-current-zero wave shape. *IET High Voltage*, pages 1–9, apr 2019. doi:10.1049/hve.2018.5103.

[3] J. Schwarz. *Berechnung von Schaltvorgängen mit einer zweifach modifizierten Mayr-Gleichung*. Dissertation, Technische Hochschule Darmstadt, 1973.

[4] R. Amsinck. Verfahren zur Ermittlung des Ausschaltverhalten bestimmender Lichtbogenkenngrössen. *ETZ-A*, 98:566, 1977.

[5] H. Rijanto. *Ein experimentelles Verfahren zur Bestimmung von Lichtbogenkenngrößen - Stromüberlagerungsverfahren*. Dissertation, Technische Universität Hannover, 1975.

[6] R. Ruppe. *Experimentelle und theoretische Untersuchungen am axial beströmten Wechselstromlichtbogen vor dem Stromnulldurchgang*. Dissertation, TH Illmenau, 1979.

[7] H. Drebenstedt et al. Ein verbessertes Verfahren zur Bestimmung der charakteristischen Funktionen des Zweipolmodells für Schaltlichtbögen. *Internat. Wiss. Kolloquium, TH Ilmenau*, 28, 1983.

[8] M. M. Walter, C. Leu, and C. M. Franck. Optimizing the Arc Characteristics for Improved Designs of Passive Resonant Hvdc-Circuit Breakers. In *XXth Symposium on Physics of Switching Arc*, pages 317–322, 2013. doi:10.3929/ethz-b-000078010.

[9] M. M. Walter. *Switching Arcs in Passive Resonance HVDC Circuit Breakers*. PhD thesis, ETH Zürich, 2013. doi:10.3929/ethz-a-010112102.

[10] O. Mayr. Beiträge zur Theorie des statischen und des dynamischen Lichtbogens. *Archiv für Elektrotechnik*, 37(12):588–608, dec 1943. doi:10.1007/BF02084317.

[11] A. N. Cassie. Arc rupture and circuit severity: a new theory. *CIGRE-Report*, 102:1–16, 1939.

[12] U. Habedank. Application of a new arc model for the evaluation of short-circuit breaking tests. *IEEE PWRD*, 8(4):1921–1925, 1993. doi:10.1109/61.248303.

[13] P. H. Schavemaker and L. van der Sluis. An improved Mayr-type arc model based on current-zero measurements [circuit breakers]. *IEEE PWRD*, 15(2):580–584, apr 2000. doi:10.1109/61.852988.

[14] M. Kapetanović. *High voltage circuit breakers*. ETF - Faculty of Eletrotechnical Engineering, Sarajevo, Sarajevo, 2011.

[15] K. Nakanishi. *Switching Phenomena in High-Voltage Circuit Breakers*. Marcel Dekker Inc., New York, NY, 1 edition, 1991.

[16] A. Khakpour et al. An Improved Arc Model Based on the Arc Diameter. *IEEE PWRD*, 31(3):1335–1341, jun 2016. doi:10.1109/TPWRD.2015.2473677.

[17] A. Ritter, L. S. J. Bort, and C. M. Franck. Five years of pulsed current testing for HVDC switchgear. In *2016 IEEE ICHVE Conference*, pages 1–4. IEEE, sep 2016. doi:10.1109/ICHVE.2016.7800658.

doi:10.14311/ppt.2019.2.180
*Plasma Physics and Technology* **6**(2):180–183, 2019 

# GENERATION OF SILVER NANOPARTICLES BY THE PIN-HOLE DC PLASMA SOURCE WITH AND WITHOUT GAS BUBBLING

Z. Kozáková*, F. Krčma, L. Čechová, S. Simić, L. Doskočil

*Faculty of Chemistry, Brno University of Technology, Purkyňova 118, 612 00 Brno, Czech Republic*

* kozakova@fch.vut.cz

**Abstract.** Silver nanoparticles were produced using the pin-hole discharge generated by dc non-pulsing high voltage directly in silver nitrate solutions. Sodium nitrate was alternatively added to increase solution conductivity and decrease input energy for the discharge breakdown. Argon or oxygen was bubbled through the discharge region. Comparative experiments were evaluated by UV-VIS spectrometry. Formation of silver nanoparticles with the average size of 100 nm was confirmed by SEM/EDS analysis.

**Keywords:** plasma in liquids, pin-hole discharges, silver nanoparticles, silver nitrate, gas bubbling, SEM/EDS analysis.

## 1. Introduction

Particles with specific size in the range between $10^{-9}$ to $10^{-7}$ m in diameter are called nanoparticles (NPs) [1, 2]. Materials in these dimensions have different properties than the bulk material and this is one of many reasons why the contemporary worldwide research is focused just on nanoparticles. Properties of NPs are dependent on their composition, size, shape, and other parameters like internal structure. The metallic nanomaterials (including the alloyed ones) [3, 4] can be synthetized as well as semiconductors [5] or organics (like graphene [6] or polymers [7]). Because of NPs unique properties, these materials are used in many application fields such as photovoltaics [8], electronics [9], diagnostics sensors [10], conductive inks [11], nanocomposites [12, 13], surface treatment [14], etc. A specific application field is in the medical therapy [15, 16] and health care. Thanks to their antibacterial effects, NPs are incorporated in biomedical devices, textiles, footwear or wound protections.

One of the means how to prepare NPs is the direct application of electrical discharge in liquids. Formed plasma channels are usually propagating through the liquid phase by the discharge itself with simultaneous discharge radiation. The discharge usually propagates from one electrode into the liquid and thus, the second electrode is formed by the gas (plasma) liquid surface. An advantage of plasma systems generated inside the liquid is the possibility to control the gaseous phase (even if some gas is bubbling into the system). Further, the whole system is efficiently self-mixed which helps to uniform the size of formed NPs. The precursors can be dissolved in the liquid or can be dissolved/sputtered from the electrode materials.

This paper deals with synthesis of silver nanoparticles by a special pin-hole electrode system using DC high voltage up to 1 kV which gave the mean power up to 150 W. Silver nitrate solutions were used as the precursor. Additionally, two gases (argon and oxygen) were bubbled through the main electrode into the plasma region in order to study effects of their flow rate on the NPs production.

## 2. Experimental setup

The electrode system of plasma reactor (Figure 1) consisted of two electrodes [17]. The main jet electrode was constructed by a tungsten wire (diameter of 1.0 mm) which was inserted into a dielectric cylindrical rod made of Macor ceramics with outer diameter of 10 mm and with one conical end. An orifice with the diameter of 1.2 mm was made in the ceramics along its longitudinal axis. A small gap of 1 mm was kept between the end of the wire electrode and the end of the dielectric rod. This configuration substantially increased electric field intensity and thus it allowed plasma ignition at relatively low applied voltage. An outer glass tube served as a holder and as a gas inlet. The second electrode was a planar plate made of platinum with dimensions of $20 \times 100\,\mathrm{mm}^2$. Both electrodes were immersed in a vessel containing 500 ml of silver nitrate solution. $AgNO_3$ concentration varied from 25 to 150 mg $l^{-1}$ (initial conductivity of 20–100 µS $cm^{-1}$). In some experiments, sodium nitrate (100 or 200 mg $l^{-1}$) was added in order to increase initial conductivity up to 300 µS $cm^{-1}$ and to decrease the input energy for the discharge breakdown. Influence of gas bubbling through the discharge region on the particle formation was studied in two gases (argon and oxygen) with flow rates in the range of 10–50 ml $min^{-1}$.

The system was supplied by dc non-pulsing high voltage up to 1 kV with the negative polarity on the main electrode. The discharge itself showed self-pulsing operation [17] giving the mean input power up to 150 W.

Treated solutions were sampled every 30 seconds during the discharge operation for UV-VIS spectrom-

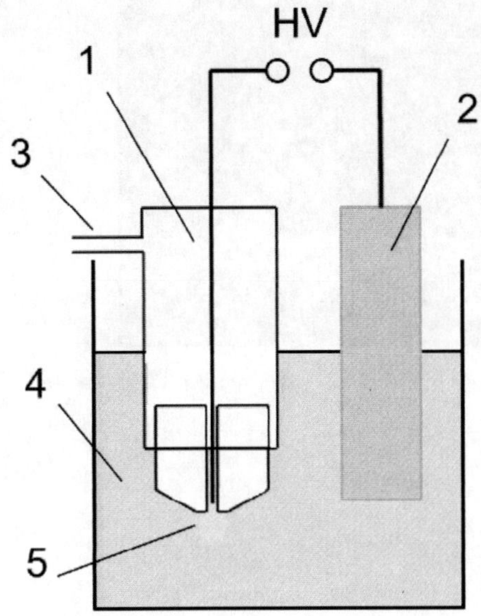


*Figure 1. Simplified scheme of plasma reactor: 1 – jet electrode, 2 – counter electrode, 3 – gas inlet, 4 – glass vessel with $AgNO_3$ solution, 5 – active plasma region.*

etry. Absorption spectra of sample solutions were recorded by the Helios Omega spectrometer in the range of 300–600 nm. Final solutions after the total two minute treatment were prepared for SEM and EDS analyses. These samples were obtained by evaporation of 20 drops of the final solutions on special aluminium holders. Detailed photos of formed nanoparticles were taken by the scanning electron microscope JEOL JSM-7600F with magnification of 3 000x and 50 000x, respectively. Material composition of nanoparticles was confirmed by the electron disperse spectrometer with the accelerating voltage of 10 kV.

## 3. Results

Formation of silver nanoparticles by the pin-hole discharge in silver nitrate solution was confirmed by SEM/EDS analysis. Figure 2(a) presents photographs in the smaller magnification of 3 000x where Ag NPs are visible as tiny dots spread among larger aggregates. These larger structures might have been formed both during the discharge treatment as well as the sample preparation (evaporation) for the SEM analysis. In the higher magnification of 50 000x (Figure 2(b)), separate Ag NPs with dimensions around 100 nm are well visible. Material of formed particles detected by EDS analysis was composed mostly of silver with some traces of aluminium (from the holder) and carbon (human contamination).

Typical absorption spectra of $AgNO_3$ solution recorded during the discharge treatment without any gas bubbling are shown in Figure 3. The initial sample of pure $AgNO_3$ solution does not absorb in the studied wavelength range around 400 nm which is the typical wavelength of silver nanoparticles absorption reported in the literature [18]. Due to the dc discharge operation, silver nanoparticles are formed in the plasma region surrounding the main jet electrode (cathode). Further, they are spread into the bulk solution by the discharge self-mixing properties. Thus, the absorbance of colloidal silver solution is increased linearly during the discharge treatment with the maximal absorption at 410 nm.

Influence of initial $AgNO_3$ concentration in the solution on formation of Ag NPs was studied in the concentration range of 25–150 $mg\,l^{-1}$. No gas was added in this series of experiments. According to the final absorbance of silver particles at 410 nm (Figure 4), it is obvious that formation of Ag NPs after two minute discharge treatment is increased up to the initial concentration of 100 $mg\,l^{-1}$. However, it is substantially decreased by higher concentration. We assume that with the increasing silver nitrate concentration, higher amount of silver particles is formed but they easily aggregate into larger structures. These heavier structures come faster under sedimentation and thus absorption of the final solution is decreased. Therefore, it is important to adjust the optimal initial concentration of the precursor $AgNO_3$ (until 100 $mg\,l^{-1}$) in order to produce homogeneous solution with silver nanoparticles.

As the initial conductivity of the primary silver nitrate solution reached only few tens of $\mu S\,cm^{-1}$ depending on the initial concentration, sodium nitrate of two concentrations was added in order to increase the initial conductivity. It is already known [19] that ignition as well as operation of dc discharges is strongly dependent on initial conductivity. Therefore, by the addition of sodium nitrate (concentration of 100 or 200 $mg\,l^{-1}$) we have reached a stable discharge operation even at 40 W instead of 100 W. However, final production of Ag NPs measured by the absorption at 410 nm seems much lower with the $NaNO_3$ addition (Figure 5).

Influence of additional gas bubbled through the discharge region on Ag NPs formation was studied for two gasses. Argon was selected as an inert representative and oxygen as a reactive gas. We had supposed that by the ionisation of argon, the transfer of plasma energy for reduction of silver ions to nanoparticles should have been more effective. However, the experimental effect was totally opposite (Figure 6). Maximal absorbance at 410 nm recorded for final $AgNO_3$ solutions after two minute discharge operation significantly decreases with the increasing flow rate of argon. The problem is probably in the formation of huge bubbles which interrupt the reduction process. The same trend was observed also for oxygen addition (Figure 7) where this suppressing effect was even stronger. The lower

*Figure 2. SEM photographs of silver nanoparticles formed after two minute discharge treatment (100 W) in $AgNO_3$ solution (100 mg $l^{-1}$); magnification: a) 3 000x, b) 50 000x.*

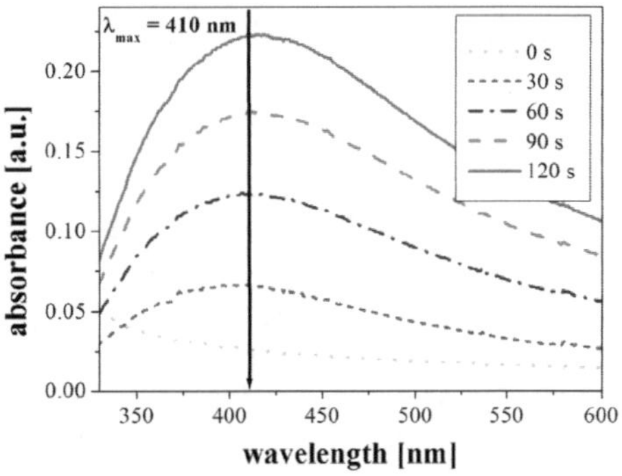


*Figure 3. Absorption spectrum of $AgNO_3$ solution (100 mg $l^{-1}$) during the dc discharge treatment (100 W) without the gas bubbling.*

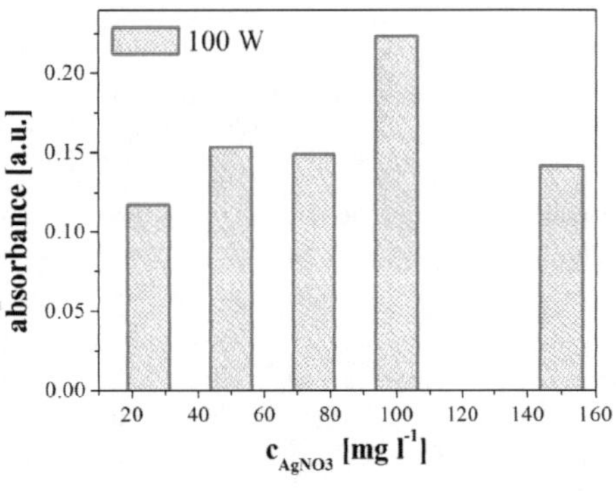


*Figure 4. Comparison of NPs absorbance (at 410 nm) after two minute discharge treatment (100 W) in $AgNO_3$ solutions with different initial concentration and without the gas bubbling.*

production of Ag NPs in presence of oxygen could be probably explained by higher variety of chemical reactions combining oxygen and water molecules which leads to the dissipation of energy to these mechanisms instead of the charge transfer needed for the silver reduction.

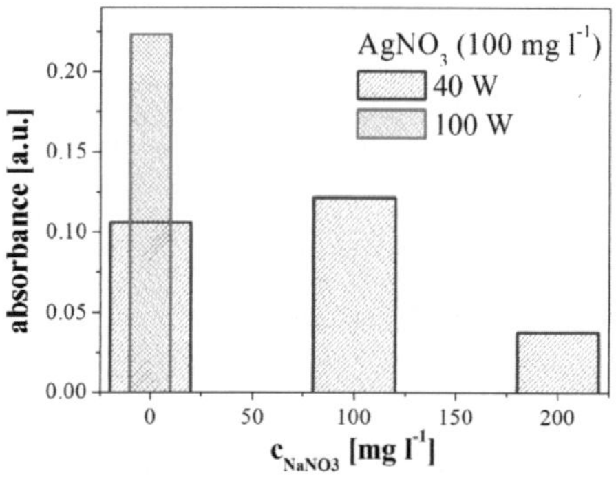


*Figure 5. Comparison of NPs absorbance (at 410 nm) after two minute discharge treatment (40 or 100 W) in $AgNO_3$ solutions (100 mg $l^{-1}$) with different $NaNO_3$ addition and without the gas bubbling.*

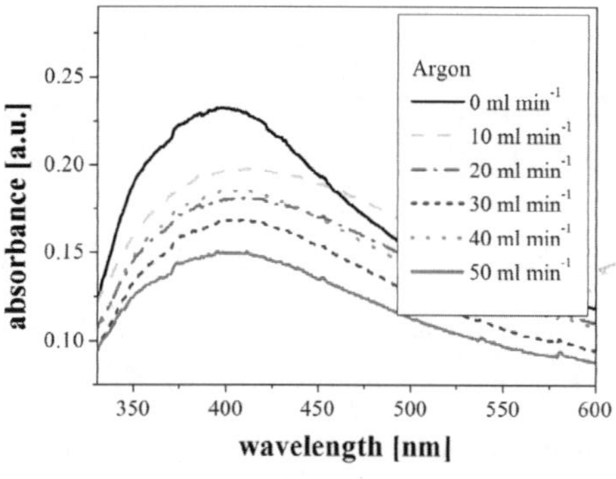


*Figure 6. Absorption spectrum of $AgNO_3$ solution (100 mg $l^{-1}$) after two minute discharge treatment (100 W) with different flow rate of argon.*

## 4. Conclusions

Silver nanoparticles were produced using the pin-hole discharge generated by dc non-pulsing high voltage directly in water solutions containing silver nitrate as the nanoparticle precursor. Formation of silver

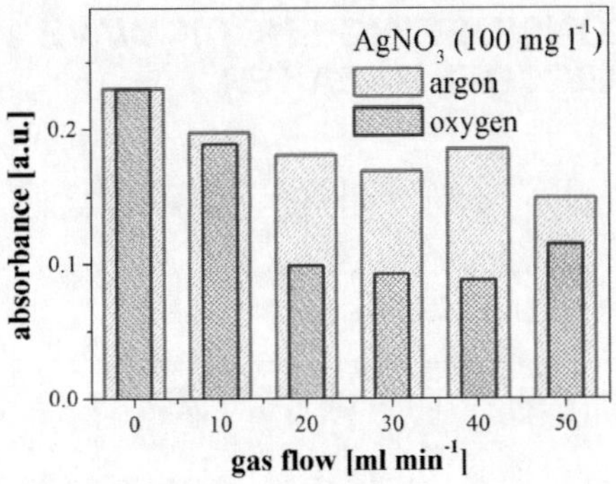


*Figure 7. Comparison of NPs absorbance (at 410 nm) after two minute discharge treatment (100 W) in $AgNO_3$ solutions (100 mg $l^{-1}$) with different flow rate of argon and oxygen.*

nanoparticles (mean size of 100 nm) was confirmed by SEM/EDS analysis. Further, UV-VIS spectrometry (absorption maximum at 410 nm) was involved for comparative experiments. Maximal production of Ag NPs was achieved at optimal $AgNO_3$ concentration of 100 mg $l^{-1}$ and applied power of 100 W. Addition of sodium nitrate, which increased the initial solution conductivity as well as decreased the minimal applied power, negatively influenced the Ag NPs formation. Also additional gas bubbling through the discharge region has a negative effect on the nanoparticle formation, both for argon and oxygen. Especially in case of oxygen, dissipation of applied energy into variable chemical reactions instead of to the reduction process was assumed.

## References

[1] M. Naito et al. *Nanoparticle technology handbook.* Second printing. Elsevier, 2012.

[2] Z. Abdullaeva. *Synthesis of nanoparticles and nanomaterials: Biological approaches.* Springer International Publishing, 2017.

[3] D. L. Fedlheim and C. Foss. *Metal nanoparticles: Synthesis, characterization, and applications.* CRC Press, 2001.

[4] G. Sharma et al. Revolution from monometallic to trimetallic nanoparticle composites, various synthesis methods and their applications: A review. *Mat. Sci. Eng. C Mater.*, 71:1216–1230, 2017. doi:10.1016/j.msec.2016.11.002.

[5] A. Rahnama and M. Gharagozlou. Preparation and properties of semiconductor CuO nanoparticles via a simple precipitation method at different reaction temperatures. *Opt. Quant. Electronics*, 44:313–322, 2012. doi:10.1007/s11082-011-9540-1.

[6] W. Choi and J. Lee. *Graphene: Synthesis and applications.* CRC Press, 2011.

[7] P. R. Sharma et al. Spherical shaped nanoparticles of cellulose and its derivatives: A short review. *Trends Carbohydrate Res.*, 7:1–5, 2015.

[8] B. Pejjai et al. Status review on earth-abundant and environmentally green Sn-X (X = Se, S) nanoparticle synthesis by solution methods for photovoltaic applications. *Int. J. Hydrogen Energy.*, 42:2790–2831, 2017.

[9] S. A. Paknejad and S. H. Mannan. Review of silver nanoparticle based die attach materials for high power/temperature applications. *Microelectron. Reliab.*, 70:1–11, 2017. doi:10.1016/j.microrel.2017.01.010.

[10] E. Roy et al. Developing electrochemical sensor for point-of-care diagnostics of oxidative stress marker using imprinted bimetallic Fe/Pd nanoparticle. *Talanta*, 132:406–415, 2015. doi:10.1016/j.talanta.2014.09.033.

[11] E. V. Agina et al. Polymer surface engineering for efficient printing of highly conductive metal nanoparticle inks. *ACS Appl. Mater. Inter.*, 7:11755–11764, 2015. doi:10.1021/am508905t.

[12] E. Nezbedová et al. Effect of particles size on mechanical properties of polypropylene particulate composites. *Int. J. Structural Integrity*, 7:690–699, 2016.

[13] M. Jouni et al. A representative and comprehensive review of the electrical and thermal properties of polymer composites with carbon nanotube and other nanoparticle fillers. *Polym. Int.*, 66:1237–1251, 2017. doi:10.1002/pi.5378.

[14] H. Teisala et al. Review on liquid flame spray in paper converting: Multifunctional superhydrophobic nanoparticle coatings. *Nord. Pulp Pap. Res. J.*, 29:747–759, 2014.

[15] N. K. Rajendran et al. A review on nanoparticle based treatment for wound healing. *J. Drug Deliv. Sci. Tec.*, 44:421–430, 2018.

[16] D. Peukert et al. Metallic nanoparticle radiosensitisation of ion radiotherapy: A review. *Phys. Medica*, 47:121–128, 2018. doi:10.1016/j.ejmp.2018.03.004.

[17] F. Krčma et al. Characterization of novel pin-hole based plasma source for generation of discharge in liquids supplied by DC non-pulsing voltage. *Plasma Sources Sci. Technol.*, 27(6):065001, 2018. doi:10.1088/1361-6595/aac521.

[18] D. Paramelle et al. A rapid method to estimate the concentration of citrate capped silver nanoparticles from UV-visible light spectra. *Analyst*, 139(19):4855–4861, 2014.

[19] F. Krčma, Z. Stará, and J. Prochazková. Diaphragm discharge in liquids: Fundamentals and applications. *J. Phys.: Conf. Ser.*, 207(1):012010, 2010. doi:10.1088/1742-6596/207/1/012010.

doi:10.14311/ppt.2019.2.184
*Plasma Physics and Technology* **6**(2):184–187, 2019 

# PERFORMANCE EVALUATION METHOD CONSIDERING SHOCK WAVE IN NOZZLE FOR HIGH-VOLTAGE CIRCUIT BREAKER

M. Terada[a, c, *], H. Urai[a], T. Nishimura[b], Y. Yokomizu[c]

[a] *Research & Development Group, Hitachi, Ltd., 7-1-1, Oomika-cho, Hitachi-shi, Ibaraki-ken, 319-1292, Japan*

[b] *Energy Business Unit, Hitachi, Ltd., 1-1-1, Kokubu-cho, Hitachi-shi, Ibaraki-ken, 316-8501, Japan*

[c] *Department of Electrical Engineering, Nagoya University, Chikusa-ku, Furo-cho, Nagoya-shi, Aichi-ken, 464-8603, Japan*

[*] masanao.terada.br@hitachi.com

**Abstract.** Interrupting-performance indexes that reflect the physical properties of gas flow are proposed for a breaker terminal fault of a high-voltage circuit breaker. A method of evaluating the shock wave and flow velocity was developed that uses a theoretical equation of supersonic flow which takes into account the nozzle's shape and stagnation pressure. The predicted flow velocities and positions at which the shock wave forms have strong correlations with the test results.

**Keywords:** supersonic flow, shock wave, flow velocity, circuit breaker.

## 1. Introduction

High-voltage circuit breaker was required to have a low operating energy for the sake of reliability and economy. Self-blast circuit breakers are widely used as they reduce the operating energy. They utilize a pressure build-up with ablation of a polytetrafluoroethylene (PTFE) insulation nozzle by arc energy, resulting in a minimization of the reaction force of mechanical compression.

Breaker terminal fault (BTF) test duties have a wide range of interrupting current. The self-blast circuit breaker utilizes ablation of the PTFE insulation nozzle by arc energy. Accordingly, the ratio of pressure rise to current is higher than that of conventional puffer-type circuit breakers. Therefore, at low currents, it is important that a small pressure rise in a pressure generating process is effectively converted into a flow generating process around zero current and hot gas is exhausted. An interrupter can be designed around a complex gas flow varying in time and space by using a computational fluid dynamics simulation technique taking into account the effects of arc heating and ablation [1–3]. Furthermore, the dielectric strength for transient recovery voltage (TRV) briefly after current interruption in a BTF can be evaluated by combining the flow simulation with an electric-field simulation [4, 5]. The dielectric strength is determined by the local density of the gas and the local electrical field. The dielectric strength is high enough if the space between electrodes is filled with a low-temperature, high-density gas. On the other hand, if the pressure is not high enough, there will be high-temperature and low-density spots between the electrodes, and in that case, the dielectric strength should be evaluated with a simulation technique.

Such a design requires a lot of trials with different structural parameters and exhaustive considerations. In this case, macroscopic evaluations focused on flow conditions will be effective. References [6, 7] evaluate the interrupting performance in correlation with the flow properties. The current study deals with supersonic flow inside the nozzle at the time around a peak value of the TRV briefly after current interruption, where performance indexes evaluating the dielectric strength are the flow velocity and shock wave position. As mentioned above, to increase the dielectric strength, the density in the nozzle should be high. One way of increasing the density is to decrease the temperature in the area between the electrodes. To do so effectively, it is important that the flow be smooth to promote the exhaust of hot gas. Thus, a stable and strong flow should be generated between the electrodes. Moreover, it is important for the shock wave to be generated on the downstream side so as to make the gas flow stable and to increase the flow velocity so as to make it powerfully flowing.

The purpose of this work is to introduce performance indexes evaluating the dielectric strength in the form of the flow velocity and the shock wave position for theoretically evaluating the physical properties of supersonic flows in nozzles in reference with the experimental data of whether the dielectric breakdown occurs or not with actual interrupting tests.

## 2. Flow conditions in region between electrodes

In the interrupting process, a supersonic flow forms in the divergence area of the nozzle and a shock wave is generated downstream of it. The gas flow is accelerated upstream of the shock wave and stalls downstream of the shock wave. Therefore, it is considered that there are two indexes related to dielectric strength briefly after current interruption. The first is the position of the shock wave. In particular, the shock wave should not form in front of the pin electrode. The second one is the maximum flow velocity in the region upstream of the shock wave.

Figure 1 shows the above ideas. Figure 1 (a) shows the case of a shock wave generated in front of the pin electrodes. In this case, the flow velocity at the tip of the pin electrode is small, resulting in hot gas remaining due to flow retention. Figure 1 (b) shows a shock wave formed to the rear of the pin electrodes. The drawing superimposes the shock wave at the front of the pin electrode, which acts as a barrier. In this case, the flow around the pin electrode is supersonic. In addition, the dielectric strength at the tip of the pin electrode is high because the area around that point in which hot gas is retained is narrow and at low temperature. Therefore, it is considered that the flow condition of Figure 1 (b) is suitable for increasing dielectric strength.

To generate a shock wave downstream of the nozzle divergence area, the pressure at the stagnation point should be increased. One way of increasing the pressure is to increase the amount of nozzle ablation and to reduce the flow resistance through the pressure build-up chamber. Another way is to narrow the nozzle divergence angle. However, the nozzle divergence angle should be enlarged to increase the flow velocity. The design of the nozzle should thus consider this trade-off relationship.

It is important for the evaluation to include the effect of varying the gas density and the temperature in combination with the electric field upstream and/or downstream of the shock wave. Figure 2 shows temperature distribution at 100 µs after current zero with CFD simulation in ANSYS Fluent which is based on the finite-volume method applied with the discrete ordinates radiation model and the realizable k-ϵ turbulence model. The nozzle ablation due to radiation heat transfer from the arc is also considered. The dielectric strength of this simulated structure is relatively low. The flow condition of the simulation result is close to the condition of the Figure 1 (a), that is, the shock wave is generated in front of the pin electrode resulting in hot gas remaining due to flow retention. One approach to design the high-voltage circuit breaker is to find the structure eliminating local weak points using with CFD simulation. Another approach is to determine the policy of the performance improvement evaluating the macroscopic performance-indexes since the interrupting process has a transient phenomena and a large variation. The target of this study is the latter approach.

## 3. Calculation method

The position of the shock wave and the flow velocity were calculated using the theoretical equation of supersonic flow. To evaluate the flow properties in terms of the pressure and the shape of the nozzle, it is considered that the pin electrode does not exist.

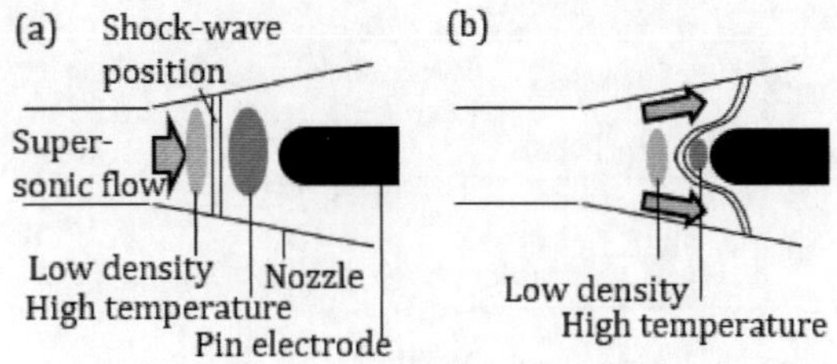


*Figure 1. Shock wave position. (a) shows a shock wave generated in front of the pin electrodes. (b) shows a shock wave generated to the rear of the pin electrodes.*

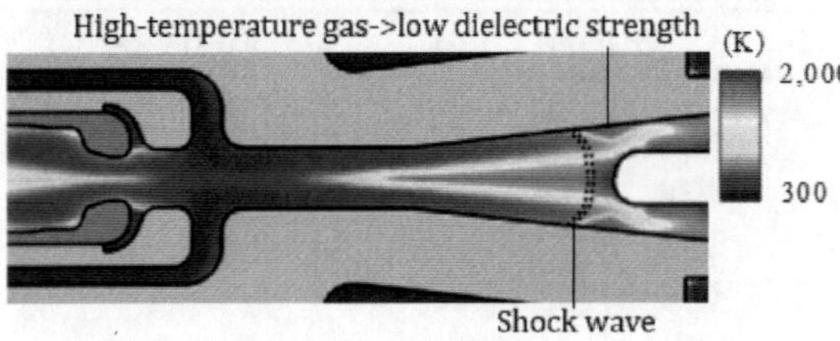


*Figure 2. Temperature distribution at 100 µs after current zero with CFD simulation.*

Figure 3 shows the calculation model for simulating supersonic flow.

The method of calculating the position of the shock wave is described below. Equations (1)–(3) show how the cross-sectional area ratio and Mach number are related to the stagnation pressure [8].

$$\frac{A_\mathrm{e}}{A^*} = \frac{\left(\frac{2}{\gamma+1}\right)^{\frac{1}{\gamma-1}} \left(\frac{P_0}{P_\mathrm{e}}\right)^{\frac{1}{\gamma}}}{\sqrt{\left(\frac{\gamma+1}{\gamma-1}\right)\left\{1-\left(\frac{P_0}{P_\mathrm{e}}\right)^{\frac{1-\gamma}{\gamma}}\right\}}} \tag{1}$$

$$M_\mathrm{e}^{\,2} = \frac{2}{\gamma-1}\left\{\left(\frac{P_0}{P_\mathrm{e}}\right)^{\frac{\gamma-1}{\gamma}} - 1\right\} \tag{2}$$

$$P_\mathrm{h} = P_0 \frac{2\gamma M_\mathrm{e}^{\,2} - (\gamma-1)}{\gamma+1}\left(1 + \frac{\gamma-1}{2} M_\mathrm{e}^{\,2}\right)^{-\frac{\gamma}{\gamma-1}} \tag{3}$$

where $P_0$, $A^*$, $A_\mathrm{e}$, $P_\mathrm{e}$, $M_\mathrm{e}$, $P_\mathrm{h}$, and $\gamma$ are the stagnation pressure, cross-sectional area of the nozzle throat, cross-sectional area of the nozzle at the shock wave's position, pressure upstream of the shock wave, Mach number upstream of the shock wave, pressure downstream of the shock wave, and ratio of specific heats, respectively. The above equations are used to calculate $A_\mathrm{e}$ at the shock wave, where $P_\mathrm{h}$ equals $P_\mathrm{b}$ witch is the back-pressure. The distance from the nozzle throat exit $L_\mathrm{s}$ to the position of the shock wave is expressed as

$$L_\mathrm{s} = \frac{\sqrt{\frac{A_\mathrm{e}}{\pi}} - \sqrt{\frac{A^*}{\pi}}}{\tan\theta} \tag{4}$$

where $\theta$ is the nozzle divergence angle.

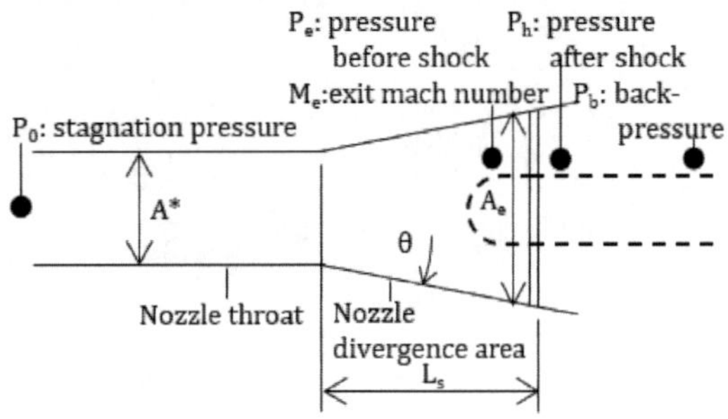


*Figure 3. Calculation model for supersonic flow.*

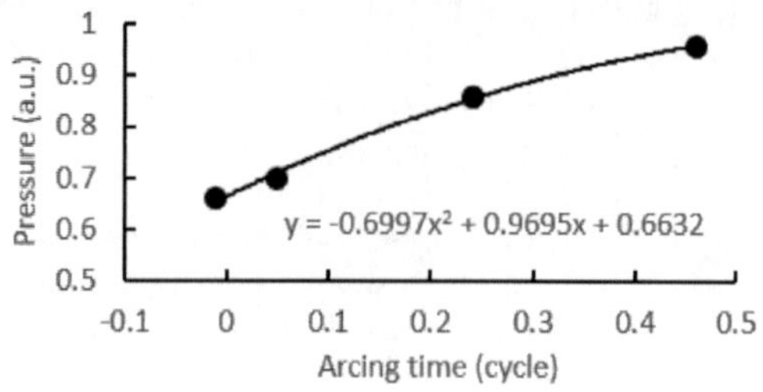


*Figure 4. Measured pressure in pressure build-up chamber.*

| No. | Th. len. (a.u.) | Div. angle (a.u.) |
|---|---|---|
| Case A | 1.0 | 1.0 |
| Case B | 0.9 | 1.0 |
| Case C | 1.0 | 1.2 |
| Case D | 0.9 | 0.9 |

*Table 1. Configurations of test structure.*

Second, the flow velocity $u_e$ under the back-pressure condition where the shock wave at the curved end of the tip of the pin is generated is expressed as

$$u_e = M_e c_e = M_e \sqrt{\gamma R T_e} \qquad (5)$$

where $M_e$, $c_e$, $T_e$, and $R$ are the Mach number upstream of the shock wave, the speed of sound, temperature at the shock wave's position, and the gas constant. The temperature at the shock wave's position is expressed as

$$T_e = \frac{T_0}{1 + \frac{\gamma - 1}{2} {M_e}^2} \qquad (6)$$

where $T_0$ is the temperature at the stagnation point. $T_0$ is constant even if the flow conditions change.

## 4. Results

The shock wave's position and the flow velocity were calculated using the method described above for BTF test structures. BTF tests were performed with a synthetic circuit which was composed with a current source supplying interrupting current and a voltage source applying TRV. The test conditions for the interrupting current and TRV were common to all cases. The correlation of these results with the test results was been evaluated.

Table 1 shows four representative test structures. Case A is a structure whose throat length and divergence angle are 1 (a.u.) and 1 (a.u.), respectively. Including the four test structures shown in Table 1, eleven test structures were built for the interrupting test. The following calculations of the shock wave position and flow velocity are for these seven cases.

In each case, several shots were performed over a range from the expected minimum arcing time to the maximum arcing time.

The stagnation pressure $P_0$ was set on the basis of the pressures measured with a piezoelectric pressure sensor in the pressure build-up chamber at current zero. The measured pressures of Case A in the pressure build-up chamber at current zero are presented in Figure 4. The horizontal axis in this figure is the arcing time, where the expected minimum arcing time is set to zero. The stagnation pressure was multiplied by 0.9 in consideration of the pressure loss from the chamber to the stagnation point. Assuming that the area of the PTFE surface close to the arc is proportional to the pressure in the chamber, the difference in throat length was determined on the basis of the pressure in Case A. The throat diameter was the same in all cases, as were the stroke characteristics of the electrode. This means that in all cases, the pin electrode was at the same position at a given arcing time.

Figures 5 and 6 show the calculated shock wave positions and flow velocities. The shock wave position was measured from the curved end of the tip of the pin electrode. The open and closed circles in the figures represent interrupting successes and failures, respectively. For the medium arcing times from 0.1 to 0.25 cycle, the cases showing positions at more than 9 mm were successful. For the short arcing times from around 0 to 0.1 cycle, the cases of more than 0.57 (a.u.) were successful. The predicted shock wave positions for the medium arcing times and predicted flow velocities for the short arcing times correlated well with the test results. However, the calculations for the arcing times longer than 0.25 cycle were not correlated with the test results. It is considered that the volume between the pin electrode and the nozzle was enlarged in the case of the longer arcing times, and this caused local weak points in the dielectric strength. The above criteria might be changing in the rated voltage. Thus, several experiments should be done before applying the method to other rated high-voltage circuit breaker.

Figures 7 (a) and 7 (b) show the position of the shock wave and the flow velocity in calculation, depending on the throat length and the divergence angle for Cases A–D. When the shock wave for a medium arcing time is located downstream, the flow velocity for the case of a short arcing time is small (Case A). In contrast, when the flow velocity for the short arcing

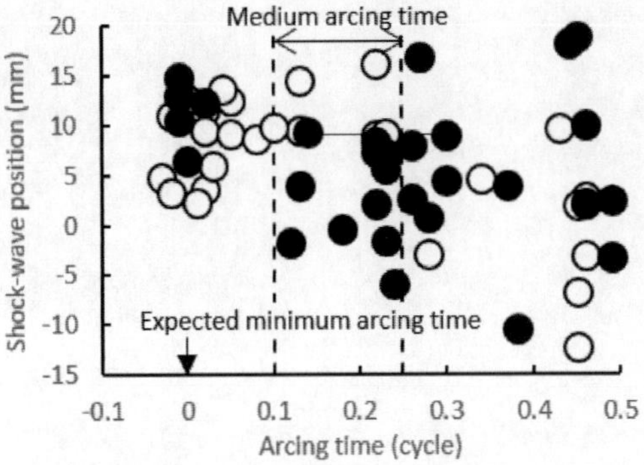


*Figure 5. Shock wave position.*

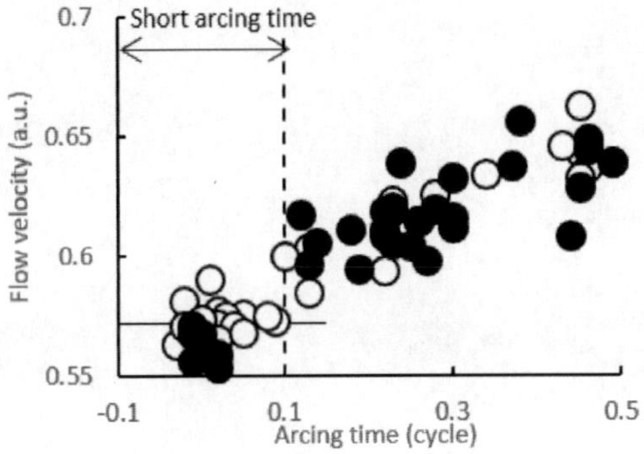


*Figure 6. Flow velocity.*

time is large, the shock wave for the medium arcing time is located upstream, in front of the tip of the pin electrode (Case B). Since the throat length of Case B is shorter than in Case A, the cross-sectional area of the tip of the pin electrode is larger and the shock wave is generated on the upstream side. As the divergence angle is widened, the flow velocity in the case of the short arcing time increases, whereas the shock wave in the case of the medium arcing time is generated on the upstream side (Case C). Cases B and C, whose shock waves are generated upstream, fail the interrupting test for the medium arcing time. On the other hand, Cases B and C whose flow velocities are large pass the interrupting test for the short arcing time. These results show that the dielectric strength from a short arcing time to a medium arcing time can be evaluated with the method described here. The dielectric strength in long arcing time should be evaluated with two dimensional CFD simulation. In the next step, the dielectric strength is evaluated using with CFD simulation.

## 5. Conclusions

Performance indexes evaluating the dielectric strength, i.e., the flow velocity and shock wave position, were introduced for theoretically evaluating the physical properties of supersonic flow in the nozzle in reference with interrupting test results. It was found that this method can be used to evaluate the dielectric strength in cases ranging from short to medium arcing times.

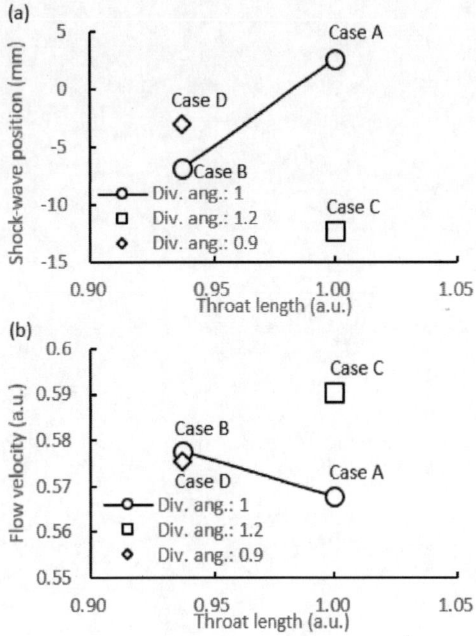


*Figure 7. Change in shock wave position (a) and flow velocity (b) depending on throat length and divergence angle.*

## References

[1] C. B. Ruchti and L. Niemeyer. Ablation Controlled Arcs. *IEEE Trans. Plasma Sci.*, 14:423–434, 1986. doi:10.1109/TPS.1986.4316570.

[2] P. Chévrier et al. Industrial applications of high-, medium- and low-voltage arc modeling. *J. Phys. D: Appl. Phys.*, 30:1346–1355, 1997. doi:10.1088/0022-3727/30/9/010.

[3] M. Seeger et al. An integral arc model for ablation controlled arcs based on cfd simulations. *J. Phys. D: Appl. Phys.*, 39:2180–2191, 2006. doi:10.1088/0022-3727/39/10/029.

[4] K. Bousoltane et al. Analysis of the breakdown in single-chamber circuit breaker. *18eme Congres Francais de Mecanique Grenoble*, 2007.

[5] D. Bešlija et al. Calculation of Steady $SF_6$ Gas Flow through a 420 kV Circuit Breaker Nozzle and Electric Field Distribution. *Journal of Energy and Power Engineering*, 8:1964–1973, 2014.

[6] K. Y. Park and M. T. C. Fang. Mathematical Modeling of $SF_6$ Puffer Circuit Breakers I: High Current Region. *IEEE Trans. on Plasma Science*, 24(2):490–502, 1996. doi:10.1109/27.510015.

[7] Y. Yokomizu et al. Dependence of Arc Interrupting Capability on Spatial Distribution of Airflow Velocity in Air-Blast Flat-Type Quenching Chamber. *IEEE Trans. on Power Delivery*, 18(1):101–106, 2003. doi:10.1109/TPWRD.2002.806687.

[8] H. W. Liepmann and A. Roshko. *Elements of Gas Dynamics*. Dover Publications, 2002.

**Gap in pagination due to unavailable papers.**

**Pages 188-193**

doi:10.14311/ppt.2019.2.194
*Plasma Physics and Technology* **6**(2):194–199, 2019 

# FROM SWITCHING ARCS TO BALL LIGHTNING TO CURING CANCER

J. J. Lowke

*CSIRO Manufacturing, PO Box 218, Lindfield NSW 2070, Sydney, Australia*

John.Lowke@csiro.au

**Abstract.** Previous modelling of switching has been through calculation of reductions in temperature of the arc at "current zero". "Enthalpy density"as a function of temperature is found to be an important property. New calculations now include an account of non-equilibrium electron density as a function of time through current zero and it is found that electron attachment rates, which are very large for $SF_6$, could be a dominant property. Modelling discharges is having other successes, for example in explaining "ball lightning"observations inside of houses and aircraft, which suddenly appear, usually at glass windows. Discharge modelling suggests these observations might be explained by the production of "singlet delta"metastable molecules of oxygen in electrical discharges in air. If metastable densities are sufficient, electrons can be produced from the detachment of negative ions to produce radiation and explain ball lightning. An exciting new development is that plasmas from electric corona in air have been found to reduce the size of cancer tumours. These excited oxygen molecules have also been proposed as having a role in this remarkable interchange between classical electrical engineering and medical science.

**Keywords:** switching arcs, current zero, $SF_6$, ball lightning, cancer.

## 1. Introduction - Computer Modelling

Naturally, of principal importance in the world, is "hardware", be it from the buildings of cities, power stations, the circuit breakers that control the power stations, or the iPhones that we all now use for everyday communications. But there is a new activity, only developed in the last 60 years of our civilisation, and that is computer modelling. There are hundreds of thousands, of people now involved in computer modelling which is also a principal activity behind a major fraction of the papers of this conference. Computer modelling has resulted in predictions of climate change, which are influencing economic policies of a significant fraction of the world. Computer modelling, based on previous representations of the world through mathematics, for example in the equations of electro magnetism, have now resulted in predictions of the very complex phenomenon of circuit interruption. Through the development of computers, predictions of the original analytic formula from mathematics have been extended to include numerical results giving details, frequently in three dimensions, of arc behaviour, be it interrupted or not, as a function of time.

Such predictions for switching arcs, based on the modelling of arc temperature, are discussed in Section 2. A new exciting development of such predictions of arc switching, is the inclusion of predictions of the electron density, discussed in Section 3. A further product of computer modelling is presented in Section 4, namely a proposed explanation of the centuries old phenomenon of Ball Lightning. This explanation results from the inclusion in the modelling of predictions of an additional particle to the particles that are usually considered, namely electrons, positive ions and negative ions. This additional particle is the metastable excited state of oxygen molecules, which have the technical name 'singlet delta'oxygen molecules. Section 2, relates modelling achievements of the past, whereas Sections 3 and 4 describe present developments of computer modelling. Section 5 outlines an area, which I am sure will result in future computer modelling, namely the use of gas discharges for the curing of cancer! It has recently been found that electrical discharge plasmas, when in contact with cancer tumours, result in a reduction in the size of the tumours, including also the possible reduction of the size of secondary tumours. Such a phenomenon raises the possibility of a completely new field: the interaction of electrical engineering plasmas with biological systems, for example through the interaction with blood and human systems of circulation.

## 2. Switching predictions from arc temperature

In a switch, or "circuit breaker", two electrodes are separated to produce an arc, which ideally is extinguished at the next current zero of the ac cycle because of the cooling of the arc during the low electric field that occurs at current zero. With increases in population and increased demands for electric power from power stations there have been increased demands on the capacity of circuit breakers. As a result there have been continuous improvements with time in, for example, the design of electrode configurations, the arc gas composition and details of imposed convective flow, all largely achieved by experiment.

With the development of computers in the 1960's, there were efforts to "model"the whole circuit inter-

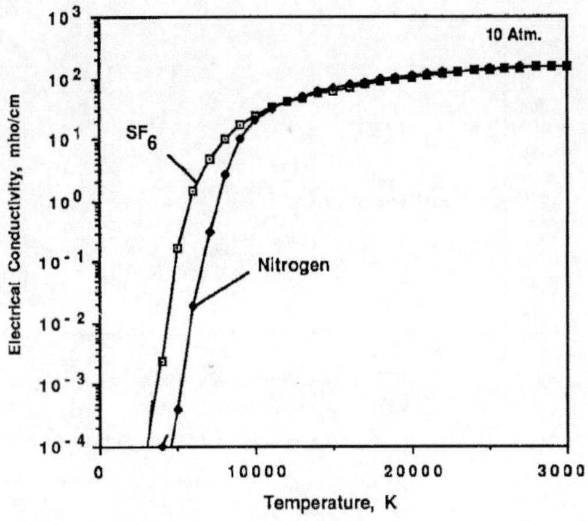


*Figure 1. Equilibrium electrical conductivity; 1 bar.*

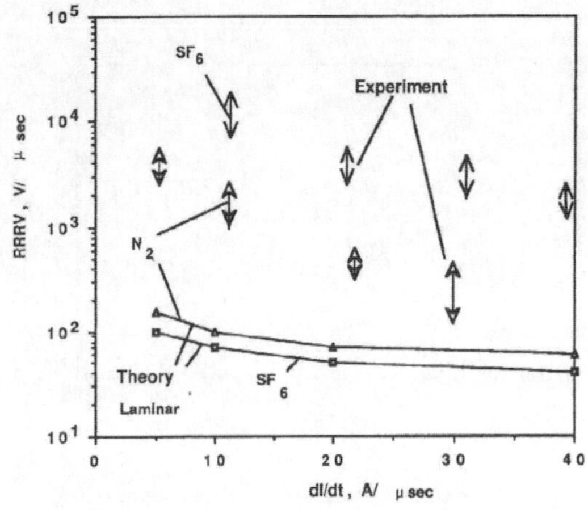


*Figure 2. Calculations of circuit interruption for laminar flow [1].*

ruption process with computers. Details of arc cooling by the processes of conduction, convection and radiation could all be modelled on a computer, as material functions of arc properties such as thermal and electrical conductivities and radiation emission coefficients were all known or could be calculated as a function of temperature and pressure. Then two-dimensional predictions of arc temperatures were made as a function of time to predict whether arc extinction would be achieved.

Figure 1 shows calculated equilibrium electrical conductivities for the two common circuit breaker gases of air and $SF_6$ as a function of pressure at 1 bar. It is seen that if cooling to temperatures below 1000 K is achieved the electrical conductivity is near zero and circuit interruption is achieved and current is not possible. Thus it was envisaged that proposed modifications in design could be modelled first theoretically on the computer and the viability of the modifications could be tested much more cheaply than by an experimental trial.

Results of such modelling based on predictions of arc temperature have been very successful in predicting arc diameters for approximately steady state arcs at high current and have been used particularly for the design of "puffer"circuit breakers, where the ablation caused by high currents could be predicted by predictions of arc diameters. These 'temperature'models are also successful in making predictions of circuit interruption, which is a significant achievement. However, early predictions shown in Fig. 2 for conditions of imposed high speed flow showed large differences between theory and experiment and also nitrogen as a better interruption gas than $SF_6$! Subsequent workers have removed the disparity of such theoretical predictions by introducing turbulence losses into the energy balance equation. But for results to give the much higher interruption ability for $SF_6$ compared with air or nitrogen, the turbulence parameters for $SF_6$ need to be significantly higher than for nitrogen or air, with no clear knowledge as to the physical reason for such a difference. One of the reasons for theory predicting superior interruption for nitrogen in Fig. 2 follows just from Fig. 1, where electrical conductivities for nitrogen are lower than for $SF_6$ at low temperatures. There are further difficulties in predicting the pressure dependence of interruption when use is just made of the energy balance equation for arc temperature. The calculations assume that the electron density is just determined by the equilibrium value for the arc temperature.

## 3. Switching arc predictions including predictions of electron densities

In this section new calculations are presented which also include a solution for the time dependence of the electron density to obtain the non-equilibrium electrical conductivity. Predictions of electrical current and its interruption depend more directly on the electron density than the arc temperature. The electron density is determined by the balance between electrons produced by ionization and losses of electrons by electron attachment to neutral molecules and also recombination of electrons and positive ions. Thus three additional completely different material properties are involved in the theoretical predictions, namely the coefficients of ionization, attachment and recombination. Electron ionization and attachment coefficients are shown in Fig. 3 and it is seen that attachment for $SF_6$ is generally more than 100 times larger for $SF_6$ than for air. Furthermore the critical electric field for which ionization is larger than attachment is about four times larger for $SF_6$.

A remarkable solution of the approximate relationship defining electron recombination as a function of time, namely $dn_e/dt = -\gamma n_e n_+ = -\gamma n_e^2$ is that for a cooling plasma after time $t$, the electron density $n_e = 1/\gamma t$, regardless of the initial electron density.

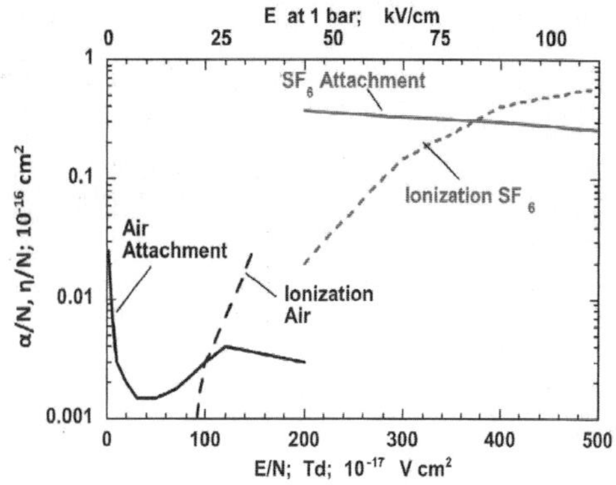


*Figure 3. Ionization and attachment coefficients; air and $SF_6$.*

This formula using a decay time of 1 µs and the large recombination coefficient, $\gamma$, of $10^{-7}\,\text{cm}^3.\text{s}^{-1}$ gives an electron density of $10^{13}\,\text{cm}^{-3}$, nowhere near the zero electron density obtained from equilibrium plasmas of low temperatures. Thus circuit interruption will be determined by whether electron attachment can convert the excess non-equilibrium electrons that remain at current zero to negative ions. This physical picture is completely different from the physical picture obtained from the calculation of arc temperatures, which is a "heat transfer"model, depending on whether there is sufficient cooling to extinguish the arc.

Predictions of circuit interruption are then quite different, as shown in Fig. 4. The predicted critical rate of rise of recovery voltage (RRRV) for $SF_6$ is $50\,\text{kV.s}^{-1}$ compared with $8\,\text{kV.s}^{-1}$ for air largely because of the higher attachment rate for $SF_6$. When the pressure of $SF_6$ is increased to 10 bar from 1 bar the RRRV increases to $200\,\text{kV.s}^{-1}$, again largely because increased attachment occurs at the higher pressure. The predictions of Figure 4 apply for a low current arc, similar to that used in welding, with very little imposed flow, and clearly predict superior interruption for $SF_6$ and increased pressure. Large circuit breakers used in power stations involve imposed gas flow at super-sonic velocities, in which case the loss term from the flow of the enthalpy of the arc will be very much larger and results may be very different from those of Fig. 4.

## 4. Explanation of Ball Lightning from $a^1\Delta_g$ oxygen

Ball lightning is occasionally seen during thunderstorms as a ball about the size of an orange, emitting light that can be easily observed, but without damage to the eyes. The balls move at about walking speed, usually about a metre above the ground and last about 10 seconds. Remarkably they have been observed most often inside of houses and even aeroplanes, frequently appearing from closed glass windows, usually without

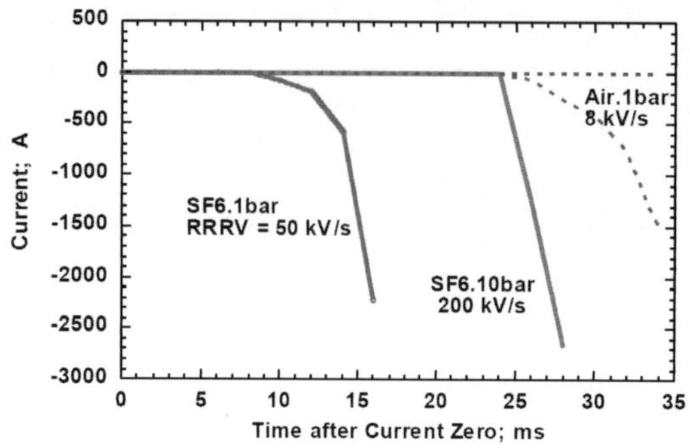


*Figure 4. Calculated rates of rise of recovery voltage; from [2].*

any remaining trace. Theories abound for the energy source, usually from something burning, such as silicon vapour, but also from microwaves, nuclear energy, dark matter, antimatter, masers and runaway electrons from outer space. In this author's view, none of these theories give a viable mechanism for producing such an energy source inside of houses or aeroplanes, almost always leaving no trace, such as an oxidation residue. The present proposal, though unlikely, does give such a model for ball-lightning, consistent with existing established concepts of electrical gas discharge behaviour.

The numerical predictions of the previous section 3, involve predictions of the behaviour of electrons, positive ions and negative ions. The new analysis involves the inclusion of an additional particle, the metastable state of oxygen, referred to as "singlet delta"oxygen molecules, represented by $a^1\Delta_g$. This molecule has well established properties, namely an energy of almost exactly 1 eV, a radiative lifetime of about an hour, a lifetime at atmospheric pressure of seconds, known cross sections as a function of energy for excitation by electrons, and also known quenching rates in air and also by water vapour. Of great significance, this metastable molecule detaches electrons from negative ions with a known detachment rate.

For most spark discharges the production of these metastables has very little effect, because the lifetime of spark discharges is of the order of micro-seconds, which is of insufficient time for significant densities of metastables to develop. However, for a discharge of continuous pulses, as occurs in negative corona discharges, it is known that sustaining electric fields are markedly reduced from $\sim 25\,\text{kV/cm}$ to $\sim 5\,\text{kV/cm}$. Such significant changes can be explained by the development of significant densities of metastables in the discharge [3]. Figure 5 shows a predicted reduction in sustaining field from these metastables and Fig. 6 shows the high densities of metastables of the order of $10^{17}\,\text{cm}^{-3}$ that are required for such an effect.

The proposed theory of ball lightning relevant for the observations inside of houses and aircraft, is in three separate stages: (1) the accumulation of ions on

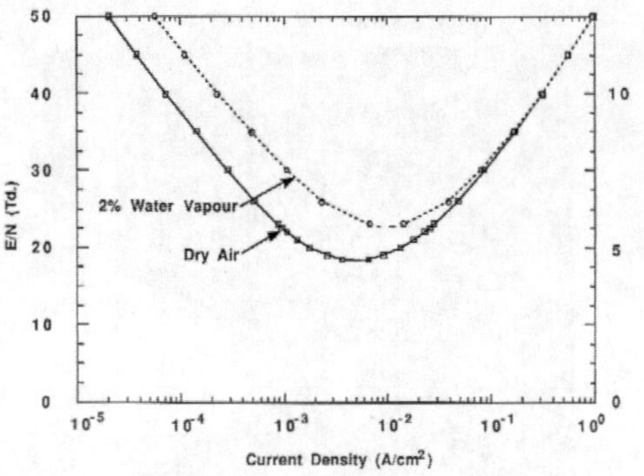


*Figure 5. Metastables enable discharge operation at 5 kV.cm$^{-1}$; from [3].*

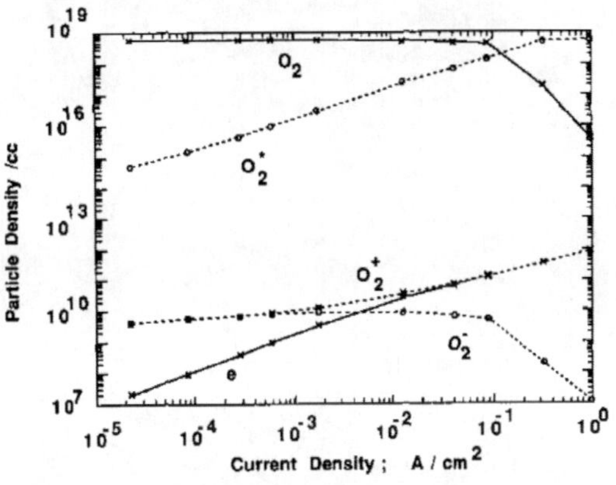


*Figure 6. Equilibrium particle densities with metastables; from [3].*

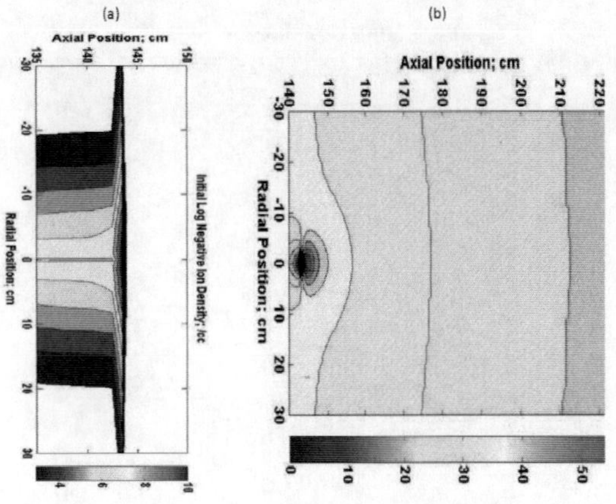


*Figure 7. Calculated charge accumulation on outside of window at 143 cm shown at (a) and the field intensification inside of the window, shown at (b) in kV.cm$^{-1}$.*

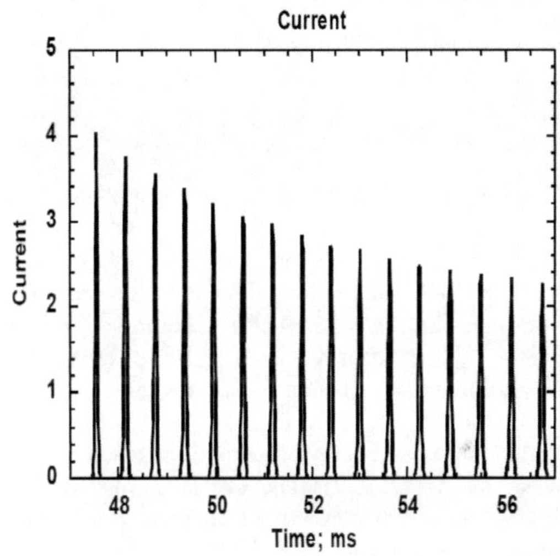


*Figure 8. Calculated current pulses for Fig. 7 discharge.*

the external surface of a glass window, (2) initiation of pulsed discharges at the inside glass surface as in [4] and (3) the accumulation of high densities of metastables to produce ball lightning.

### 4.1. Accumulation of ions on the glass window

It is proposed that a stream of background atmospheric ions remaining from stalled stepped leaders of lightning are driven by electric fields to accumulate on the insulating surface of the outside of a glass window. Such calculated densities of the order of $10^{10}$ cm$^{-3}$ are shown in Fig. 7a. Their effect is to double the background field on the other side of the glass, as shown in Fig. 7b. This mechanism was used to describe the initiation of discharges in the paper of [4].

### 4.2. Discharge pulses at inside glass surface

Fields greater than the breakdown electric field of 25 kV/cm are sufficient to initiate a discharge. Such discharges at low voltages are generally pulsed, known as Trichel pulses for negative potentials and a Hermstein discharge for positive potentials. Each discharge pulse will produce metastable molecules. Fig. 8 shows calculated electron density and thus current pulses in arbitrary units resulting from the dispersion of positive charges at the negative glass after each pulse.

### 4.3. Accumulation of large metastable densities

Metastables oxygen molecules will accumulate from each discharge pulse. From the discharge frequencies of Fig. 8, there would be ~ 10,000 pulses in 10 seconds. Calculations show, if metastable densities reach the order of $10^{17}$ cm$^{-3}$, consistent with the steady state calculations of Fig. 6, electron densities from the detachment of negative ions by the metastables become of the same order or greater than the ion densities. These large electron densities then produce a completely changed distribution of the electric field, with the electric field within the ball becoming almost zero. Radiation will be emitted from the ball with a

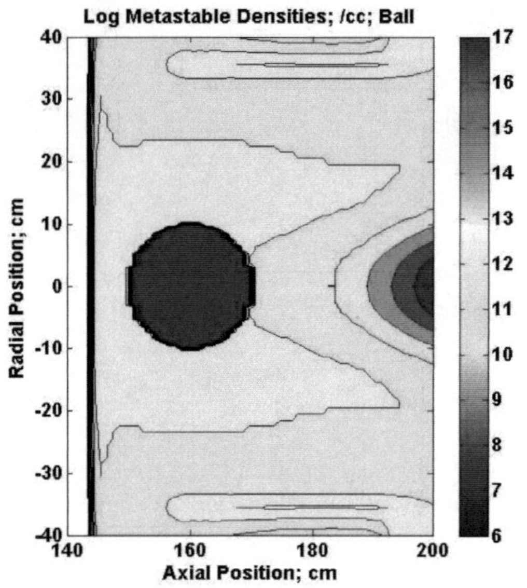


*Figure 9. Calculated metastable density after insertion into the calculations of a ball of metastables of density $10^{17}$ $cm^{-3}$.*

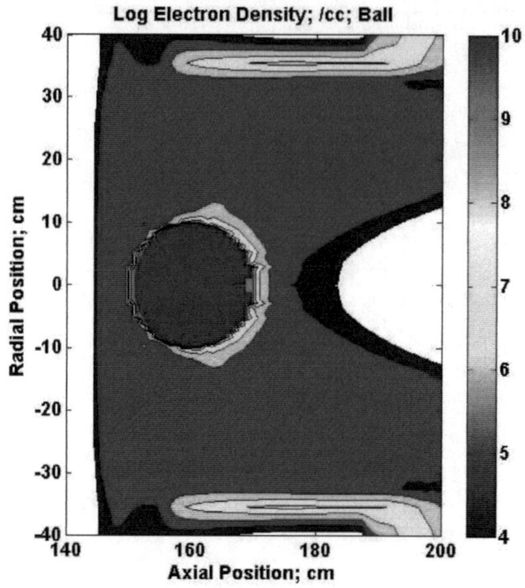


*Figure 10. Calculated electron density after insertion into the calculations of a ball of metastables of density $10^{17}$ $cm^{-3}$.*

lifetime dependent on the detachment coefficient of the metastables and the negative ion density.

Figures 9, 10 and 11 show calculated densities of metastables, electrons and negative ions after insertion into the calculations of a ball of metastables of density $10^{17}\,\mathrm{cm}^{-3}$. The ball is assumed to be spherical due to the influence of diffusion of the metastables. The electron density then rises to $\sim 10^{10}\,\mathrm{cm}^{-3}$, as shown in Figure 10, because of detachment of the negative ions by the metastables. The negative ion densities show a deficiency because of this detachment by the metastables, as shown in Fig. 11.

Figure 12 shows a remarkable reduction of the electric field to near zero within the ball and its surroundings, because of the high electrical conductivity of the electrons that are produced when the metastable density attains the density of the ions.

## 5. Discharge plasmas to cure cancer?

Traditional treatments of cancer in radiation therapy involve the bombardment of the cancer tumour with high energy particles. These high energy particles destroy the tumour cells which are impacted and also any healthy tissue that is also impacted. Recently evidence has accumulated that the impact of cancer tumours by discharges in air cause a reduction of tumour size. Not only is just the region directly impact by the plasma affected, but the whole tumour is reduced and there is even a suggestion of secondary tumours being influenced. Experimental efforts are being made to manufacture plasma sources that are safe for medical use. In particular oxygen metastables are being considered as possibly being the dominant active component. Efforts are being made to produce such plasmas that are not an electrical safety hazard and also produce significant concentrations of the oxygen metastables [5].

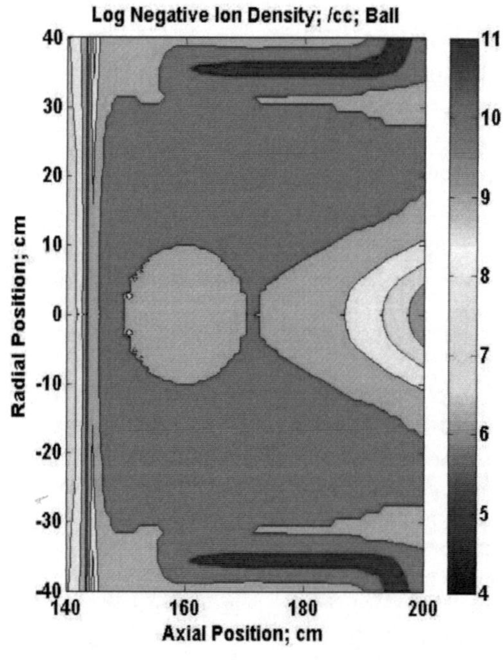


*Figure 11. Calculated density of negative ions after insertion into the calculations of a ball of metastables of density $10^{17}$ $cm^{-3}$.*

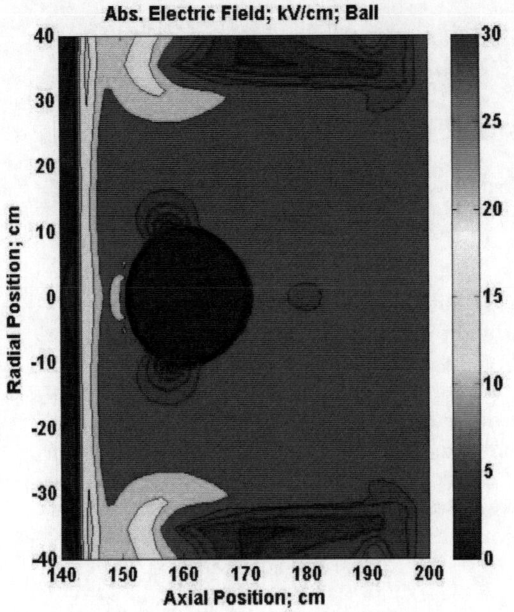


*Figure 12. Electric field within the ball and its surroundings.*

## Acknowledgements

The author is grateful for comments given by Dr. A. B. Murphy of CSIRO throughout the course of this work.

## References

[1] J. J. Lowke. Progress in arc theory, Proceedings of Sixth Int. Conf. on Switching Arc Phenomena, Lodz, Poland, Sept 12-15, 1989. In *Sbornik, Bulletin Tech University Brno*, pages 11–30, 1992.

[2] J. J. Lowke. Calculations of circuit interruption after current zero: Predictions of electron densities rather than temperatures. In *Proc. 22nd Inter. Conf. on Gas Discharges and their Applications*, volume 2, pages 347–350, 2018.

[3] J. J. Lowke. Theory of electrical breakdown in air - the role of metastable oxygen molecules. *J. Phys. D: Appl. Phys.*, 25(2):202–210, 1992. doi:10.1088/0022-3727/25/2/012/meta.

[4] J. J. Lowke, D. Smith, K. E. Nelson, C. R. W., and A. B. Murphy. Birth of ball lightning. *J. Geophys. Res.*, 117:D19107, 2012. doi:10.1029/2012JD017921.

[5] J. S. Sousa, G. Bauville, and V. Puech. Arrays of microplasmas for the controlled production of tunable high fluxes of reactive oxygen species at atmospheric pressure. *Plasma Sources Sci. Technol.*, 22:035012, 2013. doi:10.1088/0963-0252/22/3/035012.

doi:10.14311/ppt.2019.2.200
*Plasma Physics and Technology* **6**(2):200–207, 2019 

# ARC MODELING IN INDUSTRIAL APPLICATIONS

CH. RÜMPLER[a,*], R. CHECHARE[b], A. ZACHARIAS[a]

[a] *Eaton Industries GmbH, Hein-Moeller-Straße 7-11, 53115 Bonn, Germany*
[b] *Eaton India Innovation Center, Magarpatta SEZ Tower B6, Pune, India-411013*
[*] ChristianRuempler@Eaton.com

**Abstract.** Simulation methods are routinely applied in the design and development process of power distribution devices. Arcing phenomena that occur during switching operations or fault events are modeled to optimize device performance and gain deeper insights into the behavior that testing cannot easily provide. In this contribution, some applications are presented in detail. The first example describes the distribution of debris that is generated inside a molded case circuit breaker (MCCB) during short–circuit interruption. A model is used to analyze the debris transport and to derive a solution to address issues caused by the debris. Second application example is a cooling device for hot plasma gases vented by circuit breakers. A model driven design process helps to define the device dimensions to achieve a safe temperature level of the exhaust gases. The third example deals with short–circuit behavior of a hollow core high voltage surge arrester, comparing model and experimental results.

**Keywords:** arc interruption modeling, molded case circuit breaker, debris transport, blow–out cooling, surge arrester.

## 1. Introduction

Modeling methods have been developed to predict the arcing behavior in power distribution devices and are nowadays applied on a regular basis in various research and development activities. A major goal of any modeling activity in industrial applications is towards the risk mitigation of a design. By means of modeling and simulation, major processes that influence the design performance should be represented by the model, although in many cases the physical processes involving arcing are still too complex to include every aspect. But with feasible assumptions, useful models can be developed that provide guidance in the design process or enable a better understanding of the arcing phenomena and interactions. With appropriate utilization of these models, one can improve the designs ahead of testing and go into the test laboratory with higher confidence to achieve successful results.

In this contribution we introduce a basic modeling approach in Sec. 2 and provide three different applications in the following sections where the modeling approach was applied. The first example in Sec. 3 describes the distribution of debris that is generated inside a low-voltage molded case circuit breaker (MCCB) during short-circuit interruption. Based on arc model results, particle tracking is used to predict the debris transport inside the breaker and to derive a solution to address issues caused by the debris. The second application example in Sec. 4 deals with a cooling device for hot plasma gases that are vented by circuit breakers during short-circuit interruption. A model driven design process helps to define the gas cooling device dimensions and to achieve a safe temperature level of the exhaust gases, in order to reduce the risks that are involved with hot and potentially still ionized venting gas. The third example in Sec. 5 describes the short-circuit behavior of a hollow core high voltage surge arrester and modeling thereof. Model results as well as experimental results are provided and compared.

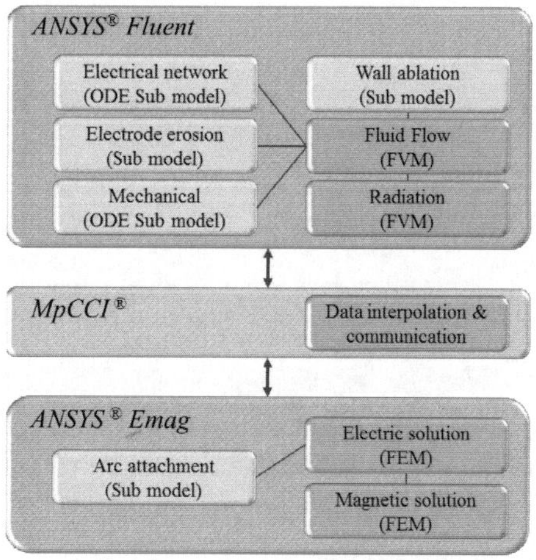


*Figure 1. Arc simulation framework.*

## 2. General arc modeling approach

In all application examples, as given in the following sections, the arc is a thermal plasma [1] where a magneto-hydrodynamic (MHD) modeling approach can be applied. Partial differential equations for fluid flow (Navier-Stokes), electromagnetic fields (Maxwell), as well as radiation transport are solved in a partitioned code-coupling approach [2]. With help of the commercial solver *ANSYS® Fluent* [3], fluid and radiation equations are solved using finite volume methods. The electric and magnetic field equations are solved by *ANSYS® Emag* solver, which is based on the finite element theory. The necessary data exchange between these solvers is done by the code coupling tool *MpCCI®* [4], as illustrated in Fig. 1. Besides the built-in functionality of the solvers, additional modules have been implemented to represent more interactions and processes in the model, which is needed to achieve a useful accuracy of the model results:

- ordinary differential equation solver (ODE solver) to describe the interaction of the arc with the electrical network,
- electrode erosion model to determine the metal vapor that is injected into the fluid region [2],
- wall ablation model for plastic materials [5],
- ODE solver for prediction of contact arm motion (driven by magnetic forces or the mechanism of a breaker),
- arc attachment model [2].

For achieving best computational performance we use the built-in parallelization capabilities of the solvers. In addition to that, we benefit in this setup from an explicit coupling approach where the codes are running in parallel as well [2], maximizing the utilization of computational resources.

## 3. Application: Debris reduction for low-voltage MCCB

### 3.1. Introduction

Short-circuit (SC) interruption testing of low-voltage molded case circuit breakers (MCCB's) has to be performed e.g. according to IEC standard 60947-2 [6]. An example for a low-voltage MCCB design with the arc chamber in focus is depicted in Fig. 2. When testing with very high prospective currents (depending on the targeted interruption ratings of the device), the current has to be limited quickly and interrupted successfully. Even with effective current limiting by means of a quick contact opening, fast arc formation, and fast rising arc voltage, the let-through currents can still be in the range of tens of kilo-amperes. High current densities flowing through the arc attachment zones can cause erosion of the electrode materials in the form of liquid droplets that are splashing away or generate metal vapor. This is happening while the arc attaches to the contacts (various contact materials), to the contact arm (usually copper), or the splitter plates (usually ferromagnetic steel).

The gas inside the arc chamber heats up quickly and the arc chamber is pressurized due to energy dissipation inside the arc. Venting openings behind the splitter plates define the main flow path from the interior of the breaker to the surroundings, while some flow is also going in the other direction towards

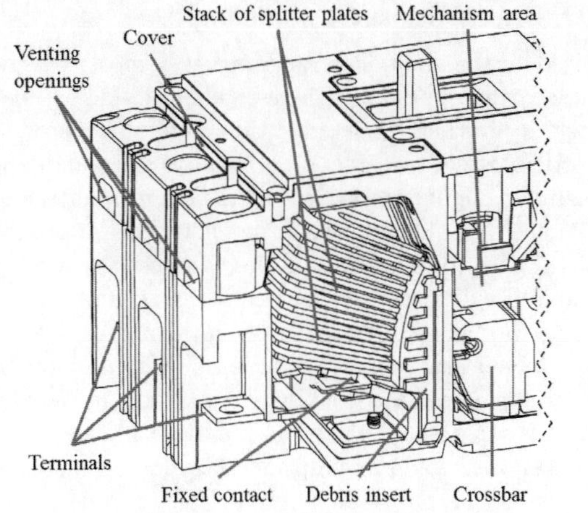


*Figure 2. Low-voltage circuit breaker with debris catching insert [7].*

the mechanism that is used to operate the crossbar and the contacts. Due to this venting path, liquid or already solidified metal droplets or particles are transported towards the crossbar and mechanism area of the breaker, potentially causing issues due to interference with moving parts. To better understand the behavior, arc modeling combined with particle tracking is performed. Based on the results, a novel design has been developed and successfully tested, as presented in the next subsections.

### 3.2. Arc model

The modeling approach, as described in Sec. 2, is used to investigate the arc behavior and flow conditions during a 70 kA/415 V/60 Hz short-circuit interruption of a low-voltage MCCB that has a rated current of 225 A. Only a single pole of the breaker is modeled and a symmetry boundary condition at the center of the pole is used to reduce the numerical effort. Model results that show the arc location and engagement with the contacts as well as the splitter plates, are depicted in Fig. 3 for two time points corresponding to smaller and larger contact opening angle.

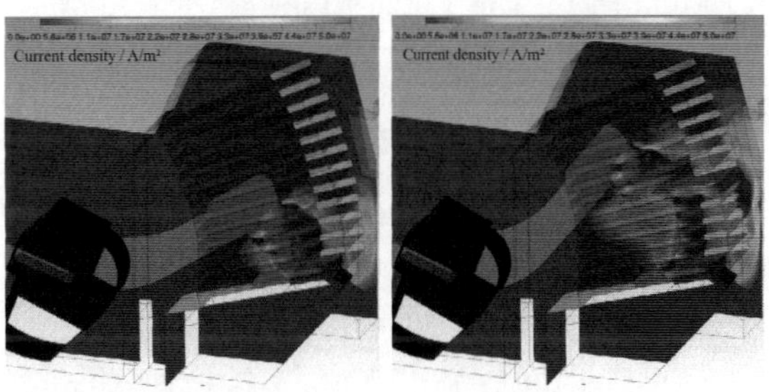


(a) t=1.0 ms (b) t=1.5 ms

*Figure 3. Arc model results showing iso-surface of current density at $5 \cdot 10^7$ $A/m^2$ for two time points.*

### 3.3. Debris transport modeling

Based on the arc model results, particle motion inside the arc chamber can be investigated. Using the calculated flow field at selected time points (frozen flow field), particle tracking is performed. Although the particles might interact with the flow in reality and change the flow conditions somehow, this approach is still able to provide an indication how the molten droplets will distribute inside the arc chamber. To define the source of the particles in the model, the arc attachment areas on any metal part – like contacts or splitter plates – are used, where the current density exceeds a threshold value. As the droplet dimensions are varying, some assumptions have to be made in the model. Droplet diameters of 0.5 mm and 1 mm are used in this study, where results are very similar for both diameters. Fig. 4 shows calculated particle tracks for 0.5 mm particle diameter at two different time points (corresponding to the arc model results depicted in Fig. 3). Once particles in the model are released and moving due to the flow, they can hit solid parts (e.g. arc chamber walls, splitter plates). In this case we use an escape boundary condition, so that the particle is removed from the calculation. With this assumption, we are focusing on the main particle transport paths where the particles can move without interferences.

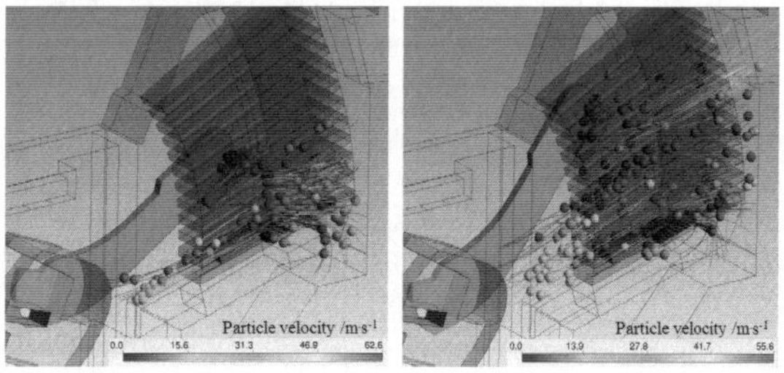


(a) t=1.0 ms (b) t=1.5 ms

*Figure 4. Visualization of modeled particle transport for two different time points.*

Because the arc is attaching on both contacts and some splitter plates at both time points, particles are generated there. Due to the strong gas stream away from the contacts towards the vent opening, majority of particles are transported into this direction. At 1.0 ms many particles hit the arc chamber wall behind the splitter plates and are removed from calculation there. But due to a recirculation zone inside the stack of splitter plates, particles are transported towards the rear of the breaker (towards the contact arm pivot point or mechanism area). The same effect is visible in the results for time point 1.5 ms. With larger contact opening angle, more particles find a free path towards the venting opening, but the recirculation pattern in the splitter plate stack transfers particles to the rear side of the breaker. As described above, this can cause issues with the mechanical operation of the breaker.

### 3.4. Solution for debris reduction

The erosion of the contacts or splitter plates cannot be avoided, therefore a method to reduce the impact of debris is needed. In order to avoid debris coming into the crossbar or mechanism area, the recirculation flow pattern can be utilized as shown in Fig. 5. By placing an insert into the arc chamber behind the splitter plate stack, as shown in the picture, the particles that are transferred backwards with the gas stream will end up in a pocket where they can accumulate. Fig. 2 shows the placing of the insert into the arc chamber as well as the pockets to accumulate debris.

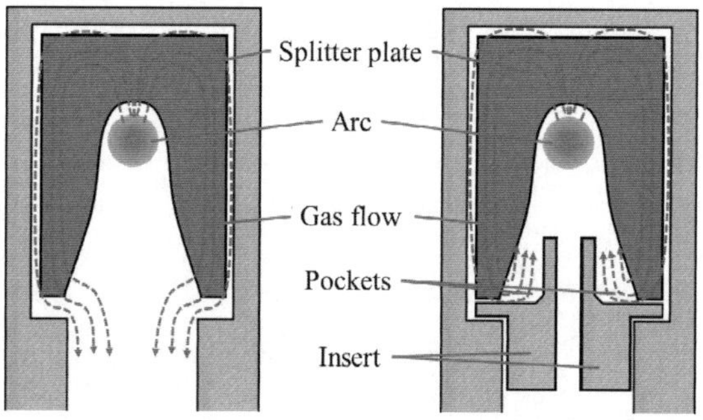


(a) Without insert (b) With insert

*Figure 5. Gas flow redirection using insert.*

To determine the effectiveness of the device, short-circuit testing with and without the insert have been performed. Some results are presented in Fig. 6. In the test without the insert, some droplet solidified and did stick below the contact arm. That led to an interference with the motion of the contact arm. Using the insert, the debris issues could successfully be reduced and the tests were successful.

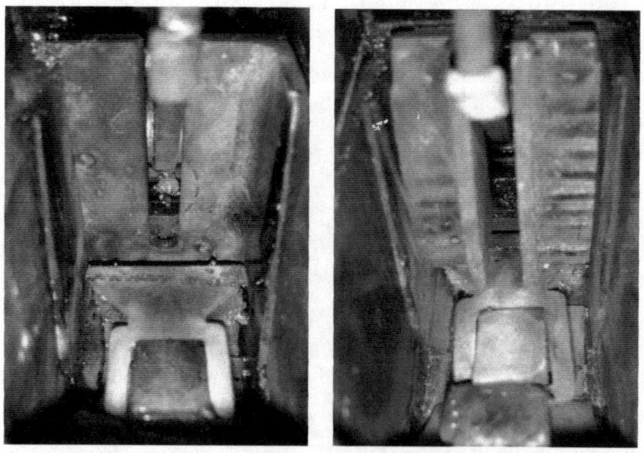

(a) Without insert (b) With insert

*Figure 6. View inside arc chambers of circuit breakers after testing (cover and stack of splitter plates removed).*

## 4. Application: Cooling device for MCCB exhaust venting

### 4.1. Introduction

During short-circuit interruption (SC), low-voltage circuit breakers vent hot plasma gas into the surroundings. To evaluate the potential harm of these gases, a metal mesh is placed around the breaker during SC testing according to IEC standard 60947-2 [6]. This metal mesh is connected to the ground via a fuse. If the vented gasses are still ionized (conductive) and the distance to the metal mesh is too short, current can flow from the breaker over the metal mesh to ground, eventually blowing the fuse. In this case the test would have failed.

To reduce this risk, a novel approach based on a heat exchanger device that is attached to the venting ports of the breaker was developed. By means of a larger number of thin metal plates with small gaps in between, the gas flow is cooled very effectively without changing the interruption performance of the breaker. The goal is to cool down the gases to a safe temperature level, so that even the risk of breakdown between conductive parts external of the breaker is eliminated.

### 4.2. Arc model

To enable the dimensioning of the cooling device, arc modeling is used to determine the venting conditions of the circuit breaker. An MCCB with rated current of 630 A has been investigated in this study that is based on a single-break design. The 3D MHD arc modeling approach, as introduced in Sec. 2, is used to calculate the AC short-circuit interruption behavior for 35 kA prospective current at line voltage of 600 V. One half of one pole of the breaker is modeled, applying a symmetry boundary condition at the center of the pole. Fig. 7 shows the temperature distribution inside the arc chamber at $t = 2.8\,\mathrm{ms}$ (symmetry plane as cut through the center of one pole). An iso-surface of temperature at 12 000 K is used to visualize the arc shape. The electrical network sub model is used to describe the interaction of the arc with the electrical network that is connected to the modeled device. Typical venting parameters like temperature and mass flow have been extracted from the model results.

### 4.3. Dimensioning of the cooling device

Modeling methods have been applied to determine optimal design parameters of the heat exchanger device. As a first step, the optimization goal needs to be determined. To avoid a breakdown between conductive parts that are in the venting path outside the breaker (e.g. live busbars), a safe venting gas temperature level has to be determined since the breakdown probability increases with higher venting gas temperatures, as shown by Tanaka [8]. The reduced critical electric field strength $(\frac{E}{N})_{\mathrm{cr}}$ of air and air-copper mixtures decreases with increasing temperature, but up to 1500 K

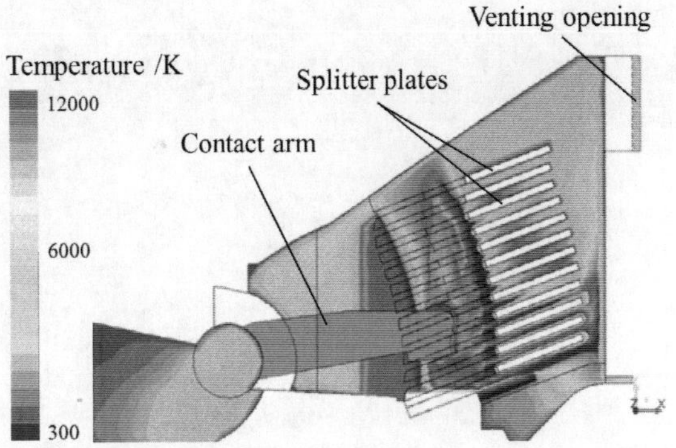


*Figure 7. Arc simulation model result (t=2.8 ms).*

the reduced critical electric field strength remains almost at the same level. Since the particle density $N$ is decreasing with rising temperature, the critical electric field $E_{\mathrm{cr}}$ is decreased by a factor of 5 when increasing the temperature from 300 K to 1500 K. The critical field of air then drops from 30 kV/cm to about 6 kV/cm (at atmospheric pressure). Metal vapor contamination does not play a role in this case, because at 1500 K all potential metal vapor that is coming from the arc chamber should be condensed. Liquid or condensed metal particles can influence the dielectric strength, but these effects are outside the scope of this study. Still some plastic decomposition products – coming from the arc interaction with plastic materials – could have an influence at this temperature level. Narayanan et al. [9] calculated the critical electric field for different air-copper-PA66 mixtures. Using these results, a value of about 2.5 kV/cm could be derived for the critical electric field.

With some typical dimensions, one can derive an approximate temperature threshold. The minimal clearance in air according to IEC standard 61439-1 [10] is 8 mm for rated impulse withstand voltage of 8 kV. Assuming that conductors have radii of 2 mm and the applied voltage is up to 1000 V, the electric field has approximately a maximum value of 2 kV/cm (uniform electric field is 1.25 kV/cm). This is still below the critical field value of 2.5 kV as shown above, so one can conclude that 1500 K is a good estimation for a safe venting gas temperature, considering that any venting gas will mix with cold gas outside the breaker reducing the gas temperature further.

The second optimization goal refers to the interruption performance of the breaker. The pressure drop created by the additional cooling device should be minimal. That ensures that the arc motion inside the breaker is not affected by the additional cooling device compared with the original breaker.

Instead of building many prototypes and running tests, a numerical "Design of Experiments" (DoE) study was performed using a simplified 2D model as shown in Fig. 8. With two symmetry boundary conditions defined, the model represents a cooling

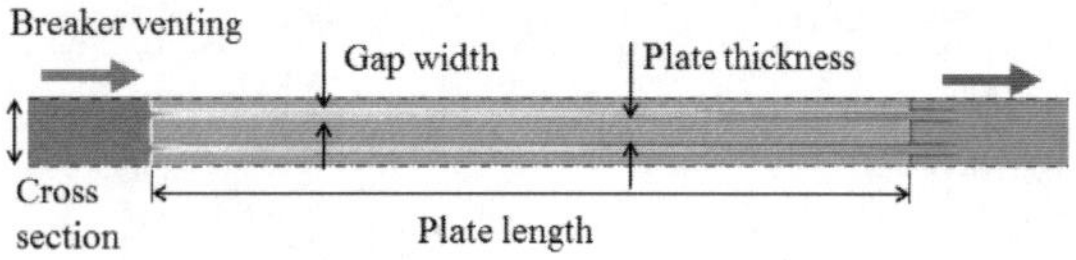


*Figure 8. Simplified 2D model used in DoE study of the cooling device.*

plate in the center, two adjacent gaps, and half of the two adjacent plates. Following design parameters describe the configuration: cross section, gap width, plate thickness, and plate length. Using mass-flow and temperature values as determined in the 3D arc simulation, a full-factorial DoE with center point is performed and evaluated where the design parameters serve as factors. Results are shown in Fig. 9, representing the normalized effects plot (ratio of effect to maximum effect) for each factor as an indicator, which design parameter or parameter combination has the most impact on the cooling device performance. To reduce the gas temperature with limited addition of pressure drop, a large total cross section is the most important design parameter. Reduction of the width of the gap between the plates also determines the cooling efficiency, but also increases the pressure drop (at fixed cross section). The length of the device can be used to control the cooling rate without changing the pressure drop, since most of the pressure drop occurs in the first section of the device.

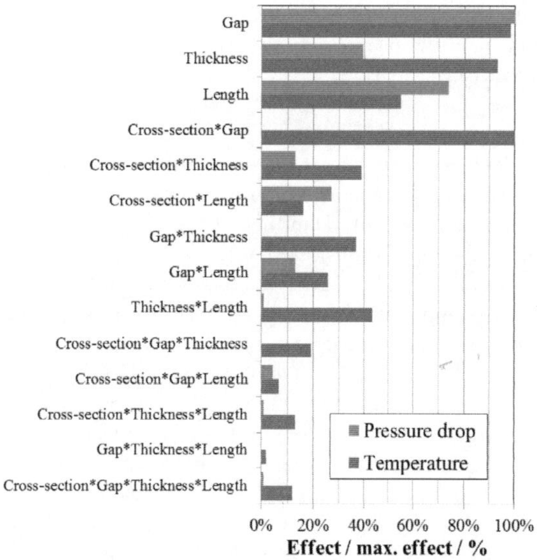


*Figure 9. DoE study results for blow-out cooling.*

## 4.4. Results

Based on the DoE study results, some prototypes with different dimensions were built and tested. In Fig. 10 a design employing 55 plates per pole with a thickness of 0.4 mm and gap of 0.2 mm is shown.

*Figure 10. Cooling device (cover removed).*

Short-circuit testing was done proving that the two performance goals can be achieved. Fig. 11 shows single pole interruption testing results for 35 kA prospective current at 600 V, comparing the breaker behavior with and without cooling device. Although the test conditions are not 100% identical (tests with and without cooling device at 50 Hz and 60 Hz circuit frequencies respectively), the interruption is very comparable. The arc voltage behaves almost identical, therefore one can conclude that the arc motion is not influenced by the cooling device.

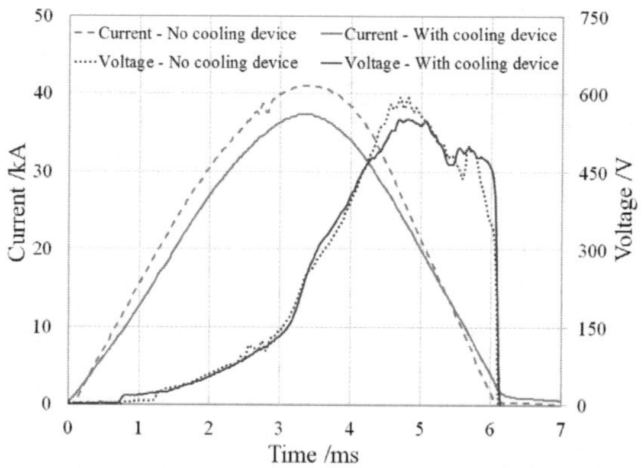


*Figure 11. Test results for 35 kA single phase interruption, without and with cooling device.*

To detect the cooling effect, a breakdown detection test setup is mounted external to the breaker directly into the venting path. It consists of a several pairs of metal rods with 8 mm distance and 2 mm radius, placed at four different distances away from the venting port of the breaker or the attached cooling device. Closest achievable distance of the first rod pair to the venting port in the setup was 10 mm. A capacitor was pre-charged to 853 V (slightly above $\sqrt{2}$ 600 V) and connected to the metal rods. If the gas stream coming from the breaker – directly or from the cooling device – is passing over the metal rods, a potential breakdown through the 8 mm gap can be detected while monitoring the voltage level. If the gas is still conductive or in case of a breakdown, the capacitor will discharge and the voltage collapses.

With aid of this setup, the risk mitigation using the cooling device could be demonstrated. In 18 different test with 72 different measured potential breakdown positions, only a single breakdown could be detected. Analyzing the breaker, it was clear that some parasitic venting not flowing through the cooling device caused that single breakdown event.

## 5. Application: Hollow core surge arrester

### 5.1. Introduction

Surge arresters employing metal-oxide varistor elements (MOV's) are widely used to protect medium and high voltage electrical distribution equipment from over voltage transients caused by internal (switching) or external events (e.g. lightning strike). MOV elements act as an electrical insulator under normal operating conditions and become electrically conductive in case the line voltage exceeds the designed threshold, creating a conductive path to ground. A short-circuit (SC) event inside the surge arrester can occur if the energy that needs to be dissipated is exceeding its capability. Surge arresters must be designed and tested as per IEC/IEEE standards for short-circuit faults to protect surrounding personnel and equipment.

In this section "Design A"type arresters according to IEC60099-4 [11] are investigated. This arrester design typically consists of metal-oxide varistor blocks that are stacked inside a hollow fiberglass reinforced polymer tube, as shown in Fig. 12. The silicone rubber housing (UltraSil™) around the tube enhances the electrical isolation even under extreme environmental conditions. The tube is mounted onto two heavy duty aluminum end castings that provide secure mounting provisions. Cast in horns (nozzles), as part of the end castings, enable directional venting of hot gases in case the built-in pressure relief device is activated due to an end-of-life event. Besides the pressure-relief functionality, the metal diaphragm as shown in Fig. 12, provides sealing of the MOV's from the environment.

In the event of a short-circuit inside the tube, a high current is flowing in a confined space forming a hot ionized plasma column inside the arrester. Heat transfer mechanisms radiation, convection, and conduction are distributing the arc energy inside the arrester cavities. Electromagnetic forces, acting on the plasma column, are responsible for further distribution of the hot ionized gas inside the tube. This results in fast heating of the air inside the surge arrester tube and a fast pressure rise that triggers the rupture of the diaphragms, once a pressure threshold is exceeded. From that point onwards hot ionized gas is venting through the nozzles to the exterior of the tube.

However, as current continues to flow through the interior of the tube, the tube pressure continues to rise gradually and drives venting of the hot gas through the nozzles. At some point in time the streams of ionized gas from top and bottom nozzles will meet and the arc

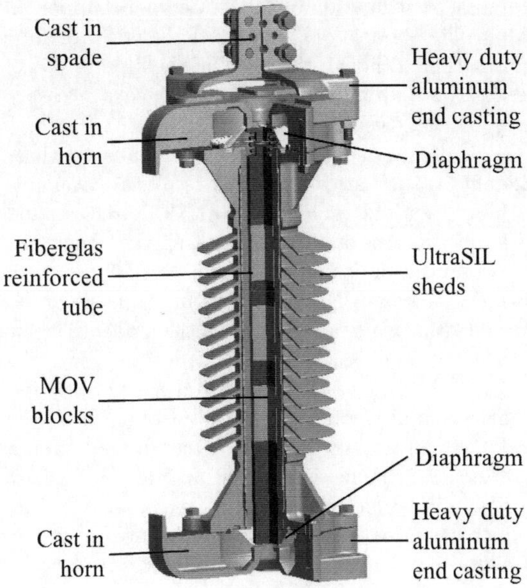


*Figure 12. Principle design of a VariSTAR™ HC station class surge arrester.*

is commutating to the exterior region. With the arc burning outside the arrester, the internal pressure rise is limited and therefore the thermal and mechanical stresses on the tube are also limited, as reported by Ozawa et al. [12]. With the help of high speed imaging, the external arc commutation phenomenon is analyzed in the SC test and compared with gas flow analysis. It was shown that a quick arc commutation from inside to outside is a key to limit pressure inside surge arrester [12].

In the certification test according to IEC60099-4 [11] a fault condition is initiated using a fuse wire that is placed in the cavity inside the tube. Testing is carried out for high, intermediate and low short-circuit currents and the arrester must satisfy various criteria in order to pass the tests. In short, no violent shattering of the structure is permitted where parts or fragments of more than 60 g leave the device during testing. The Arrester shall be able to self-extinguish open flames within two minutes at the end of the test.

Since prototyping and SC testing of surge arresters is tedious and expensive, a simulation model is developed to enhance the understanding of the physical phenomena during interruption to be able to reduce the development time.

### 5.2. Model approach

The 3D coupled multi-physics model approach introduced in Sec. 2 is applied to the surge arrester geometry. The simulation domain represents the interior of the arrester as well as some air volume outside the device to be able to capture arc commutation events. Due to the high energy density in the tube during short-circuit testing, high pressure values are expected and some deformation of the device will happen in reality. The model does not reflect the effect of tube deformation, although the effect of the pressure relief due to the diaphragm is captured once a pressure threshold is exceeded.

Calculations are done to represent and analyze three different load scenarios where test results are available (with peak current values of $I_p$=82.5 kA, 98.6 kA, and 126 kA) with focus on the initial time period where the arc commutation is happening (full test duration of 200 ms is not modeled). The current traces recorded during testing were used as load in the model to calculate the flow and electromagnetic processes. Since the complex phenomena that occur during melting and vaporization of the fuse wire are out of scope of the model, the initial condition is defined by patching a small hot arc channel inside the surge arrester where the fuse wire would be located. This enables the start of the calculation run, but the time point that is represented by the model compared to the real test needs to be determined. The arc voltage trace is used to define the time offset for the simulation, because the rapid decay of the measured voltage during the initial time period indicates the vaporization of the fuse wire and establishment of an arc discharge in the tube (see Fig. 13, Fig. 14, and Fig. 15).

### 5.3. Model results

Arc voltage and arc commutation time for $I_p$=82.5 kA, 98.6 kA, and 126 kA peak current tests are compared with simulation data in Fig. 13, Fig. 14, and Fig. 15 respectively. As indicated before, the predicted arc voltage deviates from the measurement in the initial time period due to the initial model settings and assumptions.

In the the first load case with $I_p$=82.5 kA, as shown in Fig. 13, the arrester tube ruptured in the test before the arc could commutate from the interior to the outside. Hot ionized gas is leaking at the middle section of the tube, as seen in the picture in Fig. 13 early in the test. The hot gas does help to form a conductive path outside the tube much faster compared to the case where the tube remains intact and the conductive channel is formed by the nozzle gas streams only. Therefore the arc voltage does decrease early on in the test. As the model cannot consider the tube rupture, the predicted arc voltage deviates from test result showing a later arc commutation at about 3.8 ms.

The second load case with $I_p$=98.6 kA is shown in Fig. 14. In this test the tube ruptured as well, but the arc did commutate after the rupture occurred. In this case the arc voltage almost remains constant due to gradual expansion of plasma inside pressurized arrester tube. During arc commutation, two hot gas streams meet outside to create a low resistance path for current flow, resulting in sudden drop in arc voltage. It should be emphasized that the tube rupture does not mean the test failed, as long as test criteria as per

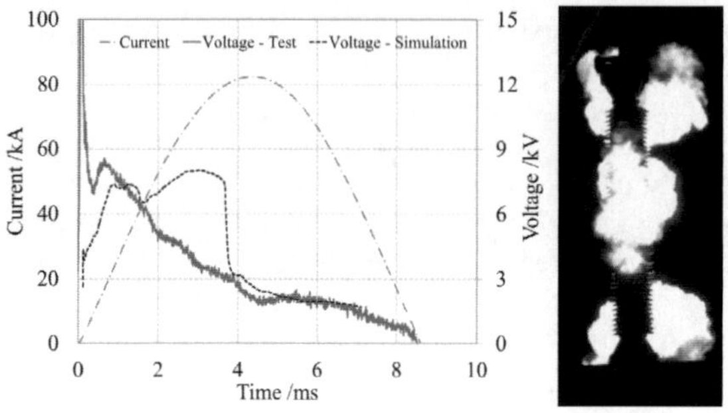


*Figure 13. $I_p$=82.5 kA peak current, tube ruptured before arc commutation. Picture taken at 1.12 ms.*

the standard are satisfied. The model does predict this commutation 1 ms earlier compared to the test. One explanation is some additional venting that was observed in the test, lowering the tube pressurization. As this venting is not considered in the model, a higher pressure inside the tube is predicted and the gases are accelerated faster through nozzle leading to an earlier arc commutation in the model.

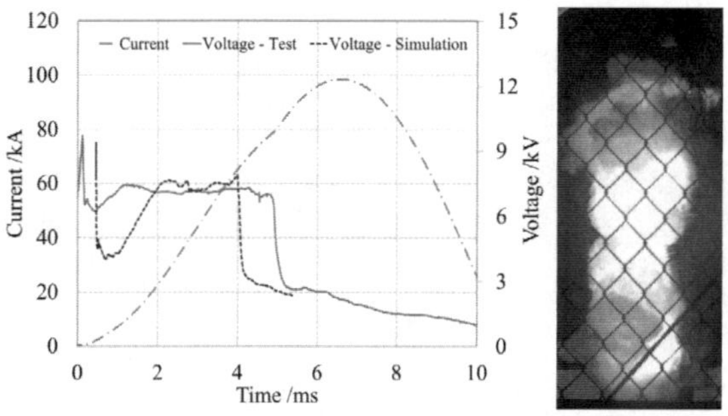


*Figure 14. $I_p$=98.6 kA peak current, tube ruptured after arc commutation. Picture taken at 4.95 ms.*

In the third load case with $I_p$=126 kA, as shown in Fig. 15, no indication for additional venting was found, also the tube did remain intact. Thus the simulation model mimics the test conditions closely and the predicted arc commutation time of 4.8 ms is in good agreement with the test result. But another effect is now visible that seems to lead to deviations of the predicted arc voltage as seen by the calculated over- and undershooting voltage results.

The interior air volume inside the surge arrester is divided into a number of channels along the length by some baffles. Depending on the current rise, the model predicts a different distribution of the arc plasma inside the tube where the available space might be more or less effectively utilized. This does lead in the $I_p$=126 kA case to a over and undershoot of the arc voltage, which can be seen in the other cases as well but not as clearly. In the test this does not seem to happen because the tube baffles might have deformed, providing more area for the arc to expand into some adjacent channels. This could be confirmed by some numerical test with bigger baffle gaps, leading to a

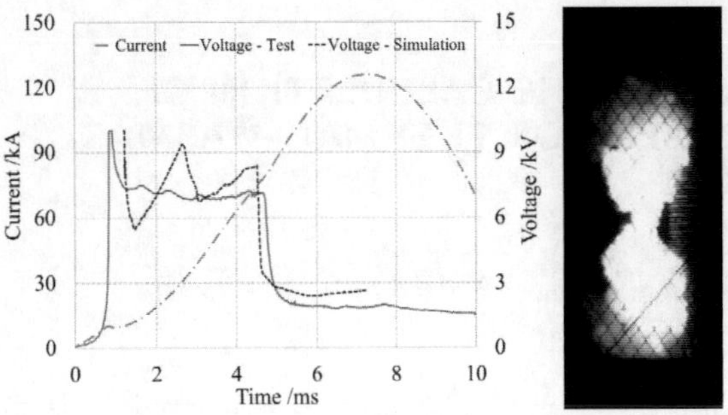


*Figure 15.* $I_p$*=126 kA peak current, tube not ruptured. Picture taken at 4.82 ms.*

flat arc voltage result (not shown here).

As discussed, the arc commutation phenomenon is recorded with a high speed camera and shown for some time points in Fig. 16. For the same time points the simulation results are shown as iso-surfaces of temperature and compared with test images for different time points in Fig. 16. The expansion of the plasma plum, as indicated by the temperature iso-values at 5 000 K and 10 000 K, does correspond well with the expansion during the test.

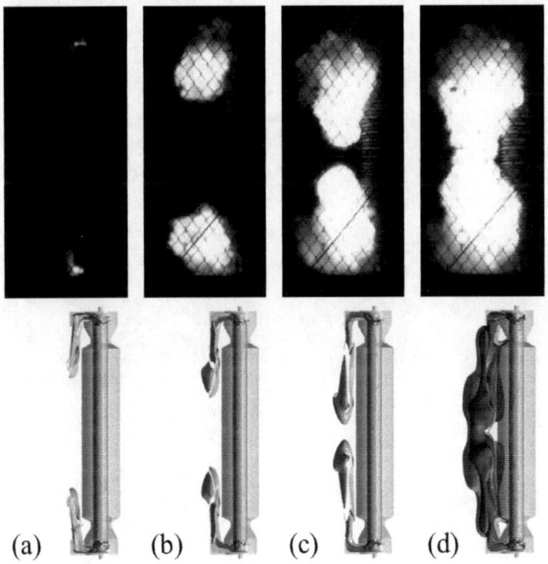


*Figure 16. Arc commutation process in experiment and simulation. Images from HS video for different time points of* $I_p$ *=126 kA fault case: (a) 3.00 ms, (b) 3.5 ms, (c) 4.29 ms, and (d) 4.88 ms. Iso-surfaces of 5 000 K and 10 000 K shown to visualize the expanding plasma jets.*

## 6. Conclusions

By means of three industrial application examples we discussed the utilization of arc modeling methods in this contribution. Although not all complex physical processes or interactions can be included in the model yet, the model results did provide useful insights for better understanding of the designs and to be a useful aid in the design process. In particular, the particle tracking results for the MCCB short-circuit application lead to the new debris catching insert design that significantly improved the short-circuit performance at high prospective current values. In case of the blow-out cooling device, a numerical DoE study did help to determine the optimal design parameters. The effectiveness of the design has been proven by various tests. In the high voltage surge arrester application, arc modeling did lead to a better understanding of the arc commutation process, which can be used in future to optimize the design further.

## References


[1] A. Gleizes, J. J. Gonzalez, and P. Freton. Thermal plasma modelling. *Journal of Physics D: Applied Physics*, 38(9):R153, 2005. doi:10.1088/0022-3727/38/9/R01.

[2] C. Rümpler. *Lichtbogensimulation für Niederspannungsschaltgeräte*. Dissertation, Technische Universität Ilmenau, 2009.

[3] ANSYS® Fluent and ANSYS® Emag. Release 19.1, ANSYS, Inc., 2018.

[4] MpCCI®-E Manual. 2018, Fraunhofer Institute SCAI, Sankt Augustin, 2018.

[5] C. Rümpler, H. Stammberger, and A. Zacharias. Low-voltage arc simulation with out-gassing polymers. In *2011 IEEE 57th Holm Conference on Electrical Contacts (Holm)*, pages 1–8, Sept 2011. doi:10.1109/HOLM.2011.6034770.

[6] Low-voltage switchgear and controlgear – Part 2: Circuit–breakers. IEC Standard 60947-2, The International Electrotechnical Commission, 2016.

[7] C. Rümpler, B. J. Schaltenbrand, A. L. Gottschalk, R. P. Malingowski, K. J. McCarthy, and X. Zhou. Electrical switching apparatus and debris barrier therefor. U.S. Patent 10 128 069, Nov 2018.

[8] Y. Tanaka. Influence of copper vapor contamination on dielectric properties of hot air at 300–3500 K in atmospheric pressure. *IEEE Transactions on Dielectrics and Electrical Insulation*, 12(3):504–512, June 2005. doi:10.1109/TDEI.2005.1453455.

[9] V. R. T. Narayanan, J. V. R. Heberlein, and C. Rümpler. The influence of metallic and plastic vapors on dielectric breakdown for low-voltage circuit breaker after current-zero. In *2013 IEEE 59th Holm Conference on Electrical Contacts (Holm 2013)*, pages 313–318, Sep 2013. doi:10.1109/HOLM.2013.6651435.

[10] Low-voltage switchgear and controlgear assemblies – Part 1: General rules. IEC Standard 61439-1, The International Electrotechnical Commission, 2011.

[11] Surge arresters – Part 4: Metal–oxide surge arresters without gaps for a.c. systems. IEC Standard 60099-4, The International Electrotechnical Commission, 2014.

[12] J. Ozawa, A. Mizukoshi, S. Maruyama, K. Nakano, K. Saito, G. St-Jean, Y. Latour, and A. Petit. Pressure relief design and performance of metal oxide surge arresters. *IEEE Transactions on Power Delivery*, 1(1):151–156, Jan 1986. doi:10.1109/TPWRD.1986.4307901.

doi:10.14311/ppt.2019.2.208
*Plasma Physics and Technology* **6**(2):208–216, 2019 

# STATE OF ART AND CHALLENGES FOR THE CALCULATION OF RADIATIVE AND TRANSPORT PROPERTIES OF THERMAL PLASMAS IN HVCB

Y. Cressault[a,*], Ph. Teulet[a], X. Baumann[a], G. Vanhulle[a], N. Kabbaj[a], F. Reichert[b], A. Petchanka[b]

[a] *(Laboratoire Plasma et Conversion d'Energie), Universite de Toulouse, CNRS, UPS, INPT, 118 route de Narbonne, F-31062 Toulouse, France*

[b] *SIEMENS AG, Nonnendammallee 104, Berlin, 13629, Germany*

[*] yann.cressault@laplace.univ-tlse.fr

**Abstract.** This paper is focused on the state-of-the-art and challenges concerning the thermophysical properties of thermal plasmas used in numerical modelling devoted to high voltage circuit breakers. For Local Thermodynamic Equilibrium (LTE) and Non-Local Thermodynamic (NLTE) and/or Chemical Equilibrium (NLCE) plasmas, the methods used to calculate the composition, thermodynamic, transport and radiative properties are presented. A review of these last data is proposed and some comparisons are given for illustrations.

**Keywords:** HVCB, chemical composition, transport coefficients, radiative properties, NLTE.

## 1. Introduction

Due to the cost of the experimental tests for HVCB, more and more numerical modelling is developed in order to compare the interruption capability of gases or mixtures with lower ODP than $SF_6$ ($CF_3I$, $CO_2$, $C_4F_7N$ with Air, $N_2$ or $O_2$), to optimize the efficiency of the process by using new designs, new electrodes or new walls, or simply to better understand the physical and chemical processes in the plasma.

To develop the simulation of a high voltage circuit breaker using magneto-hydro-dynamic models (MHD), it is necessary to elaborate databanks of fundamental properties: the thermodynamic properties, the transport coefficients, and the radiative properties, these data being used in the different equations of conservation: mass conservation, momentum conservation and energy conservation. Depending on the region of the plasma (near the electrodes, near the walls, in the surrounding gases, in the core of the plasma), depending on the pressure and the temperature, depending on the intensity or power (ignition of the arc or its cooling and extinction), presence of turbulence or not, the Local Thermodynamic Equilibrium (LTE) hypothesis can be assumed or not when non-equilibrium effects appears. Consequently, the numerical simulations have to be performed considering nonequilibrium phenomena, which are done using the two-temperature assumption. In this we usually distinguish the electron kinetic temperature $T_e$ which characterizes the high temperatures area of the plasma, while the heavy particles kinetic temperature $T_h$ describes low temperature regions.

This paper tries to make a state of the art and to present the challenges on the calculation of the properties of plasmas used in HVCB, in the case of LTE and non-LTE assumptions. The first part is devoted to properties of plasmas in LTE conditions with a summarize of the gases studied as alternative gases or mixtures to $SF_6$ and of the methods used to estimate the composition, the thermodynamic, transport and radiative properties. As the expressions available in the literature are accepted by the community in the case of LTE assumption, we will remember the recent works done on new mixtures and radiative properties. The second part is focused on the 2T-properties of plasmas in non-local thermodynamic equilibrium (nLTE) and/or non-local chemical equilibrium (nLCE). Few works have been done on this topic and lot of questions remain. We make a brief overview of the recent papers dealing with the 2T-properties applied to HVCB and finish the paper by indicating some challenges to solve.

## 2. Plasma in LTE conditions

For the HVCB applications, the first studies were devoted to the description of the heat zone, corresponding to the ignition of the arc in the first instants after the opening of the contacts. The plasma is assumed to be in Local Thermodynamic Equilibrium, the thermal properties are calculated with or without metallic vapors coming from the erosion of the electrodes (Cu, W, CuW), with or without thermoplastic vapors coming from the erosion of the nozzle ($C_2F_4$). The LTE properties have been calculated for lots of pure gases or mixtures ($SF_6$, $N_2$, Air, $CO_2$, $CF_3I$ [1, 2], $CF_4$, $C_3F_8$, $C_4F_8$, c-$C_4F_8$, $C_4F_7N$, $C_4F_8N_2$, $C_5F_{10}$, $C_6F_{12}O$, $CF_3NSF_2$, $SF_6$-$N_2$, $SF_6$-$CO_2$, $SF_6$-$CF_4$, $C_2F_4$-$CO_2$, $C_4F_8$-$CO_2$ [3–5], $CO_2$-$N_2$-Cu [6], $SF_6$-Cu [7–9], $C_5F_{10}O$-$CO_2$-$O_2$ [10]. The plasma composition is the first and unavoidable step allowing the computation of these properties, and is obtained from the law of mass action and the chemical base concept defined by Godin and Trépanier [11]. This technique is derived from the laws of thermodynamics and condensed phases are sometimes considered in the calculation of the population densities. Virial [12] and Debye [13] pressure corrections are often considered at high pressure ($P > 10$ bars) where the perfect gas law has to be replaced by a real gas law.

The thermodynamic properties (mass density, enthalpy, and specific heat at constant pressure) are directly deduced from the plasma compositions. Some results are given in [2–9] and behaviors are explained versus temperature, mixtures and pressure. The transport coefficients (viscosity, thermal and electrical conductivities) are obtained according to the Chapman-Enskog method and a previous calculation of collision integrals depending on the interaction potentials chosen to characterize the different collisions between particles (neutral-neutral, ion-neutral, electron-neutral and charged-charged collisions). The study of the collision integrals constitutes the most important part of the calculation of the transport coefficients and are often responsible for the differences observed between the authors [2–13].

The radiative losses play an important role in the HVCB arc plasmas: this is the main term of energy losses in the central region of the arc, the absorption in the surrounding plasma and cold gas leads to an increase of the pressure, it is responsible of the PTFE ablation influencing the behavior and the extinction of the plasma. The divergence of the flux is obtained by solving the radiative transfer equation which depends on the wavelengths, directions, and the geometry of the plasma. Due to the complexity of this equation, several simplifying approaches have been developed during the last decades such as the method of the net emission coefficient (NEC) initially proposed by Lowke and first calculated for $SF_6$ by Liebermann and Lowke [14, 15], then by Aubrecht and Barlova or Gleizes et al. [16, 17]. In order to treat efficiently the radiative transfer in the middle or cold regions, the P1-Model or discrete ordinates methods are often preferred using grey body approximation or Mean Absorption Coefficients (MACs) [18–22]. Then, due to rise of the computer's capacity, some works have been developed on pure $SF_6$ in order to perform an exact calculation of the radiative transfer equation with a very fine description of the spectrum, evaluating the accuracy and the validity of the simplified methods by comparisons. A quick review has been presented during the FSO conference in 2011 by A. Gleizes et al. [23]. Since, more studies on the radiative transfer equation have been done for pure gases ($SF_6$, $N_2$, Air, $CO_2$), binary mixtures ($SF_6$-$C_2F_4$, $C_4F_8$-$CO_2$, $CO_2$-Cu [3, 24]) and ternary mixtures ($SF_6$-$C_2F_4$-Cu/W) in 1D-2D-3D geometry with imposed temperatures profiles [25–28]. An example is given in Figure 1 for the divergence of the flux for a $SF_6$ pure plasma at 1bar, obtained with different approached methods [1].

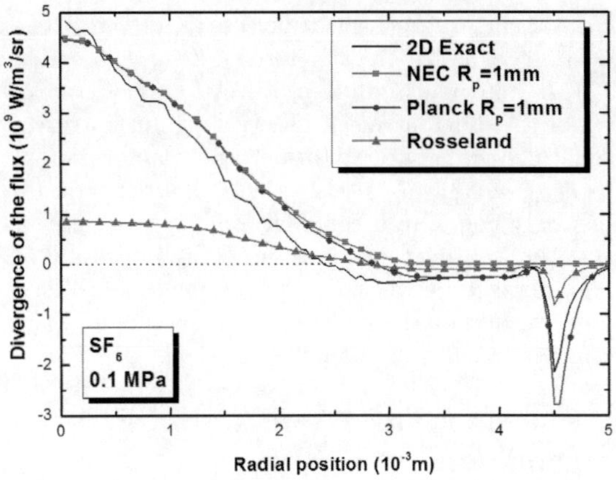


*Figure 1. Divergence of the flux for a $SF_6$ pure plasma at 0.1 MPa, obtained with different approached methods [1].*

From these studies, some conclusions can be done: (1) for the equilibrium composition and thermodynamic functions, the methods are well-known and valid, maybe questionable for plasmas at very high pressure and low temperature; (2) for transport coefficients, the interaction potentials have a significant influence on the values obtained and validation is done through other parameters such as the pressure for the thermal conductivity. A very small concentration of metallic vapors (>1%) strongly influences the electrical conductivity whereas viscosity and thermal conductivity are modified with concentration of metals or PTFE higher than 20–25%. Consequently, it is important to consider those coefficients in the conservation equations if we want to study the plasma regions where the ablation/erosion of walls and electrodes is important; (3) for the radiative transfer, some differences can be observed on the NEC when estimated using the escape factor approximation instead of the line by line method (more than 40% of difference can be observed at very high pressure > 100 bar [25]).

In LTE conditions, the challenges on thermal properties are mainly "driven"by the use of new alternative gases to $SF_6$ ($CO_2$, $CF_3I$, c-$C_4F_8$, $C_4F_7N$, $C_5F_{10}O$, $C_6F_{12}O$, $CF_3NSF_2$, pure gases or mixed with $CO_2$, $CF_4$, $C_2F_4$, $O_2$, $N_2$, air) or by the use of new materials for the manufacturing of electrodes or nozzles (Cu, W, CuW which have an important impact on the electrical conductivity and the radiation even if in few concentration). Some numerical modelling have been developed to study the influence of the erosion of electrodes or ablation of nozzle on the arc behavior from its ignition to its extinction [29–34]. The radiative transfer calculation with simplified methods can be improved following four ways: a better treatment of the absorption coefficients for molecules [26]; a careful definition of the spectral intervals for the calculation of the Mean Absorption Coefficients as in their works of Jan et al., Norborg and Iordanidis or Kloc et al. [26, 27, 35]; the elaboration of MACs tables from a mix of the different mean functions (Planck averaging at high temperature considering the absorption of the lines, and mean natural value at low temperature with only continuum contribution) [26]; the use of an optimized function based on the minimization of the difference between the exact calculation of the radiative flux or the divergence of the flux and the results obtained with simplified methods [36–38].

## 3. Plasma in nLTE and/or nLCE conditions

In many plasma processes, departures from thermal or chemical equilibrium exist on the edges of the plasma and close to the walls and the electrodes. For example, in the case of breaking devices and during the current zero phase, the electron number density is not very significant, the energy transfer between the electrons and the heavy particles is not enough efficient to preserve thermal equilibrium. As a consequence, the electron kinetic temperature $T_\mathrm{e}$ is higher than the heavy particles temperature $T_\mathrm{h}$, and two-temperature models must be developed. The recent papers propose two equations of energy conservation: one for the electrons and the other for the heavy species [39]. These models require the computation of thermodynamic, transport and radiative properties as a function of $T_\mathrm{e}$ and $\theta = T_\mathrm{e}/T_\mathrm{h}$ and the first step to obtain these data is, as in the case of LTE, the calculation of the 2T plasma composition.

There are several available theories in the literature to calculate the densities of species for non-equilibrium plasmas: (i) the minimization of a thermodynamic function (Gibbs'free enthalpy for plasma at constant pressure or Helmholtz's free energy at constant volume), (ii) the method based upon the 2T mass action law (LMA) assuming the chemical equilibrium and coupled with a set of equations for the conservation of pressure, electrical charge and mixture proportions and (iii) collisional radiative models (CR) or a chemical kinetic model (CK) coupled with an equation for the conservation of pressure. These methods give the same results in LTE but present strong differences if the plasma is not in equilibrium, depending on the temperatures affected the processes: the translatologie kinetic temperatures of electrons and heavy particles $T_\mathrm{e}$ and $T_\mathrm{h}$, the temperatures characteristic of the population of the various internal energy modes i.e. excitation temperature $T_\mathrm{ex}$ associated to the population of excited electronic states (atomic and molecular), the vibration temperature $T_\mathrm{v}$ linked to the distribution of vibrational levels and the rotation temperature $T_\mathrm{r}$ related to the population of rotational states. The LMA method is relatively easy to implement in numerical codes, and requires few data (partition functions, enthalpies of formation at 0 K of each chemical species and mass of species). The collisional-radiative (CR)

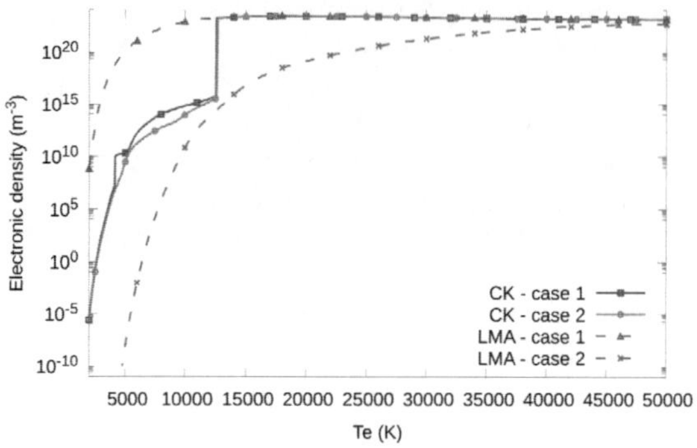


*Figure 2. Electron number densities versus electronic temperature $T_e$ (CK: chemical kinetic model, LMA : law of mass action). Pure $SF_6$, $\theta = 5$, $P = 1$ bar.*

and chemical kinetic (CK) models have advantages to reduce simplifying assumptions in the calculation. Moreover, the CK model allows studying the temporal chemical dynamic of the system. Nevertheless, a large databank of reaction rates or cross sections for each chemical process are needed. Relations between cross sections and reaction rates are well known for plasmas in thermal equilibrium conditions but are more complex for non-LTE plasmas. Indeed, electrons and heavy particles involved in chemical reactions have not the same temperature ($T_\mathrm{e}$ and $T_\mathrm{h}$) and are not described statistically by the same distribution functions. The temporal evolution of densities for each chemical species is given by a set of master equations:

$$\frac{\partial n_i}{\partial t} = C_i - n_i D_i \tag{1}$$

where $C_i$ and $D_i$ are respectively the creation and destruction factors for the specie $i$. They depend on the densities of species participating at the creation/destruction of specie $i$ through the chemical reaction and on the reaction rates of chemical reactions involved. Using relations between forward and reverse reaction rates, we used the library DVODE to solve the system formed (Eq. 1). The recent works of the team compared these methods in the case of pure $SF_6$ [40–42] and mixture $SF_6$-$C_2F_4$. We used two different assumptions concerning the population of internal energy modes: case (1) with $T_\mathrm{ex} = T_\mathrm{e}$, $T_\mathrm{v} = T_\mathrm{e}$ and $T_\mathrm{r} = T_\mathrm{h}$ more adapted to ionized gases (high temperatures), and case (2) $T_\mathrm{ex} = T_\mathrm{h}$, $T_\mathrm{v} = T_\mathrm{h}$ and $T_\mathrm{r} = T_\mathrm{h}$ more adapted to molecular gases (low temperatures). The figure 2 compares the electron number densities versus the electronic temperature $T_\mathrm{e}$ ($\theta = T_\mathrm{e}/T_\mathrm{h} = 5$), for pure $SF_6$ at 1 bar, and obtained from the CK and LMA methods.

The thermodynamic properties of 2T plasma are deduced from relations given Colombo et al. [43, 44] which propose to calculate a unique 2T specific heat $C_p$ as the derivative of the total enthalpy $H_\mathrm{tot} = H_\mathrm{e} + H_\mathrm{h}$ compared to the electronic temperature $T_\mathrm{e}$:

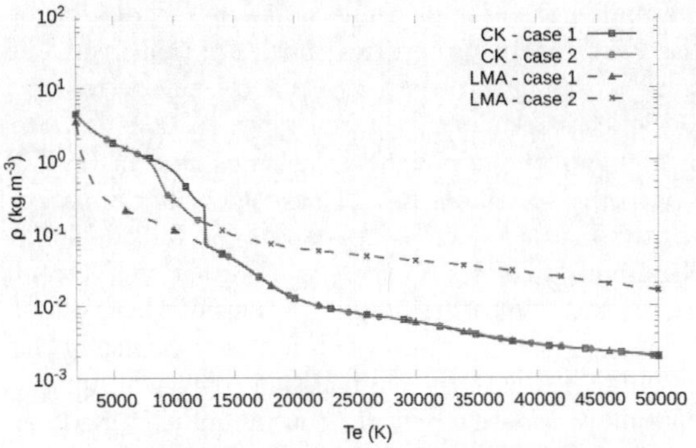


*Figure 3. Mass density ρ versus electronic temperature $T_e$ for pure $SF_6$ plasma, for $\theta = 5$ and pressure of 1 bar.*

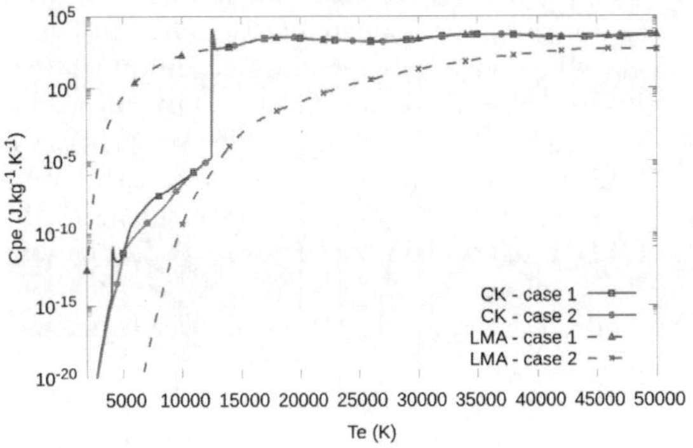


*Figure 4. Specific heat at constant pressure of electrons $C_{Pe}$ versus electronic temperature $T_e$ for pure $SF_6$, for $\theta = 5$ and pressure of 1 bar.*

$$C_P(\theta, T_e) = \left(\frac{\partial H_{\text{tot}}}{\partial T_e}\right)_P \quad (2)$$

As energetic contributions of electrons and heavies are separated in numerical codes through two coupled energy conservation equations, it is necessary to calculate independently the enthalpy of electron $H_e$ and heavy particles $H_h$, the specific heats of electrons $C_{Pe}$ and heavy species $C_{Ph}$:

$$C_{Pe}(\theta, T_e) = \left(\frac{\partial H_e}{\partial T_e}\right)_P \quad (3)$$

$$C_{Ph}(\theta, T_e) = \left(\frac{\partial H_h}{\partial T_h}\right)_P \quad (4)$$

This separation of the specific heat in two distinct contributions for electrons and heavy species is also the approach retained by Ghorui et al. [45, 46] or Zhang et al. [47]. The figure 3 shows the mass density of pure $SF_6$ obtained from the LMA and CK for the two cases, $\theta = 5$ and 1 bar. The figures 4 and 5 present the two specific heats in the same conditions. The law of mass action is in dashed lines and the chemical kinetic method in full lines.

With the CK model, there is a transition between both cases of the LMA. We can conclude that the

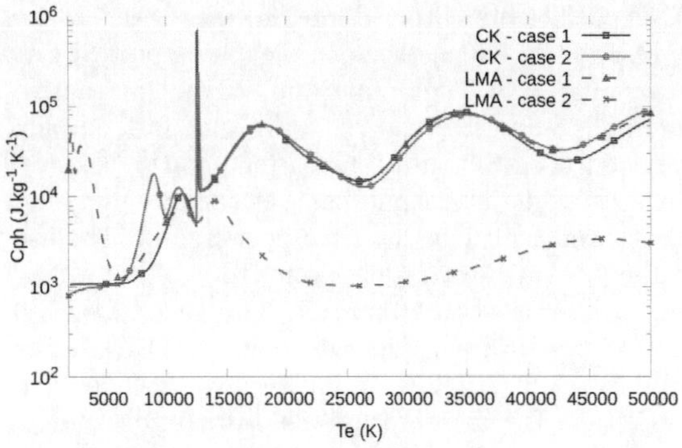


*Figure 5. Specific heat at constant pressure of heavy particles $C_{Ph}$ versus electronic temperature $T_e$ for pure $SF_6$, for $\theta = 5$ and pressure of 1 bar.*

LMA taken in the case 1 is accurate for the high temperatures and in the case 2 for the low temperature range. In high temperatures, there is lot of electrons and the inelastic processes are dominated by electron collisions and have to be calculated at $T_e$ (case 1). In low temperatures, the heavy particles chemical processes are dominant and have to be calculated at $T_h$ (case 2). There is a range of temperature where the LMA is not exact and the CK is necessary to calculate the composition.

Whereas the community agrees the calculation of the thermodynamic properties, translation thermal conductivities of electrons and heavy particles, viscosity and electrical conductivity, there is no consensus for the calculation of the reactive thermal conductivity. The theory of Hirschfelder et al. [12] or Bonnefoi et al. [48, 49] consider separately the electrons and the heavy particles whereas the theory of Rat showed that these two theories did not consider correctly not only the diffusion in 2T plasmas but also the coupling between electrons and heavy particles. The Rat's theory leads to new definitions of collision integrals, transport coefficients and consequently new terms in energy fluxes. Ghorui et al. [45, 46] in 2008 and Colombo et al. in 2009 [43, 44] showed in their works that the method of Devoto is sufficient to determine the transport properties (electrical conductivity, thermal conductivity and viscosity) in 2T assumption only if diffusion coefficients defined by Ramshaw were used. From the last works published on 2T-properties for several gases applied in various applications (Bonnefoi et al. [48, 49] for Ar-H, Mexmain [50] for Ar-O, McCourt [51] for polyatomic mixtures, Aubreton et al. [52–54] for Ar-O2, Ar-He or pure N2, Capitelli et al. [55, 56] with a "State-to-State" model applied for pure $N_2$ and $H_2$ plasmas, Guo et al. [57], Zhang et al. [58] and Yang et al. [59] on helium, argon and $SF_6$-$N_2$ mixtures) we can consider as accepted that the translation and internal thermal conductivities of heavy particles and the viscosity can be calculated at $T_h$, the translation thermal conductivity of electrons and the electrical conductivity at $T_e$. The figure 6 compare these coefficients obtained for the cases 1 and 2, the LMA and CR/CK methods, in the case of pure $SF_6$ at 1 bar and $\theta = 5$. The computation of the 2T reactive thermal conductivity at high pressure for complex gases such as $SF_6$ using multi-temperature expressions from the literature (such as the one provided by Ghorui et al. [45, 46]) gave poor results. The use of the Ghorui's expressions seems to be accurate for pressure lower than 16 bar and pure gases ($O_2$, $N_2$) or mixtures with few chemical reactions ($O_2$-$N_2$). For pure $SF_6$ and $SF_6$-$C_2F_4$ mixtures, we extended the Butler and Brokaw theory to 2T assumption. We divided the total reactive thermal conductivity into two contributions, one for the heavy particles (considering neutral atoms and molecules) and one for the electron part (considering electrons and ions):

$$\kappa_{\text{reac}}^{\text{h}} = -\frac{1}{RT_{\text{h}}^2} \frac{\begin{vmatrix} 0 & \Delta h_1 & \cdots & \Delta h_{N_\text{e}} \\ \Delta h_1 & A_{11} & \cdots & A_{N_\text{e}1} \\ \vdots & \vdots & \ddots & \vdots \\ \Delta h_{N_\text{e}} & A_{1N_\text{e}} & \cdots & A_{N_\text{e}N_\text{e}} \end{vmatrix}}{\begin{vmatrix} A_{11} & \cdots & A_{N_\text{e}1} \\ \vdots & \ddots & \vdots \\ A_{1N_\text{e}} & \cdots & A_{N_\text{e}N_\text{e}} \end{vmatrix}} \tag{5}$$

$$\kappa_{\text{reac}}^{\text{e}} = -\frac{1}{RT_{\text{e}}^2} \frac{\begin{vmatrix} 0 & \Delta h_1 & \cdots & \Delta h_{N_\text{e}} \\ \Delta h_1 & A_{11} & \cdots & A_{N_\text{e}1} \\ \vdots & \vdots & \ddots & \vdots \\ \Delta h_{N_\text{e}} & A_{1N_\text{e}} & \cdots & A_{N_\text{e}N_\text{e}} \end{vmatrix}}{\begin{vmatrix} A_{11} & \cdots & A_{N_\text{e}1} \\ \vdots & \ddots & \vdots \\ A_{1N_\text{e}} & \cdots & A_{N_\text{e}N_\text{e}} \end{vmatrix}} \tag{6}$$

with

$$A_{ij} = \frac{RT_{ij}}{P} \sum_{k=2}^{N_\text{e}-1} \sum_{l=k+1}^{N_\text{e}} \frac{x_k x_l}{D_{kl}^b} \left( \frac{\alpha(i,k)}{x_k} - \frac{\alpha(i,l)}{x_l} \right) \times \left( \frac{\alpha(j,k)}{x_k} - \frac{\alpha(j,l)}{x_l} \right) \tag{7}$$

$T_{ij}$ is equivalent to $T_\text{e}$ if one of the two particles in collision is an electron, $T_\text{h}$ in the other cases. From the original expression of Butler and Brokaw, the temperature $T$ was replaced by $T_\text{e}$, $T_\text{h}$ or $T_{ij}$ in relations (4). This choice has been validated by comparisons of shape of the reaction thermal conductivities versus temperature, $\theta$ and pressure. As examples, the results of the electrical conductivity, viscosity, translational, reactive and total thermal conductivities of electron and heavy particles are presented in figure 6 for 50%$SF_6$-50%$C_2F_4$ at 1 bar and $\theta = 5$, obtained from CK and LMA methods. For the case 1, the reaction term for heavy particles increases with $\theta$, the contribution of electrons decreases. For the case 2, the values of $\kappa_{\text{reac}}^{\text{h}}$ are shifted toward high temperatures and the peaks of $\kappa_{\text{reac}}^{\text{e}}$ are diminished and shifted to higher temperatures [60].

Finally, the last part concerns the calculation of the 2T radiative properties. For radiation in nLTE, some works have been done on atmospheric reentry by Soucasse et al. [61], or Scoggings [62] but few are available in the literature for plasmas used in HVCB. Contrarily to studies in LTE, the specialita emission and absorption coefficients cannot be linked in the Kirshhoff law through the Planck function. Recent papers propose to write two equations of conservation of energy: one for the electron with divergence of the flux obtained with the net emission coefficient approximation considering only the continuum part, another one for the heavy species with divergence of the flux obtained with the same approximation considering only the contribution of the lines. Consequently, the radiation has to be separated into two parts: one coming from atomic (radiative recombination, radiative attachment, Bremsstrahlung) and molecular continua and calculated at the temperature $T_\text{e}$, and the other one coming from the atomic lines and calculated at $T_\text{ex}$ (consequently $T_\text{e}$ or $T_\text{h}$ depending on the case 1 or 2), the molecular lines being neglected at this step. The broadenings have been calculated assuming $T_\text{h}$ for the FWHM due to the Doppler effect, $T_\text{h}$ for the van der Waals effect, and $T_\text{e}$ for the Stark perturbation. The correction proposed by Griem has been taken into consideration by calculating the parameter $R$ with $T_\text{h}$. Then, two net emission coefficients can be estimated:

$$\varepsilon_N^{\text{cont}} = \int_0^{+\infty} \varepsilon_\lambda^{\text{cont}}(\lambda, T_\text{e}) e^{-R_\text{p}\kappa'_{\text{tot}}(\lambda, T_\text{e})} \text{d}\lambda \tag{8}$$

$$\varepsilon_N^{\text{lines}} = \int_0^{+\infty} \varepsilon_\lambda^{\text{lines}}(\lambda, T_\text{ex}) e^{-R_\text{p}\kappa'_{\text{tot}}(\lambda, T_\text{ex})} \text{d}\lambda \tag{9}$$

Figure 7 shows the influence of $\theta$ on the NECs for the two cases. It means in our case that NEC for continuum is always dependent on $T_\text{e}$ but NEC for atomic lines depend on $T_\text{e}$ (case 1) or $T_\text{h}$ (case 2). As the NEC for the continuum only depends on $T_\text{e}$, the factor $\theta$ has no influence. This is not the same conclusion for the NEC of the atomic lines since $T_\text{ex}$ is equal to $T_\text{e}$ for the case 1 ($T_\text{ex} = T_\text{e}$, $T_\text{v} = T_\text{e}$ and $T_\text{r} = T_\text{h}$) or $T_\text{h}$ for the case 2 ($T_\text{ex} = T_\text{h}$, $T_\text{v} = T_\text{h}$ and $T_\text{r} = T_\text{h}$).

## 4. Conclusion

We have presented the methods commonly used to determine the thermophysical properties (composition, thermodynamic, transport and radiative properties) for LTE and NLTE/NLCE plasmas existing in HVCB. For LTE plasmas, the methods are well-known, as the influence of the metallic vapors on the properties: a small proportion of metallic vapors (inferior to 1%) strongly impacts the electrical conductivity and the radiation, higher proportions is required (more than 20%) to modify significantly the thermal conductivity and the viscosity. The challenges will be not only the calculation of these properties for new materials

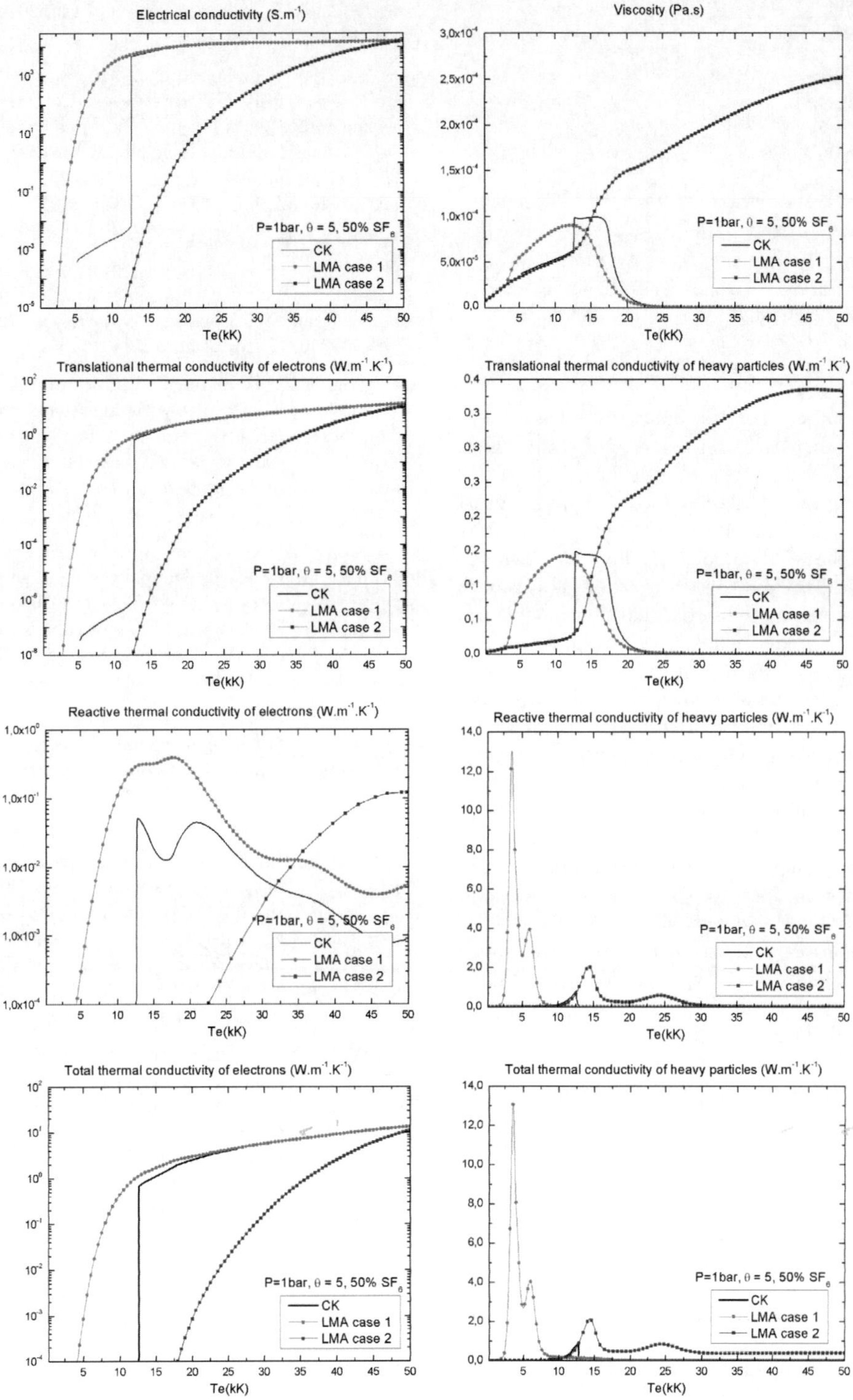


*Figure 6. 2T transport coefficients of 50%$SF_6$-50%$C_2F_4$ (mass proportions) obtained with LMA, and CK methods, for $\theta = 5$ and 1 bar.*

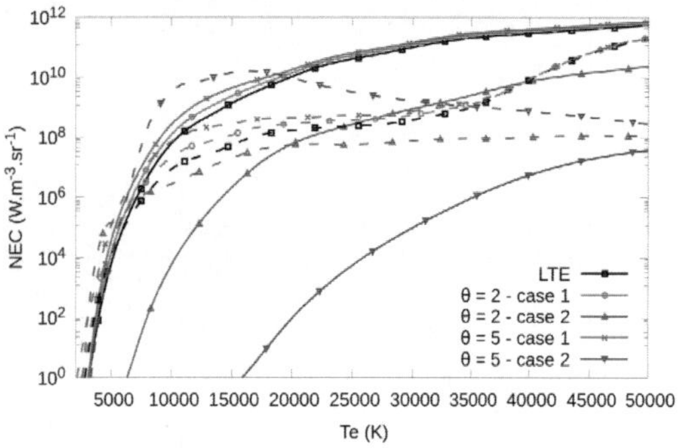


*Figure 7. Influence of $\theta$ on the two NECs versus electron temperature for a pure $SF_6$ plasma at 1 bar and for $R_p = 1\,mm$ : full lines for the atomic lines, dashed lines for the continuum.*

(used for electrodes or nozzles) but also the treatment of the RTE solving based on fine discretization and better approached methods. For NLTE/NLCE plasmas in HVCB, lots of challenges persist and must be solved rapidly in order to contribute to the improvement of the MHD modelling. For the plasmas compositions, it will be necessary to develop more precise models such as collisional radiative models or chemical kinetic models which are (for the moment) complex to achieve due to the lack of the reaction rates or collisional cross-sections. For the transport properties, the community must consider the thermal reactional conductivity and answer to the questions: should we consider only one coefficient (only for heavy particles for example), two contributions, a total thermal conductivity and how to calculate them under these conditions of non-thermal/chemical equilibrium? For the radiative properties, some works have been developed on $CO_2$ or air for re-entry atmosphere phenomena, syngas or combustion researches but few data are available for HVCB with or without metallic vapors due to the numerous chemical reactions. As the Planck and Kirchhoff laws are not accurate in non-equilibrium conditions, it is essential to treat differently the emission and absorption phenomena. The greatest difficulty lies in estimating correctly the populations energy levels, the emission and the absorption being proportional to these populations. The treatment of the atomic lines will be one of the main locks to raised (population of levels and broadenings being governed by different temperatures). New models will need to be developed to consider radiative phenomena separately: Hybrid Statistical Narrow Band model (HSNB) combining Statistical Narrow-Band model (SNB) for the optically thick molecular systems and Narrow Band means for the emission and absorption coefficients of the optically thin molecular systems and continua, a line by line (LBL) calculation for the bound-bound atomic lines [62]. Finally, these new methods and databank have to be validated with the comparison of MHD modeling results and experimental works.

## References


[1] Y. Cressault. Basic knowledge on radiative and transport properties to begin in thermal plasmas modelling. *AIP Advances*, 5(5):057112, 2015. doi:10.1063/1.4920939.

[2] Y. Cressault, V. Connord, H. Hingana, P. Teulet, and A. Gleizes. Transport properties of $CF_3I$ thermal plasmas mixed with $CO_2$, air or $N_2$ as an alternative to $SF_6$ plasmas in high-voltage circuit breakers. *Journal of Physics D: Applied Physics*, 44(49):495202, 2011. doi:10.1088/0022-3727/44/49/495202.

[3] L. Zhong, Y. Cressault, and P. Teulet. Thermophysical and radiation properties of high-temperature $C_4F_8$-$CO_2$ mixtures to replace $SF_6$ in high-voltage circuit breakers. *Physics of Plasmas*, 25(3):033502, 2018. doi:10.1063/1.5012850.

[4] P. André and Z. Koalaga. Composition of a thermal plasma formed from ptfe with copper in non-oxidant atmosphere, Part I: Definition of a test case with the $SF_6$. *High Temperature Material Processes: An International Quarterly of High-Technology Plasma Processes*, 14(3), 2010. doi:10.1615/HighTempMatProc.v14.i3.70.

[5] A. Yang, Y. Liu, B. Sun, X. Wang, Y. Cressault, L. Zhong, M. Rong, Y. Wu, and C. Niu. Thermodynamic properties and transport coefficients of high-temperature $CO_2$ thermal plasmas mixed with $C_2F_4$. *Journal of Physics D: Applied Physics*, 48(49):495202, 2015. doi:10.1088/0022-3727/48/49/495202.

[6] L. Zhong, X. Wang, M. Rong, and Y. Cressault. Effects of copper vapour on thermophysical properties of $CO_2$-$N_2$ plasma. *The European Physical Journal D*, 70(11):233, 2016. doi:10.1140/epjd/e2016-70241-3.

[7] M. Rong, L. Zhong, Y. Cressault, A. Gleizes, X. Wang, F. Chen, and H. Zheng. Thermophysical properties of $SF_6$–Cu mixtures at temperatures of 300–30 000 K and pressures of 0.01–1.0 MPa: part 1. Equilibrium compositions and thermodynamic properties considering condensed phases. *Journal of Physics D: Applied Physics*, 47(49):495202, 2014. doi:10.1088/0022-3727/47/49/495202.

[8] X. Wang, L. Zhong, Y. Cressault, A. Gleizes, and M. Rong. Thermophysical properties of $SF_6$-Cu mixtures at temperatures of 300–30,000 k and pressures of 0.01–1.0 MPa: part 2. Collision integrals and transport coefficients. *Journal of Physics D: Applied Physics*, 47(49):495201, 2014. doi:10.1088/0022-3727/47/49/495201.

[9] B. Chervy, A. Gleizes, and M. Razafinimanana. Thermodynamic properties and transport coefficients in $SF_6$-Cu mixtures at temperatures of 300–30000 K and pressures of 0.1–1 MPa. *Journal of Physics D: Applied Physics*, 27(6):1193, 1994. doi:10.1088/0022-3727/27/6/017.

[10] L. Zhong, M. Rong, X. Wang, J. Wu, G. Han, G. Han, Y. Lu, A. Yang, and Y. Wu. Compositions, thermodynamic properties, and transport coefficients of high-temperature $C_5F_{10}O$ mixed with $CO_2$ and $O_2$ as substitutes for $SF_6$ to reduce global warming potential. *AIP Advances*, 7(7):075003, 2017. doi:10.1063/1.4993305.

[11] D. Godin and J. Trépanier. A robust and efficient method for the computation of equilibrium composition in gaseous mixtures. *Plasma chemistry and plasma processing*, 24(3), 2004. doi:10.1007/s11090-004-2279-8.

[12] J. Hirschfelder, C. Curtis, and R. Bird. *Molecular Theory of Gases and Liquids*. Second printing. New York: Wiley, 1964.

[13] M. Boulos. *Thermal plasmas: fundamental and applications vol 1*. First printing. Springer-Verlag New York, 1994.

[14] J. Lowke. Predictions of arc temperature profiles using approximate emission coefficients for radiation losses. *Journal of Quantitative Spectroscopy and Radiative Transfer*, 14(2), 1974. doi:10.1016/0022-4073(74)90004-1.

[15] R. Liebermann and J. Lowke. Radiation emission coefficients for sulfur hexafluoride arc plasma. *JQSRT*, 16(253), 1976. doi:10.1016/0022-4073(76)90067-4.

[16] V. Aubrecht and M. Bartlova. Net emission coefficients of radiation in air and $SF_6$ thermal plasmas. *Plasma Chemistry and Plasma Processing*, 29(2):131, 2009. doi:10.1007/s11090-008-9163-x.

[17] A. Gleizes, B. Rahmani, J. Gonzalez, and B. Liani. Calculation of net emission coefficient in $N_2$, $SF_6$ and $SF_6$-$N_2$ arc plasmas. *Journal of Physics D: Applied Physics*, 24(8):1300, 1991. doi:10.1088/0022-3727/24/8/011.

[18] R. Siegel and J. Howell. *Thermal radiation heat transfer*. Third printing. Taylor and Francis, New York and London, 1992.

[19] M. Modest. *Radiative Heat Transfer*. Second printing. Academic Press California, London, Massachusetts, 2003.

[20] G. Raynal and A. Gleizes. Radiative transfer calculation in $SF_6$ arc plasmas using partial characteristics. *Plasma Sources and Technology*, 4(1):152, 1995. doi:10.1088/0963-0252/4/1/017.

[21] V. Aubrecht and J. Lowke. Calculations of radiation transfer in $SF_6$ plasmas using the method of partial characteristics. *Journal of Physics D: Applied Physics*, 27(10):2066, 1994. doi:10.1088/0022-3727/27/10/013.

[22] N. Bogatyreva, M. Bartlova, and V. Aubrecht. Mean absorption coefficients of air plasmas. *Journal of Physics: Conference Series*, 275(1), 2011. doi:10.1088/1742-6596/275/1/012009.

[23] A. Gleizes, H. Randrianandraina, and Y. Cressault. Improvements of radiative transfer calculation for $SF_6$ thermal plasmas. In *XIXth Symposium on Physics of Switching arc*, page 27, Brno, Czech Republic, 5-9 September 2011.

[24] T. Billoux, Y. Cressault, V. F. Boretskij, A. N. Veklich, and A. Gleizes. Net emission coefficient of $CO_2$-Cu thermal plasmas: role of copper and molecules. *J. Phys.: Conf. Ser.*, 406:012027, 2012. doi:10.1088/1742-6596/406/1/012027.

[25] H. Z. Randrianandraina, Y. Cressault, and A. Gleizes. Improvements of radiative transfer calculation for $SF_6$ thermal plasmas. *J. Phys. D: Appl. Phys.*, 44(19):194012, 2011. doi:10.1088/0022-3727/44/19/194012.

[26] C. Jan, Y. Cressault, A. Gleizes, and K. Bousoltane. Calculation of radiative properties of $SF_6$-$C_2F_4$ thermal plasmas - application to radiative transfer in high-voltage circuit breakers modelling. *J. Phys. D: Appl. Phys.*, 47(1):015204, 2013. doi:10.1088/0022-3727/47/1/015204.

[27] H. Nordborg and A. A. Lordanidis. Self-consistent radiation based modelling of electric arcs: I. Efficient radiation approximations. *Journal of Physics D: Applied Physics*, 41(13), 2008. doi:10.1088/0022-3727/41/13/135205.

[28] L. Hermette, Y. Cressault, A. Gleizes, C. Jan, and K. Bousoltane. Radiative properties of $SF_6$-$C_2F_4$-Cu mixtures in high voltage circuit breakers arc plasmas: net emission coefficient and mixing rules. In *XXth International Conference on Gas Discharge and their Applications*, page 229, Orléans, France, 6-11 July 2014.

[29] J. L. Zhang, A. B. Murphy, W. Hall, and M. T. C. Fang. Computational investigation of arc behavior in an auto-expansion circuit breaker contaminated by ablated nozzle vapor. *IEEE Transactions on Plasma Science*, 30(2):706–719, 2002. doi:10.1109/TPS.2002.1024273.

[30] Q. Ma, M. Rong, A. B. Murphy, Y. Wu, and T. Xu. Simulation Study of the Influence of Wall Ablation on Arc Behavior in a Low-Voltage Circuit Breaker. *IEEE Trans. Plasma Sci.*, 37(1):261–269, 2009. doi:10.1109/TPS.2008.2007733.

[31] J. Yan. Inclusion of nozzle dimensional change due to PTFE ablation in switching arc simulation. In *XXIIth International Conference on Gas Discharge and their Applications*, Novi-Sad, Serbia, 2-7 September 2018.

[32] P. Robin-Jouan. Ablation modelling in high voltage circuit-breakers: impact on the arc extinction. In *XXth International Conference on Gas Discharge and their Applications*, Orléans, France, 6-11 July 2014.

[33] S. Arabi, J. Trepanier, R. Camarero, and A. Vassilev. Modeling nozzle geometry changes due to the ablation in high-voltage circuit breakers. In *XXth International Conference on Gas Discharge and their Applications*, page 479, Orléans, France, 6-11 July 2014.

[34] A. Petchanka, F. Reichert, J. Gonzalez, and P. Freton. Improved modeling of ablation process in high voltage circuit breakers for switching ar simulation. In *XXth International Conference on Gas Discharge and their Applications*, page 155, Orléans, France, 6-11 July 2014.

[35] P. Kloc, V. Aubrecht, M. Bartlova, O. Coufal, and C. Rümpler. On the Selection of Integration Intervals for the Calculation of Mean Absorption Coefficients. *Plasma Chem Plasma Process*, 35(6):1097–1110, 2015. doi:10.1007/s11090-015-9648-3.

[36] P. Kloc, V. Aubrecht, and M. Bartlova. Numerically optimized band boundaries of Planck mean absorption coefficients in air plasma. *J. Phys. D: Appl. Phys.*, 50(30):305201, 2017. doi:10.1088/1361-6463/aa7627.

[37] P. Kloc, V. Aubrecht, and M. Bartlova. Objective function for numerical mean absorption bands optimization. *Plasma Physics and Technology*, 4(3):269, 2017. doi:10.14311/ppt.2017.3.269.

[38] N. Kabbaj, Y. Cressault, P. Teulet, and F. Reichert. Numerical optimization of mean absorption coefficient in air using Planck modified mean function. In *XVth Hig-Tech Plasma Processes Conference*, Toulouse, France, 2-6 July 2018.

[39] J. P. Trelles. Advances and challenges in computational fluid dynamics of atmospheric pressure plasmas. *Plasma Sources Sci. Technol.*, 27(9):093001, September 2018. doi:10.1088/1361-6595/aac9fa.

[40] A. Gleizes, F. Mbolidi, and A. A. M. Habib. Kinetic model of a decaying $SF_6$ plasma over the temperature range 12000 K to 3000 K. *Plasma Sources Sci. Technol.*, 2(3):173–179, 1993. doi:10.1088/0963-0252/2/3/007.

[41] X. Baumann, Y. Cressault, P. Teulet, F. Reichert, and A. Petchanka. First approach in the calculation of the radiative properties of multi-temperature $SF_6$ plasmas at 1 bar. In *XXIIIrd International Symposium on Plasma Chemistry*, Montréal, Canada, July 2017.

[42] X. Baumann. *Etude de la composition chimique et des propriétés thermodynamiques et radiatives d'un plasma de $SF_6$-$C_2F_4$ en condition de déséquilibre thermique.* PhD thesis, Université Paul Sabatier, France, 2019.

[43] V. Colombo, E. Ghedini, and P. Sanibondi. Thermodynamic and transport properties in non-equilibrium argon, oxygen and nitrogen thermal plasmas. *Progress in Nuclear Energy*, 50(8):921–933, 2008. doi:10.1016/j.pnucene.2008.06.002.

[44] V. Colombo, E. Ghedini, and P. Sanibondi. Two-temperature thermodynamic and transport properties of argon-hydrogen and nitrogen-hydrogen plasmas. *J. Phys. D: Appl. Phys.*, 42(5):055213, 2009. doi:10.1088/0022-3727/42/5/055213.

[45] S. Ghorui, J. V. R. Heberlein, and E. Pfender. Thermodynamic and Transport Properties of Two-temperature Oxygen Plasmas. *Plasma Chem Plasma Process*, 27(3):267, 2007. doi:10.1007/s11090-007-9053-7.

[46] S. Ghorui, J. V. R. Heberlein, and E. Pfender. Thermodynamic and Transport Properties of Two-Temperature Nitrogen-Oxygen Plasma. *Plasma Chem Plasma Process*, 28(4):553–582, 2008. doi:10.1007/s11090-008-9141-3.

[47] X.-N. Zhang, H.-P. Li, A. B. Murphy, and W.-D. Xia. A numerical model of non-equilibrium thermal plasmas. I. Transport properties. *Physics of Plasmas*, 20(3):033508, 2013. doi:10.1063/1.4794969.

[48] C. Bonnefois. *Contribution à l'étude des méthodes de résolution de l'équation de Boltzmann dans un plasma à deux températures : exemple le mélange argon-hydrogène.* PhD thesis, Limoges University, France, 1983.

[49] C. Bonnefoi, J. Aubreton, and J. M. Mexmain. New approach, taking into account elastic and inelastic processes for transport properties of a two temperature plasma. *Zeitschrift für Naturforschung*, 40:885, 1985. doi:10.1515/zna-1985-0904.

[50] J. Mexmain. *Coefficients de transport dans un plasma à deux températures: application au mélange argon-oxygène.* PhD thesis, 1983.

[51] F. McCourt. *Nonequilibrium Phenomena in Polyatomic Gases: Dilute Gases.* Vol. 1. Oxford University Press, USA, 1990.

[52] J. Aubreton, C. Bonnefoi, and J. M. Mexmain. Calcul de propriétés thermodynamiques et des coefficients de transport dans un plasma Ar-$O_2$ en non-équilibre thermodynamique et à la pression atmosphérique. *Rev. Phys. Appl. (Paris)*, 21(6):365–376, 1986. doi:10.1051/rphysap:01986002106036500.

[53] J. Aubreton, M. F. Elchinger, and P. Fauchais. New Method to Calculate Thermodynamic and Transport Properties of a Multi-Temperature Plasma: Application to $N_2$ Plasma. *Plasma Chemistry and Plasma Processing*, 18(1):1–27, 1998. doi:10.1023/A:1021785125690.

[54] J. Aubreton, M. F. Elchinger, P. Fauchais, V. Rat, and P. André. Thermodynamic and transport properties of a ternary Ar-$H_2$-He mixture out of equilibrium up to 30 000 K at atmospheric pressure. *J. Phys. D: Appl. Phys.*, 37(16):2232–2246, 2004. doi:10.1088/0022-3727/37/16/004.

[55] M. Capitelli, I. Armenise, D. Bruno, M. Cacciatore, R. Celiberto, G. Colonna, O. D. Pascale, P. Diomede, F. Esposito, C. Gorse, K. Hassouni, A. Laricchiuta, S. Longo, D. Pagano, D. Pietanza, and M. Rutigliano. Non-equilibrium plasma kinetics: a state-to-state approach. *Plasma Sources Sci. Technol.*, 16(1):S30–S44, 2007. doi:10.1088/0963-0252/16/1/S03.

[56] M. Capitelli, I. Armenise, E. Bisceglie, D. Bruno, R. Celiberto, G. Colonna, G. D'Ammando, O. De Pascale, F. Esposito, C. Gorse, V. Laporta, and A. Laricchiuta. Thermodynamics, Transport and Kinetics of Equilibrium and Non-Equilibrium Plasmas: A State-to-State Approach. *Plasma Chem Plasma Process*, 32(3):427–450, 2012. doi:10.1007/s11090-011-9339-7.

[57] X. Guo, A. B. Murphy, and X. Li. Thermodynamic properties and transport coefficients of two-temperature helium thermal plasmas. *J. Phys. D: Appl. Phys.*, 50(12):125202, 2017. doi:10.1088/1361-6463/aa5d25.

[58] X. N. Zhang, H. P. Li, A. B. Murphy, and W. D. Xia. Comparison of the transport properties of two-temperature argon plasmas calculated using different methods. *Plasma Sources Sci. Technol.*, 24(3):035011, 2015. doi:10.1088/0963-0252/24/3/035011.

[59] F. Yang, Z. Chen, Y. Wu, and al. Two-temperature transport coefficients of $SF_6$-$N_2$ plasma. *Phys. Plasmas*, 22(10), 2015. doi:10.1063/1.4932073.

[60] G. Vanhulle, Y. Cressault, P. Teulet, F. Reichert, and A. Petchanka. Transport properties of multi-temperature $SF_6$ plasmas: influence of assumptions done in the plasma composition calculation. In *XXIIIrd International Symposium on Plasma Chemistry*, Montréal, Canada, 2017.

[61] L. Soucasse, P. Riviere, and A. Soufiani. Monte Carlo methods for radiative transfer in quasi-isothermal participating media. *Journal of Quantitative Spectroscopy and Radiative Transfer*, 128:34, 2014. doi:10.1016/j.jqsrt.2012.07.008.

[62] M. Scoggins. *Development of numerical methods and study of coupled flow, radiation, and ablation phenomena for atmospheric entry.* PhD thesis, Univsersity Paris-Saclay, 2017.

doi:10.14311/ppt.2019.3.217
*Plasma Physics and Technology* **6**(3):217–222, 2019 

# LINEAR PINCH EQUILIBRIUM OF NON-NEUTRAL PLASMA REVISITED:PHENOMENOLOGICAL CONSEQUENCES OF A NUMERICAL ACCURACY PROBLEM

F. L. Braga *, D. N. Soares

*Instituto Federal de Educação, Ciências e Tecnologia do Espírito Santo, Campus Cariacica, Coordenadoria de Física, ES, Brazil.*

* filipe.braga@ifes.edu.br

**Abstract.** Weibel in 1959 under considerations of a collisionless non-neutral cylindrical plasma column studied a linear pinch confinement equilibrium. As reported here, due to non-linearity of the ordinary differential equations obtained for the electrostatic and magnetostatic fields is possible to demonstrate that the confining features previously obtained are extremely dependent on the initial conditions, and the arrangement of two parameters ($\beta$ - the ratio between ion and electron mass; $M/KT$ - ratio between relativistic rest energy associated with the pair electron-ion and thermal energy $kT$) related to the plasma column characteristics. We investigated in this paper the plasma column behavior (confining or non-confining) under modifications of that set of parameters. We detected a set of parameters values that imposes a confining configuration with an electronic skin effect on the plasma column, not yet reported or discussed in the literature.

**Keywords:** Locally Non-neutral Plasma, Linear Pinch, Skin Effect, Numerical Accuracy.

## 1. Introduction

Theoretical studies related to the equilibrium of plasma columns confining using mathematical and physical complex configurations of magnetostatic fields as non-axisymmetric magnetic fields in toroidal systems reported by [1, 2], bicuspid arrangements investigated by [3], or self-organizing structuring as observed by [4] have been extensively studied through literature. In each one of these cases, the nonlinearity and the number of differential equations that govern the physical environment are higher when compared for example with a linear pinch configuration. In spite of its simplicity, this setting has been investigated and reported previously in the literature by [5–10], and others. A significant part of these studies was intrinsically related to magnetic confining of fusion plasma [11], and in general, the phenomenological scenario obtained for the plasma column exhibit a generous amount of physical effects and other characteristics, even in simple cases of the linear pinch.

In the present article, we focus on investigating the results proposed by [10]. Beyond the ideal properties of Weibel model, the applicability and, probably, the mathematical clarity of his work, make it useful for other studies [12–14]. Any of that articles intended to reevaluate the numerical results and investigate the plasma column phenomenological behavior at the linear pinch configuration analyzed as we developed in this work.

The novel skin effect predicted by numerical solutions of Weibel model here observed seems to be of great interest for dusty plasma studies [15]. Especially the ones related to technological plasma and basic concepts articles [16–18], where the plasma column is typically non-neutral, and the mass ratio of the species (ion-dusty particle/electron) in general are sufficiently big.

We organized the article as the following: in section 2 we set up the main characteristics associated with Weibel theoretical model, in section 3 we developed and discussed the numerical solutions for the proposed model, and as an intent to explain the skin effect observed, from the perspective of an interplay between electromagnetic energy and thermal energy, we performed an extensive search for confining and non-confining configurations in order to numerical confirm this proposition, and finished the article summarizing all the results in section 4.

## 2. Linear Pinch Equations

The physical system considered was a plasma column with ions (or ionized particles) and electrons in a collisionless stationary configuration. Ions and electrons masses and electric charges were denoted by $(m_+, +e)$ and $(m_-, -e)$ respectively, being $e$ the elemental charge. All particles were moving along the $z$-axis as proposed in the diagram scheme in Figure 1. Ions and electrons move oppositely.

The electric and magnetic fields were represented by their potentials $\phi(\vec{q})$- electrostatic potential, and $\vec{A}(\vec{q})$ - magnetic vector potential, that only depends on coordinates $\vec{q}$. Each particle under the influence of these fields was governed by the following Hamiltonian

$$H_\pm = \frac{1}{2m_\pm}\sum_{i=1}^{3}(p_i \mp eA_i)^2 \pm e\phi, \qquad (1)$$

where $i = 1, 2, 3$ represents the three coordinates of vector $\vec{q}$ (position), and $\vec{p}$ was the particles linear momentum.

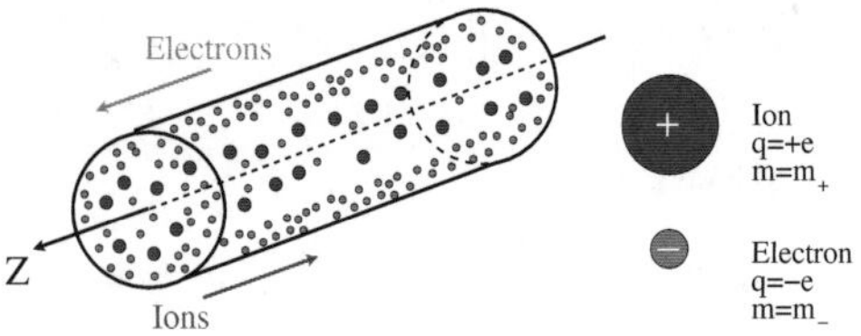


*Figure 1. Schematic diagram of the plasma column analyzed. m± and ±e are the mass and electric charge of particles considered.*

Under consideration of thermodynamic equilibrium the particles probability distribution function should follow the Boltzmann distribution

$$F_{\pm} = const \cdot \mathrm{e}^{\frac{-H_{\pm}(p_i, q_i)}{kT}}, \quad (2)$$

being *const* a normalization factor, $k$ the Boltzmann constant and $T$ the system temperature. From the statistical point of view, the ion/electron densities and the corresponding current densities could be cast as

$$n_{\pm}(\vec{q}) = \int F_{\pm}(\vec{p}, \vec{q}) d^3\vec{p}, \quad (3)$$

$$\vec{J}_{\pm}(\vec{q}) = \pm \frac{e}{m_{\pm}} \int (\vec{p} - e\vec{A}) F_{\pm}(\vec{p}, \vec{q}) d^3\vec{p}. \quad (4)$$

Using the same framework of Weibel [10] we used cylindrical system of coordinates and considered that time oscillations of the electromagnetic fields were small enough, then, they could be considered stationary, on other words $\frac{\partial \vec{E}}{\partial t} = -\frac{\partial \vec{B}}{\partial t} = 0$. In addition, we considered that the problem cylindrical symmetry imposes that $n_{\pm}(r)$, then, $\rho(r)$ - free electric charge density . And $\vec{J} = J_z(r)\hat{e}_z$ - free current density, then, $\vec{A} = A_z(r)\hat{e}_z$.

As proposed by [10], avoiding collision effects, it is possible to modulate $F_{\pm}$ by a Dirac delta function $\delta(p_z)$. The physical consequence associated with that modulation is the imposition of an equilibrium distribution function, related to purely magnetostatic configurations where no net current can be observed, despite the maintenance of local non-neutrality.

Using the following abbreviations and change of variables,

$$\frac{1}{m_{\pm}} = \frac{1}{2M}(1 \mp \beta), \quad (a_+ a_-)^{\frac{1}{2}} = n_0, \quad r = (\frac{M}{e^2 n_0})^{\frac{1}{2}} x,$$

$$\frac{eA_z(r)}{2(MkT)^{\frac{1}{2}}} = u(x), \quad \frac{e\phi(r)}{kT} - \frac{1}{2}\ln(\frac{a_+}{a_-}) = y(x),$$

where $a_{\pm}$ represents the normalization constant of each distribution $F_{\pm}$. Here $y$ and $u$ represent respectively the dimensionless electric potential and magnetic vector potential. $x$ is a dimensionless radial coordinate. $n_0$ is a typical plasma particle density at the cylinder center -taking into account ion and electrons. $M = \frac{m_+ m_-}{m_+ + m_-}$ is a reduced mass for the species analyzed, and $\beta = \frac{1 - \frac{m_-}{m_+}}{1 + \frac{m_-}{m_+}}$ is the dimensionless difference of mass between the ions and electrons. It was considered here that $c$ (light speed) is unitary – vide reference [10]. Then, densities became

$$\rho = en_0 \left[ \mathrm{e}^{-u^2 + u^2\beta - y} - \mathrm{e}^{-u^2 - u^2\beta + y} \right],$$

$$J_z = \frac{e^2 A_z n_0}{2M} [(1-\beta)\mathrm{e}^{-u^2 + u^2\beta - y} + (1+\beta)\mathrm{e}^{-u^2 - u^2\beta + y}]$$

and the potentials were given by

$$\phi = \left[ y(x) + \frac{1}{2}\ln\left(\frac{a_+}{a_-}\right) \right] \frac{kT}{e}, \quad A_z = \frac{2\sqrt{MkT}}{e} u(x).$$

As result the Maxwell equations for the potentials became

$$y''(x) + \frac{y'(x)}{x} = \frac{M}{kT}\mathrm{e}^{-u^2} \left[ e^{-u^2\beta + y} - \mathrm{e}^{u^2\beta - y} \right], \quad (5)$$

$$u''(x) + \frac{u'(x)}{x} = u(x)\mathrm{e}^{-u^2} \left[ \frac{(1-\beta)}{2}\mathrm{e}^{u^2\beta - y} + \frac{(1+\beta)}{2}\mathrm{e}^{-u^2\beta + y} \right] \quad (6)$$

the prime on each function denotes derivative with respect to the variable $x$, *e.g.* $u' = \frac{du}{dx}$. In the work developed by [10] the initial conditions were set as

$$u(0) = \left[ \frac{1}{2}\ln\left( \frac{n_0^2}{n_+(0)n_-(0)} \right) \right]^{1/2} = u_0, \; u'(0) = 0, \quad (7)$$

$$y(0) = \beta u(0) - \frac{1}{2}\ln\left(\frac{n_+(0)}{n_-(0)}\right) = y_0, \; y'(0) = 0. \quad (8)$$

In order to determine the plasma column equilibrium configuration - confining or non-confining, it was necessary to solve equations (5) and (6). As depicted by [10], the entire system is dependent on four parameters, $n_{\pm}(0)/n_0$, $\beta$ and $M/kT$. From the mathematical point of view it would be dependent on the values of $\{u(0), y(0), M/kT, \beta\}$, on other words, plasma column confining would depend explicitly on potentials at the system of coordinates origin $(u(0), y(0))$, ratio between relativistic rest energy associated with the pair electron-ion and thermal energy $kT$ $(M/kT)$ and the ratio between ion and electron mass $(\beta)$.

## 3. Results

In order to reproduce Weibel's numerical results using the non-linear differential equations (5) and (6), we performed a manual search to determine the initial value for the $x$ coordinate, since it was not explicitly mentioned on reference [10]. With the ODE2 and IC2 functions from WxMaxima software we performed various integrations of the set of differential equations with different integration starting point $x_0$, imposing the initial conditions proposed by [10] ($u(x_0) = u_0, y(x_0) = y_0, u'(x_0) = 0, y'(x_0) = 0$). In spite of several tests evaluated, during that manual search, decreasing the values from the unity up to $x_0 = 10^{-6}$, we were unable to reproduce the previous smooth numerical results obtained by Weibel [10]. In all tests, the obtained electronic densities profiles were exhibiting a singularity near the plasma edge.

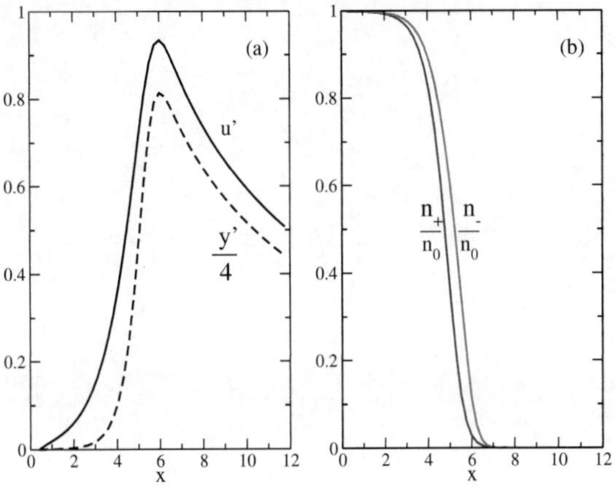


*Figure 2. Determination of the (a) dimensionless field $u'$ and $y'/4$ and (b) the normalized ion and electron densities $n_\pm/n_0$ both for the boundary conditions ($u(0) = 0.05, y(0) = 0.00279376, y(12) = 17.9961, u(12) = 5.83457$) and parameters ($\beta = 0.999455, M/kT = 10$) for a grid with 50 nodes.*

For reproduction of Weibel's profile, to clarify how he achieved them, we tried to use a finite difference method [19], changing the initial condition problem to a boundary condition problem. For that purpose we get Fig. 2 from the article [10] and using the software Digitizer we determined approximately the values of $n_\pm/n_0$ for the coordinate $x = 12.0$ and determined the respectively values of $y(12)$ and $u(12)$ solving the equations system related to $n_\pm/n_0$.

With the boundary conditions ($u(0) = 0.05$, $y(0) = 0.00279376$) – taken from [10], ($y(12) = 17.9961$, $u(12) = 5.83457$) – determined by the procedure described above, and parameters ($\beta = 0.999455$, $M/kT = 10$)), we performed two integrations of equations (5) and (6) both of them using the BC2 function from Wxmaxima software.

The first one reproduced the profiles obtained by [10] as observed in Figure 2 considering equidistant grid with 50 nodes. The obtained profiles were smooth, and especially the electronic density profile near the plasma edge decreased without any peak.

The second integration confirmed the results firstly obtained using the Wxmaxima ODE2 function as shown in Figure 3 showing a peaked electronic density near the plasma edge, but that time using a grid with 100 nodes.

The patterns that are shown in Figure 3 were precisely the same obtained by the direct integration of the initial condition problem but using the ODE2 and IC2 functions of WxMaxima.

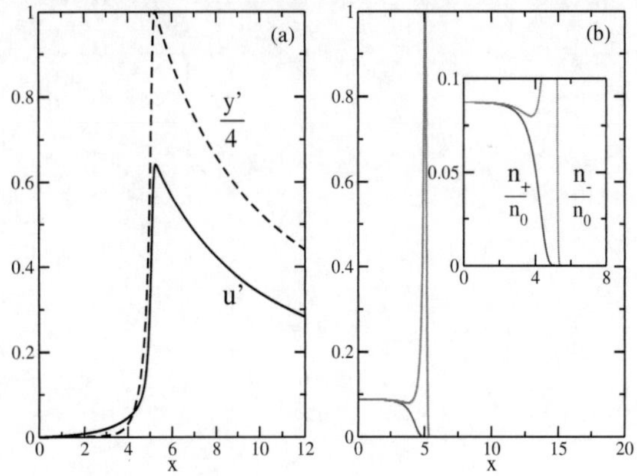


*Figure 3. Determination of the normalized graphics with respect to its maximum values (a) dimensionless field $u'$ and $y'$ and (b) the normalized ion and electron densities $n_\pm/n_0$ both for parameters ($u(x_0) = 0.05, y(x_0) = 0.00279376, \beta = 0.999455, M/kT = 10$) with initial point of integration $x_0 = 10^{-6}$, and using the finite difference method for the boundary conditions ($u(0) = 0.05, y(0) = 0.00279376, y(12) = 17.9961, u(12) = 5.83457$) for a grid with 100 nodes.*

The profiles dependency with the grid node number was numerically investigated for different grid's nodes number values (30, 50, 100 and 1000). And as observed in Figure 4 for grids larger than 100 the precision required for the integration purpose was achieved, and for grid's smaller than 50 the integration returns infinity fields profiles unpredicted by the asymptotic analysis proposed by Weibel [10].

As Weibel do not report many details associated with the computational procedure used on reference [10], we can not confirm, but probably due to hardware constraints, the numerical accuracy requirements needed to analyze the equation with greater precision were not met. And the different phenomenological results obtained here by two different and highly precise integration techniques are plausible and were not predicted before and are discussed below. As one can see given that Weibel's numerical results are recovered in the limit of low precision for the techniques here used, it indicates that the numerical accuracy was responsable for Webiel's graphical results and consequent interpretation.

Physically, the new integration results obtained showed that in spite of the plasma column local nonneutrality, visible by the fact that the difference between $n_+/n_0$ and $n_-/n_0$ at the same position $x$ are not null, the confining configuration is achieved - densities decrease for large values of $x$. The $\beta$ value close to unity shows that ion mass is greater than electron mass - acceptable for example on fusion plasma approximations. The $M/kT = 10$ value represents that relativistic rest energy is ten times greater than system thermal energy. For an electron-proton plasma $M/kT = 10$ implies a temperature $T \approx 10^{12}K \approx 100GeV$, and for an electron-positron plasma $T \approx 10^8 K \approx 100KeV$.

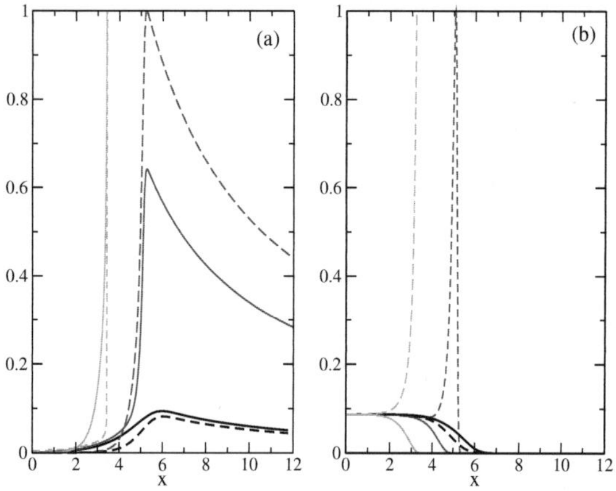

*Figure 4. Different grid nodes number simulations (orange curves - 30 nodes, black curves- 50 nodes and violet curves - 100 and 1000 nodes) for the boundary conditions ($u(0) = 0.05, y(0) = 0.00279376, y(12) = 17.9961, u(12) = 5.83457$) and parameters ($\beta = 0.999455, M/kT = 10$) representing the normalized (a) dimensionless field $u'$ (filled lines) and $y'/4$ (dashed lines) and (b) the normalized ion and electron densities $n_+/n_0$ (filled lines) and $n_-/n_0$ (dashed lines).*

On the profile observed in Figure 3 (b) is possible to detect that the normalized ion distribution $n_+/n_0$ has been shielded by the normalized electron distribution $n_-/n_0$. Hence, forming a cylindrical plasma column with a positive core and negative boundary. The peaked electron density $n_-/n_0$ near the plasma boundary (a region where the particle densities tend to zero) can be interpreted as a skin effect [20], since the entire proposed model lies on a superconductor approximation, given the plasma column is collisionless.

As observed in the field profiles, there is an inversion on the predominance of dimensionless magnetic and electric fields ($u'$ and $y'$) - Figure 3 (a) when compared to results obtained by [10] as depicted in Figure 2(a). Electromagnetic fields inversion (signal inversion or intensity inversion) have been previously reported in the literature for charged fluids [21], and for fusion plasma [22], always directly related to spatial modifications on the species densities.

Motivated by the skin effect, that was not reported in literature, we performed a scan of the parameters space ($\beta$, $M/kT$) for fixed initial potentials ($u(x_0)$, $y(x_0)$). And by direct integration of equations (5) and (6), determined which set of parameters were associated with confining and non-confining equilibrium.

The confining condition was determined by the analysis of $n_\pm/n_0$ and the signals of theirs first and second derivatives ($n'_\pm/n_0$, $n''_\pm/n_0$) at larger values of $x$ - ($x \geq 14$). If the normalized densities of both species were smaller than a minimum value $n_{min} = 10^{-2}$, in addition with $n'_\pm/n_0 < 0$ and $n''_\pm/n_0 > 0$, we accepted that a confining configuration was reached.

The result obtained by this scan can be observed in Figure 5. The filled line delimits the parameters region related to confining (Region I) and non-confining (Region II e III) configurations.

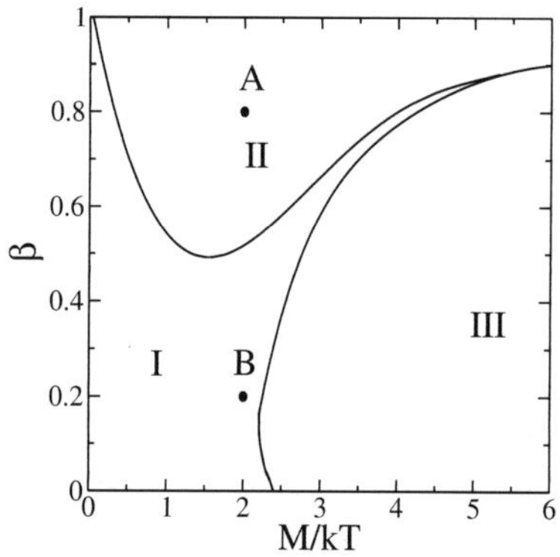


*Figure 5. Determination of the confining and non-confining configurations of parameters $\beta$ and $M/kT$ for the same initial conditions $u(0) = 0.05$, $y(0) = 0.00279376$. In the map, region I is associated with confining and region II e III with the non-confining.*

The interval of parameters $0 \leq \beta \leq 1$ is associated with the limit configurations $m_+ \gg m_- \Rightarrow \beta = 1$, $m_+ = m_- \Rightarrow \beta = 0$. From the mass point of view, this parameter allows the application of Weibel model to analyze dusty plasma [15, 16], considering extremely heavy ions when compared to electrons. For $M/kT$ we set up the interval $0 \leq M/kT \leq 6$ since that was the region where the system changes in a non-trivial way from confining to non-confining configurations. If one considered the case of an electron-positron plasma $M/kT = 6$ would be related to temperatures $T \approx 1.8 \times 10^9 K \approx 150MeV$, and $M/kT \approx 0$ must be related with $T > 10^{12}K \approx 100MeV$.

In order to get a *ansatz* about the physical explanation for this stable confining behavior, we calculate the ratio $E_{EM}/E_T$ between the system electromagnetic energy $E_{EM}$ and thermal energy $E_T$. To determine electromagnetic energy we directly integrate the fields densities

$$E_{EM} = \frac{1}{8\pi} \int [(-\nabla\Phi)^2 + (\nabla \times \vec{A})^2] rLdrd\phi,$$

and thermal energy by integration of an ideal gas approximation for energy

$$E_T = \frac{3}{2} \int (n_+ + n_-)kTrLdrd\phi.$$

Both considering a cylinder of height $L$ and radius $R_0$ (here representing the plasma boundary).

Using the same change of variables proposed by [10], is possible to determine that

$$E_{EM} = \frac{L(kT)^2}{4e^2}\int_0^b \left[\left(\frac{\partial y}{\partial x}\right)^2 + \frac{4M}{kT}\left(\frac{\partial u}{\partial x}\right)^2\right] x dx$$
$$E_T = 3kT\pi L n_0 \left(\frac{M}{e^2 n0}\right)\int_0^b \left(\frac{n_+}{n_0} + \frac{n_-}{n_0}\right) x dx$$

been, $b = R0\left(\frac{e^2 n0}{M}\right)^{\frac{1}{2}}$, so the ratio $E_{EM}/E_t$ became

$$\frac{E_{EM}}{E_T} = \frac{1}{12\pi}\frac{kT}{M}\frac{I}{II}, \quad (9)$$

where,

$$I = \int_0^{12}\left[\left(\frac{\partial y}{\partial x}\right)^2 + \frac{4M}{kT}\left(\frac{\partial u}{\partial x}\right)^2\right] x dx, \quad (10)$$
$$II = \int_0^{12}\left(\frac{n_+}{n_0} + \frac{n_-}{n_0}\right) x dx. \quad (11)$$

The integration was performed from $x_0 = 10^{-6}$ through $x = 14$ using the integration tool of Xmgrace software. The comparison between two arrangements with the same value of $M/kT$, one with confining and other with non-confining configuration, can be observed in Table 1. As quoted, confining equilibrium (Weibel case and Set B) was obtained when thermal energy is smaller than electromagnetic energy, hence $E_{EM}/E_T > 1$. In other words, the behavior of thermal expansion emulating an ideal gas is suppressed by the electrostatic and magnetostatic fields pressure, and at non-confining equilibrium (Set A), the different feature dominates, characterizing an interplay between this two effects - thermal expansion given and electromagnetic confining. That interplay can be depicted by the transition line at the parameter space.

With the proposed heuristic stability analysis is not possible to confirm the interpretation given here, a more accurate stability analysis for example using time-dependent pertubation theory must be performed in order to solve this remaining task.

| Set | $\beta$ | $M/kT$ | $E_{EM}/E_T$ | Status |
|---|---|---|---|---|
| - | 0.999455 | 10.0 | $\sim 3.16$ | C |
| A | 0.800000 | 2.00 | $\sim 0.24$ | NC |
| B | 0.200000 | 2.00 | $\sim 4.00$ | C |

*Table 1. Table of ratio between electromagnetic energy and thermal energy for compared configurations with confining (C) and non-confining (NC) equilibrium.*

## 4. Conclusions

We theoretical and numerically demonstrated that the cylindrical plasma model on a collisionless condition with ions and electrons moving along the $z$ axis, as proposed by Weibel in 1959, is dependent on the arrangement of the initial conditions for electrostatic and magnetostatic fields and a set of two parameters related to i) mass ratio between ions and electrons and ii) energy ratio between relativistic rest energy associated with the pair electron-ion and thermal energy. As observed, on the confining configurations is possible to detect a peaked electron density profile near the plasma boundary flowing around a low mobility ion core, emulating electronic skin effect on a conductor material, given the collisionless condition imposed on the model. The nonlinearity of the transition behavior from confining to non-confining at the parameters spaces ($\beta$,$M/kT$) was detected. And as proposed by the heuristic analysis probably it is related to an interplay between the electromagnetic energy (confining) and the thermal energy (expansion – non-confining) of the plasma column.


## Acknowledgements

We would like to thanks, Instituto Federal de Educação Ciências e Tecnologia do Espírito Santo, for the opportunity.


## References


[1] P. Helander. Theory of plasma confinement in non-axisymmetric magnetic fields. *Rep. Prog. Phys.*, 77(8):087001, 2014. doi:10.1088/0034-4885/77/8/087001.

[2] M. D. Kruskal and R. M. Kulsrud. Equilibrium of a Magnetically Confined Plasma in a Toroid. *Phys. Fluids*, 1(4):265–274, 1958. doi:10.1063/1.1705884.

[3] D. Mascali, G. Torrisi, L. Neri, G. Sorbello, G. Castro, L. Celona, and S. Gammino. 3D-full wave and kinetics numerical modelling of electron cyclotron resonance ion sources plasma: steps towards self-consistency. *Eur. Phys. J. D*, 69(1), 2015. doi:10.1140/epjd/e2014-50168-5.

[4] C. B. Smiet, S. Candelaresi, A. Thompson, J. Swearngin, J. W. Dalhuisen, and D. Bouwmeester. Self-Organizing Knotted Magnetic Structures in Plasma. *Phys. Rev. Lett.*, 115:095001, Aug 2015. doi:10.1103/PhysRevLett.115.095001.

[5] W. A. Newcomb. Hydromagnetic stability of a diffuse linear pinch. *Ann. Phys.*, 10(2):232–267, 1960. doi:10.1016/0003-4916(60)90023-3.

[6] J. Koliner, M. Cianciosa, J. Boguski, J. Anderson, J. Hanson, B. Chapman, D. Brower, D. Den Hartog, W. Ding, J. Duff, et al. Three dimensional equilibrium solutions for a current-carrying reversed-field pinch plasma with a close-fitting conducting shell. *Physics of Plasmas*, 23(3):032508, 2016.

[7] U. Shumlak, B. Nelson, E. Claveau, E. Forbes, R. Golingo, M. Hughes, R. Oberto, M. Ross, and T. Weber. Increasing plasma parameters using sheared flow stabilization of a z-pinch. *Physics of Plasmas*, 24(5):055702, 2017.

[8] E. Kroupp, E. Stambulchik, A. Starobinets, D. Osin, V. Fisher, D. Alumot, Y. Maron, S. Davidovits, N. Fisch, and A. Fruchtman. Turbulent stagnation in a z-pinch plasma. *Physical Review E*, 97(1):013202, 2018.

[9] J. Goedbloed. Stabilization of magnetohydrodynamic instabilities by force-free magnetic fields. *Physica*, 53(4):501–534, 1971. doi:10.1016/0031-8914(71)90113-3.

[10] E. S. Weibel. On the Confinement of a Plasma by Magnetostatic Fields. *Phys.Fluids*, 2(1):52–56, 1959. doi:10.1063/1.1724391.

[11] F. F. Chen and M. D. Smith. *Plasma*. John Wiley and Sons, Inc., 2005. doi:10.1002/0471743984.vse9673.

[12] G. Schmidt and D. Finkelstein. Magnetically Confined Plasma with a Maxwellian Core. *Phys. Rev.*, 126:1611–1615, Jun 1962. doi:10.1103/PhysRev.126.1611.

[13] C. C. Pian and A. W. McClaine. Techniques for the solution of MHD generator flows. *Comput. Fluids*, 12(4):319–338, 1984. doi:10.1016/0045-7930(84)90013-6.

[14] P. Gratreau and P. Giupponi. Vlasov equilibria of cylindrical relativistic electron beams of arbitrary high intensity. *Phys. Fluids*, 20(3):487–493, 1977. doi:10.1063/1.861887.

[15] B. M. Annaratone, W. Jacob, C. Arnas, and G. E. Morfill. Critical review of complex plasma (dusty plasma) diagnostics and manipulation techniques for the fusion community and others. *IEEE Transactions on Plasma Science*, 37(1):270–280, Jan 2009. doi:10.1109/TPS.2008.2006269.

[16] P. K. Shukla and A. A. Mamun. Introduction to dusty plasma physics. *Plasma Physics and Controlled Fusion*, 44(3):395, 2002. doi:10.1088/0741-3335/44/3/701.

[17] A. A. Fridman, L. Boufendi, T. Hbid, B. V. Potapkin, and A. Bouchoule. Dusty plasma formation: Physics and critical phenomena. theoretical approach. *Journal of Applied Physics*, 79(3):1303–1314, 1996. doi:10.1063/1.361026.

[18] S. Mayout, L. A. Gougam, and M. Tribeche. Effects of ionization and ion loss on dust ion-acoustic solitary waves in a collisional dusty plasma with suprathermal electrons. *Physics of Plasmas*, 23(3):033701, 2016. doi:10.1063/1.4942935.

[19] W. H. Press, S. A. Teukolsky, W. T. Vetterling, and B. P. Flannery. *Numerical Recipes 3rd Edition: The Art of Scientific Computing*. Cambridge University Press, New York, NY, USA, 3 edition, 2007.

[20] D. Griffiths. *Introduction to Electrodynamics*. Prentice Hall, 1999.

[21] D. Forcella, J. Zaanen, D. Valentinis, and D. Van Der Marel. Electromagnetic properties of viscous charged fluids. *Physical Review B*, 90(3):035143, 2014.

[22] F. Halpern, P. Ricci, S. Jolliet, J. Loizu, J. Morales, A. Mosetto, F. Musil, F. Riva, T.-M. Tran, and C. Wersal. The gbs code for tokamak scrape-off layer simulations. *Journal of Computational Physics*, 315:388–408, 2016.

doi:10.14311/ppt.2019.3.223
*Plasma Physics and Technology* **6**(3):223–226, 2019 

# SWITCHING BEHAVIOR OF A SERIES CONNECTION OF A VACUUM INTERRUPTER AND A GAS CIRCUIT BREAKER

N. Götte[a,*], M. Bendig[a], T. Krampert[a], P.G. Nikolic[b]

[a] *Institute for High Voltage Technology, RWTH Aachen University, 52056 Aachen, Germany*
[b] *Siemens AG, Corporate Technology, 91058 Erlangen, Germany*
[*] goette@ifht.rwth-aachen.de

**Abstract.** After being in the focus of sciences' and industry's research and development activities for many years, the investigation of possible $SF_6$ gas-alternatives has been even more intensified after the revision of the European F-Gas regulation 517/2014. As natural gases yield a significantly lower dielectric strength in comparison to $SF_6$, new challenges arise for the design of high voltage switchgear. Vacuum interrupters are environmentally friendly, reliable and able to withstand steep rising transient recovery voltages. In the last years, first installations of switchgear based on vacuum switching technology in sub-transmission level are in operation. One option for the realization of a $SF_6$ free high voltage switchgear for transmission level is the combination of a gas circuit breaker filled with an atmospheric gas with a vacuum interrupter in a hybrid switchgear. In this contribution the voltage distribution and switching behavior of a hybrid circuit breaker is experimentally investigated.

**Keywords:** vacuum circuit breaker, gas circuit breaker, hybrid circuit breaker, post-arc current, recovery voltage.

## 1. Introduction

Nowadays sulphur hexafluoride ($SF_6$) is the state-of-the-art insulating and arc-quenching medium used in gas-insulated switchgear, enabling safe current interruption and high dielectric strength. At the same time, if released to the environment, $SF_6$ is a potent greenhouse gas with a high global warming potential. After being on the agenda of research and development activities of science and industry for many years, the revision of the European regulation on fluorinated gases in 2014 even more turned the focus on the properties of possible gas-substitutes for $SF_6$. In the field of medium-voltage technology, vacuum circuit breakers (VCB) have almost completely replaced other breaking principles. These circuit breakers are reliable, environmentally friendly and low-maintenance. In recent years, there are efforts to further develop the vacuum technology to make it suitable for the high voltage level. On one hand VCB can deal with very high rate-of-rise-of-recovery voltage (RRRV) and the highest short-circuit currents, on the other hand, dielectric performance of vacuum is not rising linearly with the contact distance as it does in gases. One possible approach for an environmentally friendly switchgear for the high voltage level is a combination of VCB and gas circuit breaker (GCB) as a series connection. The interest in a hybrid circuit breaker concept has existed for almost 30 years [1], but never became a commercially available switchgear due to the success of the $SF_6$ circuit breaker.

In this contribution interaction and switching behavior of a $CO_2$-filled GCB and a VCB in a series connection around and after CZ is analyzed. It is known that the postarc current of the VCB has a great impact on voltage distribution of the series connection after CZ [2, 3]. Therefore, this work focuses on postarc current of the VCB and the relevance for the series connection and the voltage distribution after CZ.

## 2. Test Setup and Procedure

The investigated series connection consists of a commercial VCB and a self-blast GCB model. The VCB is driven by a spring mechanism and has radial-magnetic-field (RMF) contacts. The GCB model is constructed modularly so the contact distance, nozzle-system, heating volume etc. can be varied. The contact system is made of tungsten-copper (WCu) and and in this work a fixed contact distance of 40 mm is used. The contacts embedded into a nozzle-system with a heating channel made of polytetrafluoroethylene (PTFE), which leads through a 90 degree bend into the heating volume. The entire circuit breaker model is rotationally symmetric. $CO_2$ is used as insulation and quenching gas. The switching arc in the GCB is initiated by a copper ignition wire with a diameter of 60 µm.

### 2.1. Test Setup

The used synthetic test circuit is shown in Figure 1 and consists of two parts, the high current and the high voltage circuit. After charging the capacitors $C_{HC}$ and $C_{HV}$ the test is initiated by the making switch (MS). A half-cycle of a sinusoidal current with a frequency of $f \approx 50$ Hz is fed into to the device-under-test (DUT). Shortly before CZ the auxiliary switch (AS) is opened, the spark gap (SG) is triggered and the injection current of the high voltage

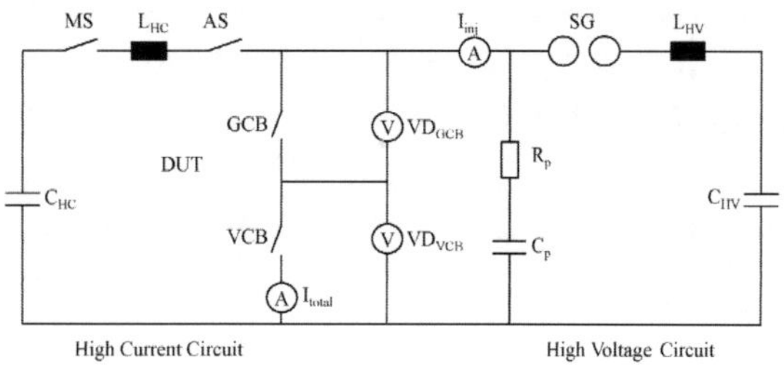


*Figure 1. Test circuit consisting of high current and high voltage circuit.*

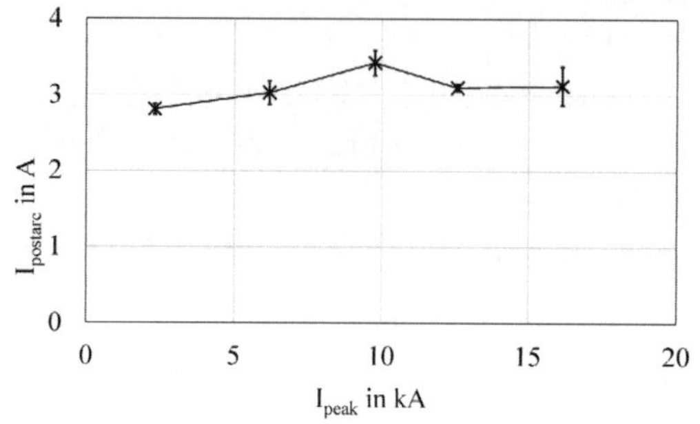


*Figure 2. Variation of the current amplitude at 10 A/µs, 6 kV/µs and 8 ms arcing time.*

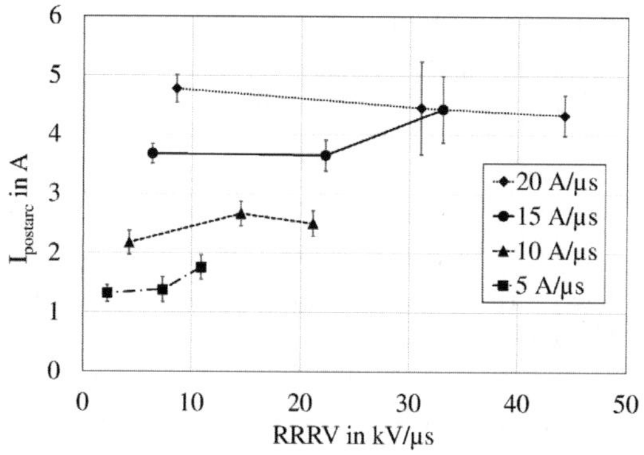


*Figure 3. Variation of current steepness and RRRV at $I_{peak} = 2.6\,kA$ and 8 ms arcing time.*

circuit is superimposed onto the test current. The current steepness in CZ is independent of the current amplitude and is set by the charging voltage of $C_{\text{HV}}$ and the resonance frequency of the high voltage circuit. The rate-of-rise-of-recovery-voltage (RRRV or $\mathrm{d}u/\mathrm{d}t_{\text{CZ}}$) is set by the parallel network $R_{\text{p}}$ and $C_{\text{p}}$. The measurement data is recorded with a high resolution (12 bit), high frequency (100 MHz) optically isolated measurement system. The postarc current is measured in the high voltage circuit ($I_{\text{inj}}$) with a coaxial shunt resistor (25 mΩ) protected by anti parallel diodes. To determine the current through the circuit breaker, the current through the stray capacitance is subtracted from the measured current value. The transient recovery voltage (TRV) over the circuit breakers are measured with two 1000:1 voltage dividers (VD), the total TRV is calculated by addition of these two signals.

## 3. Experimental Results

### 3.1. Vacuum circuit breaker

To analyze the influence of postarc current on the series connection the behavior of VCB in single operation is analyzed. Using the synthetic test circuit the current amplitude, the current steepness at CZ and the voltage steepness after CZ are set independently to determine the impact on the postarc current. With the gained information a direct comparison with the series connection is possible.

#### 3.1.1. Current amplitude

The current amplitude is varied in the range from 2 kA to 16 kA peak current. All other parameters, especially the current steepness (10 A/µs), RRRV (6 kV/µs) and arcing time (8 ms) is set constant.

Figure 2 shows the postarc current amplitude in dependence of the current amplitude. In contrast to [4] no direct correlation with the current amplitude is visible. As only currents up to 16 kA are investigated, the influence of contracted arc mode is supposed to be low.

#### 3.1.2. Current steepness and RRRV

The variation of the current steepness, RRRV and the resulting postarc currents are shown in Figure 3. For each data point at least five measurements were carried out. To achieve different $\mathrm{d}u/\mathrm{d}t_{\text{CZ}}$ values at a specific $\mathrm{d}i/\mathrm{d}t$ the parallel resistance of the parallel network $R_{\text{p}}$ is varied in three steps. It is evident that the peak value of postarc current has a dependency with the set $\mathrm{d}i/\mathrm{d}t$ in CZ. A change of $\mathrm{d}u/\mathrm{d}t_{\text{CZ}}$ shows no change on the postarc current peak value of the VCB. Similar results regarding the current steepness and the postarc current in CZ are shown in [1].

#### 3.1.3. Contact distance

From theory and other investigations it is known that the contact-distance has a direct correlation with the peak value of the postarc current [5]. The opening of the VCB is recorded using a high speed camera and synchronized with the set arcing time. In Figure 4 the measured relation between the contact distance at CZ and the measured maximum value of the postarc current is shown. A dependence between the maximum value of the postarc current and the contact distance is visible.

### 3.2. Series Connection

For the investigations of the series connection different types of voltage distributions after CZ are observed. The most basic type is the capacitive voltage distribution. It is observed at voltage pulses without a prior

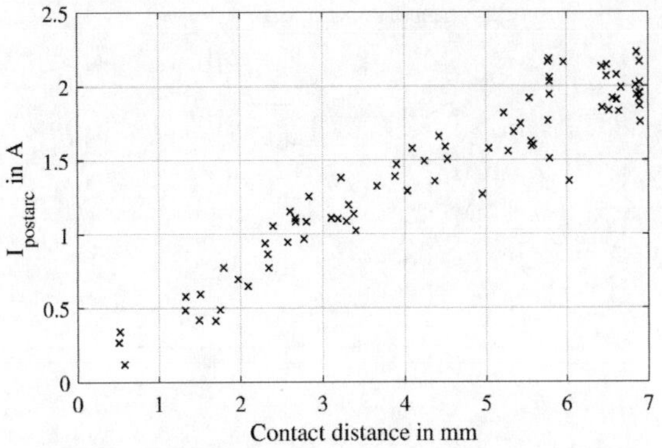


*Figure 4. Variation of the contact distance of the VCB at $I_{peak} = 2.6\,kA$, $di/dt = 10\,A/\mu s$ and $du/dt = 14\,kV/\mu s$.*

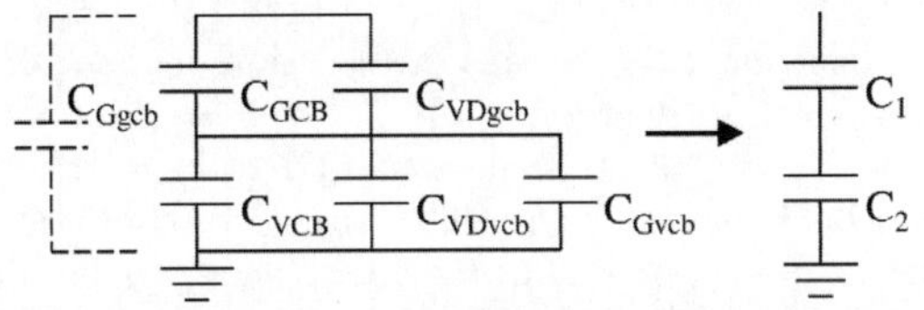


*Figure 5. Relevant capacities for the voltage distribution in the test circuit.*

current interruption, for example lightning impulses with open contacts.

During the interruption process in the hybrid circuit breaker, the current flows through both breakers. After CZ the differences between the postarc currents in the gas- and vacuum circuit breaker plays a major role in the voltage distribution. Typically, the postarc current of the GCB is in the range of several milliamperes, the postarc current of the VCB is in the range of several amperes. Since the current in a series connection must flow through both breakers, two scenarios are possible. The postarc current of the VCB forces a current flow through the GCB and the switching gap in the GCB stays conductive. The second possibility is that the postarc current of the VCB is suppressed by the GCB. Depended on the specific values of the postarc current a transition phase between these two operations is conceivable.

### 3.2.1. Capacitive Voltage distribution

The voltage distribution between both breakers during a switching pulse without prior current interruption depends only on the capacitive ratio. The stray capacitance of VCB is higher than that to the GCB due to the smaller contact distance and larger contact surface of the VCB. In Figure 5 the relevant capacities for the voltage distribution are given: the capacities of the breakers, the capacities of the voltage divider and the ground capacity in parallel to the circuit breaker placed on the ground side. The ground capacity in parallel to the breaker on the high voltage side (cf. Fig. 5: $C_{Ggcb}$) has no effect on the voltage distribution. For the given results in this investigation the division ratio is about $\frac{C_1}{C_2} \approx \frac{13\,\text{pF}}{100\,\text{pF}} = 0.13$. In the test setup this division ratio can easily be varied by changing the breakers on high voltage and ground side, the connection of the voltage divider or the distances to ground. Moreover, external grading capacities can be added to change the voltage distribution. To evaluate the influence of postarc current on the voltage distribution, the uninfluenced division ratio of the setup is necessary.

In Figure 6 two voltage distributions are compared. In both experiments the TRV shape is the same. In the left one the TRV is applied without a prior current interruption. Therefore the voltage is divided mainly by the capacitive division ratio and the TRV rises for the most part across the GCB. In the right part the postarc current of the VCB leads to a failure in the GCB and in the first microseconds the TRV rises across the VCB.

### 3.2.2. Influence of Postarc Current

In Figure 7 two experiments of a current interruption with the series connection are presented. In Figure 7 a) a high current- and voltage steepness is set. A postarc current of around 6 A flows through the series connection and the GCB fails. After 10–15 µs the voltage distribution changes into a capacitive voltage division. In Figure 7 b) the current steepness $di/dt_{CZ}$ is set to 3.13 A/µs and the voltage steepness at the beginning of the TRV is 0.9 kV/µs. At this ratings the GCB is able to suppress the postarc current in the VCB and the TRV rises in the first 10 µs exclusively over the GCB. In some experiments with these parameters, late (up to several microseconds) breakdowns of the VCB are observed. In all cases the GCB is able to withstand the whole TRV because the dielectric recovery of the $CO_2$ gas is fast enough and no failure of the hybrid circuit breaker appeared. In single operation of the VCB no breakdown in that voltage range was recorded.

## 4. Discussion

To evaluate the operating principle of a gas and vacuum circuit breaker in a series connection the VCB is investigated separately. The postarc current of the VCB and the critical $di/dt$ of the GCB at a specific switching case has the greatest influence on the voltage distribution. For the used RMF-type VCB the contact distance and the $di/dt_{CZ}$ changes the values of the postarc current.

When the postarc current of the VCB is suppressed by the GCB, late breakdowns of the VCB occur. It is suspected, that the suppression of the post arc current directly relates to the VCBs dielectric recovery. In theory the charged particles in the contact gap of the VCB are removed by the postarc current with the help of the TRV. When this process is suppressed by

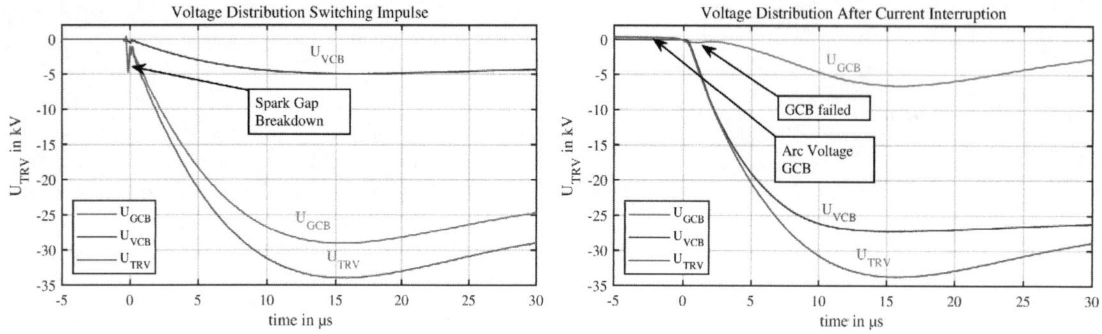


*Figure 6. Comparison of voltage distribution. Left: Without current interruption. Capacitive voltage distribution. Right: Voltage distribution after current interruption. Failed interruption in the GCB due to the postarc current of the VCB.*

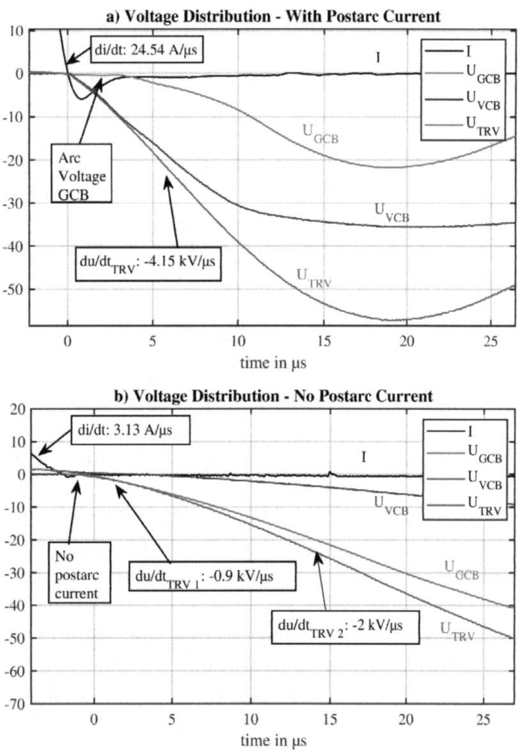


*Figure 7. a) Voltage distribution and postarc. GCB failed during postarc phase. b) Voltage distribution, postarc shut off by GCB.*

the GCB, the dielectric performance of the VCB is decreased.

## 5. Conclusions

During the interruption process of a hybrid circuit breaker, two different scenarios can be described.

1. For high current steepness in CZ or low interruption capability of the GCB, it is not able to interrupt the current. Consequently, the GCB stays conductive during the postarc phase of the VCB. The arc quenching takes place in the VCB and nearly the whole TRV rises across its contacts. After the postarc phase, in the transition phase, the distribution ratio changes with the speed of dielectric recovery of the GCB. In the last phase the voltage distribution depends only on the capacitive ratio.

2. For low current steepness in CZ or high interruption performance of the GCB, the postarc current of the VCB is suppressed by the GCB. The arc quenching takes place in the GCB. The charged particles between the VCB contacts can only be discharged very slowly and the TRV rises almost exclusively across the GCB. Late breakdowns appear in the VCB.

For the voltage distribution and the interruption process of a hybrid circuit breaker, the postarc current of the VCB plays a major role. The probability of a failure of the GCB and therefore a rise of the TRV in the first microseconds above the VCB increases with the peak value of the postarc current.

## References


[1] K. Natsui et al. Voltage distribution characteristics of series connected $SF_6$ gas and vacuum interrupters immediately after large AC current interruption. *IEEE Transactions on Power Delivery*, 3(1):241–248, 1988. doi:10.1109/61.4251.

[2] R. P. P. Smeets, V. Kertesz, D. Dufournet, D. Penache, and M. Schlaug. Interaction of a vacuum arc with an $SF_6$ arc in a hybrid circuit breaker during high-current interruption. *IEEE Transactions on Plasma Science*, 35(4):933–938, 2007. doi:10.1109/TPS.2007.896919.

[3] X. Cheng et al. Dynamic dielectric recovery synergy of hybrid circuit breaker with $CO_2$ gas and vacuum interrupters in series. *IEEE Transactions on Plasma Science*, 45(10):2885–2892, 2017. doi:10.1109/TPS.2017.2743691.

[4] N. Ide, O. Tanaka, S. Yanabu, S. Kaneko, S. Okabe, and Y. Matsui. Interruption characteristics of double-break vacuum circuit breakers. *IEEE Transactions on Dielectrics and Electrical Insulation*, 15(4):1065–1072, 2008. doi:10.1109/TDEI.2008.4591229.

[5] K. Steinke, M. Lindmayer, and K.-D. Weltmann. Post-arc currents of vacuum interrupters with axial magnetic field contacts under high current and voltage stress. In *Proceedings ISDEIV*, pages 475–480, 2000. doi:10.1109/DEIV.2000.879030.

doi:10.14311/ppt.2019.3.227
*Plasma Physics and Technology* **6**(3):227–230, 2019 

# AXIAL BLAST DISCHARGE CHAMBER WITH MOVING ELECTRODE

M.E. Pinchuk[a,*], A.V. Budin[a], N.K. Kurakina[a,b], A.G. Leks[a]

[a] *Institute for Electrophysics and Electric Power of Russian Academy of Sciences, Dvortsovaya nab., 18, St.-Petersburg, 191186, Russia*

[b] *Peter the Great Saint-Petersburg Polytechnic University, Polytechnicheskaya 29, St.-Petersburg 195251, Russia*

[*] pinchme@mail.ru

**Abstract.** The paper presents some results concerning electrophysical and gas-dynamics parameters of high-curent arc in axial blast discharge chamber. The experimental stand and numerical model were modified for axial gas flow type. Some design changes are described in the paper. The experiments were carried out for gas pressures of 1.0–6.0 MPa with current amplitude of 25–150 kA. The current half-period was of 1.0–10.0 ms. The contacts moved apart to the distance of 3–4 cm due to gas pressure boost in the chamber. OpenFOAM package with the library swak4foam was used for numerical simulation.

**Keywords:** high-current high-pressure arc, high-current circuit breaker.

## 1. Introduction

Existing systems of energy generation and distribution face necessity of breaking current from tens up to several hundreds kiloampers [1, 2]. The process of load switch-off is accompanied by an electric arc burning in all powerful switching off electric devices [3]. Now $SF_6$ is the basic electrical insulation and arc-interruption media in powerful high-voltage and generator circuit breakers [4]. But $SF_6$ is a potent greenhouse gas [4]. And general world position on the reduction of $SF_6$ gas usage [4–6] demands development of new approaches in electric-power industry, namely, a wide search of new arc-interruption environments, and also new insulation and electrode materials suitable to work in these environments at high current amplitudes.

Besides needs of the electric power industry, research of powerful arc of variable length and methods of its quenching has the great importance for the development of various technologies of lightning protection [7–9], plasma chemistry [10, 11], aerodynamics [12, 13] and others. The carrying out of experimental researches of the high current discharges at high pressure requires the creation of complicated testing systems [14].

Earlier in our laboratory, the stand for research of lightning protection systems [15] and the stand for studying arc and erosive processes in high-current gas blast breakers [16, 17] had been built. In the development of researches [16–19] for this theme it was carried out the modernization of the systems. And it was designed a new axial blast discharge chamber. Additionally, with the development of experimental techniques the numerical model [20, 21] was developed for calculations of various processes in the given system. A brief description of the axial blast chamber and preliminary results are presented in the paper.

## 2. Electrical power system and stand modernization

The energy source of set-up is the modular capacitive system with maximal storage energy of 6 MJ, maximal charging voltage of 10 kV and capacity of 0.11 F. The capacitive system is divided into six identical independent modules which can be connect to loading as simultaneously, and with a delay. For the formation of an oscillation circuit the inductive reactors of 1.5, 50 and 250 μH, reeled up on textolite drums by PIGR-16 wire (produced by GK Sevkabel, St.Petersburg, Russia) are used. The discharge current is measured by Rogovsky coil. The voltage drop across an arc was measured using a high-ohmic resistive divider with transformer decoupling. For registration of pulse pressure in the discharge chamber the piezoelectric gauge of membrane type T500 or T200 (NTIIM, Nizhny Tagil, Russia) were used.

The duration of the current period in the contour was changed by choosing a number of charged modules and appropriate inductive reactor. Current curves for two modules and the specified inductive reactors are shown in figure 1.

The experimental stand was modified for axial blast discharge chamber. Air or nitrogen were used as working gas. The design of the discharge chamber is presented in figure 2.

The housing of chamber (1) is a thick-walled vessel made from the stainless steel, inside it there is a mobile acetal resin piston (2). On the piston it is mounted the contact holder (3) on which end the contact tip (4) is screwed. The opposite end of the holder incorporates with the ground conductor line by means of flexible copper wires laid inside of polycarbonate tube (5). The ring-shaped contact (6) is fasten on the current input (7), isolated from the chamber case and connected with the high voltage conductor line.

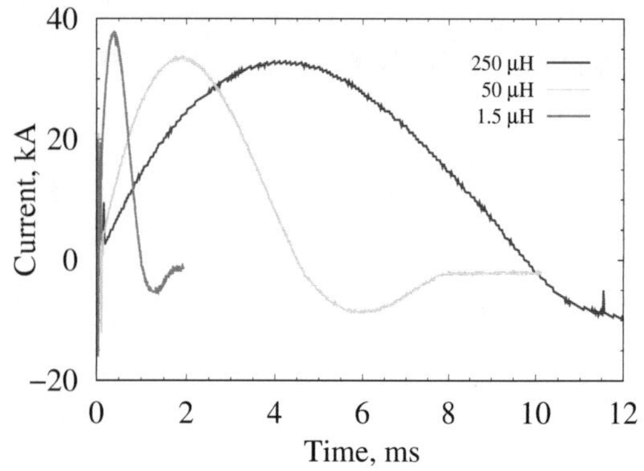


*Figure 1. Short circuit current for capacity of* 38 *mF and inductance of* 1.5, 50 *and* 250 *µH.*

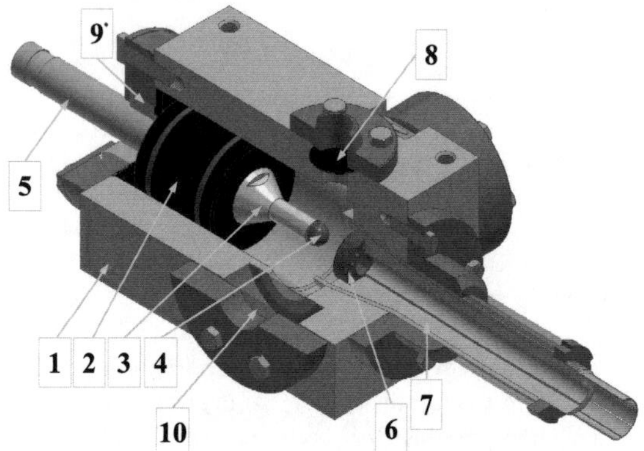


*Figure 2. Axial blast discharge chamber: (1) housing; (2) piston; (3) anode holder; (4), (6) contacts; (5) polycarbonate tube; (7) current input; (8) diaphragm; (9) fluoroplastic cartridge; (10) diagnostic window.*

Copper and MoWCu were used as materials for ring contact. And copper, CuFe and CuW for rod contact.

During the initial moment the rod contact enters into a cavity of current input on depth of 4 cm inside ring contact. Movement of the piston is carried out by means of compressed gas. Gas leaves from the discharge chamber into receiver volume after break of a diaphragm (8). The cavity behind the piston plays a role of the shock-absorber since gas is squeezed out from it through a narrow slit between a tube (5) and the fluoroplastic cartridge (9). Windows (10) are purposed for optical diagnostics.

The photo of the discharge chamber, mounted on the stand, is shown in figure 3.

Functioning of system occurs in the following sequence. The energy source (capacitor battery) is charged up to a necessary voltage. Then working gas with the help of the remote controlled valve enters into the high pressure cylinder through gas pipe line galvanically untied from installation. At the set pressure the polyester diaphragm tear, and gas starts to enter into the discharge volume. Under action of gas pressure the piston (2) begins movement. On the end

*Figure 3. Discharge chamber on the experimental stand.*

of the tube (6), made of shock-resistant polycarbonate, the plate with slits is attached. At passage of these slits by the photo diode it lights up, and on pulses from the photo diode the speed of the piston is determined. The end of the plate tears the thin wire tense across its movement, and the forming circuit gives out a signal on the discharge of the capacitor battery and start diagnostic and data-acquisition equipment. Ignition time for switching a current is chosen so that the discharge began before the moment of contacts disconnection. Thus, the discharge current starts at normally closed contacts. After contacts disconnection an electric arc, which length increases in process of the piston movement, ignites between them. Piston speed equal to contact separation speed was of 15–35 m/s at the 4th millisecond.

Filling the discharge chamber with gas before ignition of the arc has been occurred until the limiting diaphragm (8) was broken at a given pressure. Then gas flowed to the receiver tank with decreasing pressure, as shown in figure 4. When the gas output was bunged (cover plug in place of diaphragm (8)) the chamber volume was filled according to figure 5. High-frequency fluctuation in experimental curve, probably, is due to the sensor natural frequency. Calculated pressure curves are shown together with the experimental pressure curves in figures 4 and 5. The pressure was calculated by the code in the computing package OpenFOAM [20, 21]. The discharge current start occurs at the 4th millisecond for the scale in figures 4 and 5.

Certain unresolved questions do not allow to show the pressure curve during discharge. Some modifications of the simulation scheme and pressure measurement system are carried out now for verification of the measurement data and comparison with simulation results.

The current and voltage across the discharge gap are shown in figure 6 for power mode with half-period of ≈ 1 ms. During the first 100 µs after the current start, a short-circuit current flows between the closed contacts. Rod contact moves along the surfaces of the ring one and this leads to intermittent contact bounce and, as a consequence, to voltage spikes. Further, the arc burns in a narrow gap between rod and ring contacts with declining voltage. After that, the rod

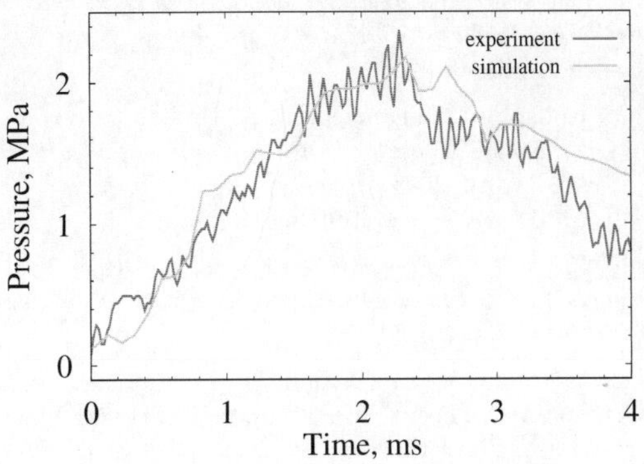


*Figure 4. Experimental (red) and calculated (green) pressure in the discharge chamber with diaphragm breaking at 2 MPa for inlet pressure of 4 MPa.*

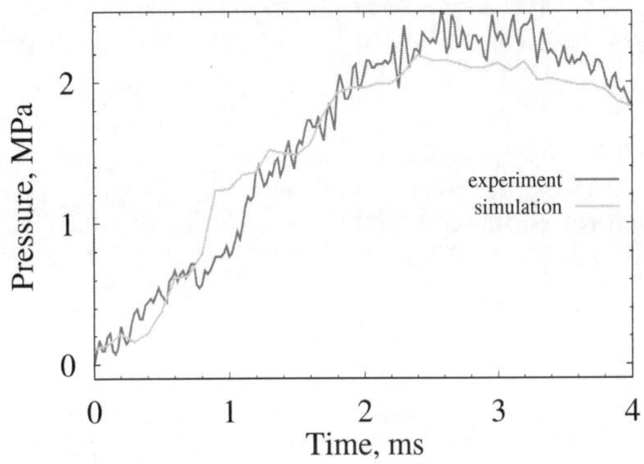


*Figure 5. Experimental (red) and calculated (green) pressure in the closed discharge chamber for inlet pressure of 4 MPa.*

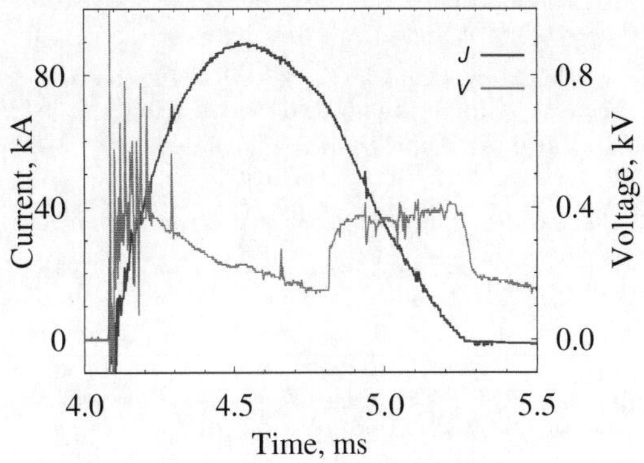


*Figure 6. Current (J) and voltage across arc gap (V) in the discharge chamber with diaphragm breaking at 2 MPa for inlet pressure of 4 MPa.*

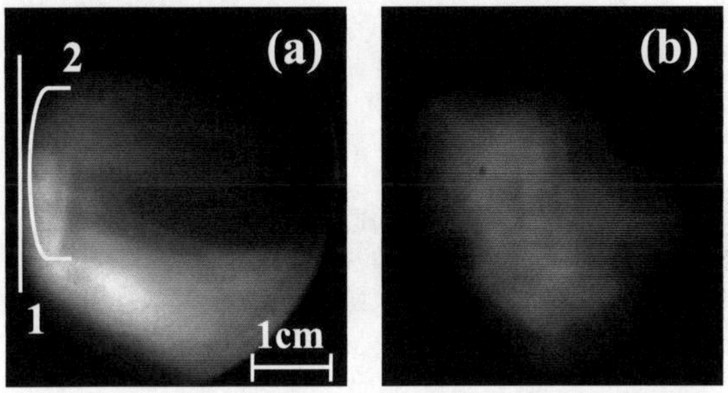


*Figure 7. High-speed images of arc between opening contacts at 4.85(a) and 5.4(b) ms with diaphragm breaking at 2 MPa for inlet pressure of 4 MPa: (1) ring contact; (2) rod contact.*

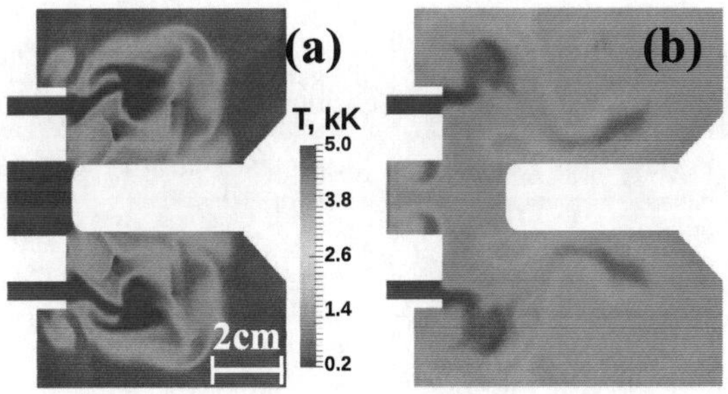


*Figure 8. Numerical simulation for temperature distribution in discharge chamber with diaphragm breaking at 2 MPa for inlet pressure of 4 MPa; (a) corresponds to ≈ 4.9 and (b) to ≈ 5.8 ms.*

contact exited ring contact hole. And this time interval of 4.8–5.3 ms with intense gas blowing through the arc is specified by considerable increase in discharge voltage value. High speed photo and calculated temperature distribution in discharge chamber for the beginning of this time interval are presented in figures 7(a) and 8(a), respectively. Intensive cooling of the gas flow through the discharge gap leads to arc quenching around current zero (figures 7(b) and 8(b)).

## 3. Conclusions

The experimental setup for simulating and investigating the processes that occur in high-current switching devices has been designed. Preliminary experiments for arcs with variable lengths diagnostics were conducted in the designed axial blast chamber. With the development of experimental techniques, the numerical model has been developed for calculations of various processes in the discharge system. To date, a qualitative agreement has been obtained between experimental data and calculations. Further improvements of the experimental and numerical tools are being made now.

## Acknowledgements

The authors acknowledge the support from Presidium RAS Program No. 7: "New developments in promising areas of energy, mechanics and robotics". Calculation results were obtained using computational resources of Peter the Great Saint-Petersburg Polytechnic University Supercomputing Centre (www.spbstu.ru).

## References

[1] V. Ignatov, M. Misrikhanov, and A. Shuntov. Concerning construction decisions for short circuit

current limitation in power system. *Proceedings of RAS. Power Engineering*, (5):94–103, 2009.

[2] A. Aleksandrov and V. Zhukov. Restruction of short circuit currents in the high voltage distribution networks of developing power systems. *Vestnik MEI*, 58(2):58–66, 2012.

[3] Current Zero Club. www.currentzeroclub.org.

[4] P. Glaubitz, S. Stangherlin, J. Biasse, F. Meyer, M. Dallet, M. Pruefert, R. Kurte, T. Saida, K. Uehara, P. Prieur, H. Ito, E. Kynast, A. Janssen, R. Smeets, and D. Dufournet. CIGRE position paper on the application of $SF_6$ in transmission and distribution networks. *Electra*, 34(274):34–39, 2014.

[5] M. Seeger, R. Smeets, et al. Recent trends in development of high voltage circuit breakers with $SF_6$ alternative gases. *Plasma Physics and Technology*, 4(1):8–12, 2017. doi:10.14311/ppt.2017.1.8.

[6] M. Eves. *A Literature review on $SF_6$ gas alternatives for use on the distribution network*. Western Power Distribution, 2018.

[7] M. E. Pinchuk, A. V. Budin, I. I. Kumkova, and A. N. Chusov. Studying energy evolution in the discharge chamber of a multichamber lightning protection system. *Technical Physics Letters*, 42(4):395–398, 2016. doi:10.1134/S1063785016040222.

[8] M. Pinchuk, A. Budin, I. Kumkova, A. Bogomaz, A. Sivaev, A. Chusov, and R. Zaynalov. Energy deposition in discharge chamber of lightning protection multichamber system. *Journal of Physics: Conference Series*, 774(1):012187, 2016. doi:10.1088/1742-6596/774/1/012187.

[9] A. Chusov, E. Rodikova, M. Murmann, H. Nordborg, and R. Fuchs. A review of progress towards simulation of arc quenching in lightning protection devices based on multi chamber systems. *Plasma Physics and Technology*, 4(3):273–276, 2017. doi:10.14311/ppt.2017.3.273.

[10] P. G. Rutberg, A. A. Safronov, A. V. Surov, A. V. Pavlov, S. D. Popov, V. A. Spodobin, and A. P. Rutberg. Spectral investigations of electric arc alternating current plasma generators with power to 600 kW. *High Temperature Material Processes: An International Quarterly of High-Technology Plasma Processes*, 13(2):195–203, 2009. doi:10.1615/HighTempMatProc.v13.i2.80.

[11] A. Surov, S. Popov, V. Popov, D. Subbotin, E. Serba, V. Spodobin, G. Nakonechny, and A. Pavlov. Multi-gas ac plasma torches for gasification of organic substances. *Fuel*, 203:1007–1014, 2017. doi:10.1016/j.fuel.2017.02.104.

[12] P. Rutberg. *Physics and Technology of High Current Discharges in Dense Gas Media and Flows*. Nova Science Publishers, 2009.

[13] V. Kolikov, A. Bogomaz, and A. Budin. *Powerful Pulsed Plasma Generators: Research and Application*. Springer Nature, 2018. doi:10.1007/978-3-319-95249-9.

[14] C. Flurscheim. *Power Circuit Breaker Theory and Design*. The Institution of Engineering and Technology, 2008. doi:10.1049/PBPO001E.

[15] A. V. Budin, M. E. Pinchuk, V. E. Pilschikov, A. G. Leks, and V. V. Leont'ev. An experimental stand for investigating protective devices for high-voltage overhead lines. *Instruments and Experimental Techniques*, 59(5):673–677, 2016. doi:10.1134/S0020441216040163.

[16] A. Budin, M. Pinchuk, V. Kuznetsov, V. Leont'ev, and N. Kurakina. Experimental stand for investigations of arc and erosion processes in high-voltage powerful breaker. *Instruments and Experimental Techniques*, 60(6):837–842, 2017. doi:10.1134/S0020441217060033.

[17] A. V. Budin, M. E. Pinchuk, V. V. Leontev, A. G. Leks, N. K. Kurakina, A. A. Kiselev, J. V. Simakova, and V. Y. Frolov. Experimental stand for investigations of insulator degradation and electrode erosion in high-current breaker. *Plasma Physics and Technology*, 4(2):120–123, 2017. doi:10.14311/ppt.2017.2.120.

[18] A. V. Budin, M. E. Pinchuk, V. E. Kuznetsov, and F. G. Rutberg. The influence of the production technology of iron-copper composite alloy on its erosion properties in a high-current high-pressure arc. *Technical Physics Letters*, 40(12):1061–1064, 2014. doi:10.1134/S1063785014120050.

[19] A. V. Budin, M. E. Pinchuk, and N. K. Kurakina. Erosion characteristics of copper-based composite electrodes in an electric arc of variable length with transverse gas blowing. *Technical Physics Letters*, 44(9):808–810, 2018. doi:10.1134/S1063785018090171.

[20] N. Kurakina, M. Pinchuk, A. Budin, A. Smirnovsky, and V. Frolov. Numerical simulation using OpenFOAM of gas-dynamics in the discharge chamber with a movable electrode. *St. Petersburg polytechnic university journal of engineering science and technology*, 24(2):69–81, 2018. doi:10.18721/JEST.240206.

[21] N. Kurakina, M. Pinchuk, A. Budin, and A. Smirnovsky. Analysis of the gas dynamics in the discharge chamber with a movable electrode. *Journal of Physics: Conference Series*, 1135(1):012094, 2018. doi:10.1088/1742-6596/1135/1/012094.

doi:10.14311/ppt.2019.3.231
*Plasma Physics and Technology* **6**(3):231–234, 2019 

# EXPERIMENTAL INVESTIGATION ON THE ARC CHARACTERISTICS AND ARC QUENCHING CAPABILITIES OF $C_5F_{10}O$-$CO_2$

Z. Guo[a], F. Tang[b], Q. Lv[b], X. Li[a,*], B. Zhang[a], S. Jia[a], R. Huang[b]

[a] *State Key Laboratory of Electrical Insulation and Power Equipment, Xi'an Jiaotong University, Xi'an, Shaanxi, 710049, China*

[b] *Shenzhen Power Supply Bureau, Shenzhen 518020*

[*] xwli@mail.xjtu.edu.cn

**Abstract.** $C_5F_{10}O$-$CO_2$ mixtures are possible alternatives to $SF_6$ - which has a high global warming potential - as the interruption medium in gas circuit breakers. This paper experimentally studies the arcing characteristics of $C_5F_{10}O$-$CO_2$ mixture, with an experimental model with viewing windows, and measures the arc voltage, current and emission spectrum. The arc evolution process is captured with a high speed camera through an inspection window. The two-dimensional distribution of arc is obtained and analyzed by the inverse transformation of Abel. The results show that, the $C_5F_{10}O$-$CO_2$ mixture arc is more volatile than $SF_6$ gas, and adding $C_5F_{10}O$ into $CO_2$ improves the stability of the arc, and significantly reduces the arc temperature.

**Keywords:** arc, $C_5F_{10}O$, inverse transformation of Abel, temperature.

## 1. Introduction

Exploring environmentally friendly gases that can replace $SF_6$ or partially replace $SF_6$ is an important research topic in this field due to the strong greenhouse effect of $SF_6$. Recently, two new fluorinated compounds, perfluoronitriles (PFN) (e.g. $C_4F_7N$ or $(CF_3)_2CFCN$), perfluoroketones (PFK) (e.g. $C_5F_{10}O$ or $CF_3COCF(CF_3)_2$), draws great attentions in leaps and are considered as very promising $SF_6$ substitutes [1, 2]. In the past few years, the investigations on the dielectric strength (DS) of gas mixtures based on $C_4F_7N$ and $C_5F_{10}O$ are carried out dramatically by various researchers. The results show that a small quantity of these gases mixed with $CO_2$ or dry air increases the DS of the gas mixture tremendously to be close to the DS of pure $SF_6$ [3, 4]. The arc interruption capability of these gas mixtures are also investigated through simulations as well as experiments [5].

This paper aims at the arcing characteristic of $C_5F_{10}O$-$CO_2$ mixtures. Based on a detachable experimental model with observation window, the arcing characteristics in the free-burning arc of $C_5F_{10}O$-$CO_2$ mixtures were studied experimentally. The emission spectrum along the radial direction of the arc was recorded by a spectrometer. The arc temperature distribution along the radial direction of the arc were obtained by Abel inverse transformation.

## 2. Experimental set up

The experiments are carried out with a detachable experimental prototype with an observation window, as shown in Fig. 1. The electrodes are made of copper. The electrode diameter is 10 mm and the distance between the electrodes is continuously adjustable. The electrode distance of the experiment is fixed to 10 mm.

*Figure 1. Geometry of the test prototype.*

In this experiment, a copper wire with a diameter of 0.1 mm is pre-connected between the electrodes to start the arc. To ensure the stability and reliability of the experiment, the copper wire is connected to the center of the two electrodes and kept as parallel as possible to the axis.

The short-circuit current is provided by a capacitor bank circuit with an oscillating frequency of 50 Hz. The sketch of the test circuit is shown in Fig. 2. The discharge circuit is connected with a thyristor switch. Since the thyristor switch only allows the forward current, a diode in parallel with the thyristor switch is adopted to obtain a cycle of current. A DG535 signal generator is used to control the thyristor switch and the operating mechanism of the GCB model in order to inject the current and trigger the spectrometer. The arc voltage and current were measured using a high voltage probe (P6015A, Tektronix) and a Rogowski

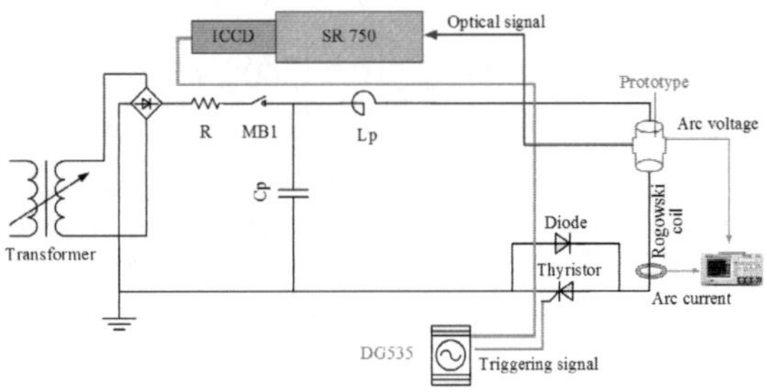


*Figure 2. Sketch of the test circuit.*

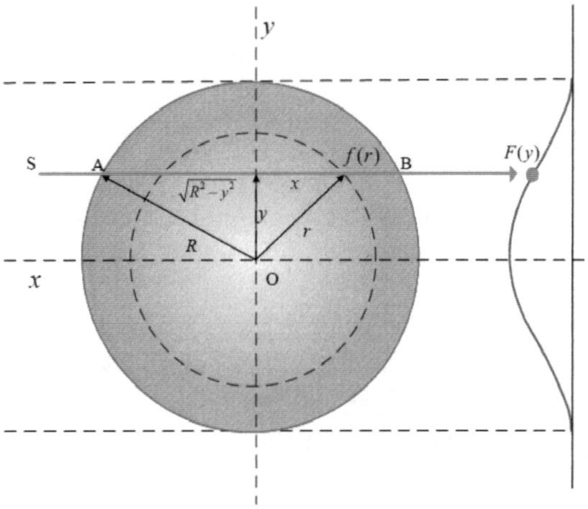


*Figure 3. Schematic diagram of Abel inverse transformation.*

coil respectively. The measurement data was recorded using a digital oscilloscope. The spectral information of the arc plasma was measured using a spectrometer (SR750, Andor) and an ICCD (iStar DH734, Andor).

In this paper, the arc temperature was evaluated by the Boltzmann slope method. Three lines of Cu I 521.8 nm, Cu I 515.32 nm and Cu I 510.55 nm and was used to obtain the copper atom excitation temperature. Assuming that the arc region is an axisymmetric region, in order to obtain the spatial distribution of the arc temperature, the radial temperature distribution of the arc is obtained by inversion of the Abel inverse transform. As shown in Fig. 3, the $f(r)$ represents the profile of the intensity of spectral line; the $r$ represents the radial distance from the arc center; the $R$ represents the radius of the arc.

As can been seen from the picture, the intensity of spectral lines is an integral of the emission coefficient along the measurement direction.

$$f(y) = \int_{-\sqrt{R^2-y^2}}^{\sqrt{R^2-y^2}} f(x)\mathrm{d}x \tag{1}$$

The distribution of emission coefficient is assumed to be symmetrical, then the relation between intensity and emission coefficient can be changed into:

$$f(y) = \int_y^R f(r)\frac{r\mathrm{d}r}{\sqrt{r^2-y^2}} \tag{2}$$

After the Abel inverse:

$$f(y) = -\frac{1}{\pi}\int_r^R f'(r)\frac{\mathrm{d}y}{\sqrt{y^2-r^2}} \tag{3}$$

$f(r)$ is expressed by three cubed spline function from discrete experiment data.

Under the LTE condition, the excitation of the atom obeys the Boltzmann equation. The arc temperature can be calculated by two different spectral line emitted from a same ion, the double-line method, and the formula is shown below.

$$\frac{I_1}{I_2} = \frac{A_1 g_1 \lambda_2}{A_2 g_2 \lambda_1}\exp\left(-\frac{E_1-E_2}{kT}\right) \tag{4}$$

where $E$ represents energy levels; $k$ represents the Boltzmann constant; $g$ represents the statistical weight; $A$ represents the transition probability; $\lambda$ represents wave length; $\varepsilon$ represents the emission coefficient, obtained from the line intensity by means of Abel inversion.

The gas scheme used in this experiment includes: $CO_2$, 28% $C_5F_{10}O$–72% $CO_2$ and $SF_6$ gas. The gas pressure is 0.1 MPa, and the expected short-circuit current is 1.6 kA. Among them, the ratio of 28% $C_5F_{10}O$–72% $CO_2$ gas mixture corresponds to a gas liquefaction temperature of -5°C at 0.1 MPa.

## 3. Experimental results

Fig. 4 shows the arc current, arc voltage and arc conductance waveforms of a 28% $C_5F_{10}O$–72% $CO_2$ mixed gas and $CO_2$ and $SF_6$ under free-burning arc situation. It can be seen from the current waveform that the peak current of $SF_6$ is the highest, and the currents of $CO_2$ and 28% $C_5F_{10}O$–72% $CO_2$ gas mixture are close. It can be seen from the voltage waveforms that the arc voltage of the $SF_6$ gas arcing process is the lowest and most stable, and the stability of $CO_2$ is the worst. There are many violent fluctuations during the second current half-wave period, while the stability of 28% $C_5F_{10}O$–72% $CO_2$ gas mixture is better than pure $CO_2$. Fig. 4(c) shows the arc conductance of several gases under free-burning arc situation. It can be seen that the arc conductance of $SF_6$ is the highest, while the arc conductance of $CO_2$ and 28% $C_5F_{10}O$–72% $CO_2$ mixed gas is relatively close.

Overall, from the arc voltage, current and conductance waveforms of $CO_2$, $SF_6$ and 28% $C_5F_{10}O$–72% $CO_2$ gas mixture, it can be seen that the arc of $SF_6$ gas is more stable, thus the arc voltage of $SF_6$ arc changes more gently, and is lower than $CO_2$ and $C_5F_{10}O$–72% $CO_2$ gas mixture, which makes the current amplitude of $SF_6$ arc higher than $CO_2$ and $C_5F_{10}O$–72% $CO_2$ gas mixture.

Fig. 5 shows the results of the spectral measurement of a 28% $C_5F_{10}O$–72% $CO_2$ gas mixture, in which the

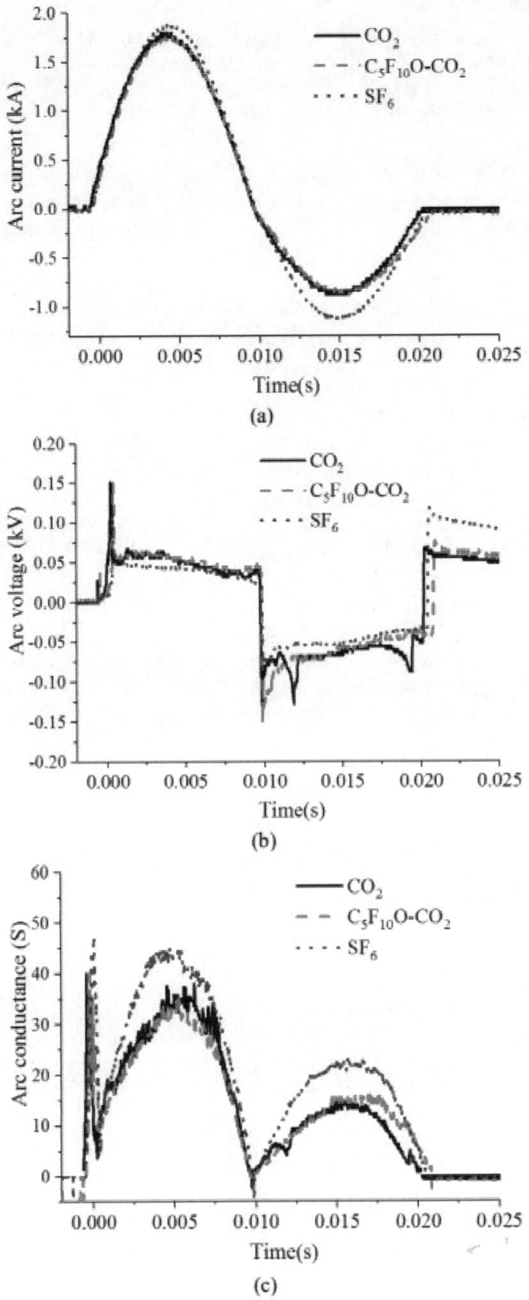


*Figure 4. Comparison of simulation result with experiment result.*

horizontal axis corresponds to the wavelength and the vertical axis corresponds to the vertical distribution. Fig. 6 shows the spectral distribution of the arc center. It can be seen that the three lines of Cu I 521.8 nm, Cu I 515.32 nm and Cu I510.55 nm are obvious.

Fig. 7 shows the radial distribution of the arc temperature at the current peak of the first half-wave in the free-burning arc of $CO_2$, 28% $C_5F_{10}O$–72% $CO_2$ and $SF_6$ gas. It can be seen that the temperature of the arc core is between 14000 K and 20000 K. The

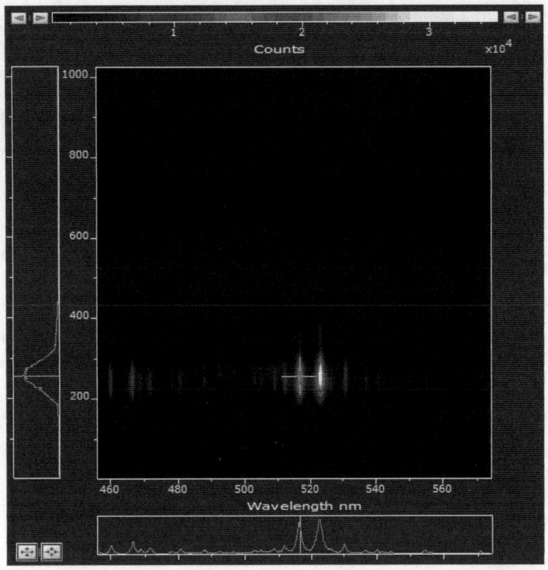


*Figure 5. The results of the spectral measurement of 28% $C_5F_{10}O$–72% $CO_2$ gas mixture.*

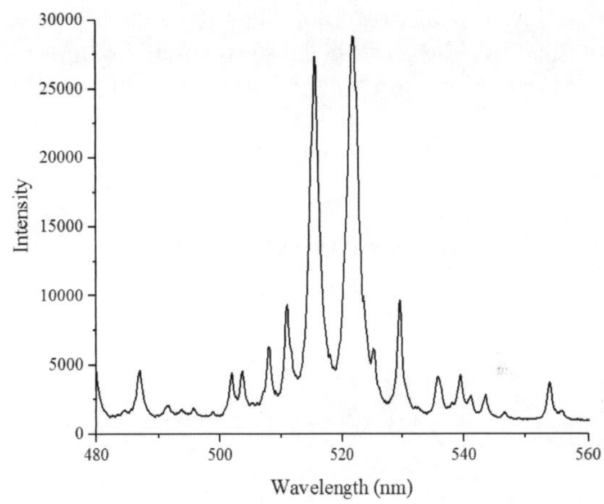


*Figure 6. The spectral distribution of the arc center.*

arc core temperature of the $CO_2$ gas arc is the highest, about 20000 K. The arc core temperature of 28% $C_5F_{10}O$–72% $CO_2$ is about 16200 K, while the $SF_6$ arc core temperature is about 14900 K, which is much lower than the other two gases. In addition, the arc diameter of $CO_2$ arcis about 13 mm, and the arc is thinner than the other two gases. The $SF_6$ and 28% $C_5F_{10}O$–72% $CO_2$ arc diameters are approximately 23 mm and 16 mm, respectively. Therefore, the addition of $C_5F_{10}O$ gas to the $CO_2$ gas can significantly lower the arc temperature and make the arc diameter larger, which corresponds to the stability of the aforementioned arc voltage.

## 4. Conclusions

Based on a detachable experimental model with observation window, the free-burning arc characteris-

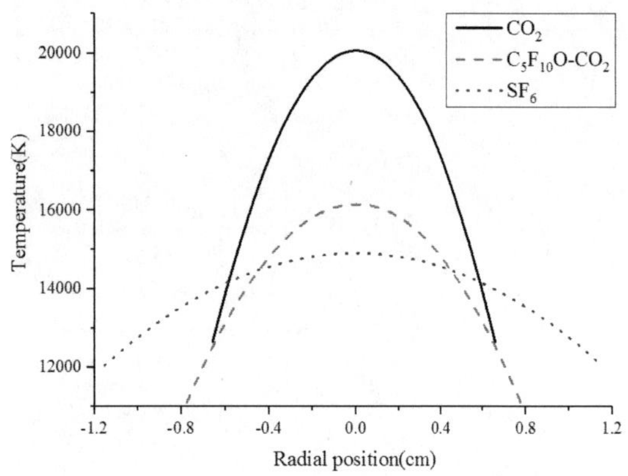


*Figure 7. The radial distribution of the arc temperature at the current peak of the first half-wave in the free-burning arc of $CO_2$, 28% $C_5F_{10}O$–72% $CO_2$ and $SF_6$ gas.*

tics of 28%$C_5F_{10}O$–72%$CO_2$ gas mixtuere and $CO_2$ were studied and compared with that of $SF_6$. The emission spectrum of the arc was measured. The two-dimensional distribution of arc temperature was obtained by Abel inverse transformation. Compared with the $SF_6$, the arc voltage fluctuation of CO2 and 28%$C_5F_{10}O$–72%$CO_2$ mixed gas is more severe and higher, and the current amplitude is smaller. After the addition of $C_5F_{10}O$ to $CO_2$ gas, the stability of the arc is improved.The addition of $C_5F_{10}O$ gas to the $CO_2$ gas can significantly reduce the arc core temperature.

## References

[1] J. K. Olthoff, R. J. Van Brunt, and L. G. Christophorou. Sulfur hexafluoride and the electric power industry. *IEEE Electr. Insul. Mag.*, 13(5):20–24, 1997. doi:10.1109/57.620514.

[2] H. Zhao, A. B. Murphy, and X. Li. $SF_6$-alternative gases for application in gas-insulated switchgear. *J. Phys. D: Appl. Phys.*, 51(15):153001, 2018. doi:10.1088/1361-6463/aab314.

[3] H. Lin, X. Li, and H. Zhao. Dielectric properties of Fluoronitriles/$CO_2$ and $SF_6$/$N_2$ mixtures as a possible $SF_6$-substitute gas. *IEEE Trans. Dielectr. Electr. Insul.*, 28(4):1332–1339, 2018. doi:10.1109/tdei.2018.007139.

[4] C. M. Franck and M. Rabie. Comparison of gases for electrical insulation: Fundamental concepts. *IEEE Trans. Dielectr. Electr. Insul.*, 25(2):649–656, 2018. doi:10.1109/TDEI.2018.006900.

[5] P. Stoller, C. Doiron, D. Tehlar, P. Simka, and N. Ranjan. Mixtures of $CO_2$ and $C_5F_{10}O$ perfluoroketone for high voltage applications. *IEEE Trans. Dielectr. Electr. Insul.*, 24(5):2712–2721, 2017. doi:10.1109/TDEI.2017.006383.

doi:10.14311/ppt.2019.3.235
*Plasma Physics and Technology* **6**(3):235–238, 2019 

# INVESTIGATION AND NUMERICAL SIMULATION OF A HIGH-CURRENT AC CIRCUIT BREAKER

I. Murashov[a,*], V. Frolov[a], A. Kvashnin[a], J. Valenta[b], D. Simek[b], L. Dostal[b], P. Kloc[b]

[a] *Institute of Energy and Transport Systems, Peter the Great St.Petersburg Polytechnic University, Polytechnicheskaya 29, 195251 St.Petersburg, Russia*

[b] *Centre for Research and Utilization of Renewable Energy, Faculty of Electrical Engineering and Communication, Brno University of Technology, Technicka 3058/12, 61600 Brno, Czech Republic*

* iuriimurashov@gmail.com

**Abstract.** The article is devoted to the study of the high-current AC circuit breaker. The results of the study are presented for various configurations of the arc divider. The study includes methods of spectral diagnostics and high-speed camera shooting synchronized with the electrical characteristics of the circuit breaker (current, voltage) in time. The obtained results allow to determine the composition of the plasma and dynamics of changes in the composition of the discharge in time. Calculation of the plasma composition and properties is made according to the obtained data, which makes it possible to take into account the products of circuit breaker materials ablation in numerical simulation. Non-stationary two-dimensional mathematical model with a moving mesh is developed. The obtained results allow to correct and verify the developed mathematical model of the circuit breaker operation. The evaluation of the arc divider influence is presented in the article.

**Keywords:** circuit breaker, plasma, simulation, mathematical model.

## 1. Introduction

The most famous and most used protection device is circuit breaker. And there is no universal method for design of a circuit breaker at the moment, but taking into account the development of computing technology, the possibility of multiphysical simulation [1] and optimization of the circuit breaker design appears. Different non-stationary mathematical models were created for analysis the circuit breaker operations previously, but all of these models did not take into account moving process of movement contact and ablation processes [1–3]. The main problem is the creation of a moving mesh, which leads to high demands on computing equipment.

There is little software which allows simulating using moving mesh and COMSOL Multiphysics is one of them. Several mesh smoothing methods are implemented in this software:

- ☐ Laplace smoothing (PDE method)
- ☐ Winslow smoothing
- ☐ Hyperelastic smoothing
- ☐ Yeoh smoothing

Laplace smoothing [4–6] introduces deformed mesh positions $x$ and $y$ as degrees of freedom in the model and it can be described in the static case by the following equation:

$$\frac{\partial^2 x}{\partial X^2} + \frac{\partial^2 x}{\partial Y^2} = 0, \tag{1}$$

where $x$ is the spatial coordinate of the spatial frame, and $X$ and $Y$ are the reference coordinates of the material frame.

*Figure 1. The circuit breaker prototype.*

And in the transient case (which is most interesting for non-stationary mathematical model of a circuit breaker operation), it can be described by the following equation:

$$\frac{\partial^2}{\partial X^2}\frac{\partial x}{\partial t} + \frac{\partial^2}{\partial Y^2}\frac{\partial x}{\partial t} = 0. \tag{2}$$

Similar equations hold for the $y$ coordinate. The speed of the movement contact is determined from the results of high-speed shooting (described below).

The circuit breaker prototype was produced for experimental studies with special windows (see Figure 1) for high-speed shooting and spectral analysis.

Different types of arc dividers (which were investigated) are presented in Figure 2.

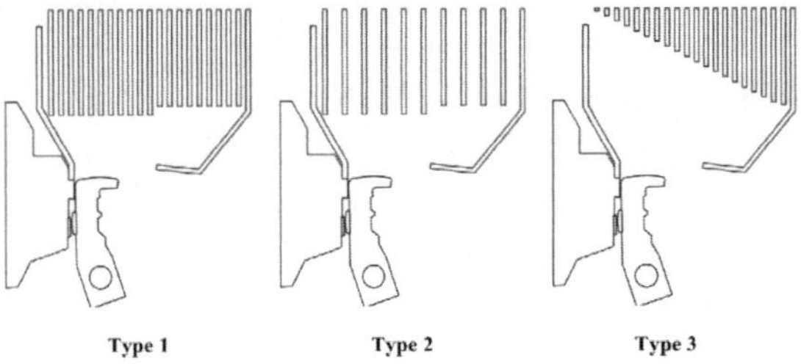


*Figure 2. Investigated types of arc dividers.*

## 2. Description of mathematical model

As mentioned earlier, the simulation of the circuit breaker operation is multiphysical task, which includes the following physical tasks: Maxwell's system of equations, momentum equation (Navier-Stokes equation) with continuity equation, heat balance equation (fluid and solid) and Laplace smoothing for moving mesh.

The circuit-breaker geometry is implemented with the ability to supplement the physical interface (for example, to solve the arc suppression problem, heating and erosion of the contact system) and the possibility of varying the position of individual components in the COMSOL Multiphysics software.

In addition, the simulation requires the dependences of the plasma thermophysical properties on temperature. The procedure for calculating the properties as a function of the plasma composition is presented in [7, 8].

The non-stationary mathematical model of the circuit-breaker operation is based on the following assumptions:

- □ Plasma is in a local thermodynamic equilibrium (LTE)
- □ Thermodynamic and transport properties depend on temperature
- □ Plasma is considered as laminar single-phase flow
- □ Near-electrode processes are not taken into account
- □ Plasma is optically thin

Basic equations describing non-stationary processes in DC circuit breaker:

□ Maxwell's system of equations:

$$\begin{cases} \nabla \times \vec{H} = \vec{J} \\ \vec{B} = \nabla \times \vec{A} \\ \vec{J} = \sigma \vec{E} + \vec{J_e} \\ \vec{E} = -\frac{\partial \vec{A}}{\partial t} \end{cases} \tag{3}$$

□ Momentum equation (Navier-Stokes equation):

$$\rho \frac{\partial \vec{v}}{\partial t} + \rho(\vec{v}\nabla)\vec{v} = \vec{F} + \nabla[-p\mathrm{I} + \mu(\nabla\vec{v} + (\nabla\vec{v})^{\mathrm{T}})] \tag{4}$$

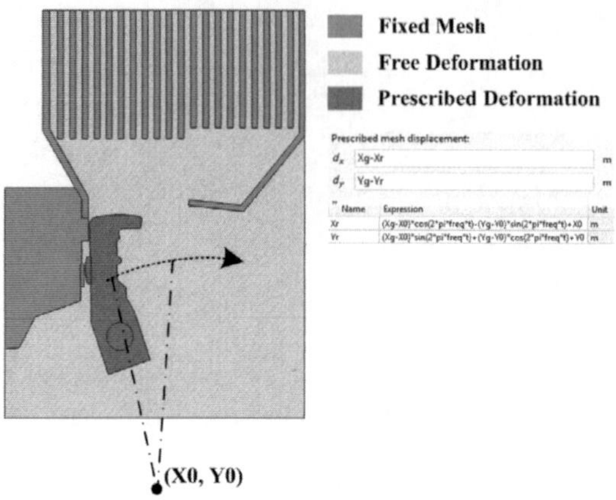


*Figure 3. Boundary conditions for "Moving Mesh" interface.*

□ Continuity equation:

$$\rho \frac{\partial \vec{v}}{\partial t} + \nabla(\rho\vec{v}) = 0 \tag{5}$$

□ Heat balance equation:

$$\rho C_p \frac{\partial T}{\partial t} + \rho C_p \vec{v} \nabla T - \lambda \nabla^2 T = Q_{\mathrm{s}} - Q_{\mathrm{rad}} \tag{6}$$

Computational domain and boundary conditions for "Moving Mesh" interface are presented in Figure 3.

The rotation is prescribed with respect to the undeformed coordinates $X_g$ and $Y_g$ using the following equations:

$$X_r = (X_g - X_0)\cos(2\pi f t) - (Y_g - Y_0)\sin(2\pi f t) + X_0, \tag{7}$$

$$Y_r = (X_g - X_0)\sin(2\pi f t) - (Y_g - Y_0)\cos(2\pi f t) + Y_0, \tag{8}$$

where $X_0$, $Y_0$ are origin coordinates; $t$ is simulation time; $f$ is determined by the velocity of the contact movement (from high-speed shooting).

The images of high-speed shooting research are shown in Figure 4. The total "switch off" time for contacts is about 20 ms and frequency $f$ is about 2 Hz (from high-speed images).

Another important aspect is the setting of material properties. The properties for most materials (see Figure 5) are well known, with the exception of the main contacts material.

The method of air properties calculating taking into account the ablation of contact materials according to the results of spectral analysis is described in [7–9].

Main contacts material is KMK-A40 in accordance with the Russian Technical Specification (TS 16-685.020 - 85). Composition of KMK-A40:

- □ Silver - 95%
- □ Carbon - 5%

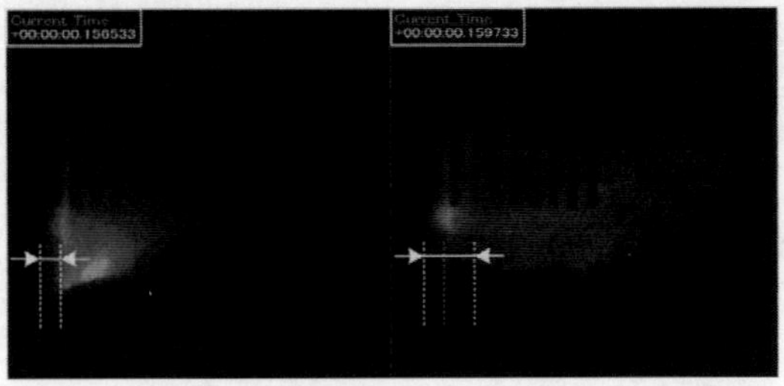

*Figure 4. The high-speed images for the movement contact velocity identification.*

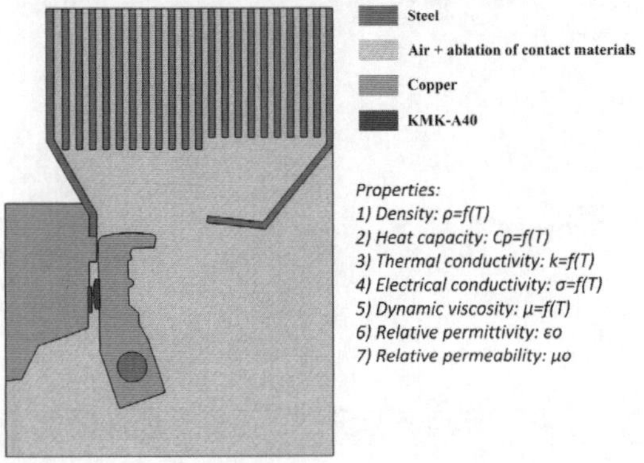


*Figure 5. Materials of computational domain.*

The calculated properties of the KMK-A40 material used in the simulation are shown in Figure 6.

The dependence of the current on time (obtained from experimental studies, see Figure 7) was used as data source for electrical task.

## 3. Simulation results

Numerical simulation was performed in Comsol Multiphysics software using the following physical interfaces:

□ Laminar Flow (spf)

□ Heat Transfer in Fluids (ht)

□ Electric Currents (ec)

□ Magnetic Fields (mf)

□ Moving Mesh

□ Multiphysics

- Nonisothermal Flow (nitf)
- Electromagnetic Heating (emh)
- Lorentz Force (lf)

Segregated solver with temperature lower limit was chosen as an algorithm for simulation of the circuit-breaker operation.

One of the main results of simulation is the temperature distribution inside the circuit breaker chamber. The temperature distributions at different times are shown in Figure 7.

The simulation results are confirmed by the results of the experimental study. New arc attachment point

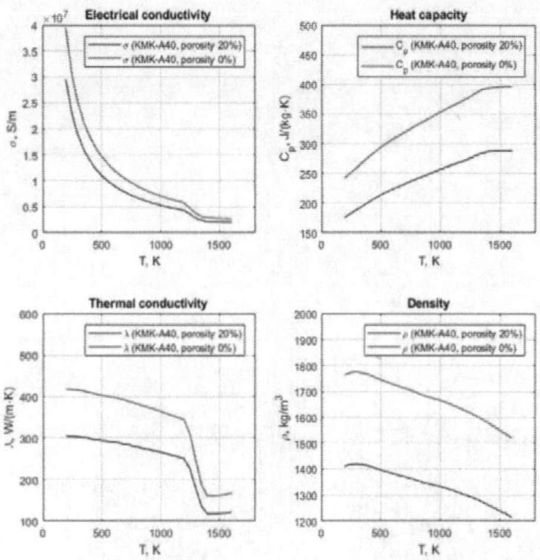


*Figure 6. KMK-A40 properties depending on temperature.*

formation is observed during the simulation a lot of times, which corresponds to the arc voltage fluctuation (from experimental study).

## 4. Results of experimental study

Experimental study included synchronized spectral analysis data, high-speed camera shooting and electrical parameters data (current and voltage waveforms). Analysis of the breaking capacity was performed on the basis of experimental study with similar electrical parameters of the power grid and the same moment of "switch off" operation. The results of experimental studies for various types of arc dividers (see Figure 2) are presented in Figure 8.

## 5. Conclusion

Non-stationary two-dimensional mathematical model with a moving mesh is developed. Laplace smoothing method was used for mesh generation. The speed of the movement contact is determined from the results of high-speed shooting (see Figure 4). KMK-A40 properties were calculated and presented (see Figure 6).

The simulation results are consistent with the experimental data. Experimental study of breaking capacity were performed for various modes and types of arc divider. The results for one of the modes are shown in Figure 8. The best result was observed when using the arc divider type 2.

### Acknowledgements

1. Research infrastructure CVVOZE Power Laboratories - project no.: LM2015092

2. CVVOZE Power Laboratories - Modernization of research infrastructure project no.: CZ.02.1.01/0.0/0.0/16_013/0001638

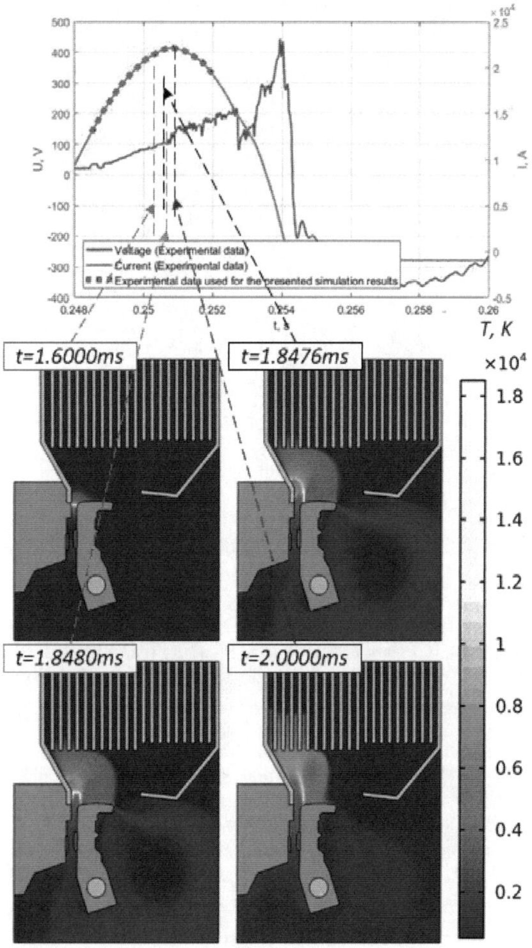


*Figure 7. Results of experimental study and numerical simulation.*

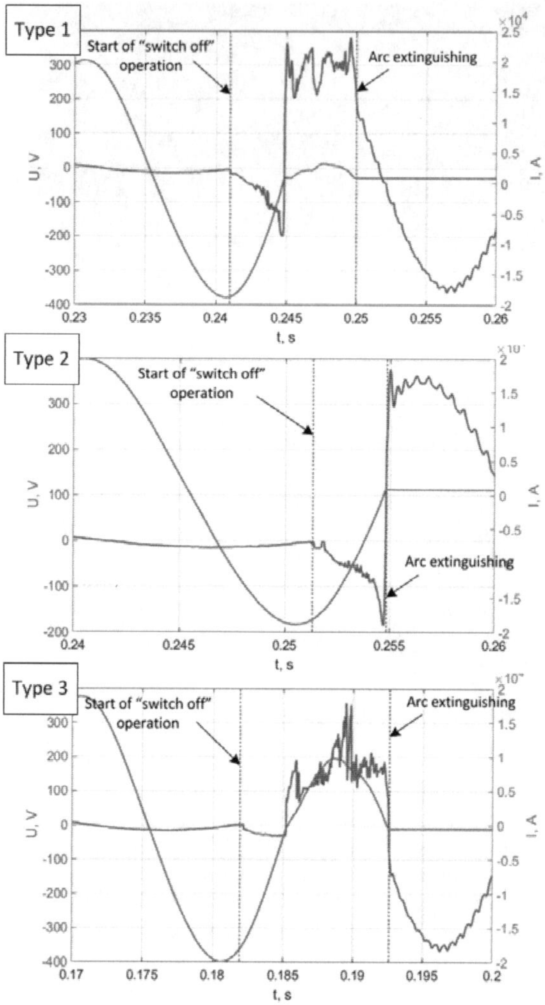


*Figure 8. Voltage and current waveforms for different types of the circuit breaker arc divider.*

## References


[1] V. Frolov, A. Kvashnin, and I. Murashov. Nonstationary mathematical model of a magnetic arc blast system. In *2018 IEEE Conference of Russian Young Researchers in Electrical and Electronic Engineering (EIConRus)*, 2018. doi:10.1109/EIConRus.2018.8317173.

[2] I. Murashov, V. Frolov, D. Ivanov, et al. Analysis of electromagnetic processes inside the arc interrupting system of a high-current circuit breaker. *Plasma Physics and Technology*, 4(2):161–164, 2017. doi:10.14311/ppt.2017.2.161.

[3] D. Chen, X. Li, L. Ji, and X. Zhou. Numerical simulation of arc motion during interruption process of low-voltage circuit breakers. In *26th International Conference on Electrical Contacts (ICEC 2012)*, 2012. doi:10.1049/cp.2012.0657.

[4] C. W. Mastin and J. F. Thompson. Quasiconformal mappings and grid generation. *SIAM J.Sci.Stat.Comput.*, 5(2):305–316, 1984. doi:10.1137/0905022.

[5] J. F. Thompson. A general three-dimensional elliptic grid generation system on a composite block structure. *Computer Methods in Applied Mechanics and Engineering*, 64:377–411, 1987. doi:10.1016/0045-7825(87)90047-8.

[6] S. P. Spekreijse. Elliptic grid generation based on Laplace equations and algebraic transformations. *J.Comput.Phys.*, 118:38–61, 1995. doi:10.1006/jcph.1995.1078.

[7] A. Chusov, G. Podporkin, M. Pinchuk, et al. Development of a physical 2-d model for arc quenching chamber of lightning protection multichamber systems. In *2016 33rd International Conference on Lightning Protection (ICLP)*, 2016. doi:10.1109/ICLP.2016.7791509.

[8] V. Y. Frolov, D. V. Ivanov, I. V. Murashov, et al. Calculation of the composition of plasma of an arc pulsed discharge in a multi-chamber arrester. *Technical Physics Letters*, 41(4):310–313, 2015. doi:10.1134/S1063785015040069.

[9] R. Kozakov, A. Khakpour, S. Gorchakow, et al. Investigation of a multi-chamber system for lightning protection at overhead power lines. *Plasma Physics and Technology*, 2(2):150–154, 2015.

doi:10.14311/ppt.2019.3.239
*Plasma Physics and Technology* **6**(3):239–242, 2019 

# INVESTIGATION OF SILICON WAFERS' INFLUENCE ON THE LOCAL MICROWAVE POWER VALUES IN A RESONATOR-TYPE PLASMATRON

S. Madveika, S. Bordusau*, M. Lushakova, O. Tsikhan

*Belarusian State University of Informatics and Radioelectronics, P.Brovki 6, 220013 Minsk, Belarus*

* bordusov@bsuir.by

**Abstract.** The article concerns the investigation results of the ø100 mm silicon wafers' influence on the microwave power value ($f_{\mathrm{MW}} = 2.45 \pm 0.05\,\mathrm{GHz}$) in local points on the axis of a reaction-discharge chamber with the volume of about $9000\,\mathrm{cm}^3$ of a resonator-type plasmatron. The experiments were carried out in the conditions of the dynamic microwave power redistribution inside a volumetric resonator by using a moving dissector. To register microwave power in the plasma volume, the method of "the active probe" was used. It has been experimentally established that the decrease of distance between the silicon wafers results in the decrease of local microwave power values between them up to 50%. The investigation results of the silicon wafers' influence on the microwave power distribution structure in the gas discharge area indicate the presence of power distribution nonuniformity in the discharge area volume.

**Keywords:** microwave plasma, silicon wafers, microwave power, plasmatron.

## 1. Introduction

New tasks and high micro- and nanoelectronics' production requirements stimulate the search for new and modernization of existing methods of a controlled plasma impact on the treated materials. Nowadays there is a great interest in the industrial application of a ultra-high-frequency (UHF) discharge, the so-called microwave discharge or microwave plasma, in the processes of electronic devices production.

The advantage of UHF range is the possibility to create a discharge device construction for excitation of plasma with a vast spectrum of parameters [1–3]. The technical methods used for excitation of microwave plasma are characteristic just for a UHF range and differ from those used at other frequencies. This is due to the fact that the wavelength is comparable with the size of the discharge device, and for their design systems with distributed parameters (waveguides, resonators, etc.) are often used.

A microwave discharge with a localized discharge region (resonator and waveguide type of discharges) has certain advantages in the solution of the problem of a plasma spatial structure influence on its chemical activity [4].

Non-equilibrium plasma of microwave discharges makes it possible to carry out plasma-chemical processes at a low gas temperature but at a higher electron temperature [5].

In a microwave discharge plasma, when the electrodynamic characteristics of a microwave applicator are changed, it is possible to control the internal discharge structure [6], which is impossible, for example, in a direct current discharge. From this point of view, volume resonators are perspective technical solutions for localization of microwave energy in a plasma volume.

However, in resonator-type plasma systems in case of high energy inputs into plasma, a large plasma non-uniformity can be formed because of significant electromagnetic field intensity gradients. It imposes certain limits on the processes of plasma-chemical treatment of semiconductor wafers with a diameter of more than 200 mm and in group treatment of smaller diameter wafers [7].

In order to reduce the microwave power non-uniformity in the plasma volume of resonator-type plasmatrons, microwave power redistribution systems are used [1]. The simplest technical solution is a rotating metal dissector placed in the resonator [8]. However, there is no information about the influence of silicon wafers with a high dielectric loss tangent on the local microwave power values in resonator-type plasmatrons with microwave power redistribution using a moving dissector.

Therefore, the investigation of the influence of the objects of treatment (their material, volume, shape, size, etc.) on the electrophysical and chemical properties of plasma as well as technological processes of treating materials with microwave plasma are of great interest.

## 2. Experimental apparatus and procedures

The investigation of the influence of silicon wafers with a high dielectric loss tangent on the local microwave power values during its dynamic redistribution in the microwave plasma was carried out in a microwave resonator-type plasma installation [9].

The principle of operation of the experimental vacuum-plasma installation is based on the transfer of microwave power from a microwave magnetron M-105 ($f_{\mathrm{MW}} = 2.45 \pm 0.05\,\mathrm{GHz}$) into an evacuated reaction-

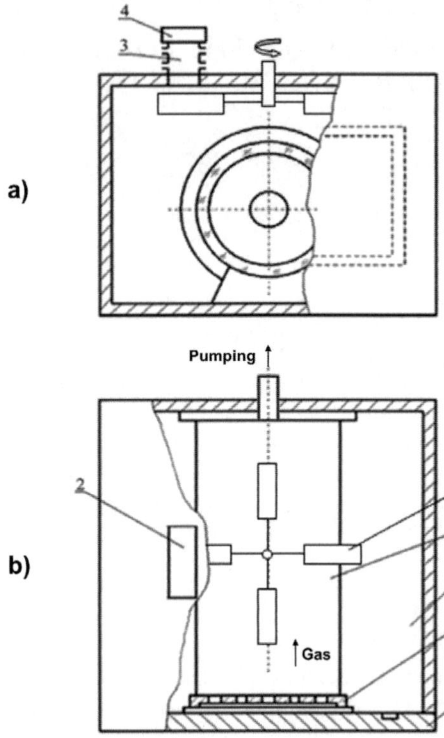


*Figure 1. Scheme of the reaction-discharge system of the microwave plasma installation (a — front view; b — top view):*
*1 — resonator; 2 — coupling aperture; 3 — waveguide; 4 — magnetron; 5 — reaction-discharge chamber; 6 — resonator's front wall; 7 — chamber's front cover; 8 — dissector.*

discharge chamber (RDC) where a low temperature gas microwave discharge is excited by a UHF electromagnetic field. Bottled 99.7% oxygen was as used as a plasma-forming gas. The operation pressure in the chamber equalled to 130 Pa.

The reaction-discharge chamber is made of quartz tube of 20 cm in diameter and 34 cm long having a total volume of about 9000 cm$^3$ and is placed in the center of a $40 \times 40 \times 30$ cm$^3$ volume resonator (see figure 1). The electromagnetic energy from the microwave magnetron is fed into the resonator through a rectangular coupling apperture in the top wall of the resonator. A metal L-form dissector is attached to the top wall inside the resonator. The dissector is placed in such a way that when it revolves its blades cross the space under the coupling apperture. The microwave magnetron in the installation worked in the pulse mode in the conditions of generating a nominal power of $P_{MW} \approx 650$ W.

The microwave power measurements at the local points on the axis of the discharge chamber with plasma and without it were performed using the "active probe" method [5, 8]. The "active probe" is a segment of a flexible coaxial cable with a wave impedance of 50 Ω, the external conductor of which is a copper pipe and the internal one is a single-core copper wire. The space between the conductors is filled with a flexible dielectric — teflon. An "active probe", which is a segment of the central internal conductor with a diameter of 1.4 mm protruding from the external conductor by 5 mm, is formed on one side of the cable. The indications of the "active probe" were registered with a microwave power meter M3-51.

Silicon wafers with a diameter of 76 and 100 mm ($\rho = 1.11$ Ω·cm) were used in the research. In each series of experiments in the discharge chamber there were two wafers of the same size placed along the gas flow parallel to each other, one on each side of the probe symmetrical to it. The distance between the wafers varied within the range of 20–100 mm.

In the first series of experiments the influence of silicon wafers on the overall structure of the microwave power distribution in plasma along the axis of the discharge chamber was studied. For this purpose the probe was moved along the discharge chamber. The investigation was performed using two silicon wafers with a diameter of 100 mm. The distance between the wafers equaled to 60 mm. The wafers were placed in the central part of the discharge chamber. In some local areas the probe passed between the silicon wafers. The indications of the "active probe" were registered along the axis every 20 mm in the direction from the chamber's front cover.

In the second series of experiments the influence of the distance between the silicon wafers (in the range of 20–100 mm) on the local values of microwave power between them was examined. The experiments were performed with silicon wafers with a diameter of 76 and 100 mm placed in the central part of the discharge chamber. The "active probe" was placed between the wafers motionless.

## 3. Results and discussion

Figure 2 shows the experimental values of microwave power meter indications in the plasma volume along the axis of the discharge chamber without wafers (figure 2a) and with two silicon wafers with a diameter of 100 mm (figure 2b), placed in the central part of the discharge chamber.

As it can be seen from the data presented in figure 2 the value of the microwave power between the silicon wafers decreases by approximately 20–30%. The changes of probe's indications in the plasma chamber's areas close to the location of silicon wafers are insignificant. At the same time, the general structure of the electromagnetic energy distribution in plasma in the microwave plasmatron's discharge chamber is maintained. The periodicity of alternations of maximum and minimum probe's indications corresponds to a half-length ($f_{MW} = 2.45 \pm 0.05$ GHz) of the electromagnetic wave exciting the plasma.

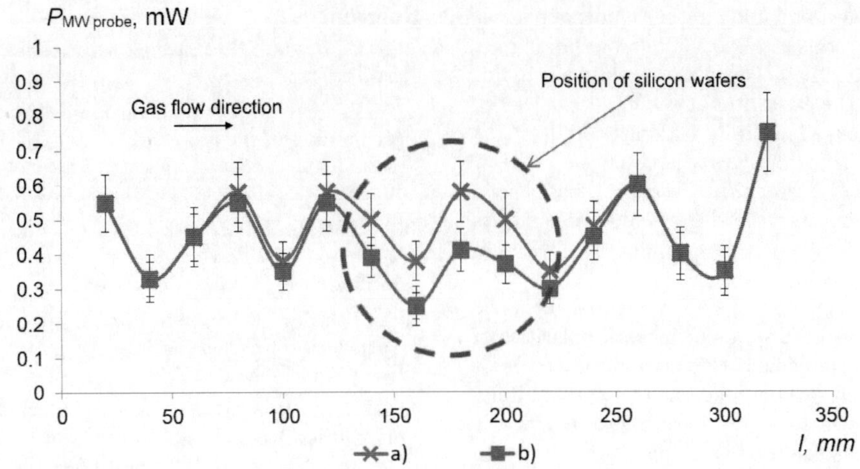


*Figure 2. Experimental values of microwave power meter indications in plasma along the axis of the discharge chamber: a — without wafers; b — with two silicon wafers with a diameter of* 100 mm *placed in the central part of the discharge chamber.*

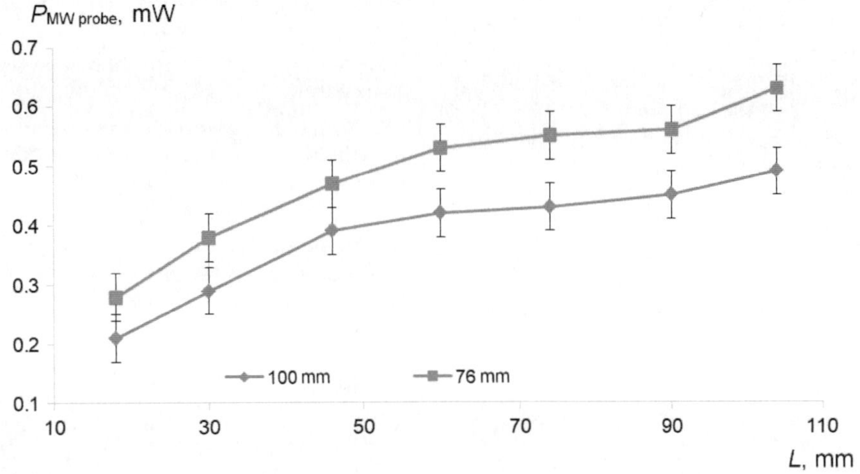


*Figure 3. Dependence of microwave power value in the plasma volume on the distance between two silicon wafers with diameters of* 76 *and* 100 *mm.*

Figure 3 shows the diagrams of the experimental dependencies of the microwave power meter's indications in the plasma volume on the distance $L$ between the pairs of silicon wafers of 76 and 100 mm in diameter.

From the dependencies presented in figure 3 it can be seen that the change of the distance between silicon wafers from 100 mm to 20 mm results in the approximate 50% decrease of the microwave power between the wafers both for the pair of wafers with 100 mm in diameter as well as for the wafers with 76 mm in diameter. The behavior of dependencies for silicon wafers of both diameters is of the same character. The presented data also show that the increase of silicon wafers' diameter results in the decrease of local microwave power values between them.

It is assumed that the effects observed in both series of experiments can be explained by partial absorption and shielding of microwave radiation by silicon wafers having high dielectric loss tangent.

Thus, the change in the position of the silicon wafers in the reaction-discharge chamber of the microwave resonator-type plasmatron, their size and mutual arrangement significantly affect the value of the microwave energy penetrating into the plasma volume.

This should be taken into account in the analysis of plasma forming processes in the conditions of microwave plasma-chemical treatment with a volumetric positioning of materials in the microwave plasmatrons' discharge chambers.

## 4. Conclusion

The performed research has shown that the insertion of silicon wafers into the discharge chamber of a microwave resonator-type plasmatron does not significantly alter the overall structure of non-uniform microwave power distribution in the volume of plasma-forming zone. In this case the spatial alternation periodicity of its maxima and minima in plasma remains the same as in the case without wafers.

It has been experimentally established that with the increase of silicon wafers' diameter the value of microwave power between them decreases. The change in the distance between the silicon wafers also results in the change of the microwave power value between them.

The performed experiments allow to supplement the model conception of physico-chemical peculiarities of microwave plasma-chemical treatment processes of materials. Particularly it concerns the effect of influence of silicon wafers on the local microwave power values in the plasma chamber volume which can also affect the degree of chemical activity of different gas discharge areas. This effect should be taken into account while designing the constructions of microwave discharge systems for technological purposes and processes of group microwave plasma-chemical treatment of electronic devices that use monocrystalline silicon wafers as substrates.

The obtained results may be used in designing of new gas discharge systems and the modernization of existing technological equipment for a group non-precision microwave plasma-chemical treatment of semiconductor wafers.

## Acknowledgements

This research has been supported by the State Program of Scientific Research "Physical materials science, new materials and technologies", assignment 3.5.08.

## References

[1] S. Bordusov. *Microwave plasma technologies in the production of electronic devices.* Minsk: Bestprint, 2002.

[2] Y. Lebedev. Microwave discharges at low pressures and peculiarities of the processes in strongly non-uniform plasma. *Plasma Sources Science and Technology*, 24(5):1–39, 2015. doi:10.1088/0963-0252/24/5/053001.

[3] V. Batenin, I. Klimovsky, G. Lisov, and V. Troitski. *Microwave Plasma Generators: Physics, Engineering, Applications.* Moscow: EnergoAtomlzdat, 1988.

[4] Y. Lebedev. The chemical activity of nonequilibrium plasma. state and perspectives of research. *Chemical Physics*, 15(5):96–100, 1996.

[5] U. Kortshagen. Electron and ion distribution functions in RF and microwave plasmas. *Plasma Sources Science and Technology*, 4(4):172–182, 1995. doi:10.1088/0963-0252/4/2/002.

[6] T. Nakano and S. Samukawa. The correlation between an electric field and the metastable chlorine ion density distributions in an ultrahigh-frequency plasma. *Japanese Journal of Applied Physics*, 37(5R):2686–2692, 1998. doi:10.1143/jjap.37.2686.

[7] A. Didenko and B. Zverev. *Microwave energetics.* M.: Nauka, 2000.

[8] S. Bordusau, S. Madveika, and A. Dostanko. Investigation of microwave energy distribution character in a resonator type plasmatron. *Plasma Physics and Technology*, 3(3):122–125, 2016.

[9] S. Bordusau. Design features of the installation and processes of microwave plasma materials processing under low vacuum. *Materials, technology, instruments*, 6(4):62–64, 2001.

doi:10.14311/ppt.2019.3.243
*Plasma Physics and Technology* **6**(3):243–246, 2019 

# ON THE INTERACTION OF A MICROWAVE EXCITED OXYGEN PLASMA WITH A JET OF PRECURSOR MATERIAL FOR DEPOSITION APPLICATIONS

R. Methling[a,*], F. Hempel[a], M. Baeva[a], T. Trautvetter[b], H. Baierl[b], R. Foest[a]

[a] *Leibniz Institute for Plasma Science and Technology (INP), Felix-Hausdorff-Str. 2, 17489 Greifswald, Germany*

[b] *Leibniz Institute of Photonic Technology (IPHT), Albert-Einstein-Str. 9, 07745 Jena, Germany*

[*] methling@inp-greifswald.de

**Abstract.** A plasma source based on a microwave discharge at atmospheric pressure is used to produce an oxygen plasma torch. A liquid precursor material is evaporated and injected into the torch through a nozzle, causing oxidization and deposition of silica at a nearby quartz substrate. The temperature generated inside the plasma source and in the plume, in the region of treatment, and at the substrate surface are key parameters, which are needed for process description and optimization of plasma-chemical reactions. Optical emission spectroscopy and thermography were applied to observe and characterize the jet behavior and composition. The experimental results are compared with self-consistent modeling.

**Keywords:** microwave plasma, plasma jet, optical emission spectroscopy, thermography.

## 1. Introduction

High performance glasses based on doped silica are essential parts of modern fiber based laser systems [1]. They provide highest precision and flexibility in system design combined with high power output and excellent beam quality. In case of pulsed operation an excellent material homogeneity is required due to high pulse peak powers.

Existing fiber preform fabrication technologies such as modified chemical vapor deposition (MCVD) gas phase or solution doping, reactive powder sintering technology (REPUSIL) or the common melt process face limitations regarding the feasibility of achievable geometrical as well as chemical dopant specifications and thus the optical performance of the resulting core material. For instance, constraints in terms of diameter of active core, homogeneity of dopant distribution within the core as well as the practicable laser active dopant concentration affect the resulting optical performance of the fabricated fiber, especially its spectral attenuation and photo darkening behavior. Furthermore, laser active dopants increase the refractive index so that a co-doping with a dopant decreasing the refractive index is required. The interaction of the two dopants usually confines, for thermodynamic reasons, the accessible core specification in terms of dopant concentration profiles, maximum dopant concentrations, and core diameter.

To overcome such thermodynamic limitations, a novel approach based on atmospheric-pressure microwave (MW) plasma has been elaborated. Previous research has demonstrated that plasma based technology at atmospheric pressure is a valuable tool for the treatment of specialty glasses and optical fibers. The use of a plasma system at atmospheric pressure offers a wide range of advantages. Atmospheric pressure plasmas are generally characterized by a high density of active species resulting in higher reaction rates which in turn are capable of significantly increasing deposition rates and efficiencies compared to low pressure plasmas. Microwave plasmas offer an even higher intrinsic electron, ion and radical density thus enhancing deposition efficiencies in comparison to other plasma sources. Furthermore, electrode-less systems such as microwave and RF plasmas circumvent possible contamination sources as encountered with e.g. DC/AC plasma jets. However, the coupling efficiency of RF energy into the plasma is considered critical. Values as low as 40 % to 50 % are reported, dropping even more significantly with high power input above 100 kW. Instead, the coupling efficiency of the incoming microwave power into the plasma can reach nearly 100 % provided that the wave guide setup is configured correctly. The plasma-based approach could offer an outstanding advantage in terms of plasma-induced kinetic control of the deposition process. It thus allows for co-doping beyond thermodynamically determined chemical equilibrium, and for overcoming diffusion limitations by the processes controlled by convection. This feature results in an overall improved outstanding homogeneity of the doped core material.

In this paper, the setup of a microwave plasma system at atmospheric pressure for the deposition of $SiO_2$ is described. Optical emission spectroscopy and thermography were applied to observe and characterize the jet behavior and composition as well as gas and target temperature. The experimental results are compared with self-consistent modeling.

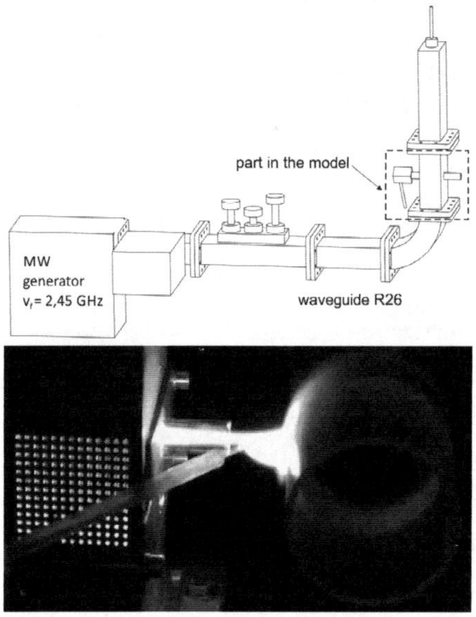


*Figure 1. Top: Schematic view of the plasma source. The plasma torch is produced by excitation of the gas flow through the quartz tube traversing the waveguide. Bottom: Photograph of the plasma torch, precursor nozzle, and rotating substrate.*

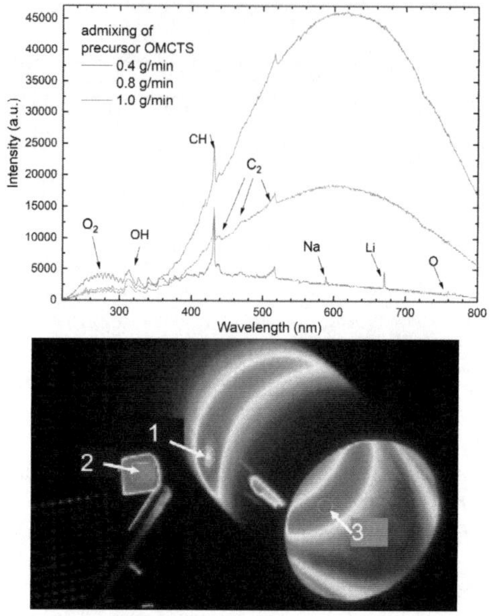


*Figure 2. Top: Overview spectrum of oxygen plasma torch with different admixtures of OMCTS precursor. Bottom: Thermography image.*

## 2. Experimental Setup

A microwave plasma torch operated at atmospheric pressure at a frequency of 2.45 GHz was used as shown in the upper part of Figure 1. The plasma source consisted of a standard wave-guide R26 and a quartz tube of 12 mm inner diameter traversing at the position of maximum electric field of the fundamental mode TE 10. Mixtures of argon and oxygen, including the pure gases, were employed as working gas. The plasma jet was directed towards a rotating substrate as shown in the lower part of Figure 1. For the generation of silica, the precursor octamethylcyclotetrasiloxane (OMCTS) was either vaporized or nebulized by means of ultrasonic impact. With some Ar flow, the precursor was fed through an additional nozzle placed outside the plasma source in an angle of about 45° towards the plasma torch. As a substrate a big quartz tube rotating perpendicular to the plasma (as shown in the photograph) was used.

Several techniques were applied for the optical analysis. The general behavior of the plasma with focus on the interaction region of the two jets and the deposition area was recorded by a high-speed video camera.

A 0.5 m imaging spectrometer (SP-2556, Roper Acton) equipped with an ICCD camera (PI-MAX4:1024i-RB, Princeton Instruments) was used to record the plasma emission along the MW jet, including positions inside the quartz tube as well as in the plasma plume. Therefore, the radiation was focused by a quartz lens onto the entrance of an UV-VIS fiber which was coupled to the spectrometer. While overview spectra as shown in the upper part of Figure 2 were taken with a grating of 50 lines per mm, higher spectral resolution ($\sim$ 0.03 nm, 2400 l/mm grating) was used for the analysis of molecular band structures.

Surface temperatures were determined by means of a thermographic camera (VarioCAM, InfraTec) with an effective resolution of 640×480 pixels. The measurement was set up in the temperature range 450–1500°C tube serving as deposition target. Time-averaged values were obtained for three positions: (1) the spot of deposition, (2) the end of the plasma torch quartz tube and (3) at the position of the local temperature maximum inside the substrate (lower part of Figure 2). The sensitivity of the measurement allowed detecting even small variations of few mm in the distance of operation.

The target was typically placed at a varying distance $L_{\text{tar}}$ from the wave guide. The position of spectroscopic investigation $x$ was related to the end of quartz tube $L_{\text{out}}$, i.e. the maximal value is at the position of the target $x = L_{\text{tar}} - L_{\text{out}}$.

## 3. Self-Consistent Model

The determination of the plasma parameters in the region of plasma-precursor interaction was of primary importance for the technological application. On the one hand, the plasma parameters in the plume depended on the conditions inside the plasma source. On the other hand, the source was hardly accessible for diagnostics. Therefore, complementary studies by modeling and experiments were carried out.

A self-consistent model was developed to describe the MW plasma source employing the plasma, electromagnetic and hydrodynamic equations[2]. It was realized on the computational platform COMSOL Multiphysics in two- and three-dimensional geometry[3].

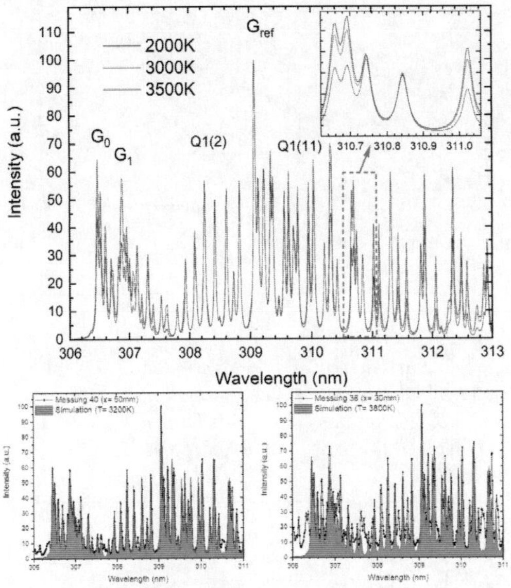


*Figure 3. OH band emission. Top: Calculated spectra for 2000 K, 3000 K, and 3500 K. Bottom: Comparison of recorded and calculated spectra with gas temperatures of about 3200 K (left) and 3800 K (right).*

The microwave entered the wave guide, propagated along it and was reflected on the short end to build a standing wave. The gas was fed into a quartz tube. The tube position was adjusted to match the maximum of the electric field in the wave guide to ignite the plasma. The model equations described in a self-consistent manner the gas flow, the plasma kinetics, the heat transfer, and the MW field in the waveguide. It was based on a hydrodynamic approach. The Navier-Stokes equations for conservation of mass and momentum, the MW field equation, and the equation for conservation species, energy of electrons and heavy particles were considered. The model avoided the assumption of quasi-neutrality and solved the Poisson equation to obtain the electrostatic potential. The results obtained in argon as a working gas[2] showed that for temperatures over ~1350 K in the plasma, the atomic ion was dominant. The molecular ion played a role in the cold region near the tube wall. Since most of the plasma flow in the source was hot, the present studies were performed in the frame of the one-ion plasma chemistry.

## 4. Gas Temperature Determination

For the evaluation of gas temperature, the rotational temperature of the OH(A-X) band with maximum intensity near 309 nm was identified by comparing recorded and calculated spectra. The calculations were performed using the simulation software Lifbase [4]. A small amount of OH, which was present from water, e.g. due to diffusion from the surrounding air into the plasma plume zone, allowed to determine the gas temperature in the most cases. For $O_2$-containing gas mixtures, an overlapping of the

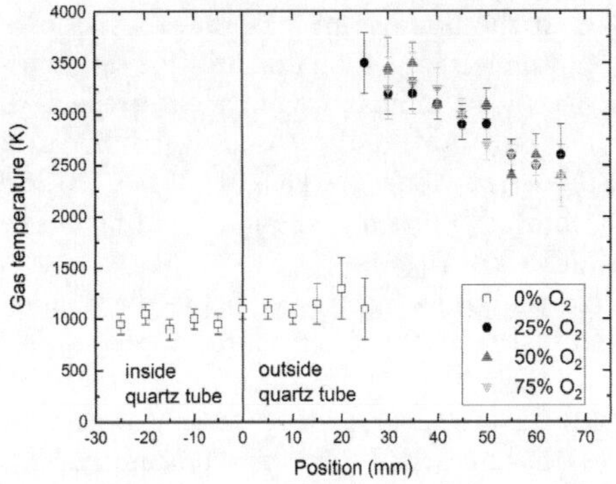


*Figure 4. Gas temperatures determined according to Figure 3 for different Ar/$O_2$ mixtures.*

OH(A-X) bands with Schumann-Runge bands of $O_2$ (about 200–450 nm with maximum intensity around 350 nm) occurred, mainly for positions near to the plasma source. Thus, the OH band structures were accessible only downstream in the plume and no temperature determinations from spectroscopy were possible at the tube end and at the place of the precursor feed. A simulation of the spectral emission pattern for three gas temperatures is shown in the upper part of Figure 3. In the lower part of the figure, two examples of comparison of measured and calculated OH-spectra are plotted for a distance of -5 mm from near the tube end in pure argon.

## 5. Results and Discussion

The gas temperature at different $x$ positions along the MW torch was determined from OH spectra and compared with values extracted from simulation with the 2D planar model. Four Ar/$O_2$ gas mixtures emitted into a substrate-free space were investigated at an incoming MW power of 1.0 kW, a gas inflow rate of 18 slm, and a tube length $L_{out}$=38 mm. The results are summarized in Figure 4. With pure argon (0% oxygen admixture) the temperature reached a maximum value of ~1500 K close to the inner tube wall on the side of the incoming microwaves, decreasing outside the tube to temperatures far below 1000 K. Interestingly, a simulated temperature decrease between the end of the wave guide (-25 mm) and the end of the quartz tube (0 mm) could not be found in the spectra. On the contrary, the gas temperatures deduced from emission spectroscopy showed a slight increase from about 1000 K inside the tube to 1200 K at a position 20 mm outside the tube. While the model predicted temperatures of about 600 K in the plume, the emission spectroscopy yielded values of 1000–1400 K. The admixture of oxygen caused a significant change in the plasma and its properties. The gas temperature could only be determined from OH emission far away from the tube end, where emission from molecular oxygen was absent. It was found to be substantially higher for any admixture of oxygen than in pure argon. This result could be probably explained with the heat released in the reactive plasma due to dissociation of oxygen molecules. Temperatures for different $O_2$ admixtures were similar within measurement accuracy. The temperature decreased nearly linear with the distance from about 3500 K ($x = 25$ mm) to 2400 K ($x = 65$ mm). No plasma emission could be observed at distances $x > 70$ mm.

Beside the above mentioned molecular bands of OH and $O_2$, several other features appeared in the overview spectra in the cases with admixture of Ar and OMCTS precursor, cf. Figure 2. Molecules originating from the carbon-containing precursor, e.g. CH and $C_2$, were detectable by their band heads. Furthermore, some atomic lines as Na and Li could be found in cases of high thermal load on the glass components of the setup while O lines only occurred for low amounts of precursor admixture. No ionic lines were observed, validating a low amount of charged particles and moderate gas temperatures.

Another specific feature of the spectra was the continuum emission above ~500 nm. Its intensity was very sensitive to the amount of OMCTS admixture, cf. overview spectra in Figure 2. Several sources are possible causing similar continuum radiation. (i) For a molecular radiation, a typical shape usually including local maxima or band heads would be expected. (ii) Recombination radiation is typically characterized by broad emission over several hundreds of nm with flat profiles without distinct lines or peaks. (iii) Thermal radiation has a spectral distribution according to the Planck function; in relevant wavelength range between VIS and NIR it is typically characterized by (nearly linear) increase towards longer wavelengths.

For advanced analysis, the influence of spectral sensitivity was eliminated by (relative) spectral calibration using a tungsten ribbon lamp. Figure 5 shows two calibrated spectra with a precursor flow of 2.0 slm Ar plus 0.8 and 1.0 g/min OMCTS. The spectral range between 550 and 800 nm was chosen only to reduce the influence from CH and $C_2$ bands (<550 nm) and reduced sensitivity causes noise (>800 nm). The curves could be fitted very well to Planck functions obtaining temperatures of 3180 K for the lower (0.8 g/min) and 2970 K for the higher amount of OMCTS (1.0 g/min). It was double-checked that no thermal radiation from the target surface was detected by the OES system and the radiation must originate from the interaction area of the two jets (oxygen plasma and precursor) near to the target. In principle, a gas jet should not emit thermal radiation. However, an explanation might be found in micro-droplets surviving from the precursor injection. Here, NIR spectroscopy could help to reveal this phenomena.

Figure 6 shows the surface temperature of the quartz substrate obtained by thermographic measurements for various distances between the end of the discharge tube and the substrate. It should be noted that differently from the results shown above, the oxygen flow through the plasma source was kept constant

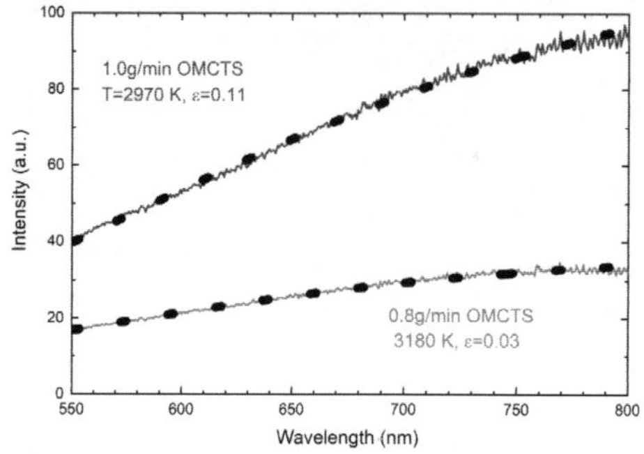


*Figure 5. Part of spectra after relative intensity calibration (1.0 kW, 10 slm $O_2$, $L_{out}$=25 mm, $L_{tar}$=45 mm, $x$=15 mm). The curves are fitted using Planck functions; temperatures of ~3000 K are obtained.*

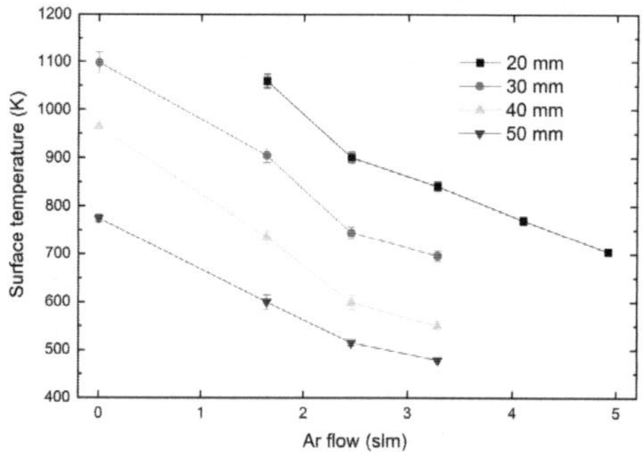


*Figure 6. Target surface temperature depending on precursor gas flow for different target distances $x$.*

and the argon was added via the precursor tube. Two well-pronounced effects were observed: On the one hand, the surface temperature was highest in pure oxygen and decreased with the increase of the Ar-concentration due to gas cooling. On the other hand, the surface temperature decreased with increasing distance due to a reduced heating by the plasma jet.

## Acknowledgements

The work was supported by grant SAW-2017-IPHT-1 of the Leibniz-Gemeinschaft. The authors acknowledge M. Stankov and J. Schäfer for helpful discussions.

## References

[1] S. Addanki, I. S. Amiri, and P. Yupapin. Review of optical fibers-introduction and applications in fiber lasers. *Results in Physics*, 10:743–750, 2018. doi:10.1016/j.rinp.2018.07.028.

[2] M. Baeva, F. Hempel, H. Baierl, T. Trautvetter, F. Foest, and D. Loffhagen. Two- and three-dimensional simulation analysis of microwave excited plasma for deposition applications: operation with argon at atmospheric pressure. *J. Phys. D: Appl. Phys.*, 51:385202, 2018. doi:10.1088/1361-6463/aad537.

[3] COMSOL. *COMSOL Multiphysics v. 5.3a*. http://www.comsol.com.

[4] J. Luque and D. R. Crosley. *SRI International Report MP 99-009, version 1.9.* LIFBASE.

doi:10.14311/ppt.2019.3.247
*Plasma Physics and Technology* **6**(3):247–250, 2019 

# ON THE ISSUE OF THE THERMAL EMISSION CATHODE RESOURCE ON THE OXIDE FILMS OF HAFNIUM AND ZIRCONIUM

N. Klochok[a,*], Yu. Popil[b], V. Chernyak[a], V. Iukhymenko[a], I. Fedirchyk[a], V. Tkach[c]

[a] *Taras Shevchenko National University of Kyiv, Volodymyrska Street 64/13, 01601, Kyiv, Ukraine*

[b] *National Technical University of Ukraine "Igor Sikorsky Kyiv Polytechnic Institute", Prospekt Peremohy 37, 03056, Kyiv, Ukraine*

[c] *V. Bakul Institute for Superhard Materials NASU, Avtozavodska Street 2, 04074, Kyiv, Ukraine*

[*] nikolo1961@gmail.com

**Abstract.** The article shows the results of research on the destruction of the thermionic cathode of an arc plasma torch when inserting hafnium and zirconium along the trace of the spot of a gas-discharge arc by means of electronic metallography with spectral analysis in the control region. The depth of penetration of the spot into the metal of the cathode insert was determined due to the discrete mechanism of action on the metal at the time of attachment. The erosion rate of zirconium oxide films was experimentally established, the extension of the operating life substantiated while mechanisms for its increase were revealed, and two types of arc contact with a hafnium or zirconium insert - the "contracted" and "diffused" ones - were offered. The theoretical operating life of the cathode is calculated.

**Keywords:** plasma, surface scanning, erosion, erosion mass, attachment spot.

## 1. Introduction

The development of modern plasma technology entails a classical contradiction between the efficiency of the thermionic cathode, its temperature and the duration of its operation - the resource, which is explained by the difference in energy between the work function and the ionization potential [1]. The ionization potential of plasma-forming gas is four times higher than the work function of copper or iron from which the electrodes are made. For example, the ionization potential of $CO_2$ is 14.3 eV [2], and the work function of copper is 4.4 eV [3]. The energy input to the spot is 4.5 eV per atom of the substance [4]. And the most effective cooling water provides heat dissipation $q_B = 1\,\mathrm{kW\,cm^{-2}}$ due to the low thermal conductivity of the metal. The life time of a spot is a long period of time, taking up approximately $10^{-3}$ s, during which the metal in the spot passes into gas (sublimation) or melts, evaporates and ionizes. The pulsed thermal effect of the spot arc leads to variable loads, and when their value exceeds the yield strength with repeated exposure, the material cracks deep. In this case, the erosion rate is much higher than the erosion rate during melting or evaporation (sublimation).

The service life of any DC electric arc plasma torch designed for the implementation of continuous technological processes is an important indicator of the reliability and technology implementation. It is determined by the service life of a plasma torch cathode - the sublimation flow rate referred to the density of the emitted current, which determines the degradation of its thermionic elements. Given the impossibility of using contact methods for studying the fastening of the attachment spot and its surface, despite publications on this topic, there is no picture of the effect of the spot on the inserted material. Therefore, in this work, we studied additional features of the destruction of the cathode of the PUN-1 plasma torch and ways of extending its service life.

## 2. Experiments and methods

The object of the study was the trace of the cathode spot of the arc on a standard insert of the PUN-1 plasma torch (Paton Electric Welding Institute for the deposition of metal films) whose capacity ranged between 20 and 10 kW. Variable values were: cathodes of various diameters with a standard zirconium or hafnium insert, as well as the discharge current from 70 to 100 A and the flow rate of plasma-forming gas of 25–46 l min$^{-1}$ air with the addition of 9 % of propane-butane, as a gas envelope, pressure ranging from 0.15 to 0.4 MPa. We used water cooling at a pressure of 0.4 MPa and temperature of 20 °C.

The study of changes in the surface microstructures of the cathode insert was conducted along the trace of the arc discharge spot, using the metallography method. The enlargement of its image from 10 to 300 times under the Metal Microscope MIG-4 and the MBE-10 digital camera, as well as the high resolving power CCDDX-4, showed two different spot mounts - the "contracted" contact with a discrete mechanism for penetrating into the metal of the hafnium and zirconium inserts to a depth of 10 to 300 microns and intense ablation. Also, it showed the "diffused" contact related to the evaporation of metal and a significant increase in the area of emission. We succeeded in detailing the nature of the destruction of the electrode surface at the attachment point with the help of the Zeiss EVO 50 XVP scanning electron microscope, which detects the minute chemical composition of the surface of the spot trace and its impurities by means of scanning spectroscopy [5].

## 3. Results and discussion

During the experiment the standard cathodes of the PUN-1 plasma torch with a hafnium and zirconium insert were used, the photos of which are shown in figure 1 (a, b, c).

During continuous operation of the PUN-1 plasma torch for 20 min and vortex gas flow in the gap of 0.5 mm between the cathode and the anode, a trace of 1 mm in diameter and 1.5 mm in depth was formed at the attachment point of the hafnium insert (see figure 1a). Regarding the zirconium cathode, which operated for 32 hours, the trace was 1.2 mm in diameter and 3.1 mm in depth. The geometry of the destruction of the cathodes is the same. When a gap between the cathode and the anode widened, which was due to a decrease in the diameter of the cathode by a factor of 1.5, the trace diameter increased to 4 mm and a depth of 2.8 mm. The nature of the destruction of the hafnium cathode insert was different. The life of PUN-1 to its complete stop, with the repeat of operating parameters, increased to 72 hours. This motivated the study of the cathode shown in figure 2 - a) the cathode 3.5 mm in diameter with a zirconium insert of 1.5 mm operated for 482 hours before a control weight measurement. The diameter of the track averaged 2.8–3.1 mm. To compare, a control measurement of the evaporation of the mass of the insert of the PUN-1 plasma torch shown in figure 3 b) and c) was taken.

As we used the Zeiss EVO 50 XVP scanning electron microscope, the results of the metallography of the trace of the arc spot on the hafnium and zirconium inserts of the PUN-1 cathode (see figure 1a and figure 1b) showed the same nature of the cathode spot fastening regardless of the operating time. As can be seen from the photographs of figure 3 a) and b), deep cracks are visible over the entire area, as well as annular, stepped and tapering in nature with a distance of 10 to 300 microns of the trace of the reference spot of the material penetrating deeply into the cathode insert, which indicates a discrete mechanism of its operation. This type of spot attachment is called "contracted". An analysis of the chemical composition of the sample with the help of X-ray radiation generated by the electron beam established that carbon, oxygen, and copper accumulate with spot work in time (see table 1) - poisoning by combustion products.

At the cathode of reduced diameter, which worked for 72 hours with a hafnium insert, along the trail of the cathode spot, by metallography, another mechanism of the spot attachment was revealed, which we called "diffused" (figure 1c). The trace over its entire area does not have large and deep cracks, the size is increased several times, and chemical analysis of the samples taken from the insert material shows the absence of "poisoning" of hafnium by combustion products. Obviously, the material of the insert into the arc is supplied by evaporation.

A decrease in the diameter of the cathode and the identification of "diffusion" attachment, with an increase in its resource, suggested an increase in the gap between the cathode and the anode, which could lead to an increase in the dynamic pressure at the point of attachment of the arc spot. For this purpose, a cathode with a diameter of 3.5 mm was manufactured with a zirconium insert of 1.5 mm (figure 1c). The metallography of the trace of the attachment of the arc spot is shown in electronic photograph on figure 4b. The analysis of the samples gives reason to believe that zirconium oxide with a melting point of 3063 K was formed at point No. 1 [6].

The weight loss of the cathode mass was $\Delta g = 0.18\,\mathrm{g}$ at the rate of mass loss $\Delta G = 1.5 \cdot 10^{-13}\,\mathrm{kg\,s^{-1}}$ [7, 8].

To check the revealed effect of reducing erosion of a zirconium insert with a diameter of 1.5 mm while reducing the speed of the eddy flow of the working medium, we measured the critical weight loss of the zirconium insert of the PUN-1 plasma torch (the gap between the cathode and anode is 0.5 mm) At a discharge current of 100 A and duration of operation 3600 s amounted to $\Delta g = 0.05\,\mathrm{g}$, which corresponds to the ablation rate $\Delta G = 0.139 \cdot 10^{-9}\,\mathrm{kg\,s^{-1}}$ and fully corresponds to the values presented in the literature [8, 9] for zirconium insert operating in an air atmosphere.

## 4. Conclusions

1. The mechanism of the action of the cathode spot of the arc gas discharge with "contracted" and "diffused" attachment to the metal surface on hafnium and zirconium inserts is presented.

2. The depth of the crater in the metal under the cathode spot reached 1.5 – 2.8 mm. The walls of the crater in the case of a "contracted" attachment are a tapering stepped terrace with a step in width of 10 – 300 microns.

3. It is shown that a decrease in the vortex flow velocity in the gap between the cathode and the anode can reduce the rate of metal entrainment from the insert crater by two orders of magnitude.

4. The theoretical life-time of the cathode assembly with a zirconium insert with a diameter of 1.5 mm with a "diffused" attachment at 70 A discharge current amounted to more than 7000 hours.

*Figure 1. Top view of the cathode PUN-1 with hafnium insert a) the cathode operated for 20 minutes, with zirconium b) it operated for 32 hours before the complete cessation of arc burning, and with hafnium c) it operated for 72 hours. The attachment location of the reference spot is indicated by the arrow.*

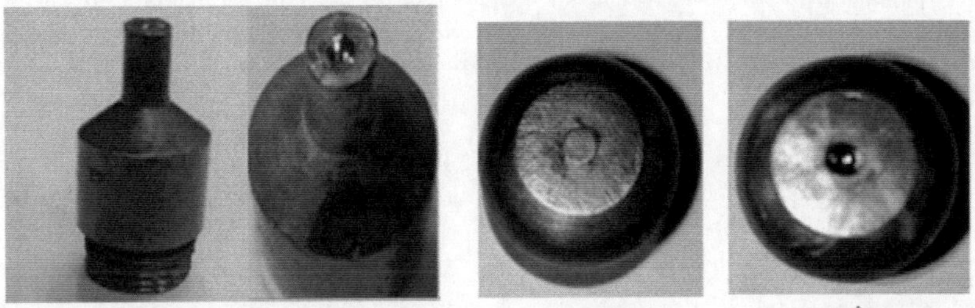


*Figure 2. a) Side view and top view of the cathode 3.5 mm in diameter with a 1.5 mm zirconium insert, which operated for 482 hours before a control measurement in a standard medium with modified parameters of the discharge, discharge current - $I_P = 70\,A$, discharge voltage - $U_P = 180\,V$, pressure = 0.15–0.25 MPa, flow rate = 25–32 l min$^{-1}$; b) Copper cathode with zirconium insert of the PUN-1 plasma torch before the control test; c) Photo of the same cathode after the test.*

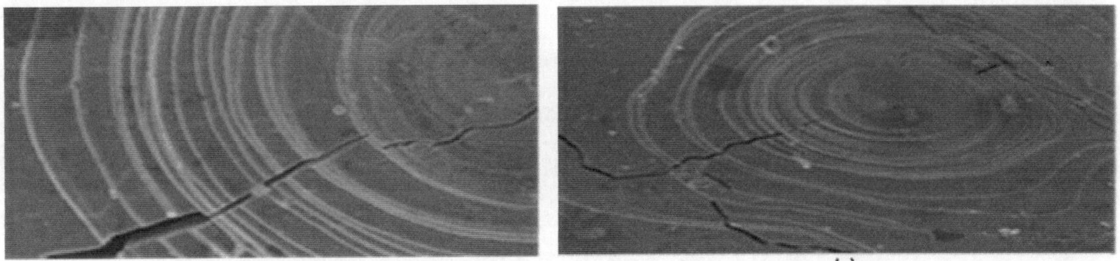


*Figure 3. a) a photograph of the trace of the cathode spot on the surface of the hafnium insert (20 minutes of work) and b) a photograph of the trace of the cathode spot on the surface of zirconium insert (32 hours of operation).*

| Spectrum | In statics | Carbon | Oxygen | Copper | Zirconium | Hafnium | Total |
|---|---|---|---|---|---|---|---|
| 1 | Yes | 18.17 | 32.73 | 0.46 | 48.31 | 0.33 | 100 |

*Table 1. The composition of the material in the trace of the cathode spot on the zirconium insert.*

| Spectrum | In statics | Carbon | Oxygen | Copper | Zirconium | Hafnium | Total |
|---|---|---|---|---|---|---|---|
| 1 | Yes | 7.33 | 29.40 | 0.46 | 62.63 | 0.16 | 100 |

*Table 2. The composition of the sampling samples from the cathode spot on the zirconium insert at the attachment point No. 1 (figure 4b).*

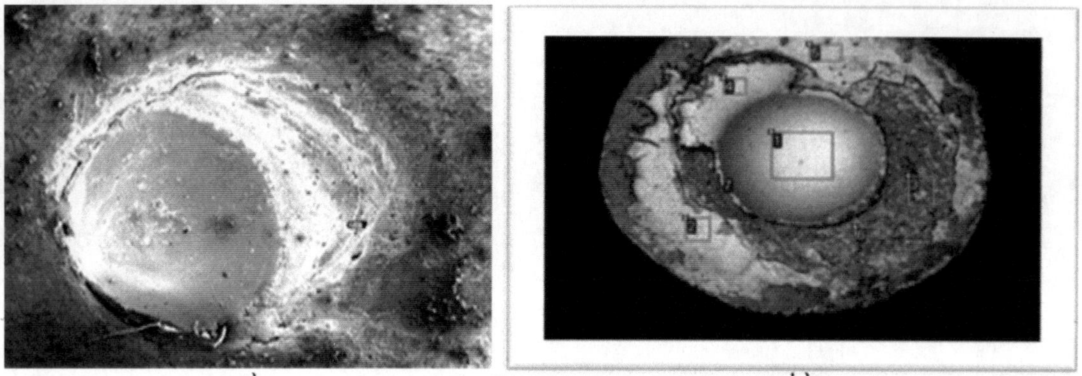

a) b)

*Figure 4. a) an electronic photograph of the trace of the cathode spot on the hafnium insert, which worked for 72 hours, with a reduced cathode diameter (see figure 1c); b) an electronic photograph of the attachment of the spot indicating the sampling location of the zirconium insert, which worked for 482 hours before weight control.*

## References


[1] A. Von Engel and M. Steenbeck. *The Physics and Technique of Electrical Discharge in Gases.* University of Washington, 1950.

[2] Welder and welding. Carbon dioxide. `http://weldering.com/uglekislyy-gaz-uglekislota-dvuokis-ugleroda`.

[3] Calculator. Work function of electrons in metals. `https://www.calc.ru/601.html`.

[4] B. I. Mikhailov. Arc spot scanning of tube electrodes in gas-vortex plasmatorches. *Thermophysics and Aeromechanics*, 15(2):307–320, Jun 2008. doi:10.1134/S0869864308020145.

[5] Institute of Solid State Physics, University of Latvia. *Scanning electron microscope Zeiss EVO 50 XVP.* `https://www.cfi.lu.lv/fileadmin/user_upload/lu_portal/projekti/cfi/Tehniskais_aprikojums/Equipment/Scanning_electron_microscope_Zeiss_EVO_50_XVP.pdf`.

[6] Wikipedia. Zirconium dioxide. `https://en.wikipedia.org/wiki/Zirconium_dioxide`.

[7] A. S. Anshakov, E. K. Urbakh, A. E. Urbakh, and V. A. Faleev. Study of thermochemical cathodes in arc plasma motors. *Thermophysics and Aeromechanics*, 12(4):685–691, 2005.

[8] V. M. Kulygin, A. V. Pereslavtsev, and S. S. Tresvyatskii. Estimation of the temporary service life of dc arc plasmatron cathode. *Technical Physics*, 62(9):1327–1331, 2017. doi:10.1134/S1063784217090146.

[9] M. F. Zhukov, A. V. Pustogarov, G. N. B. Dandaron, and A. N. Timoshevskii. *Thermochemical cathodes.* Institute of Thermo-physics of Academy of Sciences of USSR, 1985.

doi:10.14311/ppt.2019.3.251
*Plasma Physics and Technology* 6(3):251–255, 2019 

# EQUILIBRUM COMPOSITION OF THERMAL PLASMA WITH COPPER AND CHROMIUM VAPOURS ADMIXTURES

M. Bartlova[a,*], O. Coufal[a], A. N. Veklich[b], J. Pokorny[a]

[a] *Faculty of Electrical Engineering and Communication, Brno University of Technology, Technicka 10, 616 00 Brno, Czech Republic*

[b] *Taras Shevchenko National University of Kyiv, 64/13 Volodymyrska str., Kyiv, Ukraine*

[*] bartlova@feec.vutbr.cz

**Abstract.** Thermal plasma of electric arc is widely used in various technological applications: welding, cutting, lamps, spraying, protection of electrical installations, etc. Process efficiency is defined by different parameters that determine arc operation and influence the energy transfer within the medium. All energy exchanges depend on the medium, which is modified by the presence of the arc and more particularly by the appearance of new species from contact erosion. Sintered Cu-Cr composites are widely used as electrical contacts for vacuum circuit breakers. These materials take advantage of the high thermal and electrical conductivity of Cu and of the refractory and oxygen getter properties of reinforcing Cr particles. The aim of this paper is to give results of the calculation of the equilibrium composition of argon and air plasma with various admixtures of Cu and Cr.

**Keywords:** thermal plasma, equilibrium composition, sintered Cu-Cr composites.

## 1. Introduction

The complex description of the events taking place in the plasma is based on experimental and/or theoretical data, the correlation of which results from a suitable theoretical model. The results of theoretical modeling are important not only because they allow to replace missing experimental data, but also because based on these, it is possible to interpret and evaluate the measured values. One of the key data that is necessary in any modeling scenario is the amount of substances present at a given plasma location. Although this composition may change over time, as other parameters change, all chemical processes occurring in the system are directed towards equilibrium and the system is considered to be in the local thermodynamic equilibrium (LTE). Thus, the initial step in plasma modeling is to determine the composition and thermodynamic properties (TDP) of the system in thermodynamic equilibrium.

Contacts are important components in switching devices. The range of requirements for the contact materials is diverse. Besides the typical contact features like high erosion resistance, high resistance to welding, low contact resistance, good arc running properties, good arc quenching behavior other physical, mechanical and chemical properties, such as high electrical and thermal conductivity, high hardness, high corrosion resistance, etc., are desired. In addition, the materials should be environmentally friendly [1].

The composites represent a group of substances whose range of properties offers particular advantages for applications as electrical contacts in switching devices in which higher currents must be controlled. The properties of the composite materials are largely determined by the properties of their components independently. Due to its outstanding properties, the contact material copper-chromium has proven particularly suitable for high breaking currents in vacuum arcs [2]. Copper is a material with very high conductivity for heat and electricity, excellent corrosion resistance, medium strength and good formability, the chromium is used as a high-melting and thus particularly erosion-resistant component. It has been confirmed by experiments that within the scope of the invention a melting material is suitable for the above use with a composition in mass fractions Cu-Cr of about 75/25 % to 40/60 % [2].

The aim of this paper is the calculation of the equilibrium composition of argon and air thermal plasmas with various admixtures of copper and chromium vapours. The knowledge of the system composition is the first step for further investigation of the influence of various gases on the contact erosion. Theoretical results were compared with experimental data obtained by optical emission spectroscopy (OES).

## 2. Standard thermodynamic functions

In order to calculate the plasma composition, it is necessary to have input data. These input data are standard thermodynamic functions (STF) of all substances present in the system under consideration, depending on the temperature for the specified standard pressure and relative to the selected reference temperature. From these STF values, partial molar quantities must be obtained, especially the chemical potentials of substances. For this, it is necessary to know also the values of standard enthalpies of formation $\Delta_f H^\circ(T_{\mathrm{ref},i})$ (HF).

STFs are usually tabulated from 298.15 K to 6 kK; HF values for 0 K or 298.15 K reference temperature are also given [3, 4]. The essential requirement is that all HF of substances contained in the system under investigation are determined for one reference temperature. With this objective, a database system "TheCoufal" [5] has been designed which stores the HF and in the form of approximation coefficients the STF of individual substances. It allows to create a consistent set of input data for calculating the composition and TDP of systems containing substances formed from elements of the set $\mathscr{E} = \{\mathrm{C, F, H, N, O, S, W, Ar, Ca, Cl, Cu, Cr, e^-}\}$ at a temperature range of 298.15 K to 50 kK and a pressure range of 0.01 bar to 100 bar.

In order for the database to be filled with data within the specified range, it is necessary to determine the temperature dependence of the STF and possibly the HF values. The STF of gaseous substances can be calculated using partition function [3, 4]

$$Q = \sum_k \exp\left(-\frac{E_k}{\mathrm{k_B}T}\right) \tag{1}$$

where $E_k$ is the $k$-th energy level, $\mathrm{k_B}$ is the Boltzmann constant and $T$ is the temperature. From the partition function the standard reduced Gibbs energy $\Phi^\circ$ can be calculated

$$\Phi^\circ(T) = \mathrm{R}\ln\frac{Q(T)}{N} \tag{2}$$

where $N$ is the number of particles, R is the molar gas constant. From $\Phi^\circ$ all other STFs can be obtained (enthalpy change $H_i^\circ(T) - H_i^\circ(0)$, entropy $S_i^\circ(T)$, and molar specific heat $C_{p,i}^\circ(T)$).

The standard chemical potential $\mu_i^\circ$ (resp. standard Gibbs energy $G_i^\circ$) of the component $C_i$ is then given as

$$\mu_i^\circ(T) \equiv G_i^\circ(T) = \Delta_f H^\circ(T_{\mathrm{ref},i}) - T\Phi_i^\circ(T) \tag{3}$$

For further applications, the calculated discrete STF values are approximated and tabulated by the values of dimensionless coefficients $\varphi_{ik}, k = -2, -1, 0, 1, 2, 3$ which determine the approximation functions. The approximation function for $\Phi^\circ(T)$ is [3]

$$A\Phi^\circ(T) = \varphi\ln t + \sum_{k=-2}^{3} \varphi_{ik} t^k, t = \frac{T}{10^4\,\mathrm{K}} \tag{4}$$

The STFs calculation is based on the knowledge of the energy levels of atoms and molecules. The straightforward method for calculating STFs is based directly on the specified energy levels, which are inserted into the formula for calculating partition function and its derivatives. These energy levels can be obtained experimentally, mainly from spectral data, but also by theoretical calculations. In the case of atoms, the only values are electron energy levels and their statistical weights. In the case of molecules, most of the vibrational and rotational levels are specified by spectroscopic constants. Individual energy levels can be calculated from these constants on the basis of a certain spectroscopic model based on the perturbation theory. The simplest model is the model of a solid rotor and a harmonic oscillator. More complex models consider the interaction of vibration and rotation, anharmonicity, centrifugal distortion, resonance, etc. At present, however, it is also possible to use theoretical values to calculate energy. In the case of vibration-rotational energies, the Schrödinger equation for nuclear motion can be solved if the surface of potential energy is specified. The potential energy surface or curve is determined by the total energy of the electron state for a given nucleus configuration and can be obtained by ab initio methods of quantum chemistry.

Atomic energy levels are available on the NIST website [6]. Spectral data relating to polyatomic molecules are for selected molecules listed on the website [6], the diatomic molecule constants are collected in the monograph [7] and are available on the website [8]. In addition, data used to calculate STF are also given in tables [3, 4].

The available data are often incomplete, and there are none for some molecules, especially ions. In such cases, it is necessary to estimate the missing data. Spectroscopic data estimates have already been made in many cases in [3, 4]. At present, it is possible to find reliable substitution of spectral data in theoretical, especially ab initio calculations of quantum chemistry.

Other data necessary to determine the composition and thermodynamic properties of the system are standard enthalpies of formation $\Delta_f H^\circ(T_{\mathrm{ref},i})$. The values of these quantities are given in tables [3, 4] for reference temperature 0 K or 298.15 K, or the default data for their determination are available on www pages [8].

## 3. Equilibrium composition

The procedure for obtaining the composition and TDP of systems in thermodynamic equilibrium can be divided into two equivalent approaches: stoichiometric and non-stoichiometric algorithms [9, 10]. In this paper, the computation is based on a non-stoichiometric algorithm based on minimizing the total Gibbs energy $G = G(\boldsymbol{n}, p, T)$ of the system [10, 11]

$$G(\boldsymbol{n}, p, T) = \sum_{i=1}^{N} n_i \mu_i \tag{5}$$

The basic equations for calculating the equilibrium composition $\boldsymbol{n} = \{n_1, n_2, n_3, \ldots, n_N\}$ of the system $S$ at given temperature $T$ and pressure $P$ are based on the minimization of the function ($n_i$ are molar amounts of substances)

$$F(\boldsymbol{n}) = \sum_{i=1}^{N} n_i \frac{\mu_i}{\mathrm{R}T} \tag{6}$$

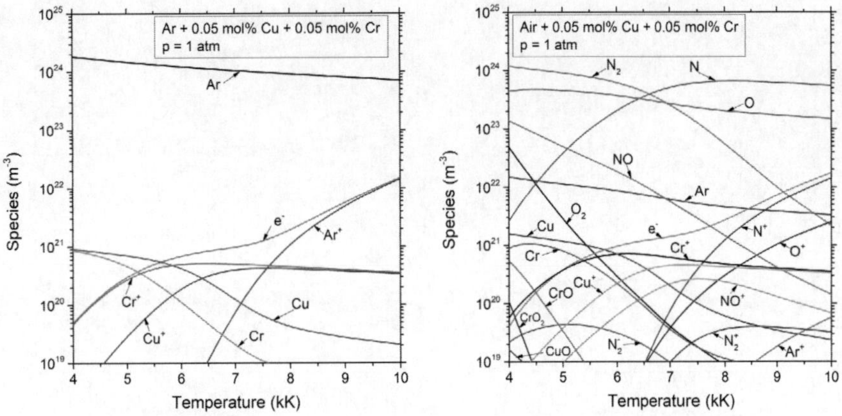


*Figure 1. Concentration of species in argon plasma with 0.05 mol % Cu + 0.05 mol % Cr (left) and air plasma with 0.05 mol % Cu + 0.05 mol % Cr (right).*

in the domain in which the mass balance conditions are satisfied

$$\sum_{i=1}^{N} n_i a_{ij} = b_j, j = 1, 2, \ldots, M \tag{7}$$

where $a_{ij}$ is an element of the constitutional matrix and $b_j$ is the total amount of the element $E_j \in S$. In an ideal system, the chemical potential $\mu_i$ is given by

$$\mu_i = \mu_i^\circ + \mathrm{R}T\ln n_i + \mathrm{R}T\ln\frac{P}{nP^\circ}, n = \sum_{i=1}^{N} n_i \tag{8}$$

where $\mu_i^\circ$ is the standard chemical potential (3), R is the molar gas constant, $P^\circ$ standard pressure ($P^\circ = 100\,\mathrm{kPa} = 1\,\mathrm{bar}$), $T$, $P$ temperature and pressure of the system. Minimizing the function $F(\boldsymbol{n})$ on the region in which the mass balance conditions apply can be done by the Lagrange method of indeterminate multipliers. By the method of indeterminate multipliers we obtain a set of $M+N$ equations, of which $N$ are nonlinear and $M$ express the mass balance conditions. The system of equations can be solved by a suitable numerical method, such as Newton-Raphson. However, it is possible to find more appropriate procedure allowing to effectively reduce the number of equations, which is important if the system contains a large amount of components. One of the most important methods for reducing the number of variables is based on minimizing the quadratic approximation of $F(\boldsymbol{n})$ [10, 11]. The function $F(\boldsymbol{n})$ develops into the second order Taylor series around the composition $\boldsymbol{n}(l)$, from which the quadratic approximation $Q(\boldsymbol{n}^{(l+1)})$ is obtained. The minimization of the quadratic approximation by the indeterminate multiplier method leads to an iteration cycle in which the system $M+1$, resp. $M$ linear equations, is solved in each step $l$. The key feature of this method is that the quadratic approximation $Q(\boldsymbol{n}^{(l+1)})$ is a convex function of the composition, which ensures the convergence of successive approximations of the composition $\boldsymbol{n}(l)$ to the minimum $\boldsymbol{n}$. The sequence of successive approximations continues until convergence is achieved.

## 4. Results and discussion

The equilibrium composition of the air plasma with admixtures of Cu and Cr and argon plasma with admixtures of Cu and Cr was calculated using the "Tmdgas" computer code [11] with input data "species" which is a part of the database system "TheCoufal" [5]. The plasma is assumed to be a gaseous medium in equilibrium and at atmospheric pressure. Air was assumed at US Standard Atmosphere [12], the volume concentration of gases in pure air was $N_2$: 78.111 vol%, $O_2$: 20.955 vol%, Ar: 0.934 vol%.

For each system three variants of the plasma composition have been calculated with the amount of copper and chromium of 0.05 mol%, 0.5 mol% and 5 mol%, respectively, i.e. in each variant there is an equal number of copper and chromium particles.

Following species were taken into account: 40 species $O_2$, $N_2$, Ar, O, $O^+$, $O^-$, $O_2^+$, $O_3$, $Ar^+$, N, $N^+$, $N_2^+$, $N_2^-$, $N_3$, $N_3^+$, NO, $NO^+$, $NO^-$, $NO_2$, $N_2O$, $N_2O^+$, Cr, $Cr^+$ ,$Cr^{+2}$, $Cr^-$, $Cr_2$, CrO , $CrO_2$, $CrO_3$, $CrO_3^-$, $Cr_2O$, $Cr_2O_2$ ,$Cr_2O_3$, CrN, Cu, $Cu^+$, $Cu^{+2}$, $Cu^-$, $Cu_2$, CuO for air plasma with Cu and Cr admixtures and 12 species Ar, $Ar^+$, Cr, $Cr^+$, $Cr^{+2}$, $Cr^-$, $Cr_2$, Cu, $Cu^+$, $Cu^{+2}$, $Cu^-$, $Cu_2$ for argon plasma with Cu and Cr admixtures. As an example, the equilibrium compositions of argon with admixture of 0.05 mol% Cu + 0.05 mol% Cr and air with admixture of 0.05 mol% Cu + 0.05 mol% Cr are, respectively, presented in Fig.1. It can be seen in both cases that for temperatures up to 6500 K electrons are produced mainly by ionization of metal atoms, especially chromium due to their relatively low ionization energy.

Different composition variants of air+Cu+Cr and Ar+Cu+Cr are presented in figure 2. Only densities of

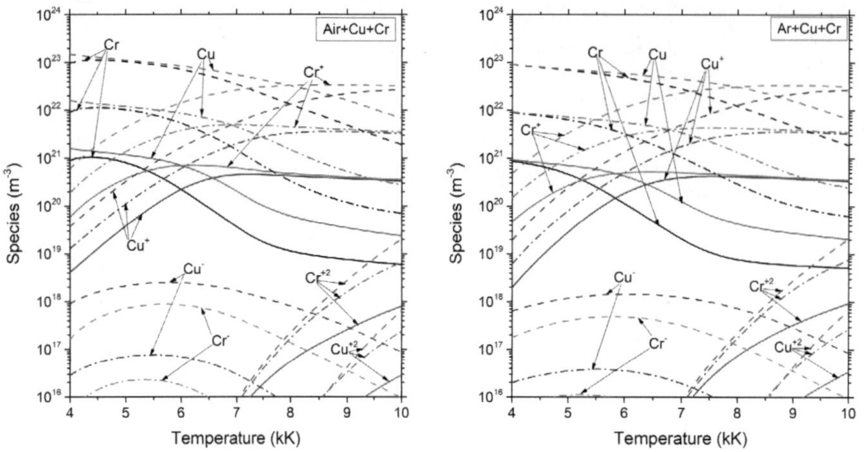


*Figure 2. Concentration of Cu and Cr atomic species at different variants of composition system of air + Cu + Cr (left) and Ar + Cu + Cr (right) at a pressure of 1 atm. Full line – 0.05 mol% Cu + 0.05 mol% Cr, dash-dot line – 0.5 mol% Cu + 0.5 mol% Cr, dashed line – 5 mol% Cu + 5 mol% Cr*

atoms and atomic ions of metal particles are given for clarity. The course of the curves is very similar which means that the inlfluence of molecular compounds of Cu and Cr that appear in the air plasma is negligible.

Theoretical calculations for air plasma with metal admixtures were compared with experimental investigations. The simple model of thermal plasma source was investigated [13]. Namely, the electric arc discharge was initiated in air between the flat end surfaces of non-cooled composite Cu-Cr electrodes. The discharge gap was 8 mm, the arc current was 3.5 A.

The radial profile of plasma temperature were determined by Boltzmann plot technique in the local thermodynamic equilibrium assumption. The emission of atomic copper spectral lines was used. We supposed that arc discharge plasma had axial symmetry. So, the Abel inversion was used to obtain the local values of the spectral line emission. Due to the instability of the discharge, statistical averaging of the recorded spatial distributions of the radiation characteristics was carried out. It was also taken into account the non-uniformity of the spectral sensitivity of the registration system.

The electron density profiles were calculated from electric conductivity obtained by solution of energy balance equation on the base of experimentally determined radial distributions of plasma temperature [14]. The profiles of temperature and electron density are given in Fig. 3. In figure 4 experimentally obtained data of electron densities were plotted as a function of temperature together with calculated electron densities in air plasma with different content of metal vapours. It can be seen that content of metal vapours in the plasma does not exceed 1 mol%. For quantitative determination of metal vapours content in plasma, additional parameter is required - the ratio between copper and chromium atoms concentration. Such parameter can be defined from the intensity ratio

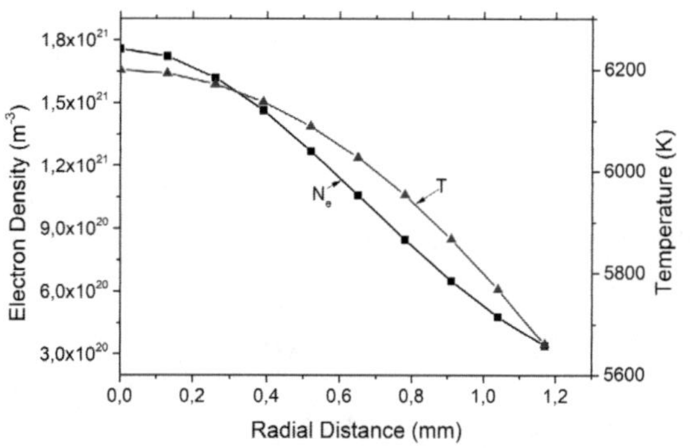


*Figure 3. Radial distribution of plasma temperature and electron density of arc discharge between Cu-Cr electrodes.*

of Cu I 515.3 nm and Cr I 540.9 nm spectral lines in an assumption of LTE. Experimentally determined temperature, electron density, and the ratio of copper and chromium concentrations were used as initial data for calculation of radial profile of plasma composition. Two various approaches were used. The simplified stoichiometric calculation assumed only following species in the plasma system: $N_2$, $O_2$, N, O, $N^+$, $O^+$, Cu, $Cu^+$, Cr, $Cr^+$. The calculation was carried out by solution of Saha equations for atoms of copper, chromium, nitrogen and oxygen, equations for dissociation of nitrogen and oxygen molecules, equation of charge neutrality, perfect gas equation, and the relation between air components [15]. The results were compared with more detailed plasma composition calculated by minimizing the Gibbs energy of the system. In Fig. 5, the comparison of Cu, Cr, $Cu^+$, $Cr^+$ concentrations is given. Both methods show good agreement in determination of metal impurities.

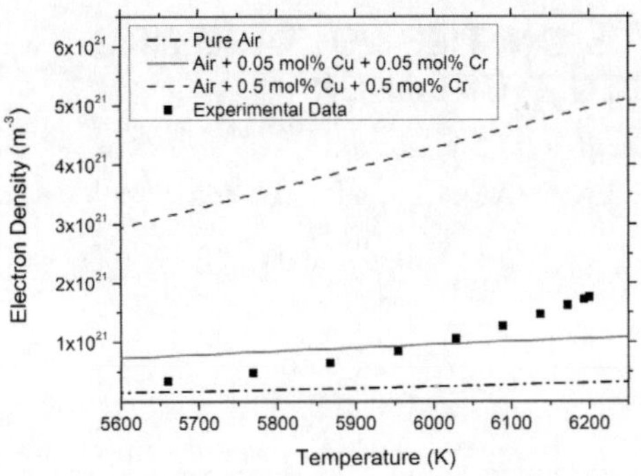


*Figure 4. Electron density as a function of temperature. Lines – calculated data for air plasma with various admixtures of Cu and Cr at the pressure of 1 atm; points – experimental data for plasma of electric arc discharge between Cu-Cr composite electrodes at current 3.5 A.*

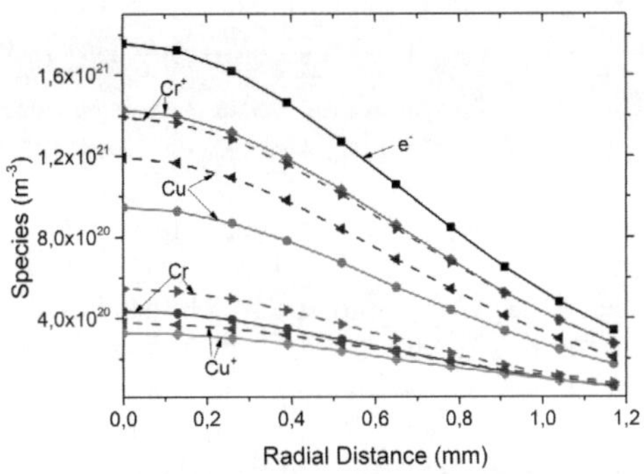


*Figure 5. Radial profiles of Cu, $Cu^+$, Cr, and $Cr^+$ densities calculated by two alternative methods. Full lines – simplified stoichiometric calculations, dash lines – minimizing the Gibbs energy.*

## 5. Conclusions

Calculations of equilibrium composition of air and argon plasmas with various admixtures of Cu and Cr have been performed. For temperatures up to 6500 K the electrical conductivity is provided mainly by ionization of metal components. This effect is evident even with a very small amount of metallic impurities. The influence of molecular compounds of Cu and Cr that appear in the air plasma is negligible. For air plasma, theoretical results were compared with experimental data for electric arc burning between composite Cu-Cr electrodes. From measured profiles of temperature, electron denisty and ratio of copper and chromium atoms the radial profile of plasma composition was calculated by two different approaches under assumption of LTE. Both methods show good agreement in determination of metal impurities.

## Acknowledgements

This work was supported from the Ministry of Education, Youth and Sports of the Czech Republic under project CZ.02.1.01/0.0/0.0/16_013/0001638 (CVVOZE Power Laboratories-Modernization of Research Infrastructure) and under joint project in the frame of research and technology collaboration between Ukraine and Czech Republic No.7AMB197UA.

## References

[1] R. A. Kornev et al. Catalytic effects of electrode materail on the silicon tetrachloride hydrogenation in RF-arc-discharge. *RSC Advances*, 6:99816, 2016. doi:10.1039/C6RA21023F.

[2] P. G. Slade. Advances in material development for high power, vacuum interrupter contacts. *IEEE Trans. on Components, Packaging and Manufacturing Technology-PartA*, 17(1):96–106, 1994. doi:10.1109/95.296374.

[3] L. V. Gurvich, I. V. Veyts, and C. B. Alcock (eds.). *Thermodynamic Properties of Individual Substances, vol. 1-4*. Hemisphere, New York, 1989, 1991.

[4] M. W. Chase (ed.). *NIST-JANAF Thermochemical Tables, vol. I and II*. J.Phys. Chem. Ref. Data, Monogr. 9. NIST, Gaithersburg, Maryland, 1998.

[5] O. Coufal, P. Sezemsky, and O. Zivny. Database system of thermodynamic properties of individual substances at high temperatures. *J. Phys. D: Appl. Phys.*, 38(8):1265–1274, 2005. doi:10.1088/0022-3727/38/8/026.

[6] NIST. Scientific and technical databases. arXiv:http://www.nist.gov/srd/index.html.

[7] K. P. Huber and G. Herzberg. *Constants of Diatomic Molecules*. Vol. IV of Molecular Spectra and Molecular Structure. Van Nostrand, Princeton, 1979.

[8] NIST Chemistry WebBook. NIST Standard Reference Database Number 69, June 2005 Release. arXiv:http://webbook.nist.gov.

[9] W. R. Smith and R. W. Missen. *Chemical Reactions Equilibrium Analysis: Theory and Algorithms*. Wiley, New York, 1982.

[10] W. B. White, S. M. Johnson, and G. B. Dantzig. Chemical equilibrium in complex mixtures. *J. Chem. Phys.*, 28(5):751–755, 1958. doi:10.1063/1.1744264.

[11] O. Coufal. Composition and thermodynamic properties of thermal plasma up to 50 kK. *J. Phys. D: Appl. Phys.*, 40(11):3371–3385, 2007. doi:10.1088/0022-3727/40/11/018.

[12] D. R. Lide. *CRC Handbook of chemistry and physics*. 88th edition. CRC Press, Boca Raton, 2007.

[13] R. V. Semenyshyn et al. Spectroscopy peculiarities of thermal plasma of electric arc discharge between electrodes with Zn admixtures. *Advances in Space Research*, 54(7):1235–1241, 2014. doi:10.1016/j.asr.2013.11.042.

[14] S. Fesenko et al. Properties of thermal air plasma with admixing of copper and carbon. *Journal of Physics: Conference Series*, 550:012008, 2014. doi:10.1088/1742-6596/550/1/012008.

[15] I. L. Babich et al. Spectroscopic data and Stark broadening of Cu I and Ag I spectral lines: Selection and analysis. *Advances in Space Research*, 54(7):1254–1263, 2014. doi:10.1016/j.asr.2013.10.034.

doi:10.14311/ppt.2019.3.256
*Plasma Physics and Technology* **6**(3):256–260, 2019 

# EXPERIMENTAL AND SIMULATION STUDIES ON THE VOLTAGE DROP OF ARC IN LOW-VOLTAGE CIRCUIT BREAKER

J. Lu[a], G. Déplaude[b], P. Freton[a], J-J. Gonzalez[a,*], P. Joyeux[b]

[a] *Laboratory on Plasma and Conversion of Energy, University of Toulouse, CNRS, INPT, UPS., 118 Route de Narbonne, F-31062 Toulouse cedex 9, France*

[b] *Hager Electro SAS, 132 Bd d'Europe, 67215 Obernai, France*

[*] gonzalez@laplace.univ-tlse.fr

**Abstract.** In low voltage circuit breaker (LVCB) apparatus, a current limitation is performed by increasing the arc voltage. This increase is mainly realized in the splitters plates of the arc chamber by additional drop voltages due to anode and cathode sheaths regions. The consideration of the voltage drops near-electrodes regions is so one of the most important mechanism to improve the description of the arc behavior in LVCB. In this paper, the arc voltage evolution has been studied by experimental and simulation by considering a simple geometry constituted by two rails runner with one or two splitters plates. One magneto hydrodynamic model in three dimensions (3D) was developed to simulate the arc motion and the arc splitting process. In order to compare with the model results, experimental tests have been carried out.

**Keywords:** Low-voltage circuit breaker, voltage drop, splitters, experiment, simulation.

## 1. Introduction

The principle of LVCB consists to create a current limitation by increasing the arc voltage. The increase of the voltage is carried out by various mechanisms: increase of the losses, arc elongation and multiplication of the voltage drops in the splitters. To quantify and better describe these mechanisms, studies are conducted through models and experiments [1–4]. The most theoretical approach studied and applied in LVCB is the one proposed by Lindmayer [2]. This method allows considering an additional drop voltage through the resistivity in the layers. Another approach consists to describe the sheath with a non-equilibrium approach [5]. Nevertheless this method is difficult to implement for a real configuration with several splitters plates. Some authors [6–10] studied numerically the splitting process. Considering a non-linear permeability of ferromagnetic material's they focus their studies on the influence of eddy currents [9] and on the influence of metal vapors coming from iron splitter plate erosion [10]. In those studies, the cathode voltage was treated by U-J curves and the anode arc root description was determinedwith the LTE-diffusion model of Lowke and Tanaka [11]. Another group has studied the arc phenomena in LVCB by an imaging system applied to a flexible test apparatus [12–15]. The studies were devoted to the influence of vent aperture size on re-ignition and arcing phenomena. They also proposed a modified method to consider the voltage due to the sheath regions. The ability to predict the arc extinction was analyzed and compared [16] with the Lindmayer approach [2]. The most complete study was proposed by Iturregi et al. [17]. They designed a 3D model in LVCB to analyze the arc behavior. Nevertheless incoherencies appear in the comparison between one and two splitters plate: the additional drop voltage due to a second splitter is not present in the total voltage evolution versus time. The method is not clearly described as the authors used directly the @Ansys module which includes the Lindmayer approach. In this paper, the process of arc splitting is analyzed by theoretical and experimental works. Experimental tests have been realized using one and two splitters: high speed camera and electrical measurements. For the model we have implemented the Lindmayer approach and used the same geometries of Iturregi et al. [17].

## 2. Numerical method

### 2.1. Hypotheses

The following assumptions have been used:

(1) The air medium plasma is assumed to be in Local Thermodynamic Equilibrium (LTE).

(2) The plasma is a Newtonian fluid and the flow is laminar.

(3) Vapors from the walls and electrodes are not taken into account.

(4) Radiation is treated using the net emission coefficient method.

### 2.2. Equations

Based on the Fluent software, we have implemented a fluid model to simulate electrical arc behavior in a LVCB. It can be characterized by the macroscopic fluid quantities: velocity, pressure, enthalpy. All these quantities are obtained from the Navier-Stokes equations with Lorentz forces and homic heating and completed with Maxwell equations to consider electromagnetism effects. The magnetic field is obtained from the current path in the runners and plasma through the vector potential resolution and a BiotSavart formulation is used at the boundaries. In order to represent the additional voltage drop for the runners and splitters, an effective electrical conductivity is defined [2] based on experimental works [6, 8]. The voltage evolution versus the current density is plotted in the Figure 1. These characteristics (a-b-c-d) suggested by Lindmayer et al. [2] in air medium with copper runners enable some tuning of the model to adjust the voltage, they represent the current ability to jump on the splitter plate.

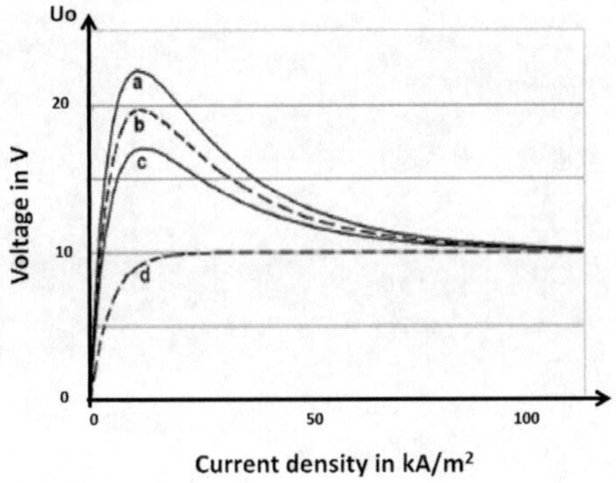


*Figure 1. Evolution of voltage drop versus current density [3]*

### 2.3. Boundary conditions

"Classical" boundary conditions are used [1]. The magnetic field is calculated from the vector potentials but the BiotSavart formulation is used to close the system resolution.

### 2.4. Geometries

The geometries for the simulation are presented in the Figure 2. The dimensions of the chambers are $40 \times 2.5 \times 11$ mm ($xyz$) and the rails are $40 \times 2.5 \times 1.5$ mm ($xyz$). It should be noted that the dimension of the splitters plate are not the same: for one splitter: $20 \times 2.5 \times 2$ mm ($xyz$) and for two splitters: $20 \times 2.5 \times 1$ mm ($xyz$). $D1 = 8$ mm represents the distance between the two runners; $D2 = 3$ mm is the distance between runners and splitters, for the third geometry $D3 = 2$ mm. These dimensions are far from real LVCB but chosen to correspond with Iturregi et al. [17]. A uniform grid is used in the geometry leading to 1.12 million cells.

## 3. Simulation results

We present the main results obtained with the 3D model applying the Lindmayer approach. The arc

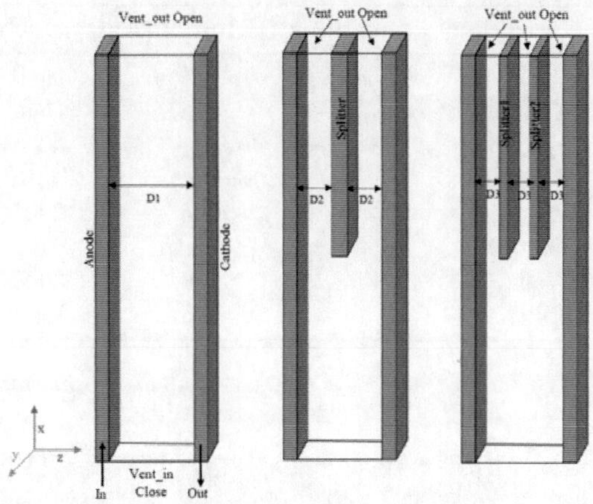


*Figure 2. Geometries used for the simulation (From left to right: 0 splitter, 1 splitter and 2 splitters).*

splitting behavior is shown through the current densities in Figure 4 for one and two splitter plates. Figure 3 presents the voltage evolution for the two cases. The calculation time step is 10 µs and the input current is 50 A DC. The curve "c" of Figure 1 was chosen to simulate the sheath contribution. There is no symmetry plane in our simulation. In order to reduce the calculation time, the arc is ignited 1.5 mm front of splitters plates.

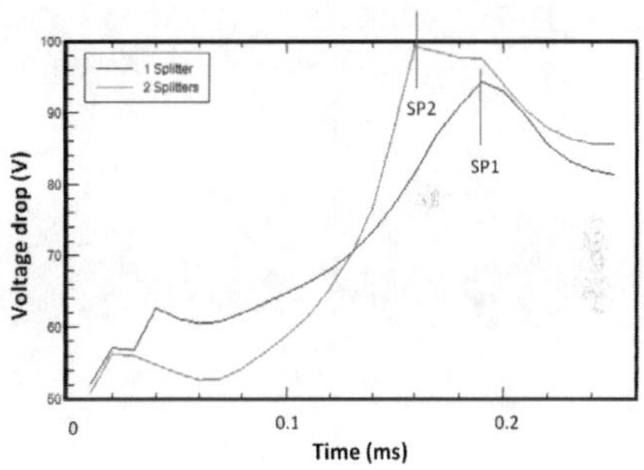


*Figure 3. Voltage drop in the simulation with 1 and 2 splitters.*

In case of one splitter: At $t = 50$ µs the arc starts to bend and moves toward the splitters. Due to the arc elongation the voltage increases progressively from 61 V at $t = 50$ µs to its maximum value 94 V at $t = 0.19$ ms. Then the voltage decreases. At $t = 0.19$ ms the drop voltage is due to the effects of the arc elongation and to the additional voltage sheathes. After time $t = 0.19$ ms, the current flows progressively into the splitter leading to a diminution of the current path and of the total voltage. The current path goes into the splitter totally at $t = 0.21$ ms. At $t = 0.25$ ms there is only one path of current and the voltage drop decreases to 81 V. The contribution of the splitter sheath on the total voltage is around 30 V in the following case. In case of two splitters, the presented times are not the same; indeed the

| Boundary conditions | Momentum | Enthalpy | Scalar Potential | Vector Potential |
|---|---|---|---|---|
| Cathode | v=0 m/s | Heat Transfer | Continuity | BiotSavart |
| Anode | v=0 m/s | Heat Transfer | Continuity | BiotSavart |
| In | / | dT/dn=0 | I=50 A | / |
| Out | v=0 m/s | dT/dn=0 | 0 V | / |
| Vent In | v=0 m/s | 300 K | dV/Dn=0 | BiotSavart |
| Vent Out | P=1 atm | Convection | dV/Dn=0 | BiotSavart |
| Walls | v=0 m/s | 300 K | dV/Dn=0 | / |
| Splitters | v=0 m/s | Heat Transfer | Continuity | BiotSavart |

*Table 1. Boundary conditions used in the simulation model.*

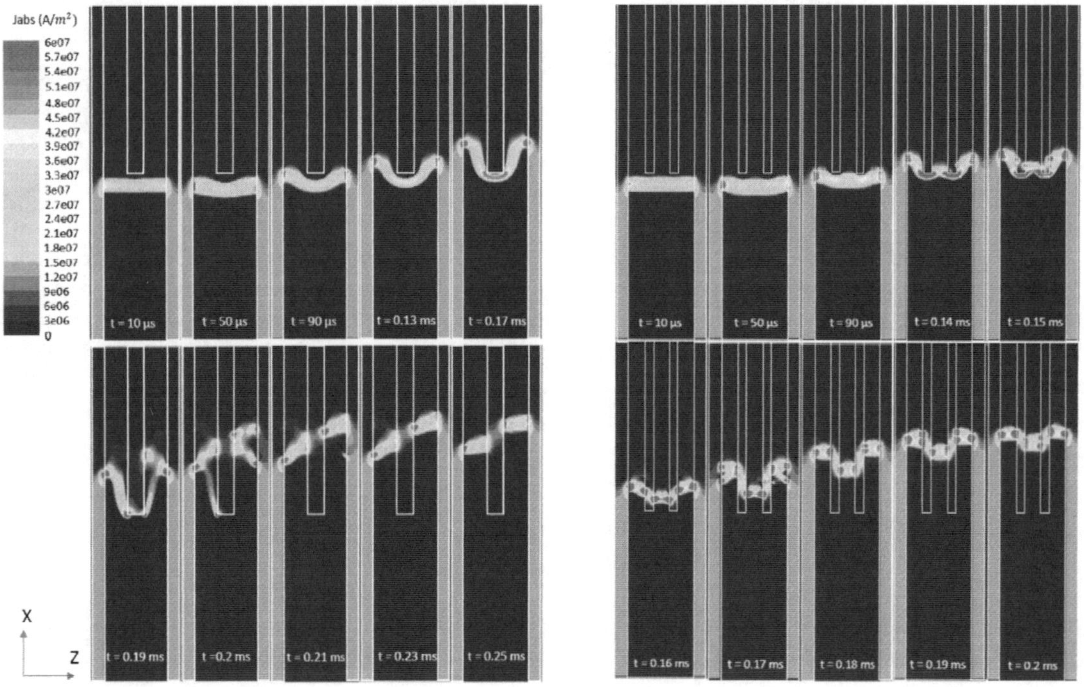


*Figure 4. Simulated current densities using curve c for one (left) and two (right) splitters.*

arc motion, the splitting process time and behavior are different. At $t = 0.14$ ms, the arc column bends around the splitter plates, the arc voltage increases up to 76.6 V. Compared to one splitter case, we can observe that at $t = 0.15$ ms the arc begins to jump on the splitters. From this time, two arc roots appear on the splitters plates, the current density through the splitter plates increases and the current path out of the splitters plates gradually fades. Compared with the case of one splitter, the arc voltage evolution is faster. At time $t = 0.25$ ms the difference on the total arc voltage between the two cases is only 4 V. The comparison is difficult to make as the current paths and arc lengths are not the same. According to the results with one splitter an additional voltage $U = 30$ V between the two cases should exist. Nevertheless the arc positions are different, the medium temperature is different and the arc behavior not the same as shown by the voltage evolutions Figure 3.

## 4. Experimental results

The dimensions of the experimental setup are the following: 126 × 15 × 20 mm ($xyz$) for the chamber, 126 × 15 × 1 mm ($xyz$) for the runners and 50 × 15 × 1 mm for the splitter plate. There is a wire located in front of the splitter to ignite the electrical arc. The experimental setup is composed by a generator which can product a prospective peak current up to 13 kA with a maximum charging voltage of 600 V in AC, a high-speed camera (Photron SA5), a differential voltage probe and a Rogowski coil for the electrical measurements. Figure 5 presents the current evolution and the total voltages versus time for zero and one splitter. The first peak of voltage at $t = 0.25$ ms characterizes the arc ignition. After that, the arc moves toward to the upper side of the chamber. During the arc motion the voltage is nearly constant 60 V until $t = 2$ ms. At this time the arc arrives in front of the splitter plate. In the case with

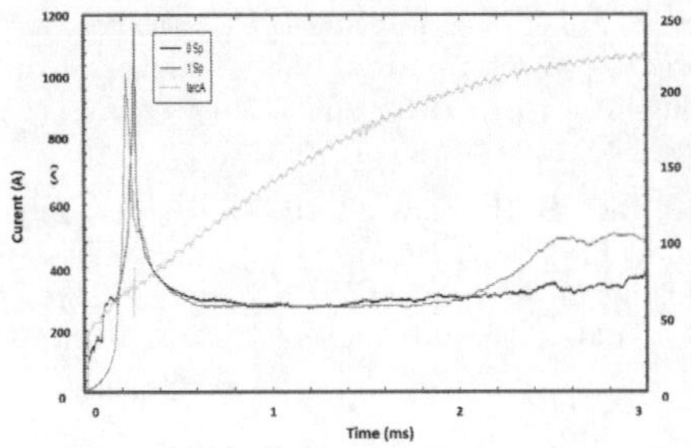


*Figure 5. Experimental voltage drop and total current. Cases with one and without splitter in the chamber.*

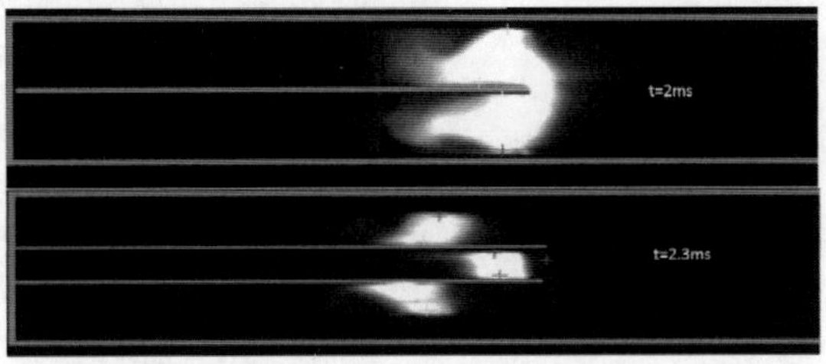


*Figure 6. Example of arc behavior in front of one and two splitters plates obtained by high-speed camera.*

one splitter, arc voltage increases by squeezing against the splitter, by touching the lower edge of the splitter until arc was totally on the splitter from $t = 2\,\text{ms}$ to $t = 3\,\text{ms}$. The voltage drop contribution dues to the splitter can be calculated by the difference between the case without splitter. The value is nearly 25 V. This sheath contribution is in the same order of magnitude than the theoretical case, nevertheless we have seen between the two theoretical cases (one and two splitters) that the comparison is difficult due to the tortuous current paths or arc behavior in the plasma.

Finally we present Figure 6 two images obtained by the high speed camera to illustrate the arc behavior observed experimentally. In the upper picture the arc is bent in front of the splitter as in the down picture the arc is segmented by the presence of the two splitters plates.

## 5. Conclusion

A 3D model was developed and the effective electrical model of conductivity suggested by Lindmayer [2] introduced in the developments to characterize the additional voltage due to the splitter plate sheathes. The feasibility is demonstrated in cases of one and two splitters plates. Interesting results are obtained on the voltage evolutions. In the same time investigations are made to characterize the arc behavior in an experimental setup by high speed camera and electrical measurements. The theoretical and experimental configurations are at this time different nevertheless the same order of magnitude was observed with an additional drop voltage around 20–30 V for each splitter plate. The next step of our study will consist to use the same geometry for the model and the experience and to perform one parametric study on the number of splitters plates.

## References

[1] J. Quéméneur, J. Lu, J.-J. Gonzalez, and P. Freton. Arc motion in low voltage circuit breaker (LVCB) experimental and theoretical approaches. *Science and Technology*, 8(2):35–45, 2018. doi:10.5923/j.scit.20180802.02.

[2] M. Lindmayer, E. Marzahn, A. Mutzke, T. Ruther, and M. Springstubbe. The process of arc splitting between metal plates in low voltage arc chutes. *IEEE Trans. CPMT*, 29(2), 2006. doi:10.1109/HOLM.2004.1353090.

[3] A. Mutzke, T. Ruther, M. Kurrat, M. Lindmayer, and E.-D. Wilkening. Modeling the arc splitting process in low-voltage arc chutes. In *53rd IEEE Holm Conf. on Electrical Contacts, Pittsburgh*, 2007. doi:10.1109/HOLM.2007.4318213.

[4] A. Mutzke, T. Ruther, M. Lindmayer, and M. Kurrat. Arc behavior in low-voltage arc chambers. *European Physical Journal-applied Physics*, 49, 2010. doi:10.1051/epjap/2010001.

[5] M. S. Benilov and A. Marotta. A model of the cathode region of atmospheric pressure arc. *J. Phys. D: Appl. Phys.*, 28(9), 1995. doi:10.1088/0022-3727/28/9/015.

[6] F. Yang, M. Rong, and Y. Wu. Numerical analysis of arc characteristics of splitting process considering ferromagnetic plate in low-voltage arc chamber. *IEEE Trans. Plasma Sci.*, 38(11), 2010. doi:10.1109/TPS.2010.2070084.

[7] Z. Sun, M. Rong, and F. Yang. Numerical modeling of arc splitting process with ferromagnetic plate. *IEEE Trans. on Plasma Sci.*, 36(4):1072–1073, 2008. doi:10.1109/TPS.2004.924559.

[8] M. Rong, F. Yang, Y. Wu, et al. Simulation of arc characteristics in miniature circuit breaker. *IEEE Trans. on Plasma Sci.*, 38(9):2306–2311, 2010. doi:10.1109/TPS.2010.2050703.

[9] F. Yang, M. Rong, Y. Wu, et al. Numerical simulation of the eddy current effects on the arc splitting process. *Plasma Science and Technology*, 14(11), 2012. doi:10.1088/1009-0630/14/11/05.

[10] F. Yang, M. Rong, Y. Wu, et al. Numerical analysis of the influence of splitter-plate erosion on an air arc in the quenching chamber of a low-voltage circuit breaker. *J. Phys. D: Appl. Phys.*, 43(43), 2010. doi:10.1088/0022-3727/43/43/434011.

[11] J.-J. Lowke and M. Tanaka. Lte-diffusion approximation for arc calculations. *J. Phys. D: Appl. Phys.*, 39(16), 2006. doi:10.1088/0022-3727/39/16/017.

[12] J. W. McBride and P. M. Weaver. Review of arcing phenomena in low voltage current limiting circuit breakers. *IEE Proc. Sci. Meas. and Tech.*, 1(23), 2001. doi:10.1049/ip-smt:20010185.

[13] J. W. McBride and D. Shin. A study of the motion of high current arcs in splitter plates using an arc imaging system. In *International Conference on Electrical Contacts, Edinburg*, 2016.

[14] J. W. McBride, A. Balestrero, L. Ghezzi, G. Tribulato, and K. J. Cross. Optical fiber imaging for high speed plasma motion diagnostics: Applied to low voltage circuit breakers. *Rev. Sci. Instrum*, 81(5), 2010. doi:10.1063/1.3428737.

[15] D. Shin, I. O. Golosnoy, and J. W. McBride. A study of arc modelling in low-voltage switching devices. In *International Conference on the Computation of Electromagnetics Fields*, 2017.

[16] D. Shin, J. W. McBride, and I. O. Golosnoy. Arc modeling to predict arc extinction in low-voltage switching devices. In *IEEE Holm Conf. on Elec. Cont.*, 2018. doi:10.1109/HOLM.2018.8611712.

[17] A. Iturregi, B. Barbu, E. Torres, F. Berger, and I. Zamora. Electric arc in low-voltage circuit breakers: Experiments and simulation. *IEEE Trans. Plasma Sci.*, 45(1):113–120, 2017. doi:10.1109/TPS.2016.2633400.

doi:10.14311/ppt.2019.3.261
*Plasma Physics and Technology* **6**(3):261–264, 2019 

# PRELIMINARY STUDY ON MAKING ARC CHARACTERISTICS OF $AgSnO_2$ CONTACT

Guo Fengyi, Zhang Yuehui*, Wang Zhiyong, Wang Xili, Gao Hongxin

*Faculty of Electrical and Control Engineering, Liaoning Technical University, Longwan South Street 188, 125000, Huludao China*

* 1004071138@qq.com

**Abstract.** In order to analyze the making arc characteristics of $AgSnO_2$ contact, experiments were carried out with a self-developed experimental equipment. It was found that contact resistance had no obvious change with the increase of the number of experiments.In the later stage of the experiment, the contact bounces occurred during the contact closing process, which not only prolonged the making arc duration, but also increased making arc energy. When the contact was eroded to a certain extent by arc, making welding occurs.

**Keywords:** $AgSnO_2$ contact, contact welding, arc energy, arcing duration.

## 1. Introduction

$AgSnO_2$ is an environmentally friendly conatct material with arc ablation resistance, wear resistance and harmlessness. $AgSnO_2$ contact is widely used in relays and other electrical switches with its excellent performance. However, $AgSnO_2$ contact can be welded under harsh electrical condition, which reduces the electrical life of relay. Studying the making arc characteristics of $AgSnO_2$ contact can improve material performance and prolong the relay electrical life.

Many scholars have done a lot of work in the research of contact welding mechanism. W. F. Rieder considered the arc whose bounce height was less than 10 µm and arc duration was less than 100 µs as short arc, and found that the welding caused by short arc was stronger than that caused by long arc [1, 2]. L. Morin found that the contact displacement curve did not change when the voltage pulse occurred. It was believed that the voltage pulse was caused by the increase of the constriction resistance [3]. L. Zhenbiao found that contact welding happens randomly in the electrical life of relays. There are not obvious changes in the making arc and the breaking arc durations before and after welding occurrence [4]. K. Hotta demonstrated that the increase of arc energy on the contact surface would increase the welding force, and the real welding area accounted for about 25% to 40% of the melting area [5]. T. Mutzel found that different base metal oxides and their total content are of significant influence for weld break forces [6]. R. Wanbin observed the substantial decrease and final disappearance of sliding distance accelerated the contact welding failure process [7]. G. Fengyi obtained arc erosion characteristic curve and mechanism of various contact materials through breaking experiments on different level of currents and micro-analysis of contact after breaking [8]. Because the factors affecting the contact welding are complex, it is difficult to study the mechanism of contact welding. Although researchers have obtained some useful experimental conclusions, the research on the phenomenon of contact welding still requires further study.

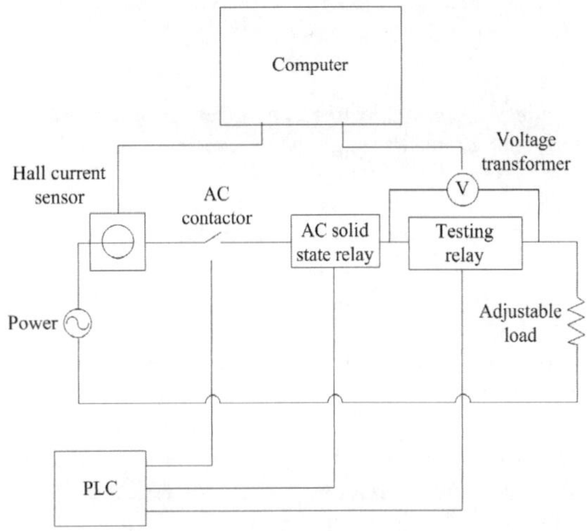


*Figure 1. Electrical schematic diagram of the experimental equipment.*

The arc experiments of $AgSnO_2$ contact were carried out using the self-developed experimental equipment. The relationship between arc erosion, making contact bounces and welding was analyzed by using experiment results.

## 2. Experimental conditions

In order to study the making arc characteristics of $AgSnO_2$ contact, a set of experimental equipment was built. It can set the turn-on and turn-off time of relay, automatically collect the current and voltage data of contact, and change the type of load.

Electrical schematic diagram of the experimental equipment is shown in Figure 1. The power supply used in the experiment is AC 238 V\50 Hz. An AC con-

| Number of experiments | Load | Voltage | Making speed |
|---|---|---|---|
| 1 – 5000 | 1.6 Ω | 238 V | 0.2 m/s |
| 5001 – 13946 | 1.6 Ω 6 mH | 238 V | 0.2 m/s |

*Table 1. Experimental parameters.*

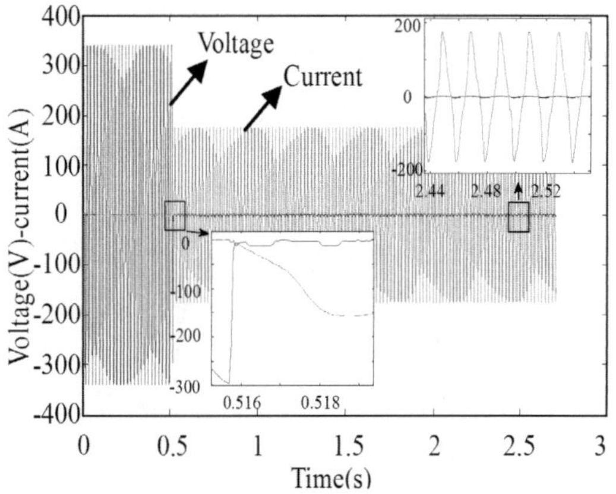


*Figure 2. Typical welding current and voltage waveforms of contacts.*

tactor was used to break the circuit in emergency. An AC solid state relay was used to realize the on-load or off-load making and breaking of the magnetic latching relay. The magnetic latching relay in this experiment was on-load making and breaking. The AC contactor, solid state relay and magnetic latching relay were controlled by programmable logic controller (PLC). The data collected by Hall current sensor and voltage transformer were transmitted to the LABVIEW through a data acquisition card.

The testing relay is a magnetic latching relay, which can break 250 V, 100 A load under normal working conditions. The diameter and thickness of the stationary and movable contacts are 5 mm and 2.2 mm, and the gap-length is 1.5 mm. The contact material is $AgSnO_2$ (90\10), which is produced by powder metallurgy with minor additives of Cu and Li. The experimental parameters are given in Table 1. In each operation cycle, the contact closure and disconnection state were 2 s and 5 s, respectively.

## 3. Experimental results and analysis

The waveforms of current and voltage in the experiments were similar to those in Figure 2 when the contact welding occurred. The relay was set to make at 0.515 s, it can be seen from the voltage waveform of the amplification diagram that the contact bounces occurred during the making process. At 2.5 s, the relay was set to break, but the current and voltage waveforms had no change. It meant that the contacts hadn't successfully broken and the making welding adhesion phenomenon occurred.

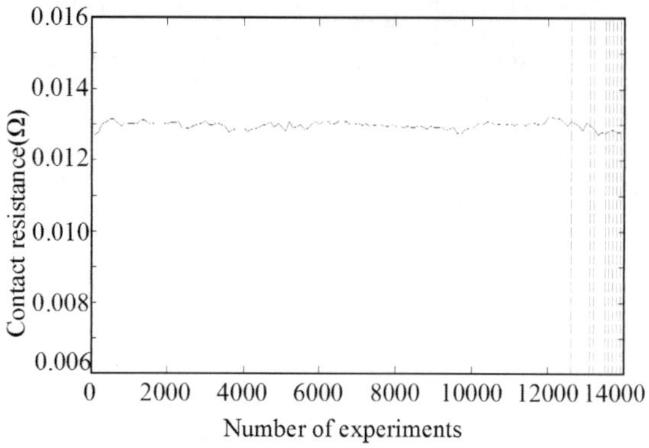


*Figure 3. The relationship between contact resistance and number of experiments.*

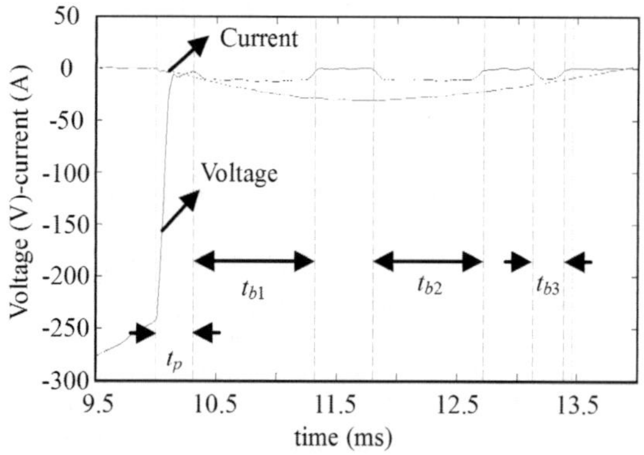


*Figure 4. Typical current and voltage waveforms of making bounce.*

### 3.1. Analysis of Contact Resistance

The contact resistance of the relay was calculated using the voltage and current waveforms. The change trend of the contact resistance with the number of experiments is shown in Figure 3. The experimental intervals of welding occurrence were marked with red dotted lines in the figure and the following experimental figures. As shown in Figure 3, there are no obvious changes in contact resistance from the beginning of relay experiment to the end of relay life, and contact resistance fluctuated around 0.013 Ω.

### 3.2. Meaning of parameters $t_b$, $t_p$, $W$

Making arc duration and bounce times can be identified by voltage waveform. In Figure 4, the interval between two vertical green dotted lines is the making arc time regions and the intervals between the red dotted lines are the bouncing intervals, which were automatically marked by Matlab. The $t_p$ and $t_b$ are the pre-breakdown arc duration and making bounce arc duration, respectively. From Figure 4, we can see that three bounces occurred. The bouncing time $t_b$ is the sum of multiple bouncing times in the making process. The making arc duration is $t = t_p + t_b$. The arc energy is $W = \sum_{n=1}^{N} u(n)\, i(n)\, \Delta t$.

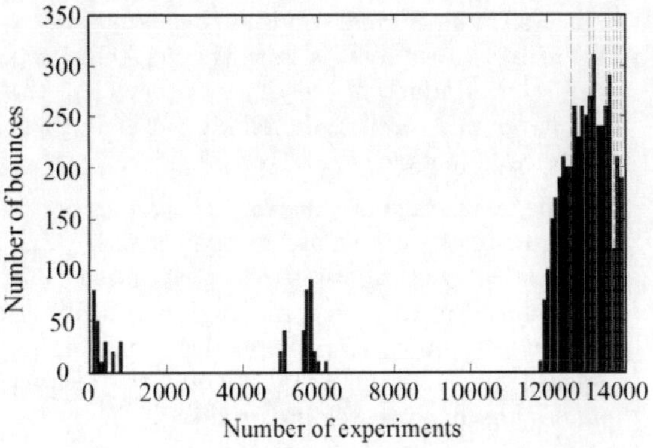


*Figure 5. The relationship between the number of making bounces and number of experiments.*

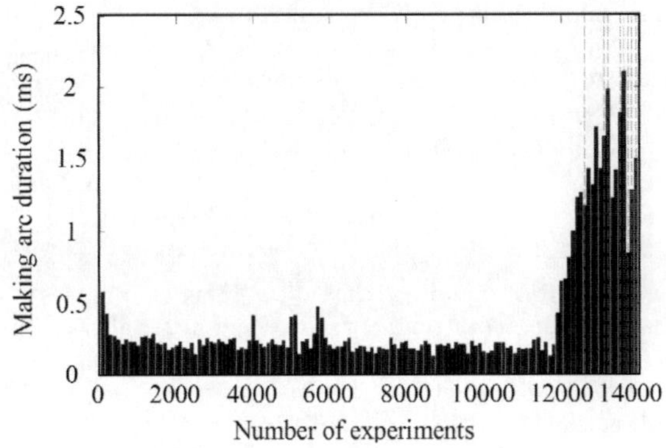


*Figure 6. The relationship between the making arc duration and number of experiments.*

### 3.3. Analysis of Contact Making Bounce

As shown in Figure 5, the ordinate is the total number of making bounces per 100 cycles. In the about 800 and 6 000 cycles, there is a few contact bounces, which may be due to the mechanical structure of the relay in the running-in stage. But after 12 000 cycles, a large number of contact bounces occurred during the making process.

At the same time, welding frequently occurred after 12 000 cycles. From Figure 5, it can be seen that a large number of bounces began to occur from 12 000 cycles, but at the time the contact welding did not occur immediately. The first contact welding adhesion occurred in 12 618 cycles, which means that only contact bounce is not enough to cause welding phenomenon.

### 3.4. Analysis of Making Arc Duration

In order to analyze the relationship between arc duration and contact welding, the trend of making arc duration with the number of experiments was plotted by analyzing current and voltage waveforms. The results are shown in Figure 6.

From Figure 6, it can be found that the making arc duration in the later stage of the experiment is obviously longer than that in the earlier stage. The

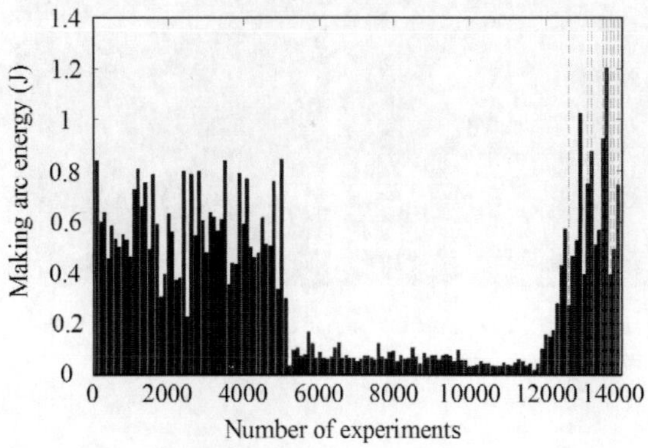


*Figure 7. The relationship between the making arc energy and number of experiments.*

making arc duration maintained at about 0.227 ms in first 12 000 cycles, and then the arcing duration rose rapidly after 12 000 cycles. The average value of the making arc duration after 12 000 cycles is 1.382 ms. The increase of making arc duration in the later stage is mainly due to the contact bounces in the making process.

### 3.5. Analysis of Making Arc Energy

From Figure 7, it can be seen that the making arc energy in the switching resistance load (the first 5 000 cycles) is much higher than the making arc energy in switching resistance-inductance load (after 5 000 cycles). Because the current can not change abruptly due to the effect of inductance, the making arc energy under resistance-inductive load is smaller. The maximum making arc energy occurred in the range of 13 600 to 13 700, and the average value of making arc energy is 1.2 J.

It can be seen that the making arc energy increased sharply after 12 000 experiments, and then welding occurred frequently. However, from Figure 7, it can be seen that the average value of making arc energy after 12 000 times is close to that of the making arc energy in first 5 000 experiments. The making arc energy in the first 5 000 experiments is 0.58 J. The average value of making arc energy in the eight weld intervals (marked with red dotted lines) in the figure is 0.707 J. There are many experiments where the making arc energy exceeded 0.707 J in the first 5 000 experiments, but there is not any contact welding. It can be found that only high making arc energy is not enough to cause contact welding.

### 3.6. Analysis of Making Welding

Figure 8 contains the images of the contacts before(left) and after(right) the experiment. In the figure, the upper contact is a movable one and the lower contact is a static one. The contacts before the experiment were copper yellow with metallic luster and smooth surface. After 13 946 cycles, serious erosion occurred on the surface of the relay contacts. The

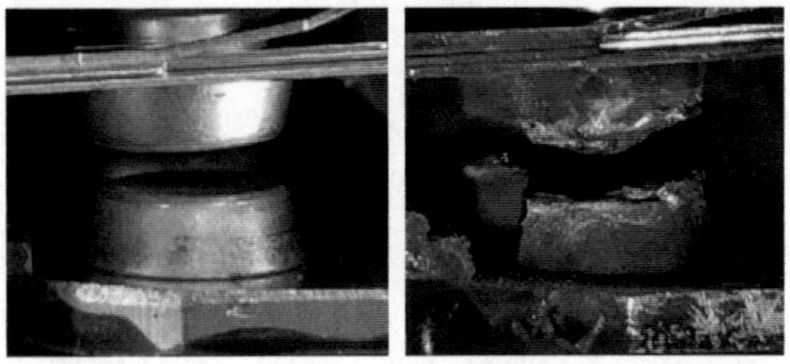

*Figure 8. The contact image before and after experiment.*

color of the contacts changed from copper yellow to carbon black. The contacts appearance in the right image in Figure 8 is not caused by one experiment, but by the accumulation of arc erosion and contacts' mechanical hammering in each experiment.

In order to obtain more experiment samples, when the contact of the relay couldn't be automatically broken because of the high welding force, the contact was forced to separate by external force. The experiment was carried out until current couldn't passes through closed contacts, which meant the relay was completely invalid.

In the initial stage of the experiment, the arc erosion contacts were produced by switching operation, in which the breaking arc played a major role. The arc changed the surface morphology of the contacts, which was harmful to the formation of stable contact. Then, when the contact surface morphology deteriorated to a certain extent, many successive moving contact making bounces occurred. The addition of making bounce arc increased the total arc energy and duration, which accelerated the deterioration of the contact surface. Under the effect of high arc temperature, the silver on the contact surface formed a melt pool. When the contact closed, the liquid silver melt pool cooled, solidified and then making weld occurred. Due to the contact surface deterioration, when the contacts closed in the later stage of cycles, conductive spots were smaller. So the current passing through the spots was high and generated a large amount of heat, thereby increasing the possibility of contact welding.

## 4. Conclusions

According to experimental results, the following conclusions can be drawn.

1. The contact resistance of the relay had no obvious change with the increase of the number of experiments.

2. In the first 12 000 cycles, the number of making bounces was relatively low. In the later stage of the experiment, the number of making bounces increased due to the deterioration of the surface contact condition. Because contact bounces prolong the making arc duration, making arc duration has the same trend as the number of making bounces. Under normal conditions, making arc energy in resistance load is much higher than that in resistance-inductance load. However, when the making bounces occurs, the making arc energy in resistance-inductance load will be higher.

3. In the later stage of experiment, the number of making bounces, making arc energy were large, and the electrical condition was extremely harsh. The arc produced in the single making process erodes the contacts more seriously. When the contact was eroded to a certain extent by arc, the contact bounces and making welding occurred.

## Acknowledgements

The project was supported by National Natural Science Foundation of China(51674136) and LiaoNing Revitalization Talents Program(XLYC1802110).

## References

[1] W. F. Rieder and A. R. Neuhaus. Contact welding influenced by anode arc and cathode arc, respectively. *IEEE Holm Conference on Electrical Contacts and the International Conference on Electrical Contacts*, 1(1):378–381, 2004. doi:10.1109/HOLM.2004.1353144.

[2] W. F. Rieder and A. R. Neuhaus. Short arc modes determining both contact welding and material transfer. *IEEE Transactions on Components and Packaging Technologies*, 30(1):9–14, 2007. doi:10.1109/TCAPT.2007.892056.

[3] L. Morin, N. B. Jemaa, and D. Jeannot. Make arc erosion and welding in the automotive area. *IEEE Transactions on Components and Packaging Technologies*, 23(2):240–246, 2000. doi:10.1109/6144.846760.

[4] L. Zhenbiao, W. Meifang, and M. Hasegawa. Effect of arcing behavior characteristics on welding resistance of relay contacts. *IEICE Transactions on Electronics*, E90-C(7):1385–1390, 2007. doi:10.1093/ietele/e90-c.7.1385.

[5] H. Katsuki and I. Takashi. Contact welding mechanism with bounce arc on Ag and Cu contacts in low-voltage switches. *IEEE Transactions on Components, Packaging and Manufacturing Technology*, 7(3):363–370, 2017. doi:10.1109/TCPMT.2017.2652458.

[6] T. Mutzel, M. Bender, and R. Niederreuther. The effect of material composition on dynamic welding of electrical contacts. *IEEE Holm Conference on Electrical Contacts*, 1(9):1–7, 2013. doi:10.1109/HOLM.2013.6651432.

[7] R. Wanbin, W. Tengyu, Z. Xu, Z. Ming, and W. Jianmin. Research on the rolling and sliding behavior of making contact and associated welding mechanism for low-current switching devices. *IEEE Transactions on Components, Packaging and Manufacturing Technology*, 9(1):18–27, 2019. doi:10.1109/TCPMT.2018.2827035.

[8] G. Fengyi, W. Guoqiang, D. Na, and F. W. Leuschner. The arc erosive characteristics and crack formation mechanisms analysis of sliver-based contact materials. *Proceedings of the CSEE*, 24(9):213–221, 2004. doi:10.13334/j.0258-8013.pcsee.2004.09.038.

doi:10.14311/ppt.2019.3.265
*Plasma Physics and Technology* **6**(3):265–269, 2019 

# APPROXIMATE DETERMINATION OF RADIATION PROPERTIES IN THE ARC PLASMAS WITH ADMIXTURES OF COPPER VAPOURS

N. Bogatyreva[a,*], M. Bartlová[a], V. Aubrecht[b], P. Kloc[b]

[a] *Department of Physics, Faculty of Electrical Engineering and Communication, Brno University of Technology, Technická 3058/10, 616 00 Brno, Czech Republic*

[b] *Department of Power Electrical and Electronic Engineering, Faculty of Electrical Engineering and Communication, Brno University of Technology, Technická 3058/12, 616 00 Brno, Czech Republic*

[*] bogatyreva@feec.vutbr.cz

**Abstract.** The aim of this paper is to evaluate radiation properties of air arc plasmas with various admixtures of copper vapours. The first order of the method of spherical harmonics ($P_1$–approximation) has been used as the way to solve the equation of radiation transfer. Calculations of the absorption coefficients for a thermal plasma have been performed as a function of the temperature and the frequency. The frequency variable in the equation of transfer was handled by means of multigroup method. Methods for prediction of the average absorption coefficients were described and compared in detail. The net emission coefficients have been determined for comparison.

**Keywords:** arc plasma, radiation transfer, absorption properties, admixtures of copper vapours.

## 1. Introduction

This paper presents the methods for research and prediction of the radiation properties in electric arc plasmas. Radiation transport of energy plays an important role in many plasma processing devices. Electric arc plasmas are utilized in a number of industrial applications. The electric arc parameters depend on the purity of gas in the chamber and the controllability of other additives in plasma. In many cases concentration of different species in the plasma is not possible to measure or measurement of different parameters will be difficult for technical realization. Then, modeling of the plasma spectrum is highly important for investigation of electric arc.

## 2. Energy transfer

Mathematical modeling is necessary to determine of radiation flux and its divergence. The solution of the equation of radiation transfer makes possible to estimate these parameters. The complete stationary equation of radiation transfer for an absorbing and emitting medium is

$$\mathbf{\Omega} \cdot \mathrm{grad}(I_\nu) = \kappa_\nu [B(\nu) - I_\nu], \qquad (1)$$

where $\mathbf{\Omega}$ is the unit direction vector, $I_\nu$ is the radiation intensity, $\kappa_\nu$ is the absorption coefficient, $B(\nu)$ is the Planck's spectral radiation intensity for equilibrium radiation, $\nu$ is the frequency [1].

The spectral intensity depends in general upon independent variables $(\mathbf{r}, \nu, \mathbf{\Omega})$. One must approximate the equation of transfer, either analytically or numerically, in order to obtain a solution. Due to the non-linearity of the equation describing the radiation field and the strong dependence of input parameters on the radiation frequency, various approximate methods are used. One of them is the method of spherical harmonics $P_N$–approximation.

The $P_1$–approximation consists of expanding the radiation intensity in spherical harmonics and including only the first order terms [2]. Under this assumption, the spectral equation of radiation transfer leads to a simple elliptic equation for radiation density $U_\nu$

$$-\mathrm{div}\left\{\frac{c}{3\kappa_\nu}\mathrm{grad}\,[U_\nu(\mathbf{r},\nu)]\right\} + \kappa_\nu c U_\nu(\mathbf{r},\nu) = \kappa_\nu 4\pi B(\nu), \qquad (2)$$

where $c$ is the speed of light, $\mathbf{r}$ is the position vector fixing the location of a point in space.

## 3. Absorption coefficients

Radiation in arc plasma depends on concentrations of all chemical species present in the plasma. Equilibrium compositions of plasma can be computed by means of several methods. In our case, plasma compositions are calculated by using the Tmdgas computer code, which is part of the database system TheCoufal [3, 4]. Our calculations take into account continuum and line radiation of atomic and ionic species and photoabsorption of diatomic molecules. It enables to obtain an approximate solution of required accuracy. One of the options considered is clean air without $CO_2$. We assume that the plasma containes some elements $O_2$, $N_2$, Ar, Cu, $e^-$ and also ions of O, N, Ar, Cu.

Comparison of temperature dependences of species concentration in the arc plasma for various admixtures of copper vapours is presented in Fig. 1. Compositions were calculated for admixtures of 5%, 50%, 95% Cu and pure air at pressure 0.1 MPa.

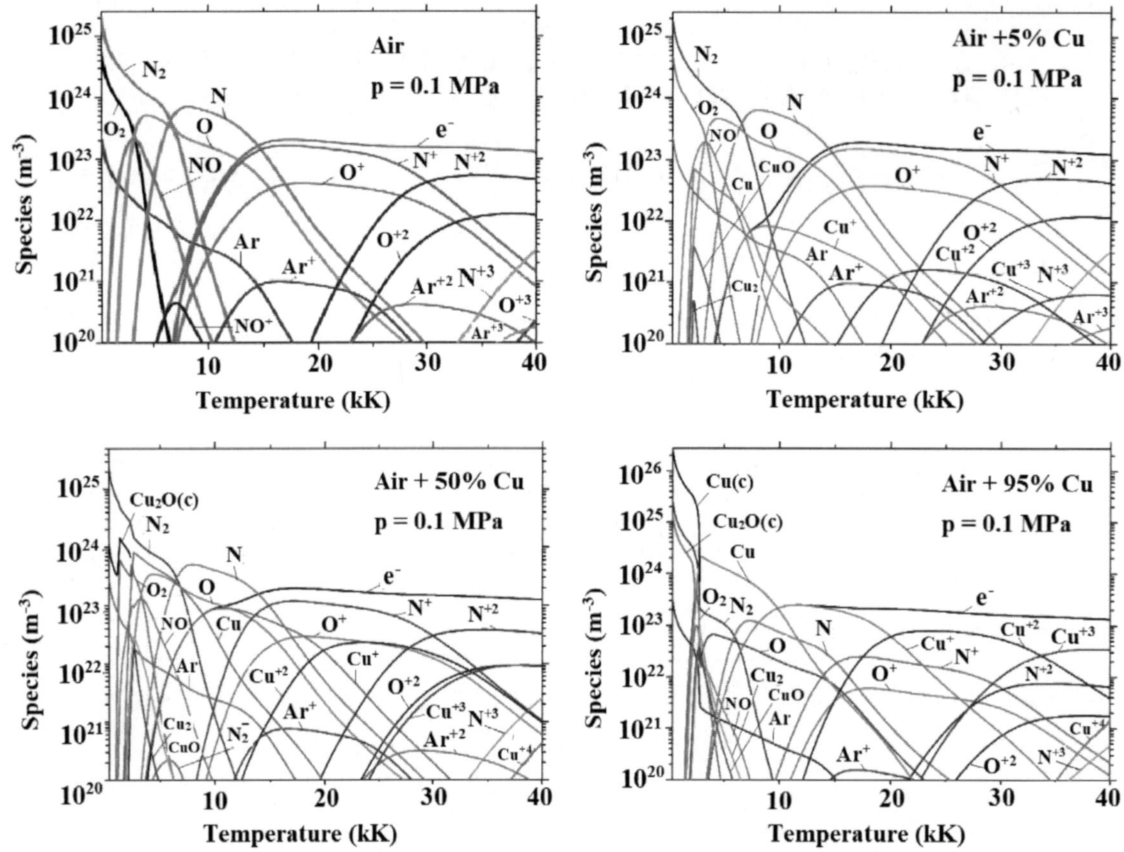


*Figure 1. Concentration of different species in the thermal plasma as a function of temperature.*

For atomic and ionic species, the absorption coefficients $\kappa_\nu$ were calculated. Both line and continuous radiation were considered in the calculation.

The total coefficient of absorption is given by the linear sum

$$\kappa_\nu(\nu, T, p) = \kappa_\nu^{\mathrm{ff}} + \kappa_\nu^{\mathrm{bf}} + \kappa_\nu^{\mathrm{bb}}. \qquad (3)$$

The indexes "ff, bf, bb" denote free-free transitions (bremsstrahlung), bound-free transitions (photo-recombination) and bound-bound transitions (discrete radiation), respectively.

The absorption coefficients of air plasma were calculated for three concentrations of copper (5%, 50%, 95%) at a pressure of 0.1 MPa. The plasma temperature was considered in the range of (300, 35 000) K. The frequency interval for calculations was (0.01 to 10) $\cdot\, 10^{15}$ Hz with increment of $10^{12}$ Hz.

Evaluated total spectral absorption coefficients of air arc plasmas with various admixtures of copper vapours at temperature 15 000 K are shown in Fig. 2. This spectrum is a complex of discrete and continuous spectra. Both dependences of the absorption coefficients on the radiation frequency and the concentrations of copper vapours can be seen in Fig. 2. The concentration of admixtures has an effect on the magnitude of the absorption coefficients. The greater influence is found in the first half of the frequency interval considered. The increase of the admixtures of copper vapours lead to an increase of the number of spectral lines.

## 4. Mean absorption coefficients

The simplification of the frequency dependence imposes a cutting of the spectrum in several frequency groups, in which the absorption coefficient is supposed to be constant with certain average value. We have calculated Planck ($\kappa_{\mathrm{P}}$) and Roseland ($\kappa_{\mathrm{R}}$) means for cutting the spectrum into 5 frequency groups. Fig. 3 shows the mean Planck and Rosseland absorption coefficients as a function of the temperature for air plasma with admixtures of 5% Cu and 95% Cu at the pressure of 0.1 MPa. Admixtures of copper vapours increase particularly the Planck absorption means since the Planck means stronger dependent upon the spectral lines.

The Planck mean is appropriate in the case of optically thin and emission dominated systems. It can be

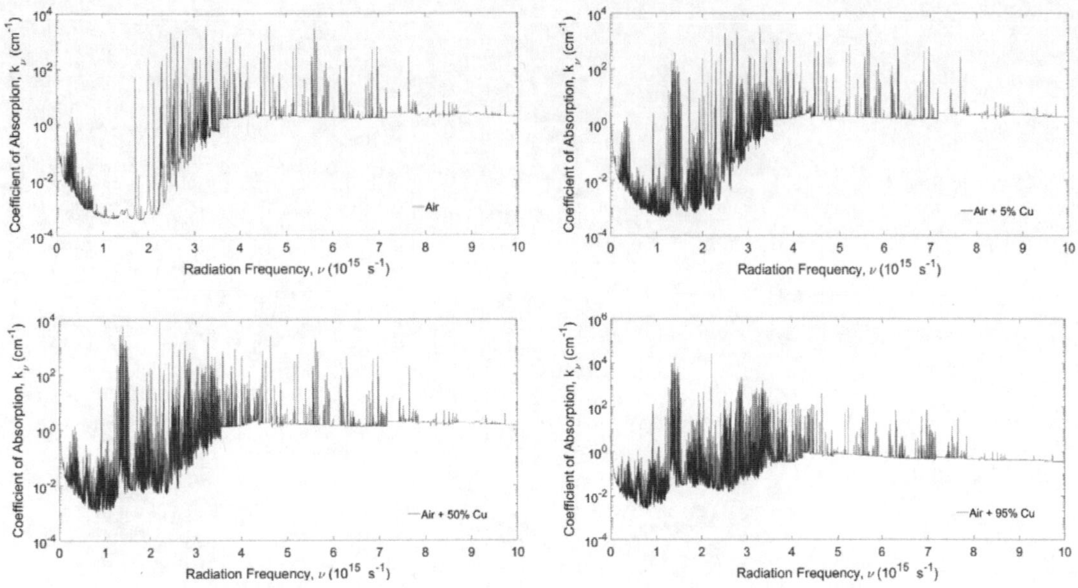


*Figure 2. Comparison of absorption coefficients in thermal plasma for different concentrations of copper.*

expressed as

$$\overline{\kappa_{\mathrm{P}}} = \frac{\displaystyle\int_{\nu_{k-1}}^{\nu_k} \kappa_\nu B(\nu,T)\mathrm{d}\nu}{\displaystyle\int_{\nu_{k-1}}^{\nu_k} B(\nu,T)\mathrm{d}\nu}. \tag{4}$$

Here, radiation intensity is given by the Planck function

$$B(\nu,T) = \frac{2h\nu^3}{c^2\left[\exp\left(\frac{h\nu}{k_BT}\right)-1\right]}, \tag{5}$$

where $h$ is the Planck constant, $k_B$ is the Boltzmann constant, $T$ is the temperature.

The Rosseland mean, also called mean free path of radiation is appropriate when the system approaches equilibrium (almost all radiation is reabsorbed). It has the form

$$\overline{\kappa_{\mathrm{R}}^{-1}} = \frac{\displaystyle\int_{\nu_{k-1}}^{\nu_k} \kappa_\nu^{-1}\frac{\mathrm{d}B(\nu,T)}{\mathrm{d}T}\mathrm{d}\nu}{\displaystyle\int_{\nu_{k-1}}^{\nu_k} \frac{\mathrm{d}B(\nu,T)}{\mathrm{d}T}\mathrm{d}\nu}. \tag{6}$$

The use of Rosseland and Planck mean absorption coefficients is only strictly appropriate in limiting circumstances. In the multigroup method, the splitting of the whole frequency interval has to be made according to frequency dependence of the absorption coefficients. For frequency groups with low values of absorption coefficients, the use of Planck mean is appropriate; for groups with high values of absorption coefficients, the Rosseland mean is more suitable. In our case, for groups 1–4 the Planck mean gives better results, in the group 5 the Rosseland mean is more appropriate.

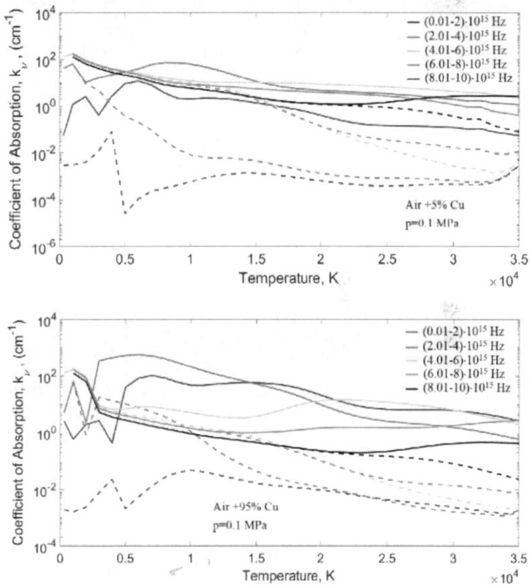


*Figure 3. Planck (full lines) and Rosseland (dotted lines) means absorption coefficients for various frequency groups as a function of temperature.*

## 5. Isothermal plasma cylinder

In the case of a cylindrically symmetrical plasma, the multigroup equation (2) depends only on one variable – the radial range $r$. Due to the fact that functions $\overline{\kappa}_n$ (mean absorption coefficient for any frequency interval) and $B_n$ (Planck function for any frequency interval) are dependent on the arc temperature and it decreases from the axis to the edge they are usually regarded as radial range functions. If we assume (to simplify the task) that the plasma is isothermal, i.e.

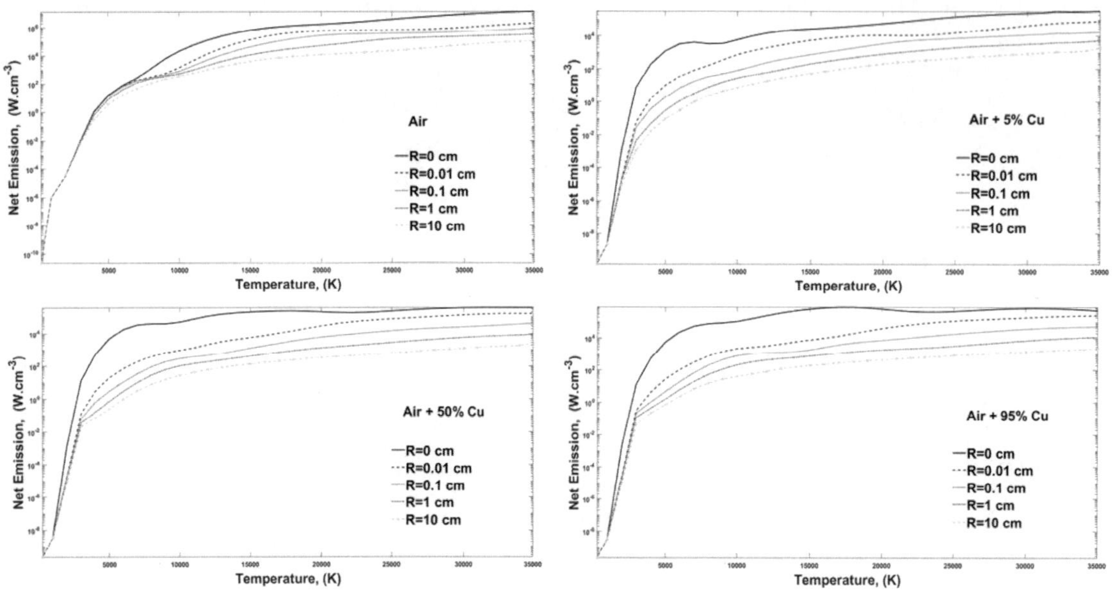


*Figure 4. Net emission coefficient of isothermal plasma cylinder as a function of temperature for various thicknesses of the plasma.*

the arc temperature remains constant, functions $\overline{\kappa}_n$ and $B_n$ also remain constant for the current solution of equation (2).

If no radiation enters from outside, the group net emission over the volume of the arc is

$$(w_{\text{avg}})_n = \frac{2\pi}{\pi R^2}\int_0^R r\,\text{div}\mathbf{W}_n(r)\,\mathrm{d}r = \frac{8\pi B_n J_1(\sqrt{3}\,\overline{\kappa}_n R)}{R[2J_1(\sqrt{3}\,\overline{\kappa}_n R) + \sqrt{3}\,J_0(\sqrt{3}\,\overline{\kappa}_n R)]}, \quad (7)$$

where $J_1(x)$ and $J_0(x)$ are the modified Bessel functions, $R$ is the arc radius, and $\text{div}\mathbf{W}_n(r)$ is the group radiation flux.

The sum over all frequency groups gives the net emission of radiation

$$\text{div}\mathbf{W}_R = w_{\text{avg}} = \sum_n (w_{\text{avg}})_n = 4\pi\varepsilon_N, \quad (8)$$

where $\varepsilon_N$ is the net emission coefficient.

The isothermal net emission coefficient corresponds to the fraction of the total power per unit volume and unit solid angle irradiated into a volume surrounding the axis of the arc plasma and escaping from the arc column after crossing a thickness $R$ of the isothermal plasma.

The net emission coefficient for the isothermal plasma cylinder with radius $R$ is given by [5]

$$\varepsilon_N = \int_0^\infty B(\nu)\kappa_\nu \exp(-\kappa_\nu R)\mathrm{d}\nu \quad (9)$$

and can be calculated with a direct integration over the real spectrum.

Comparison of the net emission coefficients for five different radii of the plasma cylinder (0 cm; 0.01 cm; 0.1 cm; 1 cm; 10 cm) at a pressure of 0.1 MPa is presented in the Fig. 4. The zero radius corresponds to the case omitting self-absorption. Besides the temperature variation, the plasma thickness has great influence on the radiation emission.

## 6. Results

The radiation properties in the arc plasmas were determinated by using the approximate method mentioned above. Plasma compositions with admixtures of 5%, 50%, 95% copper vapours were prepared at a pressure of 0.1 MPa in the temperature interval (300, 35 000) K. Absorption coefficients for a plasma with these admixtures of Cu were calculated and the spectra were plotted in a wide frequency range. Mean absorption coefficients were obtained by employing Planck and Rosseland means. Then, net emission coefficients were computed for different radii of the isothermal plasma cylinder. Based on the determined radiation parameters it is possible to analyze the influence for different admixtures of copper vapours (5%, 50%, 95%) in the arc plasmas. Results of our work will be useful for practical purposes, for example in the construction of circuit breakers.

Many authors have calculated the parameters for air plasma with admixtures of copper vapours. Most of them perform computations for 1%, 10% or 100% of Cu. Our calculations were carried out for other concentrations. Results of our work were compared with values of net emission coefficients of Cressault [6] and Aubrecht [7]. Our method allows us to get a

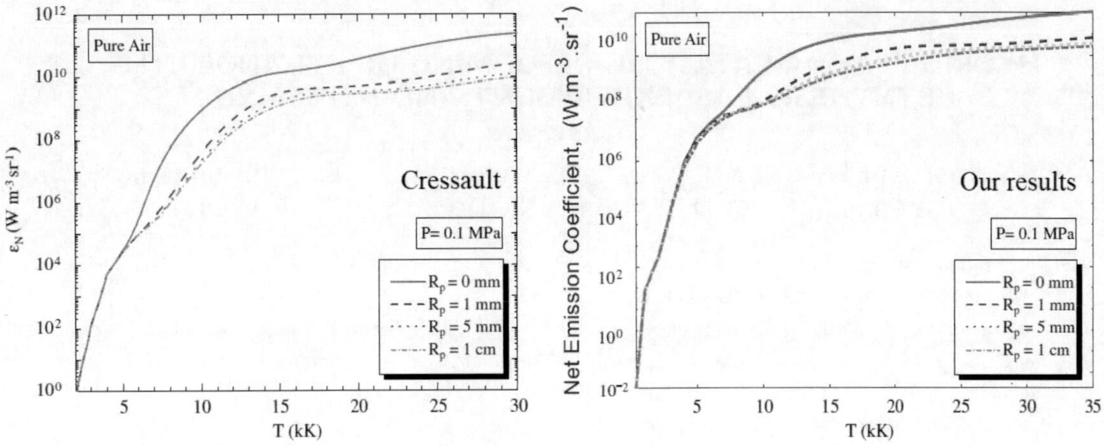


*Figure 5. Comparison of net emission coefficient in air plasma.*

satisfactory agreement with the calculations of other authors (Fig. 5, Fig. 6). Discrepancies between the results can be explained by the different approximate methods of calculation.

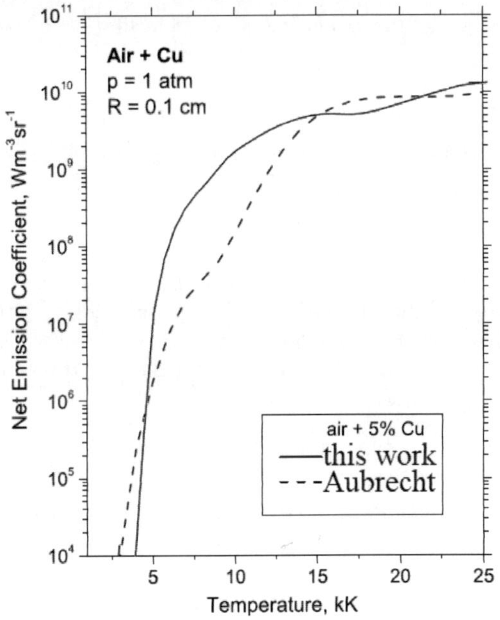


*Figure 6. Net emission coefficient in air plasma with 5% Cu.*

## 7. Conclusions

The Rosseland and Planck mean absorption coefficients can differ by several orders of magnitude. Discrete radiation influences significantly the values of Planck means; on the other hand, Rosseland means ignore the role of lines mainly at lower frequencies. Admixture of Cu vapour in air arc plasma contributes to the increase of the net emission. The contribution depends on the Cu vapour concentrations. The choice of mixtures was guided by the amount of evaporated electrode material in self-blast circuit breakers with copper electrodes.

## Acknowledgements

This research work was carried out in the Centre for Research and Utilization of Renewable Energy (CVVOZE). The authors gratefully acknowledge financial support from the Ministry of Education, Youth and Sports of the Czech Republic under NPU I programme (project No. LO1210 Energy for Sustainable Development) and OP VVV Programme (project No. CZ. 02.1.01/0.0/0.0/16o13/0001638 CVVOZE Power Laboratories-Modernization of Research Infrastructure).

## References

[1] M. I. Boulos, P. Fouchais, and E. Pfender. *Thermal plasmas: fundamentals and applications.* 1. Plenum Press, New York, 1994.

[2] B. N. Chetverushkin. *Matematicheskoe modelirovanie zadach dinamiki izluchajuschego gaza.* Nauka, Moskva, 1985.

[3] O. Coufal. Database system of thermodynamic properties of individual substances at high temperatures. *J.Phys.D: Appl.Phys.*, 38(8):1265–1274, 2005. doi:10.1088/0022-3727/38/8/026.

[4] O. Coufal. Tmdgas. arXiv:http://www.feec.vutbr.cz/~coufal.

[5] J. Lowke. Prediction of arc temperature profiles using approximate emission coefficients for radiation losses. *J. Quant. Spectroscop. Radiat. Transfer*, 14(2):111–122, 1974. doi:10.1016/0022-4073(74)90004-1.

[6] Y. Cressault, R. Hannachi, P. Teulet, A. Gleizes, J.-P. Gonnet, and J.-Y. Battandier. Influence of metallic vapours on the properties of air thermal plasmas. *Plasma Sources Sci. Technol.*, 17(3):35016, 2008. doi:10.1088/0963-0252/17/3/035016.

[7] V. Aubrecht, M. Bartlová, and O. Coufal. Radiative emission from air thermal plasmas with vapour of Cu or W. *J. Phys. D: Appl. Phys.*, 43(43):434007, 2010. doi:10.1088/0022-3727/43/43/434007.

doi:10.14311/ppt.2019.3.270
*Plasma Physics and Technology* **6**(3):270–273, 2019 

# HYBRID PLASMA-CATALYTIC REFORMING OF ETHANOL INTO SYNTHESIS GAS: EXPERIMENT AND MODELING

O. Nedybaliuk[a,*], I. Fedirchyk[a], V. Chernyak[a], T. Tereshchenko[a], O. Tsymbaliuk[a], V. Demchina[b], M. Bogaenko[c], V. Popkov[c]

[a] *Faculty of Radio Physics, Electronics and Computer Systems, Taras Shevchenko National University of Kyiv, 64/13 Volodymyrska Street, 01601 Kyiv, Ukraine*

[b] *The Gas Institute of the National Academy of Sciences of Ukraine, 39 Degtyarivska Street, 03113 Kyiv, Ukraine*

[c] *The Institute for the Problems in Material Science of the National Academy of Sciences of Ukraine, 3 Krzhizhanovsky Street, 03142 Kyiv, Ukraine*

[*] oanedybaliuk@gmail.com

**Abstract.** Understanding of the plasma-assisted reforming of hydrocarbons requires a combined application of the experimental studies of reforming systems and the kinetics modeling of reforming processes. Experiments were conducted on a system with a wide-aperture rotating gliding discharge with atmospheric air used as a plasma gas. Reforming parameters essential for the kinetics modelling of the reforming process were obtained. The influence of water addition method on the product composition of plasma-catalytic ethanol reforming was investigated.

**Keywords:** plasma-catalytic reforming, ethanol, synthesis gas, oxidative steam reforming.

## 1. Introduction

Synthesis gas produced during the reforming is commonly characterised by the volumetric concentration of $H_2$ and the ratio between $H_2$ and CO in the reforming products. Addition of water as a reforming reactant has shown to provide an additional source of H, O and OH species that initiate reforming reactions [1, 2] and can lead to better conversion efficiency and higher hydrogen content in the reforming products. In addition, the hydrogen content in the produced gas can be increased by converting CO using water via the water-gas shift reaction:

$$CO + H_2O \rightarrow H_2 + CO_2 \tag{1}$$

There are several ways of introducing water into the plasma-catalytic reforming systems, either directly into a discharge as a separate reactant or into a reaction chamber in a mixture with the hydrocarbon. The extent to which the method of water introduction impacts synthesis gas composition and $H_2$ yield during plasma-catalytic reforming remains under question. Understanding this topic requires a combination of experimental research and kinetics modelling of the processes inside the reforming system based on the operating parameters obtained during the experiments.

This work is focused on investigating the changes in the composition of the gas produced during the plasma-catalytic reforming of ethanol, which are caused by adding water as a reactant.

## 2. Experimental description

Plasma-catalytic reforming of ethanol was conducted via the partial oxidation reforming process (equation 2) and the oxidative steam reforming process (equation 3).

$$C_2H_5OH + 0.5\,O_2 \rightarrow 3\,H_2 + 2\,CO \tag{2}$$

$$C_2H_5OH + 0.5\,O_2 + 2\,H_2O \rightarrow 5\,H_2 + 2\,CO_2 \tag{3}$$

Figure 1 shows the scheme of the experimental setup used for the plasma-catalytic reforming of ethanol into synthesis gas. Reforming system used during experiments was created at the Taras Shevchenko National University of Kyiv. The system consisted of discharge chamber and reaction chamber, which were connected with each other. During the experiments, the food-grade ethanol was used for the conversion, atmospheric air was used as a source of oxygen and distilled water was used as a source of water.

All reactants were entering the system tangentially to the walls of discharge and reaction chambers, which led to the generation of vortex flows of reactants in both chambers. Air was supplied into the system using Fiac Cosmos air compressor. The total required airflow was separated and injected partly through the discharge chamber and partly through the reaction chamber. The airflow into each chamber was controlled using Dwyer RMA-22 SSV flowmeters. A flow of gas inside the discharge chamber caused a rotation of discharge channel. The part of air introduced into the discharge chamber was used to produce air plasma, which contained chemically active ions, free radicals and excited particles. Produced species were injected into the reaction chamber as a torch of rotating gliding discharge. The rest of the airflow was mixed with the ethanol or ethanol-water mixture, which was injected into the system at a specific rate using a syringe pump, and introduced into the reaction chamber.

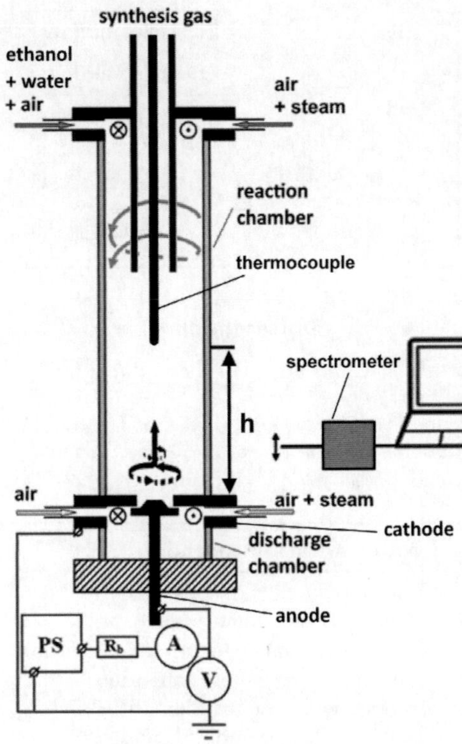


*Figure 1. Scheme of experimental setup.*

Plasma-catalytic reforming of ethanol was conducted in the cylindrical stainless steel or quartz reaction chamber 100 mm long and with 36 mm inner diameter. Type K thermocouple with 0.75 % T accuracy sleeve was introduced inside the ceramic into the reaction chamber via the outlet for the reforming products. This thermocouple measured the temperature inside the reaction chamber at the distance from the entrance into the reaction chamber from the discharge chamber ($h$) alongside the axis of the chamber.

Discharge system comprised central high-voltage stainless steel anode and grounded circular stainless steel cathode. Cathode had an aperture with 20 mm diameter. The distance between the electrodes was set to 1 mm. Rotating gliding discharge was powered using BP-100 power supply, which combined with the 33 kΩ ballast resistance $R_b$ was able to apply up to 7 kV voltage to the discharge gap. During the reforming, the current of rotating gliding discharge was set to and maintained at specific value. The accuracy of current measurement was ±5 mA and the accuracy of voltage measurement was ±0.1 kV.

The introduction of the distilled water into the reforming system was done either as steam mixed with air and injected into the discharge or reaction chamber or as a liquid ethanol-water mixture introduced into reaction chamber. In case of the former, required steam flow was obtained by supplying 90 W of power to Kärcher SC1 steam generator .

Gaseous products obtained during the reforming were cooled down to the room temperature using a water cooling system. This led to the condensation of excess water, leftover ethanol and heavy reforming products and their separation from the synthesis gas. Resulting gas was sampled and the volumetric concentrations of its components were determined via the gas chromatography on Agilent 6890 N gas chromatograph. Gas chromatograph used HP-MOLSIV and HP-PLOT Q columns thermostated in a temperature range from +4 to +450 °C, thermal conductivity detector at 200 °C and Ar carrier gas. Obtained volumetric concentrations were used to determine the specific gravity of produced synthesis gas in relation to air. The flow of produced gas was measured using Dwyer RMA-23 SSV flowmeter. This flowmeter was calibrated using air, therefore, calculated specific gravity of synthesis gas was used in the formula provided by the flowmeter manufacturer [3] to convert obtained flow measurements into the correct values of the synthesis gas flows.

Additional information about the optical emission spectroscopy of the radiation emitted out of the quartz reaction chamber in this system and rotational and vibrational temperatures of OH molecules inside the reaction chamber can be found in works [4, 5].

## 3. Results and Discussion

Temperatures inside the reaction chamber were measured for the 10 l min$^{-1}$ airflow into the discharge chamber and 5 l min$^{-1}$ airflow into the reaction chamber at distances $h$ every 5 mm starting from 10 mm up to 65 mm. The range of the discharge current during the temperature measurements was from 20 mA up to 70 mA, with a 10 mA step. Average discharge voltage was 2.2 kV in the case of 20 mA and 1.6 kV for other currents. Figure 2 presents the obtained dependencies of the temperatures measured by the thermocouple at different distances $h$ inside the reaction chamber on the discharge current.

Figure 2 shows that the temperature inside the reaction chamber increases with the rise of discharge current. This can be connected with the increased discharge power, which led to the higher specific energy input per mole of injected air. At the 20 mA discharge current, the temperature difference between the extreme positions of the thermocouples ($h = 10$ mm and $h = 65$ mm) is approximately 250 K, while at 70 mA this temperature difference increases to 500 K. Therefore, the temperature gradient between the hottest area of the reaction chamber, which is at the bottom of the chamber, and the reaction chamber exit increases with the rise of the discharge current and power. The gas temperature at the reaction chamber exit is approximately 1.5 times lower than near the bottom of the chamber. The temperature decrease can be caused by the mixing between the 10 l min$^{-1}$ airflow injected into the discharge chamber and heated by the discharge and the 5 l min$^{-1}$ airflow injected into the reaction chamber at room temperature.

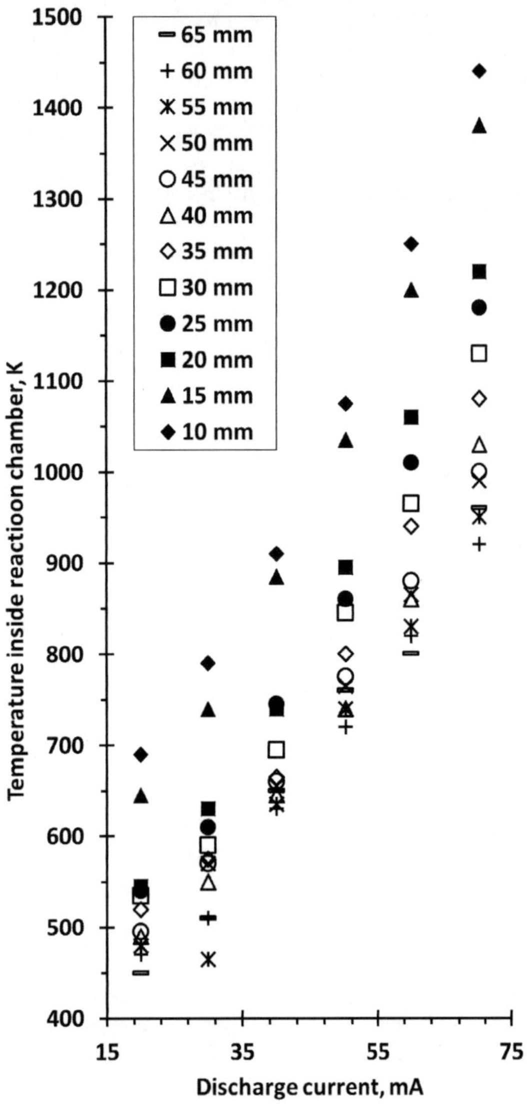


*Figure 2. Temperatures at the specific distances h alongside the reaction chamber axis measured at the different discharge currents. Airflow into discharge chamber is 10 l min$^{-1}$, airflow into reaction chamber is 5 l min$^{-1}$.*

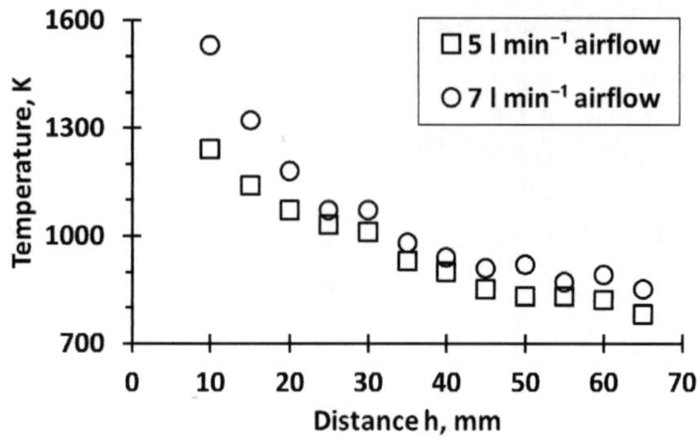


*Figure 3. Distributions of temperature at distance h alongside the axis of the reaction chamber at 10 l min$^{-1}$ airflow into discharge chamber, 5 l min$^{-1}$ and 7 l min$^{-1}$ airflows into the reaction chamber, discharge current is 60 mA and voltage is 1.4 kV.*

| Sample | Water addition method |
|---|---|
| 1 | Without water addition |
| 2 | 2 g min$^{-1}$ steam flow added into discharge chamber together with airflow |
| 3 | 1 ml min$^{-1}$ water flow added into reaction chamber together with ethanol as 1:11 water-ethanol mixture |
| 4 | 2 ml min$^{-1}$ water flow added into reaction chamber together with ethanol as 2:11 water-ethanol mixture |

*Table 1. Connections between the collected samples and water addition methods.*

The distributions of the gas temperature at the distance $h$ along the reaction chamber axis measured by the thermocouple at 10 l min$^{-1}$ airflow into the discharge chamber and 5 l min$^{-1}$ and 7 l min$^{-1}$ airflows into the reaction chamber are shown at figure 3. Temperatures were measured for 60 mA discharge current and 1.4 kV discharge voltage.

A slight rise of the temperature at the reaction chamber axis is becoming observable with the increase of the airflow into the reaction chamber. This increase is more noticeable closer to the discharge system at $h = 10$ mm than near the reaction chamber exit at $h = 65$ mm. The rise of the airflow into the reaction chamber leads to the contraction of the discharge area inside the reaction chamber and to its restriction to the chamber axis, away from the reaction chamber wall.

Several methods were used to introduce water into the reaction chamber during the investigation of the influence of water addition on the plasma-catalytic reforming of ethanol. Water was injected either as a steam-air mixture directly into the discharge or reaction chamber or introduced into the reaction chamber as liquid in ethanol-water mixture. In all cases the input ethanol flow was set to 648 ml h$^{-1}$. The airflow into the discharge chamber was kept at 10 l min$^{-1}$ and the airflow into the reaction chamber was set to 5 l min$^{-1}$. Gas produced during the reforming was sampled for 4 different methods of water addition into the reaction chamber, which are presented in table 1. Figure 4 shows the volumetric concentrations of species in samples, which were obtained using gas chromatography.

The analysis of samples composition provided some insights into the impact of water addition method onto the reforming process. The addition of 2 g min$^{-1}$ water steam flow into the discharge chamber (sample 2) led to the increase of the ratio between $H_2$ and CO in the produced gas from 0.95 (sample 1) to 1.4. In addition, the volumetric concentrations of $H_2$ and CO decreased, while several more complex hydrocarbons,

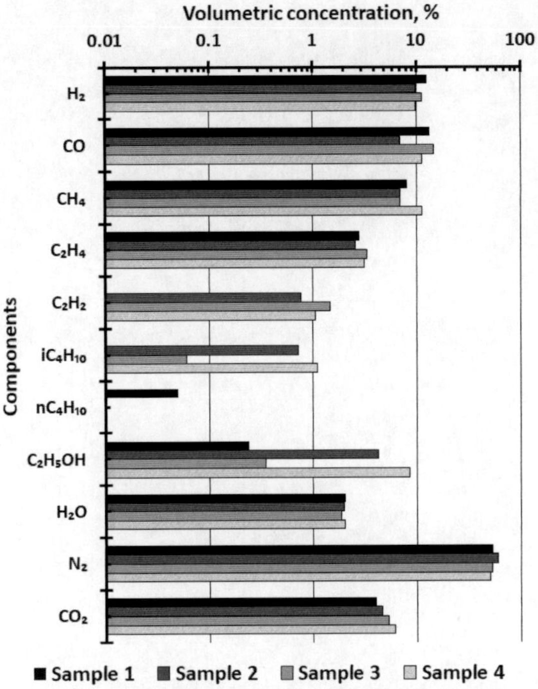


*Figure 4. Volumetric concentration of species in gaseous reforming products obtained at 10 l min$^{-1}$ airflow into discharge chamber, 5 l min$^{-1}$ airflow into the reaction chamber, 648 ml h$^{-1}$ ethanol flow into the reaction chamber, 60 mA discharge current and 1.4 kV voltage.*

such as $C_2H_2$ and $iC_4H_{10}$, appeared in the products. In case of introducing water in a mixture with ethanol (samples 3 and 4), the ratio between $H_2$ and CO drops compared to the sample 1 to 0.77 for sample 3 but begins to recover with the increase of water percentage in a mixture and reaches 0.87 for sample 4. The volumetric concentrations of $H_2$ and CO decrease, $C_2H_2$ and $iC_4H_{10}$ appear in the products, and the volumetric concentration of $CH_4$ increased in the case of water-ethanol mixture with the higher amount of water (sample 4). The introduction of water into both chambers led to the increase of $CO_2$ volumetric concentration in the reforming products. The rise of $CO_2$ concentration can be caused by the water-gas shift reaction (equation 1) or complete oxidation of hydrocarbons during reforming.

## 4. Conclusions

The highest measured gas temperature inside the reaction chamber is localized at the bottom of the reaction chamber near the exit from the discharge chamber. The temperature measured at the reaction chamber outlet is 1.5 times lower than the temperature near the discharge chamber. This temperature decrease may be caused by the addition of the air that was injected directly into the reaction chamber to the air that entered the reaction chamber from the discharge chamber. The rise of discharge current and the consequent increase of the discharge power leads to the increase of the gas temperature along the reaction chamber axis. The rise of the airflow injected into the reaction chamber causes the increase of the gas temperature on the reaction chamber axis. This temperature increase can be caused by the airflow injected into the reaction chamber, which constricted the volume filled by the discharge and localized it closer to the reaction chamber axis.

The addition of water into the plasma-catalytic system during ethanol reforming leads to the increase of the volumetric concentrations of $C_2H_2$, $iC_4H_{10}$ and $CO_2$. Addition of 2 g min$^{-1}$ flow of water steam into the discharge chamber rises the ratio between $H_2$ and CO by 1.5 times in comparison to the reforming without water. The addition of 2 g min$^{-1}$ flow of water in a mixture with injected ethanol provides the same ratio between $H_2$ and CO as the reforming of ethanol without water. This shows that activation of water in the discharge before its injection into the reaction chamber is preferable for the reforming process over the direct addition of water into the reaction chamber.

The impact of water activation on the reforming products can be connected to the increased production of H, O and OH species inside the discharge plasma, which are necessary for the initiation of the reforming reaction chains. However, a detailed modelling of the reaction kinetics is required for the understanding of this mechanism, which will be conducted in the future studies using the reforming parameters and plasma properties obtained in this study.

## Acknowledgements

This work was supported by the state grant of the Ministry of Education and Science of Ukraine 0118U002057.

## References

[1] A. I. Shchedrin, D. S. Levko, V. Y. Chernyak, V. V. Yukhimenko, and V. V. Naumov. Effect of air on the concentration of molecular hydrogen in the conversion of ethanol by a nonequilibrium gas-discharge plasma. *JETP Letters*, 88(2):99–102, 2008. doi:10.1134/S0021364008140063.

[2] A. Frassoldati, A. Cuoci, T. Faravelli, and E. Ranzi. Kinetic Modeling of the Oxidation of Ethanol and Gasoline Surrogate Mixtures. *Combustion Science and Technology*, 182(4-6):653–667, 2010. doi:10.1080/00102200903466368.

[3] Dwyer Instruments. Variable Area Flow Corrections. [2018-11-16]. URL: http://blog.dwyer-inst.com/2017/09/06/variable-area-flow-corrections/{#}sthash.syxIW2q1.dpbs.

[4] I. I. Fedirchyk, O. A. Nedybaliuk, V. Y. Chernyak, and V. P. Demchina. Plasma-catalytic reforming of rich ethanol-air mixtures. *Problems of Atomic Science and Technology*, (1):164–167, 2019.

[5] O. A. Nedybaliuk, I. I. Fedirchyk, and V. Y. Chernyak. Plasma-catalytic system with wide-aperture rotating gliding discharge. *Problems of Atomic Science and Technology*, (6):214–217, 2018.

# AUTHOR INDEX

# AUTHOR INDEX